Fields Institute Communications

VOLUME 76

The Fields Institute is a centre for research in the mathematical sciences, located in Toronto, Canada. The Institutes mission is to advance global mathematical activity in the areas of research, education and innovation. The Fields Institute is supported by the Ontario Ministry of Training, Colleges and Universities, the Natural Sciences and Engineering Research Council of Canada, and seven Principal Sponsoring Universities in Ontario (Carleton, McMaster, Ottawa, Queen's, Toronto, Waterloo, Western and York), as well as by a growing list of Affiliate Universities in Canada, the U.S. and Europe, and several commercial and industrial partners.

More information about this series at http://www.springer.com/series/10503

Donald Dawson • Rafal Kulik
Mohamedou Ould Haye • Barbara Szyszkowicz
Yiqiang Zhao

Editors

Asymptotic Laws and Methods in Stochastics

A Volume in Honour of Miklós Csörgő

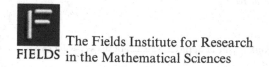

The Fields Institute for Research
in the Mathematical Sciences

Editors
Donald Dawson
School of Mathematics and Statistics
Carleton University
Ottawa, ON, Canada

Mohamedou Ould Haye
School of Mathematics and Statistics
Carleton University
Ottawa, ON, Canada

Yiqiang Zhao
School of Mathematics and Statistics
Carleton University
Ottawa, ON, Canada

Rafal Kulik
Department of Mathematics
   and Statistics
University of Ottawa
Ottawa, ON, Canada

Barbara Szyszkowicz
School of Mathematics and Statistics
Carleton University
Ottawa, ON, Canada

ISSN 1069-5265
Fields Institute Communications
ISBN 978-1-4939-5011-9
DOI 10.1007/978-1-4939-3076-0

ISSN 2194-1564 (electronic)

ISBN 978-1-4939-3076-0 (eBook)

Mathematics Subject Classification (2010): 60-02, 62-02, 60F05, 60F15, 60F17, 60G15, 60G17, 60G50, 60G55, 60J55, 60J65, 60K37, 62G30, 62M10
Springer New York Heidelberg Dordrecht London
© Springer Science+Business Media New York 2015
Softcover re-print of the Hardcover 1st edition 2015

Cover illustration: Drawing of J.C. Fields by Keith Yeomans

Printed on acid-free paper

Springer Science+Business Media LLC New York is part of Springer Science+Business Media (www.springer.com)

Carleton hosting international
mathematics symposium July
3–6 to honour 50 years of
Prof. Miklós Csörgő's research

Preface

A Fields Institute International Symposium on Asymptotic Methods in Stochastics was organized and held in honour of Miklós Csörgő's work on the occasion of his 80th birthday at Carleton University, Ottawa, Canada, July 3–6, July 2012. The symposium was hosted and sponsored by the School of Mathematics and Statistics, Carleton University, and co-sponsored by the Fields Institute for Research in Mathematical Sciences.

The symposium attracted more than 70 participants from around the world, including many graduate students and postdoctoral fellows. It is with great sadness that we are to write here that in January 2014, one of the participants, **Marc Yor**, passed away. We recall the happy days we were lucky to spend with him here, while he was attending our conference. We are very pleased that in collaboration with Francis Hirsch and Bernard Roynette, he also contributed a paper for publication in this volume. Unfortunately, we cannot any more thank him for his eminent participation in our symposium, where he also gave a talk on peacocks and associated martingales.

The opening address was given by **Don Dawson**, "Path properties of fifty years of research in Probability and Statistics: a tribute to Miklós Csörgő," that was followed by Miklós presenting his 50- year involvement in *Asymptotic Methods in Stochastics* in a historical context.

In this regard we wish to mention that there were two previous conferences, both held at Carleton University, in celebration of Miklós Csörgő's contributions to Probability and Statistics on the respective occasions of his 65th and 70th birthdays. *The first one*, ICAMPS '97 (International Conference on Asymptotic Methods in Probability and Statistics, 8–13 July 1997), was organized by Barbara Szyszkowicz, who also edited the proceedings volume of this conference (cf. [**V1**] in **Publications of Miklós Csörgő**; bold-face letters and/or numbers in square brackets will throughout refer to the latter list of publications). *The second one*, ICAMS '02 (International Conference on Asymptotic Methods in Stochastics, 23–25 May 2002), was organized by Lajos Horváth and Barbara Szyszkowicz, and, just like the preisent symposium, it was also co-sponsored by Tthe Fields Institute. For the proceedings volume of ICAMS'02, we refer to [**V2**], that is, Volume 44 of Fields Institute Communications, as well as to the there indicated Fields Institute website: www.fields.utoronto.ca/publications/supplements/, where the editors of the latter volume, Lajos Horváth and Barbara Szyszkowicz, also have a 69-page résumé of Miklós' work over the past forty or so years at that time, titled "Path Properties of Forty Years of Research in Probability and Statistics: In Conversation with Miklós Csörgő". This article with its 311 references, together with Miklós' list of publications at that time, is also available as no. 400 – 2004 of the Technical Report Series of LRSP. It can also be accessed on the LRSP website: http://www.lrsp.carleton.ca/trs/trs.html.

We much appreciate having been given the opportunity by the Editorial Board of Publications of the Fields Institute to include in this volume Miklós' above-mentioned list of publications (cf. **Table of Contents**). The Editors also have a 45-page resume, titled "**A Review of Miklós Csörgő's Mathematical Biography**", that can be accessed on the Fields Institute website www.fields.utoronto.ca/publications/supplements/. Unfortunately, due to space limitations, we could not include in this collection our expository style review of SELECTED PATH PROPERTIES OF 50+ YEARS OF RESEARCH IN PROBABILITY AND STATISTICS: IN CONVERSATION WITH MIKLÓS CSÖRGŐ.

In the abstract of his talk at the conference, "Almost exact simulations using Characteristic Functions", **Don McLeish** nicely relates asymptotics, numerical methods and simulations as tools of approximation in Probability and Statistics. We quote the first part of his abstract here:

Asymptotic statistics explores questions like when and how do functions of observed data behave like functions of normal random variables? and much of the work of Miklós Csörgő and his coauthors can be described analogously as when and how do functionals of an observed path behave like those of corresponding Gaussian processes? . For much of the past century, asymptotics provided the main approximation tool in probability and statistics. Although it is now supplemented with other approximation tools such as numerical methods and simulation, asymptotics remains a key to understanding the behaviour of random phenomena.

The following participants presented 30-minute talks at the conference: Raluca Balan, István Berkes, David Brillinger, Alexander Bulinski, Murray Burke, Endre Csáki, Herold Dehling, Dianliang Deng, Richard Dudley, Shui Feng, Antónia Földes, Peter W. Glynn, Edit Gombay, Karl Grill, Lajos Horváth, Gail B. Ivanoff, Jana Jurečková, Reg Kulperger, Deli Li, Zhengyan Lin, Peter March, Yuliya Martsynyuk, Don McLeish, Masoud Nasari, Emmanuel Parzen, Magda Peligrad, Jon N.K. Rao, Bruno Rémillard, Pál Révész, Murray Rosenblatt, Susana Rubin-Bleuer, Thomas Salisbury, Qi-Man Shao, Zhan Shi, Josef G. Steinebach, Qiying Wang, Martin Wendler, Wei-Biao Wu, Marc Yor, and Hao Yu.

We are pleased to publish this collection of twenty articles in the *Fields Institute Communications* series by Springer, and it is our pleasure to dedicate this volume to Miklós Csörgő as a token of respect and appreciation of his work in Probability and Statistics by all the contributors to this volume, and all the participants in our 2012 Fields Institute International Symposium. We are grateful to the contributors for submitting their papers for publication in this volume, as well as to the referees for their valuable time and enhancing work on it. All papers have been refereed, and accordingly revised if so requested by the editors. We wish to record here our sincere thanks to everyone for their extra time, care and collaboration throughout this elaborate process. The papers in this volume contain up-to-date surveys and original results at the leading edge of research in their topics written by eminent international experts. They are grouped into seven sections whose headings are indicative of their respective main themes that also reflect Miklós' wide-ranging research areas in Probability and Statistics. Except for Section 2, the listing of the articles in each is in the alphabetical order resulting from that of their authors. The reason for making an exemption from this "rule" in Section 2 is that the Csáki et al. paper there also provides a general footing for the results that are proved in Révész's exposition right after.

In **Section 1**, **Miklós Csörgő** and **Zhishui Hu** present, in a historical context, and then establish, a weak convergence theorem for self-normalized partial sums processes of independent identically distributed summands when the latter belong to the domain of attraction of a stable law with index $\alpha \in (0, 2]$. In particular, Theorem 1.1 of this paper identifies the limiting distribution in Theorem 1.1 of Chistyakov and Götze (cf. 2. in References therein) under the same necessary and sufficient conditions in terms of weak convergence in $D[0, 1]$. Initiated by his primary contributions [97] (with Lajos Horváth), [190] and [191] (both with Barbara Szyszkowicz and Qiying Wang), self-normalization and Studentization have become an important global research area of Miklós Csörgő and his collaborators (cf., e.g., [192], [204], [205], [216], [217], [220], [221], [222], [223] and [224]). In the introduction of their paper in this section, **Dianliang Deng** and **Zhitao Hu** present an up-to-date survey of results dealing with the precise asymptotics for the deviation probabilities of self-normalized sums and continue with establishing integrated precise asymptotics results for the general deviation probabilities of multidimensionally indexed self-normalized sums. **Magda Peligrad** and **Hailin Sang** deal with asymptotic results for linear processes in general and, in the latter context, review some recent developments, including the central limit theorem (CLT), functional CLT and their self-normalized forms for partial sums. They study these in terms of independent and identically distributed summands (cf. 16. in References therein) and, via self-normalization, for short memory linear processes as well, as, e.g., in 14. in References therein. Self-normalized CLT and self-normalized functional CLT are also covered for long memory linear processes with regularly varying coefficients (cf. 15. in References therein).

In **Section 2**, **Endre Csáki**, **Antónia Földes** and **Pál Révész** survey their joint work with Miklós on anisotropic random walks on the two-dimensional square lattice $\mathbb{Z}^2$ of the plane (cf. [210], [213], [215], [218], and [219]). Such random walks possibly have unequal symmetric horizontal and vertical step probabilities, so that these step probabilities can only depend on the value of the vertical coordinate. In particular, if such a random walk is situated at the site on the horizontal line $y = j \in \mathbb{Z}$, then, at the next step, it moves with probability p_j to either vertical neighbour and with probability $1/2 - p_j$, to either horizontal neighbour. It is assumed throughout that $0 < p_j \leq 1/2$ and $\min_{j \in \mathbb{Z}} p_j < 1/2$.

The case $p_j = 1/2$ for some j means that the horizontal line $y = j$ is missing, a possible lack of complete connectivity. The initial motivation for studying such two-dimensional random walks on amnisotropic lattice has originated from the so-called transport phenomena of statistical physics (cf. 12., 14., 15. and 16. in References therein), where having $p_j = 1/2, j = \pm1, \pm2, \ldots$, but $p_0 = 1/4$, the so-called random walk on the two-dimensional comb, i.e., when all the horizontal lines of the x axis are removed, is also of interest (cf. 1., 2., 5. and 29. in References therein). In his paper, **Pál Révész** continues the investigation of the latter comb-random walk, and also that of a random walk on a half-plane half-comb lattice (cf. [**218**]), and concludes a result for each on the area of the largest square they respectively succeed in covering at time n. **Gail B. Ivanoff** reviews martingale techniques that play a fundamental role in the analysis of point processes on $[0, \infty)$, and revives the question of applying martingale methods to point processes in higher dimensions. In particular, she revisits the question of a compensator being defined for a planar point process in such a way that it exists, it is unique and it characterizes the distribution of the point process. She proceeds to establish a two-dimensional analogue of Jacod's characterization of the law of a point process via a regenerative formula for its compensator and also poses some related open questions.

In **Section 3**, the paper of **Alexander Bulinski** deals with high -dimensional data that can be viewed as a set of values of some factors and a binary response variable. For example, in medical studies the response variable can describe the state of a patient's health that may depend only on some parts of the factors. An important problem is to determine collections of significant factors. In 3. of References of the paper, Bulinski establishes the basis for the application of the multifactor dimensionality reduction (MDR) method in this regard, when one uses an arbitrary penalty function to describe the prediction error of the binary response variable by means of a function of the factors. The goal of thie present paper is to conclude multidimensional CLT's for statistics that justify the optimal choice of a subcollection of the explanatory variables. Statistical variants of these CLT's involving self-normalization are also explored. The paper of **Deli Li**, **Yongcheng Qi** and **Andrew Rosalsky** is devoted to extending recent theorems of Hechner, and Hechner and Heinkel (cf. 6. and 7. in References therein) dealing with sums of independent Banach space valued random variables. The proof of the main result, Theorem 3 in this paper, is based on new versions of the classical Lévy, Ottaviani, and Hoffmann-Jorgensen inequalities (cf. 11., 3. and 8., respectively, in References therein) that were recently obtained by Li and Rosalsky (cf. 13. in References of the paper). In her second paper in this volume, **Magda Peligrad** surveys the almost sure CLT and its functional form for stationary and ergodic processes. Her survey addresses the question of limit theorems, started at a point, for almost all points. These types of results are also known under the name of quenched limit theorems, or almost sure conditional invariance principles. All these results have in common is that they are obtained via a martingale approximation in the almost sure sense. As applications of the surveyed results, several classes of stochastic processes are shown to satisfy quenched CLT and quenched invariance principles, namely, classes of mixing sequences, shift processes, reversible Markov

Chains and Metropolis Hastings algorithms. In his paper, **Qiying Wang** revisits, with some improvements, his recent extended martingale limit theorem (MLT) and, for a certain class of martingales, concludes that the convergence in probability of the conditional variance condition in the classical MLT can be reduced to the less restrictive convergence in distribution condition for the conditional variance (cf. 7. in References therein). The aim of this paper is to show that the latter extended MLT can be used to investigate a specification test for a nonlinear cointegrating regression model with a stationary error process and a nonstationary regressor. This, in turn, leads to a neat proof for the main result in Wang and Phillips of 11. in References.

Anchored by his 1997 book with Lajos Horváth (cf. [**A5**]), change-point analysis has been an important research area of Miklós and his collaborators for almost three decades now (cf. [**93**], [**94**], [**105**], [**106**], [**110**], [**147**], [**148**], [**155**], [**175**], [**198**], [**204**] and [**223**]). The three papers in **Section 4** present recent advances in the field. **Alina Bazarova**, **István Berkes** and **Lajos Horváth** develop two types of tests to detect changes in the location parameters of dependent observations with infinite variances. In particular, autoregressive processes of order one with independent innovations in the domain of attraction of a stable law of index $\alpha \in (0, 2)$ are considered, and, for testing the null hypothesis of the stability of the location parameter versus the at most one-change alternative, they construct a suitably trimmed CUSUM process via removing the d largest observations from the sample. They recall (cf. 8. in References therein) that the thus adjusted CUSUM process converges weakly to a Brownian bridge, if $d = d(n) \to \infty$ fast enough but so that $d(n)/n \to 0$, as $n \to \infty$. However the normalizing sequence depends heavily on unknown parameters. In view of this, two types of test statistics are studied, namely, maximally selected CUSUM statistics whose long run variance is estimated by kernel estimators, and ratio statistics that do not depend on the long run variances whose estimation is thus avoided. **Herold Dehling**, **Roland Fried**, **Isabel Garcia**, and **Martin Wendler** study the detection of change-points in time series. Instead of using the classical CUSUM statistic for detection of jumps in the mean that is known to be sensitive to outliers, a robust test based on the Wilcoxon two-sample test statistic is proposed. The asymptotic distribution of the proposed test can be derived from a functional central limit theorem for a two-sample U-statistics -dependent data that in the case of independent data was studied by Csörgő and Horváth (cf. 5. in References therein, [**106**] in Miklós' list). In their present paper, their result is extended to short-range-dependent data, namely, data that can be represented as functionals of a mixing process. Similar results were obtained for long-range-dependent data by Dehling, Rooch and Taqqu (cf. 10. in References therein). Further to [**106**], we mention [**204**], where the projection variate is assumed to be in the domain of attraction of the normal law, possibly with infinite variance. **Edit Gombay** deals with retrospective change-point detection in a series of observations generated by a binary time series model with link functions other than the logit link function that was considered by Fokianos, Gombay and Hussein in 5. of References therein that appeared in 2014. It is shown that the results in the latter work carry over if, instead of the logit link function, one uses the probit, the log-log, and complementary log-log link functions in the binary regression model. Some of the technical details omitted in 5. are also detailed in their present paper.

In **Section 5**, **Kilani Ghoudi** and **Bruno Rémillard** investigate the asymptotic behaviour of multivariate serial empirical and copula processes based on residuals of autoregressive-moving-average (ARMA) models. Motivated by Genest et al. 14. as in References therein, multivariate empirical processes based on squared and other functions of residuals are also investigated. Under the additional assumption of symmetry about zero of the innovations, it is shown that the limiting processes are parameter-free. This, in turn, leads to developing distribution-free nonparametric tests for a change-point in the distribution of the innovations, tests of goodness-of-fit for the law of innovations, and tests of independence for m consecutive innovations. Simulations are also carried out to assess the finite-sample properties of the proposed tests and to provide tables of critical values. **Murray Rosenblatt** presents a historical overview of the evolution of a notion of strong mixing as a measure of short-range dependence and a sufficient condition for a CLT. He also discusses a characterization of strong mixing for stationary Gaussian sequences, as well as examples of long-range dependence leading to limit theorems with nonnormal limiting distributions. Results concerning the finite Fourier transform are noted, and a number of open questions are considered. We also note in passing that the articles [197], [200], [206], [207], [214] and [227] in Miklós' list of publications deal with empirical and partial sums processes that are based on short and long memory sequences of random variables.

In **Section 6**, the paper by **Hongwei Dai**, **Donald Dawson** and **Yiqiang Zhao** extends the classical kernel method employed for two-dimensional discrete random walks with reflecting boundaries. The main focus of the paper is to provide a survey on how one can extend the latter kernel method to study asymptotic properties of stationary measures for continuous random walks. The semimartingale reflecting Brownian motion is taken as a concrete example to detail all key steps in the extension in hand that is seen to be completely parallel to the method for discrete random walks. The key components in the analysis for a boundary measure, including analytic continuation, interlace between the two boundary measures, and singularity analysis, allow the authors to completely characterize the tail behaviour of the boundary measure via a Tauberian-like theorem. In their paper, **Peter Glynn** and **Rob Wang** develop central limit theorems and large deviation results for additive functionals of reflecting diffusion processes that incorporate the cumulative amount of boundary reflection that has occurred. In particular, applying stochastic calculus and martingale ideas, partial differential equations are derived from which the central limit and law of large numbers behaviour for additive functionals involving boundary terms can be computed. The corresponding large deviation theory for such additive functionals is then also developed. For papers on additive functionals in Miklós' list of publications, we refer to [134], [152], [187] and [212]. Paper [173] in the same list contains a self-contained background on stochastic analysis, Itô calculus included. The paper by **Francis Hirsch, Bernard Roynette** and **Marc Yor** studies peacock processes which play an important role in mathematical finance. A deep theorem of Kellerer (cf. 9. in References therein) asserts the existence of peacock processes as a Markov martingale with given marginals, assumed to increase in the convex order. The paper in hand revisits Kellerer's theorem with a proof, in the light of the papers 5. and 8. in its References by Hirsch-Roynette and G. Lowter, respectively, and presents, without proof, results of 6., 7. and 8. by G. Lowter, which complete and make Kellerer's theorem more precise on some points. Many other references around Kellerer's theorem can be found in 4. of References of the paper.

Mayer Alvo's paper in **Section 7** deals with applying empirical likelihood methods to various problems in two-way layouts involving the use of ranks. Specifically, it is shown that the resulting test statistics are asymptotically equivalent to well-known statistics such as the Friedman test for concordance. It is also shown that empirical likelihood methods can be applied to the two-sample problem, as well as to various block design situations. In her paper **Jana Jurečková** highlights asymptotic behaviour in statistical estimation via describing some of the most distinctive differences between the asymptotic and finite-sample properties of estimators, mainly of robust ones. The latter are, in general, believed to be resistant to heavy-tailed distributions, but they can themselves be heavy-tailed. Indeed, as pointed out by Jurečková, many are not finite-sample admissible for any distribution, though they are asymptoticallly admissible. Hence, and also in view of some other examples she deals with in her paper, she rightly argues that before taking a recourse to asymptotics, we should analyzse the finite-sample behaviour of an estimator, whenever possible.

The Fields Institute announcement of our Ssymposium was also noticed by Dr. László Pordány, Aambassador for Hungary in Canada (2012). Seeing the programme, he wrote to Miklós, conveying his wish to receive the Hungarian participants of the conference in his ambassadorial residence. We, in turn, reciprocated with an invitation to His Excellency to attend, and also participate in, the opening of the symposium, that he gracefully accepted. Following the first -day programme, in the evening, Ambassador László Pordány and Mrs. Mária Csikós welcomed the Hungarian participants at the Aambassador's residence, and His Excellency used the occasion to speak *In Memoriam* **Sándor Csörgő** (Egerfarmos, 16 July 16, 1947 – Szeged, 14 February 14, 2008). We most sincerely thank Dr. László Pordány for the eminent role he played in making the first day of our conference especially memorable.

The occasion of presenting this volume also gives us the opportunity to sincerely thank the Fields Institute for Research in Mathematical Sciences for their financial support of our symposium. We hope very much that this volume, and the national and international success of our conference itself, will have justified their much appreciated trust in us.

Last, but not least, we most sincerely wish to thank Gillian Murray, the coordinator of our manifold LRSP (Laboratory for Research in Statistics and Probability) activities for more than three decades, for her help in preparing this volume, in collaboration with Rafal Kulik and Barbara Szyszkowicz, for publication, while in retirement now.

In conclusion, we also want to express our appreciation to the Editorial Board of the Fields Institute for their approval of the publication of these proceedings in their Communications series; to Carl R. Riehm, the Managing Editor of Publications, for his kind attention to, and sincere interest in, the publication of this volume, and to the Publications Manager, Debbie Iscoe, for her cooperation and expert help in its preparation for Springer. We hope very much that the readers will find this collection of papers, and our introductory comments on them, informative and also of interest in their studies and research work in Stochastics.

Ottawa, ON, Canada   Donald Dawson
                                                                       Rafal Kulik
                                                                Mohamedou Ould Haye
                                                                Barbara Szyszkowicz
                                                                     Yiqiang Zhao

References

1. *Asymptotic Methods in Probability and Statistics: A volume in Honour of Miklós Csörgő* (B. Szyszkowicz, Ed.). Elsevier, Amsterdam 1998.
2. Path Properties of Forty Years in Probability and Statistics: in Conversation with Miklós Csörgő, the Editors: Lajos Horváth and Barbara Szyszkowicz. *Technical Report Series of LRSP, No. 400, Carleton University-University of Ottawa 2004.*
3. *Asymptotic Methods in Stochastics: Festschrift for Miklós Csörgő* (L. Horváth, B. Szyszkowicz, Eds.). Fields Institute Communications, Volume 44, AMS 2004.

   - Fields Institute - Supplements:
     www.fields.utoronto.ca/publications/supplements/
   - *Path Properties of Forty Years of Research in Probability and Statistics. In Conversation with Miklós Csörgő*
   - *Miklós Csörgő List of Publications.*

Contents

Part I
Limit Theorems for Self-normalized Processes

Weak Convergence of Self-normalized Partial Sums Processes

Miklós Csörgő and Zhishui Hu

1 Introduction

Throughout this paper $\{X, X_n, n \geq 1\}$ denotes a sequence of independent and identically distributed (i.i.d.) non-degenerate random variables. Put $S_0 = 0$, and

$$S_n = \sum_{i=1}^n X_i, \quad \bar{X}_n = S_n/n, \quad V_n^2 = \sum_{i=1}^n X_i^2, \quad n \geq 1.$$

The quotient S_n/V_n may be viewed as a self-normalized sum. When $V_n = 0$ and hence $S_n = 0$, we define S_n/V_n to be zero. In terms of S_n/V_n, the classical Student statistic T_n is of the form

$$
\begin{aligned}
T_n(X) &= \frac{(1/\sqrt{n}) \sum_{i=1}^n X_i}{\left((1/(n-1)) \sum_{i=1}^n (X_i - \bar{X}_n)^2 \right)^{1/2}} \\
&= \frac{S_n/V_n}{\sqrt{(n - (S_n/V_n)^2)/(n-1)}}.
\end{aligned}
\tag{1}
$$

M. Csörgő
School of Mathematics and Statistics, Carleton University,
1125 Colonel By Drive, Ottawa, ON K1S 5B6, Canada
e-mail: mcsorgo@math.carleton.ca

Z. Hu (✉)
Department of Statistics and Finance, School of Management,
University of Science and Technology of China, Hefei, Anhui 230026, China
e-mail: huzs@ustc.edu.cn

© Springer Science+Business Media New York 2015
D. Dawson et al. (eds.), *Asymptotic Laws and Methods in Stochastics*,
Fields Institute Communications 76, DOI 10.1007/978-1-4939-3076-0_1

If T_n or S_n/V_n has an asymptotic distribution, then so does the other, and they coincide [cf. Efron [10]]. Throughout, $\overset{d}{\to}$ will indicate convergence in distribution, or weak convergence, in a given context, while $\overset{d}{=}$ will stand for equality in distribution.

The identification of possible limit distributions of normalized sums $Z_n = (S_n - A_n)/B_n$ for suitably chosen real constants $B_n > 0$ and A_n, the description of necessary and sufficient conditions for the distribution function of X such that the distributions of Z_n converge to a limit, were some of the fundamental problems in the classical theory of limit distributions for identically distributed summands [cf. Gnedenko and Kolmogorov [13]]. It is now well-known that Z_n has a non-degenerate asymptotic distribution for some suitably chosen real constants A_n and $B_n > 0$ if and only if X is in the domain of attraction of a stable law with index $\alpha \in (0, 2]$. When $\alpha = 2$, this is equivalent to $\ell(x) := EX^2 I(|X| \le x)$ being a slowly varying function as $x \to \infty$, one of the necessary and sufficient analytic conditions for $Z_n \overset{d}{\to} N(0, 1)$, $n \to \infty$ [cf. Theorem 1a in Feller [11], p. 313], i.e., for X to be in the domain of attraction of the normal law, written $X \in \mathrm{DAN}$. In this case A_n can be taken as nEX and $B_n = n^{1/2}\ell_X(n)$ with some function $\ell_X(n)$ that is slowly varying at infinity and determined by the distribution of X. Moreover, $\ell_X(n) = \sqrt{\mathrm{Var}\,(X)} > 0$ if $\mathrm{Var}\,(X) < \infty$, and $\ell_X(n) \nearrow \infty$ if $\mathrm{Var}\,(X) = \infty$. Also, X has moments of all orders less than 2, and variance of X is positive, but need not be finite. The function $\ell(x) = EX^2 I(|X| \le x)$ being slowly varying at ∞ is equivalent to having $x^2 P(|X| > x) = o(\ell(x))$ as $x \to \infty$, and thus also to having $Z_n \overset{d}{\to} N(0, 1)$ as $n \to \infty$. In a somewhat similar vein, Z_n having a non-degenerate limiting distribution when X is in the domain of attraction of a stable law with index $\alpha \in (0, 2)$ is equivalent to

$$1 - F(x) + F(-x) \sim \frac{2-\alpha}{\alpha} x^{-\alpha} h(x)$$

and

$$\frac{1 - F(x)}{1 - F(x) + F(-x)} \to p, \quad \frac{F(-x)}{1 - F(x) + F(-x)} \to q$$

as $x \to +\infty$, where $p, q \ge 0, p + q = 1$ and $h(x)$ is slowly varying at $+\infty$ [cf. Theorem 1a in Feller [11], p. 313]. Also, X has moments of all orders less than $\alpha \in (0, 2)$. The normalizing constants A_n and B_n, in turn, are determined in a rather complicated way by the slowly varying function h.

Now, in view of the results of Giné, Götze and Mason [12] and Chistyakov and Götze [2], the problem of finding suitable constants for Z_n having a non-degenerate limit in distribution when X is in the domain of attraction of a stable law with index $\alpha \in (0, 2]$ is eliminated via establishing the convergence in distribution of the self-normalized sums S_n/V_n or, equivalently, that of Student's statistic T_n, to a non-degenerate limit under the same necessary and sufficient conditions for X.

For X symmetric, Griffin and Mason [14] attribute to Roy Erickson a proof of the fact that having $S_n/V_n \xrightarrow{d} N(0,1)$, as $n \to \infty$, does imply that $X \in DAN$. Giné, Götze and Mason [12] proved the first such result for the general case of not necessarily symmetric random variables (cf. their Theorem 3.3), which reads as follows.

Theorem A. *The following two statements are equivalent:*

(a) $X \in DAN$ and $EX = 0$;
(b) $S_n/V_n \xrightarrow{d} N(0,1)$, $n \to \infty$.

Chistyakov and Götze [2], in turn, established the following global result (cf. their Theorem 1.1.) when X has a stable law with index $\alpha \in (0,2]$.

Theorem B. *The self-normalized sums S_n/V_n converge weakly as $n \to \infty$ to a random variable Z such that $P(|Z| = 1) < 1$ if and only if*

(i) X is in the domain of attraction of a stable law with index $\alpha \in (0,2]$;
(ii) $EX = 0$ if $1 < \alpha \le 2$;
(iii) if $\alpha = 1$, then X is in the domain of attraction of Cauchy's law and Feller's condition holds, that is, $\lim_{n\to\infty} nE\sin(X/a_n)$ exists and is finite, where $a_n = \inf\{x > 0 : nx^{-2}EX^2I(|X| < x) \le 1\}$.

Moreover, Chistyakov and Götze [2] also proved (cf. their Theorem 1.2) that the self-normalized sums S_n/V_n converge weakly to a *degenerate limit* Z if and only if $P(|X| > x)$ is a slowly varying function at $+\infty$.

Also, in comparison to the Giné et al. [12] result of Theorem A above that concludes the asymptotic *standard* normality of the sequence of self-normalized sums S_n/V_n if and only if $X \in DAN$ and $EX = 0$, Theorem 1.4 of Chistyakov and Götze [2] shows that S_n/V_n is asymptotically normal if and only if S_n/V_n is asymptotically *standard* normal.

We note in passing that Theorem 3.3 of Giné et al. [12] (cf. Theorem A) and the just mentioned Theorem 1.4 of Chistyakov and Götze [2] confirm the long-standing conjecture of Logan, Mallows, Rice and Shepp [21] (LMRS for short), stating in particular that " S_n/V_n is asymptotically normal if (and perhaps only if) X is in the domain of attraction of the normal law" (and X is centered). And in addition "It seems worthy of conjecture that the only possible nontrivial limiting distributions of S_n/V_n are those obtained when X follows a stable law". Theorems 1.1 and 1.2 of Chistyakov and Götze [2] (cf. Theorem B above and the paragraph right after) show that this second part of the long-standing LMRS conjecture also holds if one interprets nontrivial limit distributions as those, that are not concentrated at the points $+1$ and -1.

The proofs of the results of Chistyakov and Götze [2] (Theorems 1.1–1.7) are very demanding. They rely heavily on auxiliary results from probability theory and complex analysis that are proved in their Sect. 3 on their own.

As noted by Chistyakov and Götze [2], the "if" part of their Theorem 1.1 (Theorem B above) follows from the results of LMRS as well, while the "if" part of their Theorem 1.2 follows from Darling [8]. As described in LMRS [cf. Lemma 2.4 in Chistyakov and Götze [2]; see also Csörgő and Horváth [3], and S. Csörgő [7]], the class of limiting distributions for $\alpha \in (0, 2)$ does not contain Gaussian ones. For more details on the lines of research that in view of LMRS have led to Theorems A and B above, we refer to the respective introductions of Giné et al. [12] and Chistyakov and Götze [2].

Further to the lines of research in hand, it has also become well established in the past twenty or so years that limit theorems for self-normalized sums S_n/V_n often require fewer, frequently much fewer, moment assumptions than those that are necessary for their classical analogues [see, e.g. Shao [27]]. All in all, the asymptotic theory of self-normalized sums has much extended the scope of the classical theory. For a global overview of these developments we refer to the papers Shao [28–30], Csörgő et al. [5], Jing et al. [16], and to the book de la Peña, Lai and Shao [9].

In view of, and inspired by, the Giné et al. [12] result of Theorem A above, Csörgő, Szyszkowicz and Wang [4] established a self-normalized version of the weak invariance principle (sup-norm approximation in probability) under the same necessary and sufficient conditions. Moreover, Csörgő et al. [6] succeed in extending the latter weak invariance principle via weighted sup-norm and L_p-approximations, $0 < p < \infty$, in probability, again under the same necessary and sufficient conditions. In particular, for dealing with sup-norm approximations, let Q be the class of positive functions $q(t)$ on $(0, 1]$, i.e., $\inf_{\delta \leq t \leq 1} q(t) > 0$ for $0 < \delta < 1$, which are nondecreasing near zero, and let

$$I(q, c) := \int_{0+}^{1} t^{-1} \exp(-cq^2(t)/t)dt, \quad 0 < c < \infty.$$

Then [cf. Corollary 3 in Csörgő et al. [6]], *on assuming that $q \in Q$, the following two statements are equivalent*:

(a) $X \in$ DAN *and* $EX = 0$;
(b) *On an appropriate probability space for* $X, X_1, X_2, \ldots$, *one can construct a standard Wiener process* $\{W(s), 0 \leq s < \infty\}$ *so that, as* $n \to \infty$,

$$\sup_{0 < t \leq 1} \left| S_{[nt]}/V_n - W(nt)/n^{1/2} \right| \Big/ q(t) = o_P(1) \qquad (2)$$

if and only if $I(q, c) < \infty$ *for all* $c > 0$.

With $q(t) = 1$ on $(0, 1]$, this is Theorem 1 of Csörgő et al. [4], and when $\sigma^2 = EX^2 < \infty$, then (2) combined with Kolmogorov's law of large numbers results in the classical weak invariance principle that in turn yields Donsker's classical functional CLT.

This work was inspired by the Chistyakov and Götze [2] result of Theorem B above. Our main aim is to identify the limiting distribution in the latter theorem under the same necessary and sufficient conditions in terms of weak convergence on $D[0, 1]$ (cf. Theorem 1). Our auxiliary Lemma 1 may be viewed as a scalar normalized version of Theorem B (Theorem 1.1 of Chistyakov and Götze [2]).

2 Main Results

An **R**-valued stochastic process $\{X(t), t \geq 0\}$ is called a *Lévy process*, if the following four conditions are satisfied:

(1) it starts at the origin, i.e. $X(0) = 0$ a.s.;
(2) it has independent increments, that is, for any choice of $n \geq 1$ and $0 \leq t_0 < t_1 < \cdots < t_n$, the random variables $X(t_0)$, $X(t_1) - X(t_0), \cdots, X(t_n) - X(t_{n-1})$ are independent;
(3) it is time homogeneous, that is, the distribution of $\{X(t+s) - X(s) : t \geq 0\}$ does not depends on s;
(4) as a function of t, $X(t, \omega)$ is a.s. right-continuous with left-hand limits.

A *Lévy process* $\{X(t), t \geq 0\}$ is called α-*stable* (with index $\alpha \in (0, 2]$) if for any $a > 0$, there exists some $c \in \mathbf{R}$ such that $\{X(at)\} \stackrel{d}{=} \{a^{1/\alpha} X(t) + ct\}$. If $\{X(t), t \geq 0\}$ is an α-*stable Lévy process*, then for any $t \geq 0$, $X(t)$ has a stable distribution. For more details about Lévy and α-stable Lévy processes, we refer to Bertoin [1] and Sato [26].

It is well known that G is a stable distribution with index $\alpha \in (0, 2]$ if and only if its characteristic function $f(t) = \int_{-\infty}^{\infty} e^{itx} dG(x)$ admits the representation (see for instance Feller [11])

$$f(t) = \begin{cases} \exp\left\{i\gamma t + c|t|^\alpha \frac{\Gamma(3-\alpha)}{\alpha(\alpha-1)}\left[\cos\frac{\pi\alpha}{2} + i(p-q)\frac{t}{|t|}\sin\frac{\pi\alpha}{2}\right]\right\}, & \text{if } \alpha \neq 1; \\ \exp\left\{i\gamma t - c|t|\left[\frac{\pi}{2} + i(p-q)\frac{t}{|t|}\log|t|\right]\right\}, & \text{if } \alpha = 1, \end{cases} \tag{3}$$

where c, p, q, γ are real constants with $c, p, q \geq 0$, $p + q = 1$. Write $G \sim S(\alpha, \gamma, c, p, q)$ and, as in Theorem B, let

$$a_n = \inf\{x > 0 : nx^{-2} EX^2 I(|X| < x) \leq 1\}.$$

The following result is our main theorem.

Theorem 1. *Let* $X, X_1, X_2, \cdots$ *be a sequence of i.i.d. non-degenerate random variables and let* $G \sim S(\alpha, \gamma, c, p, q)$. *If* X *is in the domain of attraction of* G *of index* $\alpha \in (0, 2]$, *with* $EX = 0$ *if* $1 < \alpha \leq 2$ *and* $\lim\limits_{n\to\infty} nE\sin(X/a_n)$ *exists and is finite if* $\alpha = 1$, *then, as* $n \to \infty$, *we have*

$$\frac{S_{[nt]}}{V_n} \stackrel{d}{\to} \frac{X(t)}{\sqrt{[X]_1}}$$

on $D[0, 1]$, *equipped with the Skorokhod J_1 topology, where $X(t)$ is an α-stable Lévy process of index $\alpha \in (0, 2]$ on $[0, 1]$, $X(1) \sim S(\alpha, \gamma', 1, p, q)$ with $\gamma' = 0$ if $\alpha \neq 1$ and $\gamma' = \lim_{n \to \infty} nE \sin(X/a_n)$ if $\alpha = 1$, and $[X]_t$ is the quadratic variation of $X(t)$.*

When $\alpha = 2$, G is a normal distribution, $X(1) \overset{d}{=} N(0, 1)$ and $[X]_1 = 1$. Consequently, $X(t)/\sqrt{[X]_1}$ is a standard Brownian motion and thus we obtain the weak convergence of $S_{[nt]}/V_n$ to a Brownian motion as in (c) of Theorem 1 of Csörgő et al. [4] [see also our lines right after (2)].

Consider now the sequence $T_{n,t}$ of Student processes in $t \in [0, 1]$ on $D[0, 1]$, defined as

$$\{T_{n,t}(X), \ 0 \leq t \leq 1\} := \left\{ \frac{(1/\sqrt{n}) \sum_{i=1}^{[nt]} X_i}{\left((1/(n-1)) \sum_{i=1}^{n}(X_i - \bar{X}_n)^2\right)^{1/2}}, \ 0 \leq t \leq 1 \right\}$$

$$= \left\{ \frac{S_{[nt]}/V_n}{\sqrt{(n - (S_n/V_n)^2)/(n-1)}}, \ 0 \leq t \leq 1 \right\}. \tag{4}$$

Clearly, $T_{n,1}(X) = T_n(X)$, with the latter as in (1). Clearly also, in view of Theorem 1, the same result continues to hold true under the same conditions for the Student process $T_{n,t}$ as well, i.e., Theorem 1 can be restated in terms of the latter process. Moreover, if $1 < \alpha \leq 2$, then $EX =: \mu$ exists and the following corollary obtains.

Corollary 1. *Let $X, X_1, X_2, \cdots$ be a sequence of i.i.d. non-degenerate random variables and let $G \sim S(\alpha, \gamma, c, p, q)$. If X is in the domain of attraction of G of index $\alpha \in (1, 2]$, then, as $n \to \infty$, we have*

$$T_{n,t}(X - \mu) = \frac{(1/\sqrt{n}) \sum_{i=1}^{[nt]}(X_i - \mu)}{\left((1/(n-1)) \sum_{i=1}^{n}(X_i - \bar{X}_n)^2\right)^{1/2}} \overset{d}{\to} \frac{X(t)}{\sqrt{[X]_1}}$$

on $D[0, 1]$, equipped with the Skorokhod J_1 topology, where $X(t)$ is an α-stable Lévy process of index $\alpha \in (1, 2]$ on $[0, 1]$, $X(1) \sim S(\alpha, 0, 1, p, q)$, and $[X]_t$ is the quadratic variation of $X(t)$.

As noted earlier, with $\alpha = 2$, $X(t)/\sqrt{[X]_1}$ is a standard Brownian motion. Moreover, in the latter case, we have $(X - \mu) \in$ DAN and this, in turn, is equivalent to having (2) with $T_{n,t}(X - \mu)$ as well, instead of $S_{[nt]}/V_n$ [cf. Corollary 3.5 in Csörgő et al. [5]].

Corollary 1 extends the feasibility of the use of the Student process $T_{n,t}(X - \mu)$ for constructing functional asymptotic confidence intervals for μ, along the lines of Martsynyuk [22, 23], beyond $X - \mu$ being in the domain of attraction of the normal law.

Via the proof of Theorem 1 we can also get a weak convergence result when X belongs to the domain of partial attraction of an infinitely divisible law (cf. Feller [11], p. 590).

Theorem 2. *Let $X(t)$ be a Lévy process with $[X]_1 \neq 0$, where $[X]_t$ is the quadratic variation of $X(t)$. If there exist some positive constants $\{b_n\}$ and some subsequence $\{m_n\}$, where $m_n \to \infty$ as $n \to \infty$, such that $S_{m_n}/b_n \overset{d}{\to} X(1)$ as $n \to \infty$, then $S_{[m_n t]}/V_{m_n} \overset{d}{\to} X(t)/\sqrt{[X]_1}$ on $D[0,1]$, equipped with the Skorokhod J_1 topology.*

As will be seen, in the proof of Theorems 1 and 2 we make use of a weak convergence result for sums of *exchangeable random variables*. For any finite or infinite sequence $\xi = (\xi_1, \xi_2, \cdots)$, we say ξ is exchangeable if

$$(\xi_{k_1}, \xi_{k_2}, \cdots) \overset{d}{=} (\xi_1, \xi_2, \cdots)$$

for any finite permutation $(k_1, k_2, \cdots)$ of **N**. A *process $X(t)$* on $[0,1]$ is *exchangeable* if it is continuous in probability with $X_0 = 0$ and has exchangeable increments over any set of disjoint intervals of equal length. Clearly, a Lévy process is exchangeable.

By using the notion of exchangeability, we can get the following corollary from the proof of Theorem 1.

Corollary 2. *Let $X, X_1, X_2, \cdots$ and G be as in Theorem 1. If X is in the domain of attraction of G of index $\alpha \in (0,2]$, with $EX = 0$ if $1 < \alpha \leq 2$ and $\lim_{n\to\infty} nE\sin(X/a_n)$ exists and is finite if $\alpha = 1$, then, as $n \to \infty$,*

$$\left(\frac{S_n}{a_n}, \frac{V_n^2}{a_n^2}, \frac{\max_{1\leq i \leq n} |X_i|}{a_n}\right) \overset{d}{\to} (X(1), [X]_1, J), \tag{5}$$

where, with $\Delta X(t) := X(t) - X(t-)$, $J = \max\{|\Delta X(t)| : 0 \leq t \leq 1\}$ is the biggest jump of $X(t)$ on $[0,1]$, where, as in Theorem 1, $X(t)$ is an α-stable Lévy process with index $\alpha \in (0,2]$ on $[0,1]$, $X(1) \sim S(\alpha, \gamma', 1, p, q)$ as specified in Theorem 1, and $[X]_t$ is the quadratic variation of $X(t)$.

We note in passing that, under the conditions of Corollary 2, the joint convergence in distribution as $n \to \infty$

$$\left(\frac{S_n}{a_n}, \frac{V_n^2}{a_n^2}\right) \overset{d}{\to} (X(1), [X]_1) \tag{6}$$

amounts to *an extension of Raikov's theorem from $X \in DAN$ to X being in the domain of attraction of G of index $\alpha \in (0,2]$.* When $\alpha = 2$, i.e., when $X \in DAN$, the statement of (6) reduces to Raikov's theorem in terms of having $\left(\frac{S_n}{a_n}, \frac{V_n^2}{a_n^2}\right) \overset{d}{\to}$ $(N(0,1), 1)$ as $n \to \infty$ (cf. Lemma 3.2 in Giné et al. [12]).

As a consequence of Corollary 2, *under the same conditions*, as $n \to \infty$, we have

$$\frac{\max_{1 \leq i \leq n} |X_i|}{S_n} \xrightarrow{d} \frac{J}{X(1)}, \tag{7}$$

and

$$\frac{\max_{1 \leq i \leq n} |X_i|}{V_n} \xrightarrow{d} \frac{J}{\sqrt{[X]_1}}. \tag{8}$$

In case of $\alpha = 2$, G is a normal distribution, $X \in DAN$ with $EX = 0$, and $X(t)/\sqrt{[X]_1}$ is a standard Brownian motion. Consequently, J in Corollary 2 is zero and, as $n \to \infty$, we arrive at the conclusion that when $X \in DAN$ and $EX = 0$, then the respective conclusions of (7) and (8) reduce to $\max_{1 \leq i \leq n} |X_i|/|S_n| \xrightarrow{P} 0$ and $\max_{1 \leq i \leq n} |X_i|/V_n \xrightarrow{P} 0$. Kesten and Maller ([20], Theorem 3.1) proved that $\max_{1 \leq i \leq n} |X_i|/|S_n| \xrightarrow{P} 0$ is equivalent to having

$$\frac{x|EXI(|X| \leq x)| + EX^2I(|X| \leq x)}{x^2 P(|X| > x)} \to \infty,$$

and O'Brien[24] showed that $\max_{1 \leq i \leq n} |X_i|/V_n \xrightarrow{P} 0$ is equivalent to $X \in DAN$.

For X in the domain of attraction of a stable law with index $\alpha \in (0, 2)$, Darling [8] studied the asymptotic behavior of $S_n / \max_{1 \leq i \leq n} |X_i|$ and derived the characteristic function of the appropriate limit distribution. Horváth and Shao [15] established a large deviation and, consequently, the law of the iterated logarithm for $S_n / \max_{1 \leq i \leq n} |X_i|$ under the same condition for X symmetric.

Proofs of Theorems 1, 2 and Corollary 2 are given in Sect. 3.

3 Proofs

Before proving Theorem 1, we conclude the following lemma.

Lemma 1. *Let $G \sim S(\alpha, \gamma, c, p, q)$ with index $\alpha \in (0, 2]$ and let Y_α be a random variable associated with this distribution. If there exist some positive constants $\{A_n\}$ satisfying $S_n/A_n \xrightarrow{d} Y_\alpha$ as $n \to \infty$, then*

(1) X is in the domain of attraction of G,
(2) $EX = 0$ if $1 < \alpha \leq 2$, and $\lim_{n \to \infty} nE \sin(X/a_n)$ exists and is finite if $\alpha = 1$.

Conversely, if the above conditions (1) and (2) hold, then

$$S_n/a_n \xrightarrow{d} Y'_\alpha, \quad \alpha \in (0, 2],$$

where Y'_α is a random variable with distribution $G' \sim S(\alpha, \gamma', 1, p, q)$, with $\gamma' = 0$ if $\alpha \neq 1$ and $\gamma' = \lim\limits_{n\to\infty} nE\sin(X/a_n)$ if $\alpha = 1$.

Proof. If $\alpha = 2$, then G is a normal distribution and the conclusion with $X \in DAN$ and $EX = 0$ is clear.

If $0 < \alpha < 2$, then X belongs to the domain of attraction of a stable law G with the characteristic function $f(t)$ as in (3) if and only if (cf. Theorem 2 in Feller [11], p. 577)

$$\ell(x) = EX^2 I(|X| \leq x) = x^{2-\alpha}L(x), \ x \to \infty,$$

and

$$\frac{P(X > x)}{P(|X| > x)} \to p, \ x \to \infty,$$

where $L(x)$ is a slowly varying function at infinity. In this case, as $n \to \infty$, we have (cf. Theorem 3 in Feller [11], p. 580)

$$\frac{S_n}{a_n} - b_n \xrightarrow{d} \tilde{Y}_\alpha \text{ with distribution } \tilde{G}, \ \alpha \in (0, 2), \tag{9}$$

where

$$b_n = \begin{cases} (n/a_n)EX, & \text{if } 1 < \alpha < 2; \\ nE\sin(X/a_n), & \text{if } \alpha = 1; \\ 0, & \text{if } 0 < \alpha < 1, \end{cases}$$

and $\tilde{G} \sim S(\alpha, 0, 1, p, q)$. Thus if (2) holds, then, as $n \to \infty$, we have

$$\frac{S_n}{a_n} \xrightarrow{d} Y'_\alpha \text{ with distribution } G', \ \alpha \in (0, 2),$$

where $G' \sim S(\alpha, \gamma', 1, p, q)$ with $\gamma' = 0$ if $\alpha \neq 1$ and $\gamma' = \lim\limits_{n\to\infty} nE\sin(X/a_n)$ if $\alpha = 1$.

If, as $n \to \infty$, there exists some positive constants $\{A_n\}$ satisfying $S_n/A_n \xrightarrow{d} Y_\alpha$ with distribution G with index $\alpha \in (0, 2)$, then (1) holds. Hence (9) is also true. Consequently, by Theorem 1.14 in Petrov [25], we have $b_n \to b$ for some real constant b, as $n \to \infty$. Thus if $\alpha = 1$, then $\lim\limits_{n\to\infty} nE\sin(X/a_n)$ exists and is finite, and if $1 < \alpha < 2$, since in this case $n/a_n = na_n^{-\alpha}L(a_n)(a_n^{\alpha-1}/L(a_n)) \sim a_n^{\alpha-1}/L(a_n) \to \infty$ as $n \to \infty$, we have $EX = 0$.

Proof of Lemma 1 is now complete. $\qquad\qquad\qquad\qquad\qquad\qquad\qquad\qquad\square$

Proof of Theorem 1. Since $X(t)$ is a Lévy process, we have the Lévy-Itô decomposition (see for instance Corollary 15.7 in Kallenberg [18])

$$X(t) = bt + \sigma W(t) + \int_0^t \int_{|x| \le 1} x(\eta - E\eta)(ds, dx) + \int_0^t \int_{|x| > 1} x\eta(ds, dx), \quad (10)$$

for some $b \in \mathbf{R}, \sigma \ge 0$, where $W(t)$ is a Brownian motion independent of η, and $\eta = \sum_t \delta_{t, \Delta X_t}$ is a Poisson process on $(0, \infty) \times (\mathbf{R} \setminus \{0\})$ with $E\eta = \lambda \otimes v$, where $\Delta X_t = X_t - X_{t-}$ is the jump of X at time t, λ is the Lebesgue measure on $(0, \infty)$ and v is some measure on $\mathbf{R} \setminus \{0\}$ with $\int (x^2 \wedge 1) v(dx) < \infty$. The quadratic variation of $X(t)$ is (cf. Corollary 26.15 in Kallenberg [18])

$$[X]_t = \sigma^2 t + \sum_{s \le t} (\Delta X_s)^2. \quad (11)$$

Noting that a Lévy Process is exchangeable, by Theorem 2.1 of Kallenberg [17] (or Theorem 16.21 in Kallenberg [18]), $X(t)$ has a version $X'(t)$, with representation

$$X'(t) = b't + \sigma' B(t) + \sum_j \beta_j (I(\tau_j \le t) - t), \quad t \in [0, 1], \quad (12)$$

in the sense of a.s. uniform convergence, where

(1) $b' = X(1)$, $\sigma' \ge 0$, $\beta_1 \le \beta_3 \le \cdots \le 0 \le \cdots \le \beta_4 \le \beta_2$ are random variables with $\sum_j \beta_j^2 < \infty$, a.s.,
(2) $B(t)$ is a Brownian bridge on $[0, 1]$,
(3) $\tau_1, \tau_2, \cdots$ are independent and uniformly distributed random variables on $[0, 1]$,

and the three groups (1)–(3) of random elements are independent. $X(t)$ has a version $X'(t)$ means that for any $t \in [0, 1]$, $X(t) = X'(t)$ a.s. But since both $X(t)$ and $X'(t)$ are right continuous, we have

$$P(X(t) = X'(t) \text{ for all } t \in [0, 1]) = 1.$$

Thus we may say that $X(t) \equiv X'(t)$ on $[0, 1]$. By (12), we get that $(\beta_1, \beta_2, \cdots)$ are the sizes of the jumps of $\{X(t), t \in [0, 1]\}$ and $(\tau_1, \tau_2, \cdots)$ are the related jump times. Thus $\eta = \sum_j \delta_{\tau_j, \beta_j}$ on $(0, 1] \times (\mathbf{R} \setminus \{0\})$ and, by (11),

$$[X]_1 = \sigma^2 + \sum_{s \le 1} (\Delta X_s)^2 = \sigma^2 + \sum_j \beta_j^2.$$

We are to see now that we also have

$$\sigma' = \sigma. \quad (13)$$

Write

$$X^n(t) = bt + \sigma W(t) + \int_0^t \int_{1/n < |x| \le 1} x(\eta - E\eta)(ds, dx) + \int_0^t \int_{|x| > 1} x\eta(ds, dx)$$

$$= \tilde{b}_n t + \sigma \tilde{B}(t) + \sum_{|\beta_j| > 1/n} \beta_j I(\tau_j \le t), \quad n \ge 1,$$

where $\tilde{b}_n = b + \sigma W(1) - \int xI(1/n < |x| \le 1)\nu(dx)$ and $\tilde{B}(t) = W(t) - tW(1)$ is a Brownian bridge. Noting that $W(1)$ and $\{\tilde{B}(t)\}$ are independent, $X^n(t)$ is also an exchangeable process for each $n \ge 1$. From the proof of Theorem 15.4 in Kallenberg [18], we have

$$E \sup_{0 \le s \le 1} (X(s) - X^n(s))^2 \to 0, \quad n \to \infty.$$

Thus, as $n \to \infty$, $X^n(t) \stackrel{d}{\to} X(t)$ on $D(0, 1)$ with the Skorokhod J_1 topology. Then, by Theorem 3.8 in Kallenberg [19], as $n \to \infty$, we have

$$\sigma^2 + \sum_{|\beta_j| > 1/n} \beta_j^2 \stackrel{d}{\longrightarrow} \sigma'^2 + \sum_j \beta_j^2.$$

Hence $\sigma'^2 = \sigma^2$, and (13) holds.

By Lemma 1, $S_n/a_n \stackrel{d}{\to} X(1)$. Hence, by Theorem 16.14 in Kallenberg [18], we have $S_{[nt]}/a_n \stackrel{d}{\to} X(t)$ on $D(0, 1)$ with the Skorokhod J_1 topology. By noting that $\{X_i/a_n, i = 1, \cdots, n\}$ are exchangeable random variables for each n, and by using Theorems 3.8 and 3.13 in Kallenberg [19], as $n \to \infty$, we have

$$\left(\frac{S_n}{a_n}, \sum_{i=1}^n \frac{X_i^2}{a_n^2}, \sum_{i=1}^n \delta_{X_i/a_n}\right) \stackrel{vd}{\longrightarrow} \left(X(1), [X]_1, \sum_j \delta_{\beta_j}\right) \text{ in } \mathbf{R} \times \mathbf{R}_+ \times \mathcal{N}(\mathbf{R} \setminus \{0\}), \quad (14)$$

where $\stackrel{vd}{\longrightarrow}$ means convergence in distribution with respect to the vague topology, and $\mathcal{N}(\mathbf{R} \setminus \{0\})$ is the space of integer-valued measures on $\mathbf{R} \setminus \{0\}$ endowed with the vague topology. Hence

$$\left(\frac{S_n}{V_n}, \sum_{i=1}^n \frac{X_i^2}{V_n^2}, \sum_{i=1}^n \delta_{X_i/V_n}\right) \stackrel{vd}{\longrightarrow} \left(\frac{X(1)}{\sqrt{[X]_1}}, 1, \sum_j \delta_{\beta_j/\sqrt{[X]_1}}\right) \text{ in } \mathbf{R} \times \mathbf{R}_+ \times \mathcal{N}(\mathbf{R} \setminus \{0\}).$$

Since $\{X_i/V_n, i = 1, \cdots, n\}$ are exchangeable for each n, by Theorems 3.8 and 3.13 in Kallenberg [19], we have

$$\frac{S_{[nt]}}{V_n} \stackrel{d}{\to} \frac{X(t)}{\sqrt{[X]_1}}$$

on $D[0, 1]$, equipped with the Skorokhod J_1 topology. $\qquad\square$

Proof of Theorem 2. It is similar to the proof of Theorem 1 with only minor changes. Hence we omit the details. □

Proof of Corollary 2. Note that (14) is equivalent to (see remarks below Theorem 2.2 of Kallenberg [17])

$$\left(\frac{S_n}{a_n}, \sum_{i=1}^{n} \frac{X_i^2}{a_n^2}, \frac{X_{n1}}{a_n}, \frac{X_{n2}}{a_n}, \cdots \right) \xrightarrow{d} (X(1), [X]_1, \beta_1, \beta_2, \cdots) \text{ in } \mathbf{R}^{\infty}, \tag{15}$$

where $X_{n1} \leq X_{n3} \leq \cdots \leq 0 \leq \cdots \leq X_{n4} \leq X_{n2}$ are obtained by ordering $\{X_i, 1 \leq i \leq n\} \cup \{\tilde{X}_i, i > n\}$ with $\tilde{X}_i \equiv 0$, $i = 1, 2, \cdots$. Now the conclusion of (5) follows directly from (15). □

Acknowledgements We wish to thank two referees for their careful reading of our manuscript. The present version reflects their much appreciated remarks and suggestions. In particular, we thank them for calling our attention to the newly added reference Kallenberg [19], and for advising us that the proof of our Theorem 2.1 needs to be done more carefully, taking into account the remarks made in this regard. The present revised version of the proof of our Theorem 2.1 is done accordingly, with our sincere thanks attached herewith.

This research was supported by an NSERC Canada Discovery Grant of Miklós Csörgő at Carleton University and, partially, also by NSFC(No.10801122), the Fundamental Research Funds for the Central Universities, obtained by Zhishui Hu.

References

1. Bertoin, J.: Lévy Processes. Cambridge University Press, Cambridge (1996)
2. Chistyakov, G.P., Götze, F.: Limit distributions of Studentized means. Ann. Probab. **32**, 28–77 (2004)
3. Csörgő, M., Horváth, L.: Asymptotic representations of self-normalized sums. Probab. Math. Stat. **9**, 15–24 (1988)
4. Csörgő, M., Szyszkowicz, B., Wang, Q.: Donsker's theorem for self-normalized partial sums processes. Ann. Probab. **31**, 1228–1240 (2003)
5. Csörgő, M., Szyszkowicz, B., Wang, Q.: On weighted approximations and strong limit theorems for self-normalized partial sums processes. In: Horváth, L., Szyszkowicz, B. (eds.) Asymptotic Methods in Stochastics. Fields Institute Communications, vol. 44, pp. 489–521. American Mathematical Society, Providence (2004)
6. Csörgő, M., Szyszkowicz, B., Wang, Q.: On weighted approximations in $D[0, 1]$ with application to self-normalized partial sum processes. Acta Math. Hung. **121**, 307–332 (2008)
7. Csörgő, S.: Notes on extreme and self-normalised sums from the domain of attraction of a stable law. J. Lond. Math. Soc. **39**, 369–384 (1989)
8. Darling, D.A.: The influence of the maximum term in the addition of independent random variables. Trans. Am. Math. Soc. **73**, 95–107 (1952)
9. de la Peña, V.H., Lai, T.L., Shao, Q.-M.: Self-normalized Processes: Limit Theory and Statistical Applications. Probability and Its Applications (New York). Springer, Berlin (2009)
10. Efron, B.: Student's t-test under symmetry conditions. J. Am. Stat. Assoc. **64**, 1278–1302 (1969)
11. Feller, W.: An Introduction to Probability Theory and Its Applications, vol. 2. Wiley, New York (1971)

12. Giné, E., Götze, F., Mason D.: When is the Student t-statistic asymptotically standard normal? Ann. Probab. **25**, 1514–1531 (1997)
13. Gnedenko, B.V., Kolmogorov, A.N.: Limit distributions for sums of independent random variables. Addison-Wesley, Cambridge (1968)
14. Griffin, P.S., Mason, D.M.: On the asymptotic normality of self-normalized sums. Proc. Camb. Philos. Soc. **109**, 597–610 (1991)
15. Horváth, L., Shao, Q.-M.: Large deviations and law of the iterated logarithm for partial sums normalized by the largest absolute observation. Ann. Probab. **24**, 1368–1387 (1996)
16. Jing, B.-Y., Shao, Q.-M., Zhou, W.: Towards a universal self-normalized moderate deviation. Trans. Am. Math. Soc. **360**, 4263–4285 (2008)
17. Kallenberg, O.: Canonical representations and convergence criteria for processes with interchangeable increments. Z. Wahrsch. Verw. Geb. **27**, 23–36 (1973)
18. Kallenberg, O.: Foundations of Modern Probability. Springer, New York (2002)
19. Kallenberg, O.: Probabilistic Symmetries and Invariance Principles. Springer, New York (2005)
20. Kesten, H., Maller, R.A.: Infinite limits and infinite limit points for random walks and trimmed sums. Ann. Probab. **22**, 1475–1513 (1994)
21. Logan, B.F., Mallows, C.L., Rice, S.O., Shepp, L.A.: Limit distributions of self-normalized sums. Ann. Probab. **1**, 788–809 (1973)
22. Martsynyuk, Yu.V.: Functional asymptotic confidence intervals for the slope in linear error-in-variables models. Acta Math. Hung. **123**, 133–168 (2009a)
23. Martsynyuk, Yu.V.: Functional asymptotic confidence intervals for a common mean of independent random variables. Electron. J. Stat. **3**, 25–40 (2009b)
24. O'Brien, G.L.: A limit theorem for sample maxima and heavy branches in Galton-Watson trees. J. Appl. Probab. **17**, 539–545 (1980)
25. Petrov, V.V.: Limit Theorems of Probability Theory, Sequences of Independent Random Variables. Clarendon, Oxford (1995)
26. Sato, K.: Lévy Processes and Infinitely Divisible Distributions. Cambridge University Press, Cambridge (1999)
27. Shao, Q.-M.: Self-normalized large deviations. Ann. Probab. **25**, 285–328 (1997)
28. Shao, Q.-M.: Recent developments on self-normalized limit theorems. In: Szyszkowicz, B. (ed.) Asymptotic Methods in Probability and Statistics. A Volume in Honour of M. Csörgő, pp. 467–480. North-Holland, Amsterdam (1998)
29. Shao, Q.-M.: Recent progress on self-normalized limit theorems. In: Lai, T.L., Yang, H., Yung, S.P. (eds.) Probability, Finance and Insurance. World Scientific, Singapore (2004)
30. Shao, Q.-M.: Stein's method, self-normalized limit theory and applications. In: Proceedings of the International Congress of Mathematicians 2010 (ICM 2010), Hyderabad, pp. 2325–2350 (2010)

Precise Asymptotics in Strong Limit Theorems for Self-normalized Sums of Multidimensionally Indexed Random Variables

Dianliang Deng and Zhitao Hu

It is a great pleasure for us to dedicate this paper in honour of Professor Miklós Csörgő's work on the occasion of his 80th birthday

1 Introduction

Let $\{X, X_n, X_\mathbf{n}; n \in Z_+, \mathbf{n} \in Z_+^d\}$ be the independent and identically distributed (i.i.d.) random variables on a probability space $(\Omega, \mathscr{F}, P)$ where Z_+ denote the set of positive integers and Z_+^d denote the positive integer d-dimensional lattice with coordinate-wise partial ordering $\leq$. The notation $\mathbf{m} \leq \mathbf{n}$, where $\mathbf{m} = (m_1, m_2, \ldots, m_d)$ and $\mathbf{n} = (n_1, n_2, \ldots, n_d)$, thus means that $m_k \leq n_k$, for $k = 1, 2, \ldots, d$ and also $|\mathbf{n}|$ denotes $\prod_{k=1}^{d} n_k$, $\mathbf{n} \to \infty$ means $n_k \to \infty$ for $k = 1, 2, \ldots, d$. Set $S_n = \sum_{i=1}^{n} X_i$, $W_n^2 = \sum_{i=1}^{n} X_i^2$, $S_\mathbf{n} = \sum_{\mathbf{k} \leq \mathbf{n}} X_\mathbf{k}$, and $W_\mathbf{n}^2 = \sum_{\mathbf{k} \leq \mathbf{n}} X_\mathbf{k}^2$ where $n \in Z_+$ and $\mathbf{n} \in Z_+^d$.

In the classical limit theory, the concentration is on the asymptotic properties for the normalized partial sum $S_n/(EW_n^2)^{1/2}$ under the finite second moment assumption. However the current concern is to study the same properties for the self-normalized sum S_n/W_n without the finite moment assumption and many results have been obtained on this topic. Griffin and Kuelbs [8] established a self-normalized law of the iterated logarithm for all distributions in the domain of attraction of a normal or stable law. Shao [23] derived the self-normalized large deviation for arbitrary random variables without any moment conditions. In addition, Slavova [25], Hall [14], and Nagaev [20] obtained the Berry-Esseen bounds. Wang and Jing [29] derived exponential nonuniform Berry-Esseen bounds. Further results for self-normalized sums include large deviation (see [4, 28]) Cramér type results (see

D. Deng (✉)
Department of Mathematics and Statistics, University of Regina,
Regina, SK S4S 0A2, Canada
e-mail: deng@uregina.ca

Z. Hu
Department of Mathematics and Information, Chang'An University,
Nan Er Huan Zhong Duan, XiAn, Shanxi, 710064, China

© Springer Science+Business Media New York 2015
D. Dawson et al. (eds.), *Asymptotic Laws and Methods in Stochastics*,
Fields Institute Communications 76, DOI 10.1007/978-1-4939-3076-0_2

[24, 27]), Darling-Erdös theorem and Donsker's theorem (see [2, 3]), Kolmogorov and Erdös test (see [29]) and the law of iterated logarithm (see [5, 9, 22]), among many others. The known results have shown that comparing the same problems in the standard normalization, the moment assumptions under self-normalization can be eliminated and fundamental properties can be maintained much better by self-normalization than deterministic normalization. Furthermore, the limit theorems of self-normalized sums have resulted in more and more attention and been widely used in statistical analysis. Griffin and Mason [10] derived the asymptotic normality. Giné, Götze and Mason [7] studied the asymptotic properties of the Student t-statistic $T_n = \frac{S_n}{W_n} \left(\frac{n-1}{n-(S_n/W_n)^2} \right)^{1/2}$. Mason and Shao [19] extended this result to the bootstrapped Student t-statistic. Most recently, Csörgő and Martsynyuk [1] established functional central limit theorems for self-normalized type versions of the vector of the introduced least squares processes for (β, α), as well as for their various marginal counterparts. They also discussed joint and marginal central limit theorems for Studentized and self-normalized type least square estimators of the slope and intercept. The results obtained in Csörgő and Martsynyuk [1] provide a source for completely data-based asymptotic confidence intervals for β and α.

However, the recent interest lies in the precise asymptotics for self-normalized sum S_n/W_n. Zhao and Tao [30] obtained the following result.

Theorem 1. *Suppose that $EX = 0$ and $l(x) = EX^2 I\{|X| \leq x\}$ is a slowly varying function at ∞. Then for any $\beta > 0$ and $\delta > \max(-1, 2/\beta - 1)$,*

$$\lim_{\epsilon \to 0^+} \epsilon^{\beta(\delta+1)-2} \sum_{n=2}^{\infty} \frac{(\log n)^{\delta-2/\beta}}{n} \times$$

$$E\left[\left(\frac{S_n}{W_n} \right)^2 I\left(\left| \frac{S_n}{W_n} \right| \geq \epsilon(\log n)^{1/\beta} \right) \right] = \frac{\beta E|N|^{\beta(\delta+1)}}{\beta(\delta+1)-2} \tag{1}$$

This theorem extends result in Liu and Lin [18] from the normalized sum to the self-normalized sum. On the other hand, Pang et. al. [21] obtained the precise asymptotics of the law of iterated logarithm (LIL) for self-normalized sums, which can be thought of as the extension of the result on the precise asymptotics of LIL obtained in Gut and Spătaru [13].

Theorem 2. *Suppose that X is symmetric with $EX = 0$ and $l(x) = EX^2 I\{|X| \leq x\}$ is a slowly varying function at ∞, satisfying $l(x) \leq c_1 \exp(c_2(\log x)^\beta)$ for some $c_1 > 0, c_2 > 0$ and $0 \leq \beta < 1$. Let $a > -1$ and $b > -1/2$. Assume that $\alpha_n(\epsilon)$ is a nonnegative function of ϵ such that*

$$\alpha_n(\epsilon) \log\log n \to \tau \quad as \quad n \to \infty \quad and \quad \epsilon \searrow \sqrt{1+a}.$$

Then

$$\lim_{\epsilon \downarrow \sqrt{a+1}} (\epsilon^2 - a - 1)^{b+1/2} \sum_{n=1}^{\infty} \frac{(\log n)^a (\log\log n)^b}{n} \times$$

$$P\left(\left| \frac{S_n}{W_n} \right| \geq (2\log\log n)^{1/2}(\epsilon + \alpha_n(\epsilon)) \right) = \exp(-2\tau\sqrt{1+a}) \frac{\Gamma(b+\frac{1}{2})}{\sqrt{\pi}(a+1)}. \tag{2}$$

In addition, Deng [6] extended Theorems 1 and 2 and derived the more general results for the precise asymptotocs in the deviation probability of self-normalized sums for the one-dimensionally indexed random variables.

Comparing with the precise asymptotics for normalized sums and self-normalized sums of one-dimensionally indexed random variables, the analogues for multidimensionally indexed random variables also resulted in the attention to the researchers. Gut and Spătaru [13] studied the precise asymptotics for normalized sums of multidimesionally indexed random variables and established precise asymptotocs for $\sum_{\mathbf{n}} |\mathbf{n}|^{r/p-2} P(|S_{\mathbf{n}}| \geq \epsilon |\mathbf{n}|^{1/p})$, and for $\sum_{\mathbf{n}} \frac{(\log |\mathbf{n}|)^{\delta}}{|\mathbf{n}|} P(|S_{\mathbf{n}}| \geq \epsilon \sqrt{|\mathbf{n}| \log |\mathbf{n}|})$, $(0 \leq \delta \leq 1)$ as $\epsilon \searrow 0$, and for $\sum_{\{\mathbf{n}:|\mathbf{n}|\geq 3\}} \frac{1}{|\mathbf{n}| \log |\mathbf{n}|} P(|S_{\mathbf{n}}| \geq \epsilon \sqrt{|\mathbf{n}| \log \log |\mathbf{n}|})$ as $\epsilon \searrow \sqrt{2(d-1)EX^2}$. One of the results obtained in Gut and Spătaru [13] is as follows.

Theorem 3. *Suppose that $EX=0$, that $E[X^2(\log(1+|X|))^{d-1}(\log\log(e+|X|))^{\delta}] < \infty$ for some $\delta > 1$, and set $EX^2 = \sigma^2$. Then,*

$$\lim_{\epsilon \searrow \sigma \sqrt{2(d-1)}} \sqrt{\epsilon^2 - 2(d-1)\sigma^2} \times$$

$$\sum_{\{\mathbf{n}:|\mathbf{n}|\geq 3\}} \frac{1}{|\mathbf{n}| \log |\mathbf{n}|} P(|S_{\mathbf{n}}| \geq \epsilon \sqrt{|\mathbf{n}| \log \log |\mathbf{n}|}) = \frac{\sigma}{(d-1)!} \sqrt{\frac{2}{d-1}}. \tag{3}$$

Meanwhile, Jiang and Yang [16] proved a result for self-normalized sums of multidimensionally indexed random variables as follows.

Theorem 4. *Assume that $EX = 0$, and $EX^2 I(|X| \leq x)$ is a slowly varying function at infinity. Then, for $0 \leq \delta \leq 1$,*

$$\lim_{\epsilon \downarrow 0} \epsilon^{2\delta+2} \sum_{\mathbf{n}} \frac{(\log |\mathbf{n}|)^{\delta}}{|\mathbf{n}|(\log |\mathbf{n}|)^{d-1}} P\left(\left|\frac{S_{\mathbf{n}}}{W_{\mathbf{n}}}\right| \geq \epsilon \sqrt{\log |\mathbf{n}|}\right) = \frac{E|N|^{2\delta+2}}{(d-1)!(1+\delta)}. \tag{4}$$

From the previous discussion, we can summarize that the common interest for the precise asymptotics of normalized sums and self-normalized sums of multidimensionally indexed random variables is to find the convergence rate for the following infinite series

$$\sum_{\mathbf{n}} h_1(|\mathbf{n}|) P(|S_{\mathbf{n}}| \geq \epsilon \sqrt{|\mathbf{n}|} \phi_1(|\mathbf{n}|)) \quad \text{for normalized sums} \tag{5}$$

and

$$\sum_{\mathbf{n}} h_2(|\mathbf{n}|) P(|S_{\mathbf{n}}| \geq \epsilon W_{\mathbf{n}} \phi_2(|\mathbf{n}|)) \quad \text{for self-normalized sums} \tag{6}$$

as $\epsilon \to \epsilon_0$ for specified functions $h_1(x)$, $h_2(x)$, $\phi_1(x)$ and $\phi_2(x)$. For example, Gut and Spătaru [13] derived the precise asymptotics of (5) for $h_1(x) = (\log x)^\delta/x$, $h_1(x) = 1/(x \log x)$; $\phi_1(x) = \sqrt{\log x}$, $\phi_1(x) = \sqrt{\log \log x}$ and Jiang and Yang [16] did that of (6) for $h_2(x) = (\log x)^\delta/(x(\log x)^{d-1})$ and $\phi_2(x) = \sqrt{\log x}$. However, there is no result on the precise asymptotics of series (5) and (6) for the general forms of functions $h_1(x), h_2(x), \phi_1(x)$ and $\phi_2(x)$. Therefore the interest of this paper is to derive the analogues of Theorem 3 and to give the extensions of Theorem 4. Although we can derive the precise asymptotics for the series (5), in what follows we will focus on the derivation of precise asymptotics for the series (6) with the general functions $h_2(x)$ and $\phi_2(x)$. In fact we will extend the foregoing results in some sense. Firstly we will investigate the precise asymptotics in the deviation probabilities of self-normalized sums for the multidimensionally indexed random variables as $\epsilon \searrow \epsilon_0$ where ϵ_0 can be 0 or greater than 0. Secondly instead of special functions such as x^r, $\log x$, $\log \log x$ in the deviation probabilities for self-normalized sums, we will consider the precise asymptotics of (6) for general functions. Thirdly, since there is no discussion on the precise asymptotics of (5) with general functions $h_1(x)$ and $\phi_1(x)$, the results obtained in this paper can also be suitable to the series (5). Therefore the present paper will give the integrated results and the theorems stated above can be considered as the special cases of our results. Moreover some novel results are derived. The remainder of this paper is organized as follows. Section 2 introduces some definitions, notation and states main results. Section 3 will give some preliminaries for the proofs of theorems, which follow in Sect. 4.

2   Main Results

In the remaining sections, suppose that $\{X, X_n, n \in Z_+^d\}$ be nondegenerate i.i.d. random variables and set $S_n = \sum_{k \leq n} X_k$ and $W_n^2 = \sum_{k \leq n} X_k^2$. Let N denote the standard normal random variable. Let $\phi(x)$, $g(x)$ be positive valued functions on $[1, \infty)$, $[\phi(1), \infty)$, respectively, and $\alpha(x)$ be a positive function on the positive finite interval $[a, b]$ such that:

(A1) $\phi(x)$ is differentiable with the positive derivative $\phi'(x)$ and $\phi(x) = o(\sqrt{x})$.
(A2) $g(x)$ is an integrable function with the anti-derivative $G(x) = \int_{\phi(1)}^x g(t)dt$;
(A3) $\lim_{x \to \epsilon_0^+} \alpha(x) = 0$ for some $\epsilon_0 \in [a, b]$.

Based on the discussion in Sect. 1 the main results are stated as follows.

Theorem 5. *Suppose that $EX^2 I(|X| \leq x)$ is a slowly varying function at ∞, $\phi(x)$, $g(x)$ and $\alpha_i(x)(i = 1, 2)$ satisfy (A1)–(A3), respectively.*

(i)   If for fixed $1/2 < \gamma < 1$, the integral

$$\alpha_1(\epsilon) \int_0^\infty G\left(\frac{x}{\epsilon}\right) \exp\left(-\frac{\gamma x^2}{2}\right) dx < +\infty \qquad (7)$$

uniformly with respect to $\epsilon \in [a, b]$, then

$$\lim_{\epsilon \to \epsilon_0^+} \alpha_1(\epsilon) \sum_{\mathbf{n}} \frac{g[\phi(|\mathbf{n}|)]\phi'(|\mathbf{n}|)}{(\log |\mathbf{n}|)^{d-1}} P\left(\left|\frac{S_\mathbf{n}}{W_\mathbf{n}}\right| > \epsilon\phi(|\mathbf{n}|)\right)$$

$$= \lim_{\epsilon \to \epsilon_0^+} \frac{\alpha_1(\epsilon)}{(d-1)!} E\left[G\left(\frac{|N|}{\epsilon}\right)\right]. \qquad (8)$$

By choosing the appropriate forms to functions $g(x), \phi(x), \alpha_1(x)$ and $\alpha_2(x)$, many known results can follow from Theorem 5.

At first, by setting $g(x) = (\log x)^{d+\tau-1}/x(d + \tau > 0), \phi(x) = x^{1/q}(q > 2)$ and $\alpha(x) = (-\log x)^{-(d+\tau)}$, we have the following corollary.

Corollary 1. *Suppose that $EX^2 I(|X| \leq x)$ is a slowly varying function at ∞. Then*

$$\lim_{\epsilon \to 0+} \left(\log \frac{1}{\epsilon}\right)^{-(d+\tau)} \sum_{\mathbf{n}} \frac{(\log |\mathbf{n}|)^\tau}{|\mathbf{n}|} P\left(\left|\frac{S_\mathbf{n}}{W_\mathbf{n}}\right| \geq \epsilon|\mathbf{n}|^{1/q}\right) = \frac{q^{d+\tau}}{(d+\tau)(d-1)!}. \qquad (9)$$

In particular, by setting $\tau = 0$ in (9), the self-normalized version of Theorem 1 in Gut and Spătaru [13] is obtained and thus this result can be thought of as the generalization of the aforementioned theorem. Next one can also obtain the self-normalized version of Theorem 4 in Gut and Spătaru [13] by choosing $g(x) = x^{2(d+\delta)-1}, \phi(x) = (\log x)^{1/2}$ and $\alpha(x) = x^{2(d+\delta)}$:

$$\lim_{\epsilon \to 0+} \epsilon^{2(d+\delta)} \sum_{\mathbf{n}:|\mathbf{n}|\geq 3} \frac{(\log |\mathbf{n}|)^\delta}{|\mathbf{n}|} P\left(\left|\frac{S_\mathbf{n}}{W_\mathbf{n}}\right| \geq \epsilon(\log |\mathbf{n}|)^{1/2}\right) = \frac{E|N|^{2(d+\delta)}}{(d-1)!(d+\delta)} \qquad (10)$$

Then Theorem 4 can be obtained by replacing δ with $\delta + 1 - d$ in (10). Moreover, if we choose $g(x) = x^{2\eta+1}, \phi(x) = (\log \log x)^{1/2}, \alpha_1(x) = \alpha_2(x) = x^{2\eta+2}$, we have that

$$\lim_{\epsilon \to 0+} \epsilon^{2\eta+2} \sum_{\mathbf{n}:|\mathbf{n}|\geq 3} \frac{(\log \log |\mathbf{n}|)^\eta}{|\mathbf{n}|(\log |\mathbf{n}|)^d} P\left(\left|\frac{S_\mathbf{n}}{W_\mathbf{n}}\right| \geq \epsilon(\log \log |\mathbf{n}|)^{1/2}\right) = \frac{E|N|^{2\eta+2}}{(d-1)!(1+\eta)}$$

$$(11)$$

More generally, by taking $g(x) = \beta x^{\beta(\delta+1)-3}, \phi(x) = (\eta\varphi(x))^{1/\beta}$ and $\alpha_1(x) = \alpha_2(x) = x^{\beta(\delta+1)-2}$, we have the succeeding corollary.

Corollary 2. *Suppose that $\varphi(x)$ is a positive valued differentiable function on $[1, \infty)$ such that $\varphi'(x) > 0$, $\lim_{x \to \infty} \varphi(x) = \infty$ and $\varphi(x) = o(x^{\beta/2})$. Then for $\beta > 0, \delta > \frac{2}{\beta} - 1$ and $\eta > 0$,*

$$\lim_{\epsilon \to 0^+} \epsilon^{\beta(\delta+1)-2} \sum_{n:|n|\geq 3} \frac{(\varphi(|n|))^{\delta-2/\beta}\varphi'(|n|)}{(\log|n|)^{d-1}} P\left(\left|\frac{S_n}{W_n}\right| \geq \epsilon(\eta\varphi(|n|))^{1/\beta}\right)$$

$$= \frac{\beta E|Z|^{\beta(\delta+1)-2}}{(d-1)!\eta^{\delta+1-2/\beta}[\beta(\delta+1)-2]}. \tag{12}$$

Now we consider deriving the analogue of Theorem 3. So far, the choice of function $g(x)$ is limited to the power functions or slow varying functions, for which, the condition (7) always holds. However, the conditions in Theorem 5 are no longer satisfied for the exponential functions. In fact, by setting $g(x) = 2x\exp(x^2)$, $\phi(x) = (\log\log x)^{1/2}$, $\alpha_1(x) = \sqrt{x^2-2}$ and $\epsilon_0 = \sqrt{2}$, we have that

$$\lim_{\epsilon \to \sqrt{2}} \sqrt{\epsilon^2-2} \sum_{|n|\geq 3} \frac{1}{|n|(\log|n|)^{d-1}} P\left(\left|\frac{S_n}{W_n}\right| > \epsilon(\log\log|n|)^{1/2}\right)$$

$$= \lim_{\epsilon \downarrow \sqrt{2}} \alpha_1(\epsilon)E\left[G\left(\frac{|N|}{\epsilon}\right)\right] = \frac{\sqrt{2}}{(d-1)!},$$

provided that (7) in Theorem 5 holds. However, (7) in Theorem 5 does not hold for the above choices of functions $g(x)$ and $\alpha(x)$. In fact, for $0 < \gamma < 1$, the integral

$$\int_0^\infty \alpha_1(\epsilon)G\left(\frac{x}{\epsilon}\right)\exp\left(-\frac{\gamma x^2}{2}\right)dx = \sqrt{\epsilon^2-2}\int_0^\infty \frac{\epsilon\sqrt{\gamma}}{\sqrt{\gamma\epsilon^2-2}}\exp\left(-\frac{\gamma y^2}{2}\right)dy$$

does not converge uniformly with respect to $\epsilon \geq \sqrt{2}$. Therefore in order to obtain the self-normalized version of Theorem 3, the stronger condition should be added on the random variables. Actually, we have the following result.

Theorem 6. *Let X be a variable with $E|X|^{2+\delta} < +\infty$ for some $0 < \delta < 1$. Suppose that (A2) and (A3) hold for $g(x)$ and $\alpha_i(\epsilon)(i = 1, 2)$, respectively, and $\phi(x)$ satisfies the following condition:*

(A1′) $\phi(x)$ is differentiable with the positive derivative $\phi'(x)$ and $\phi(x) = O(x^{\frac{\delta}{4+2\delta}})(\delta > 0)$.

Then (8) holds provided that

$$\alpha_1(\epsilon)E\left[G\left(\frac{|N|}{\epsilon}\right)\right] < +\infty \tag{13}$$

uniformly with respect to $\epsilon \in [a, b]$.

Now many specified results can also be obtained by choosing different forms to $g(x), \phi(x), \alpha_1(x)$ and $\alpha_2(x)$. Under the condition that $E|X|^{2+\delta} < +\infty(\delta > 0)$, we have for $m > -1, d + \tau > 0$ that

$$\lim_{\epsilon \downarrow \sqrt{2(d+\tau)}} (\epsilon^2 - 2(d+\tau))^{\frac{2m+1}{2}} \sum_{|\mathbf{n}| \geq 3} \frac{(\log |\mathbf{n}|)^{\tau} (\log \log |\mathbf{n}|)^m}{|\mathbf{n}|} \times$$

$$P\left(\left|\frac{S_\mathbf{n}}{W_\mathbf{n}}\right| \geq \epsilon (\log \log(|\mathbf{n}|))^{1/2}\right) = \frac{2^{\frac{2m+1}{2}} \Gamma(\frac{2m+1}{2})}{(d-1)! \sqrt{(d+\tau)\pi}}, \quad (14)$$

$$\lim_{\epsilon \downarrow 0} \epsilon^{2m} \exp\left(-\frac{(d+\tau)^2}{2\epsilon^2}\right) \sum_{|\mathbf{n}| \geq 3} \frac{(\log |\mathbf{n}|)^{\tau} (\log \log |\mathbf{n}|)^m}{|\mathbf{n}|} \times$$

$$P\left(\left|\frac{S_\mathbf{n}}{W_\mathbf{n}}\right| \geq \epsilon \log \log |\mathbf{n}|\right) = \frac{2(d+\tau)^{m-1}}{(d-1)!}, \quad (15)$$

and

$$\lim_{\epsilon \downarrow 0} \epsilon^{2(2m+1)} \exp\left(-\frac{\tau^2}{2\epsilon^2}\right) \sum_{n \geq 3} \frac{(\log \log |\mathbf{n}|)^m \exp[\tau (\log \log |\mathbf{n}|)^{1/2}]}{|\mathbf{n}|(\log |\mathbf{n}|)^d} \times$$

$$P\left(\left|\frac{S_\mathbf{n}}{W_\mathbf{n}}\right| \geq \epsilon (\log \log |\mathbf{n}|)^{1/2}\right) = \frac{4\tau^{2m}}{(d-1)!} \quad (16)$$

In particular, by choosing $\tau = -1$ and $m = 0$ in (14), the self-normalized version of Theorem 3 can be obtained:

$$\lim_{\epsilon \downarrow \sqrt{2(d-1)}} \sqrt{\epsilon^2 - 2(d-1)} \sum_{|\mathbf{n}| \geq 3} \frac{1}{|\mathbf{n}| \log |\mathbf{n}|} P\left(\left|\frac{S_\mathbf{n}}{W_\mathbf{n}}\right| \geq \epsilon (\log \log(|\mathbf{n}|))^{1/2}\right)$$

$$= \frac{1}{(d-1)!} \sqrt{\frac{2}{(d-1)}} \quad (17)$$

Now for $\mu > -1, \nu > 0$ and $\eta > 0$, by taking $g(x) = x^\mu \exp(\nu x^2)$, $\phi(x) = (\eta \varphi(x))^{1/2}$, $\alpha_1(x) = (x^2 - 2\nu)^{\frac{\mu}{2}}$ and $\alpha_2(x) = (x^2 - 2\nu)^{\frac{\mu+2}{2}}$ in Theorem 6, the subsequent corollary follows.

Corollary 3. *Let X be a random variable with $E|X|^{2+\delta} < +\infty (\delta > 0)$. Suppose that $\varphi(x)$ is a positive valued differentiable function on $[1, \infty)$ such that $\varphi'(x) > 0$, $\lim_{x \to \infty} \varphi(x) = \infty$ and $\varphi(x) = O(x^{\delta/(1+\delta)})$. Then*

$$\lim_{\epsilon \downarrow \sqrt{2\nu}} (\epsilon^2 - 2\nu)^{\frac{\mu}{2}} \sum_{|\mathbf{n}| \geq 3} \frac{\exp(\eta \nu \varphi(|\mathbf{n}|))[\varphi(|\mathbf{n}|)]^{\frac{\mu-1}{2}} \varphi'(|\mathbf{n}|)}{(\log |\mathbf{n}|)^{d-1}} \times$$

$$P\left(\left|\frac{S_\mathbf{n}}{W_\mathbf{n}}\right| \geq \epsilon (\eta \varphi(|\mathbf{n}|))^{1/2}\right) = \frac{2^{\frac{\mu}{2}} \Gamma(\frac{\mu}{2})}{(d-1)! \sqrt{\nu \pi} \eta^{\frac{\mu+1}{2}}}.$$

Also, the following corollary can be derived by taking $g(x) = x^\mu \exp(\nu x)$, $\alpha_1(x) = \exp(-\nu^2/2x^2)x^{2\mu}$ and $\alpha_2(x) = \exp(-\nu^2/2x^2)x^{2(\mu+1)}$ in Theorem 6.

Corollary 4. *Let X be a random variable with $E|X|^{2+\delta} < +\infty$. Suppose that $\phi(x)$ is a positive valued differentiable function on $[1,\infty)$ such that $\phi'(x) > 0$, $\lim_{x\to\infty} \phi(x) = \infty$ and $\phi(x) = O(x^{\delta/(2+2\delta)})$. Then for $\nu > 0$,*

$$\lim_{\epsilon\downarrow 0} \epsilon^{2\mu} \exp\left(-\frac{\nu^2}{2\epsilon^2}\right) \sum_n \frac{(\phi(|n|))^\mu \exp(\nu\phi(|n|))\phi'(|n|)}{(\log|n|)^{d-1}} P\left(\left|\frac{S_n}{W_n}\right| \geq \epsilon\phi(|n|)\right) = \frac{2\nu^{\mu-1}}{(d-1)!}.$$

So far, we obtain the precise asymptotics for the probability deviation series of self-normalized sums of multidimensionally indexed random variable under the moment condition that $E|X|^{2+\delta} < +\infty$ for $\delta > 0$. However the analogue of Theorem 1 cannot be derived under the same moment condition. If strong conditions are added on the random variable X, the precise asymptotics in complete moment convergence for self-normalized sums of multidimensionally indexed random variables can be obtained.

Theorem 7. *Let X be a symmetric random variable with $E|X|^3 < +\infty$. Suppose that (A1)–(A3) hold for $\phi(x)$, $g(x)$ and $\alpha_2(\epsilon)$. Then*

$$\lim_{\epsilon\to\epsilon_0^+} \alpha_2(\epsilon) \sum_n \frac{g(\phi(|n|))\phi'(|n|)}{(\log|n|)^{d-1}} E\left[\left(\frac{S_n}{W_n}\right)^2 I\left(\left|\frac{S_n}{W_n}\right| \geq \epsilon\phi(|n|)\right)\right]$$

$$= \lim_{\epsilon\to\epsilon_0^+} \frac{\alpha_2(\epsilon)}{(d-1)!} E\left[N^2 G\left(\frac{|N|}{\epsilon}\right)\right].$$

provided that

$$\alpha_2(\epsilon)E\left[N^2 G\left(\frac{|N|}{\epsilon}\right)\right] < +\infty \tag{18}$$

uniformly with respect to $\epsilon \in [a, b]$, respectively.

For the different choices of functions $g(x), \phi(x)$ and $\alpha_2(x)$, the analogues of previous results can also be obtained. Moreover replacing $\epsilon\phi(|n|)$ by $\epsilon\phi(|n|) + \kappa(|n|)$ and $\epsilon\phi(|n|) + \kappa(\epsilon, |n|)$ in (8), the self-normalized versions of Theorem 2 can be derived for the multidimensionally indexed random variables.

Theorem 8. *Suppose that the same conditions as that in Theorem 5 or Theorem 6 hold.*

(i) *If $\kappa(x)$ is a nonnegative function of x such that $\kappa(x) = O(1/\phi(x))$, then*

$$\lim_{\epsilon\to 0^+} \alpha_1(\epsilon) \sum_n \frac{g[\phi(|n|)]\phi'(|n|)}{(\log|n|)^{d-1}} P\left(\left|\frac{S_n}{W_n}\right| > \epsilon\phi(|n|) + \kappa(|n|)\right)$$

$$= \lim_{\epsilon\to\epsilon_0^+} \alpha_1(\epsilon)E\left[G\left(\frac{|N|}{\epsilon}\right)\right].$$

(ii) If $\kappa(\epsilon, x)$ is a nonnegative function with respect to ϵ and x such that

$$\kappa(\epsilon, x)\phi(x) \to \rho \text{ as } x \to \infty \text{ and } \epsilon \to \epsilon_0 > 0,$$

then

$$\lim_{\epsilon \to \epsilon_0^+} \alpha_1(\epsilon) \sum_n \frac{g[\phi(|n|)]\phi'(|n|)}{(\log |n|)^{d-1}} P\left(\left|\frac{S_n}{W_n}\right| > \epsilon\phi(|n|) + \kappa(\epsilon, |n|)\right)$$

$$= \exp(-\epsilon_0 \rho) \lim_{\epsilon \to \epsilon_0^+} \alpha_1(\epsilon) E\left[G\left(\frac{|N|}{\epsilon}\right)\right].$$

Theorem 9. *Suppose that the same conditions as that in Theorem 7 hold.*

(i) If $\kappa(x)$ is a nonnegative function of x such that $\kappa(x) = O(1/\phi(x))$, then

$$\lim_{\epsilon \to 0^+} \alpha_2(\epsilon) \sum_n \frac{g[\phi(|n|)]\phi'(|n|)}{(\log |n|)^{d-1}} E\left[\left(\frac{S_n}{W_n}\right)^2 I\left(\left|\frac{S_n}{W_n}\right| > \epsilon\phi(|n|) + \kappa(|n|)\right)\right]$$

$$= \lim_{\epsilon \to \epsilon_0^+} \alpha_2(\epsilon) E\left[N^2 G\left(\frac{|N|}{\epsilon}\right)\right].$$

(ii) If $\kappa(\epsilon, x)$ is a nonnegative function with respect to ϵ and x such that

$$\kappa(\epsilon, x)\phi(x) \to \rho \text{ as } x \to \infty \text{ and } \epsilon \to \epsilon_0 > 0,$$

then

$$\lim_{\epsilon \to \epsilon_0^+} \alpha_2(\epsilon) \sum_n \frac{g[\phi(|n|)]\phi'(|n|)}{(\log |n|)^{d-1}} E\left[\left(\frac{S_n}{W_n}\right)^2 I\left(\left|\frac{S_n}{W_n}\right| > \epsilon\phi(|n|) + \kappa(\epsilon, |n|)\right)\right]$$

$$= \exp(-\epsilon_0 \rho) \lim_{\epsilon \to \epsilon_0^+} \alpha_2(\epsilon) E\left[G\left(\frac{|N|}{\epsilon}\right)\right].$$

Also, one can obtain many results by choosing different forms for functions $g(x)$, $\phi(x)$, $\kappa(x)$ and $\kappa(\epsilon, x)$. Finally we end this section by a specific result on the precise asymptotic in the complete moment convergence.

Corollary 5. *Let $\tau > 0$, $m > 1/2$ and $\kappa(\epsilon, x)$ be a nonnegative function of ϵ and x such that*

$$\kappa(\epsilon, x)(2 \log \log \log x)^{1/2} \to \rho \text{ as } x \to \infty \text{ and } \epsilon \downarrow \sqrt{\tau}.$$

Then, under the same conditions as in Theorem 9,

$$\lim_{\epsilon \downarrow \sqrt{\tau}} (\epsilon^2 - \tau)^m \sum_{\mathbf{n}} \frac{(\log \log |\mathbf{n}|)^{\tau-1} (\log \log \log |\mathbf{n}|)^{m-\frac{3}{2}}}{|\mathbf{n}|(\log |\mathbf{n}|)^d} E\left[\left(\frac{S_\mathbf{n}}{W_\mathbf{n}}\right)^2 \times \right.$$

$$\left. I\left(\left|\frac{S_\mathbf{n}}{W_\mathbf{n}}\right| \geq \epsilon(2 \log \log \log |\mathbf{n}|)^{1/2} + \kappa(\epsilon, |\mathbf{n}|)\right)\right] = 2 \exp\{-\sqrt{\tau}\rho\} \frac{\sqrt{a}\Gamma(m)}{(d-1)!\sqrt{\pi}}.$$

3 Preliminaries

Again suppose that $\{X_\mathbf{k}, \mathbf{k} \in Z_+^d\}$ are random variables and $\{S_\mathbf{n}, \mathbf{n} \in Z_+^d\}$ are their partial sums. Note that $S_\mathbf{n}$ is simply a sum of $|\mathbf{n}|$ random variables. Let

$$d(j) = \text{card}\{\mathbf{k} : |\mathbf{k}| = j\}, \quad \text{and} \quad M(j) = \text{card}\{\mathbf{k} : |\mathbf{k}| \leq j\}.$$

From Hardy and Wright [15], the following asymptotics hold:

$$M(j) \sim \frac{j(\log j)^{d-1}}{(d-1)!} \quad \text{as} \quad j \to \infty$$

and

$$d(j) = o(j^\delta) \quad \text{for any } \delta > 0 \quad \text{as} \quad j \to \infty.$$

Further, since all terms in the sums we are considering are nonnegative, the order of summation can be changed as follows (see Gut [11, 12]).

$$\sum_{\mathbf{n}} \cdots = \sum_{j=1}^{\infty} \sum_{|\mathbf{n}|=j} \cdots$$

In particular, if the functions involving $\mathbf{n}$ only depend on the value of $|\mathbf{n}|$, the second summation can be further simplified. For example,

$$\sum_{\mathbf{n}} g[\phi(|\mathbf{n}|)]\phi'(|\mathbf{n}|)P(|S_\mathbf{n}| \geq \epsilon W_\mathbf{n}\phi(|\mathbf{n}|))$$

$$= \sum_{j=1}^{\infty} \sum_{|\mathbf{n}|=j} g[\phi(|\mathbf{n}|)]\phi'(|\mathbf{n}|)P(|S_\mathbf{n}| \geq \epsilon W_\mathbf{n}\phi(|\mathbf{n}|))$$

$$= \sum_{j=1}^{\infty} d(j)g[\phi(j)]\phi'(j)P(|S_{\pi(j)}| \geq \epsilon W_{\pi(j)}\phi(j))$$

where $\pi(j) = (j, 1, \ldots, 1) \in Z_+^d$ and $S_{\pi(j)} = \sum_{i=1}^{j} X_{\pi(i)}$.

Now we give a proposition which is crucial in the proofs of theorems for the multidimensionally indexed random variables.

Proposition 1. *Let $\alpha(\epsilon)$ be a function satisfying (A3) in Sect. 2, $\{a(j, \epsilon), j \geq 1\}$ be a nondecreasing sequence of functions such that $\lim_{j \to \infty} a(j, \epsilon) = 0$ for all $\epsilon \in [a, b]$. Suppose that the infinite series $\sum_{j=1}^{\infty} d(j)a(j, \epsilon)$ converges for $\epsilon \in [a, b]$. Then,*

$$\lim_{\epsilon \to \epsilon_0^+} \alpha(\epsilon) \sum_{j \geq 1} d(j)a(j, \epsilon) = \lim_{\epsilon \to \epsilon_0^+} \alpha(\epsilon) \sum_{j \geq 1} (\log j)^{d-1} a(j, \epsilon)/(d-1)!$$

The proof of Proposition 1 can be obtained from the following lemma (see Spătaru [27]).

Lemma 1. *Let $\{a(j), j \geq 1\}$ be a nondecreasing sequence such that $\lim_{j \to \infty} a(j) = 0$, and let $0 < \delta < 1$. Then, there exists $k_0 = k_0(\delta)$ such that*

$$C_1 + \frac{1-\delta}{(d-1)!} \sum_{j \geq k_0} (\log j)^{d-1} a(j) \leq \sum_{j \geq 1} d(j)a(j) \geq C_1 + \frac{1+\delta}{(d-1)!} \sum_{j \geq k_0} (\log j)^{d-1} a(j).$$

Since $S_{|n|}$ and S_n, $W_{|n|}$ and W_n have the identical distributions, respectively, one can obtain the multidimensional-index versions of several inequalities for self-normalized sums of random variables, which will be used in the proofs of our theorems.

Lemma 2 (Shao [24]). *Let $\{X, X_k, k \in Z_+^d\}$ be i.i.d. random variables with $EX = 0$ and $EX^2 I(|X| \leq x)$ is slowly varying as $x \to \infty$. Then for arbitrary $1/2 < \gamma < 1$, there exist $0 < \delta < 1, x_0 > 1$ and n_0 such that for any $|n| \geq n_0$ and $x_0 < x < \delta\sqrt{|n|}$,*

$$P(S_n/W_n \geq x) \leq \exp\left(-\frac{\gamma x^2}{2}\right).$$

Lemma 3 (Wang and Jing [29]). *Let $\{X_k, k \in Z_+^d\}$ be a sequence of symmetric random variables with $E|X_k|^3 < +\infty$ for $k \in Z_+^d$. Then for all $n \in Z_+^d$ and $x \in R$,*

$$|P(S_n/W_n \geq x) - P(N \geq x)| \leq A \min\{(1 + |x|^3)L_{3n}, 1\}e^{-x^2/2}$$

where $L_{3n} = \sum_{|k| \leq |n|} E|X_k|^3 / (\sum_{|k| \leq |n|} EX_k^2)^{3/2}$.

Lemma 4 (Jingh et al. [17]). *Let $\{X_k, k \in Z_+^d\}$ be a sequence of random variables with $EX_k = 0$ and $E|X_k|^{2+\delta} < +\infty$ for $0 < \delta \leq 1$. Then for all $n \in Z_+^d$ and $x \in R$,*

$$|P(S_n/W_n \geq x) - P(N \geq x)| \leq A(1 + x)^{1+\delta} e^{-x^2/2}/d_{n,\delta}^{2+\delta}$$

holds for $0 \leq x \leq d_{n,\delta}$ where $d_{n,\delta} = (\sum_{|k| \leq |n|} EX_k^2)^{1/2}/(\sum_{|k| \leq |n|} E|X_k|^{2+\delta})^{1/(2+\delta)}$.

4 Proofs of Main Results

We will first present a lemma on the standard normal random variable. Similar to Lemma 3.1 in Deng [6], the following lemma follows.

Lemma 5. *Suppose that* $g(x), \phi(x)$ *and* $\alpha_i(x)(i = 1, 2)$ *satisfy the conditions (A1)–(A3), respectively. Then we have*

$$\lim_{\epsilon \to \epsilon_0^+} \alpha_1(\epsilon) \sum_n g[\phi(|n|)]\phi'(|n|)(\log |n|)1 - dP(|N| > \epsilon\phi(|n|))$$

$$= \lim_{\epsilon \to \epsilon_0^+} \frac{\alpha_1(\epsilon)}{(d-1)!} E\left[G\left(\frac{|N|}{\epsilon} \right) \right], \tag{19}$$

and

$$\lim_{\epsilon \to \epsilon_0^+} \alpha_2(\epsilon) \sum_n g[\phi(|n|)]\phi'(|n|)(\log |n|)^{1-d} \int_{\epsilon\phi(|n|)}^{\infty} 2xP(|N| > x)dx$$

$$= \lim_{\epsilon \to \epsilon_0^+} \frac{\epsilon^2 \alpha_2(\epsilon)}{(d-1)!} E\left[G_1\left(\frac{|N|}{\epsilon} \right) \right], \tag{20}$$

where $G_1(x) = \int_{\phi(1)}^{x} 2uG(u)du$.

4.1 The Proof of Theorem 5

From Lemma 5, proving (8) is equivalent to proving

$$\lim_{\epsilon \to \epsilon_0^+} \alpha_1(\epsilon) \sum_n \frac{g[\phi(|\mathbf{n}|)]\phi'(|\mathbf{n}|)}{(\log |\mathbf{n}|)^{d-1}} \left| P\left(\left| \frac{S_\mathbf{n}}{W_\mathbf{n}} \right| > \epsilon\phi(|\mathbf{n}|) \right) - P(|N| > \epsilon\phi(|\mathbf{n}|)) \right| = 0 \tag{21}$$

Now, for a fixed $M > 0$, set $K = K(M, \epsilon) = \phi^{-1}(M/\epsilon)$. Then,

$$\lim_{\epsilon \to \epsilon_0^+} \alpha_1(\epsilon) \sum_n \frac{g[\phi(|\mathbf{n}|)]\phi'(|\mathbf{n}|)}{(\log |\mathbf{n}|)^{d-1}} \left| P\left(\left| \frac{S_\mathbf{n}}{W_\mathbf{n}} \right| > \epsilon\phi(|\mathbf{n}|) \right) - P(|N| > \epsilon\phi(|\mathbf{n}|)) \right|$$

$$= \lim_{\epsilon \to \epsilon_0^+} \alpha_1(\epsilon) \sum_{|\mathbf{n}| \le K} \frac{g[\phi(|\mathbf{n}|)]\phi'(|\mathbf{n}|)}{(\log |\mathbf{n}|)^{d-1}} \left| P\left(\left| \frac{S_\mathbf{n}}{W_\mathbf{n}} \right| > \epsilon\phi(|\mathbf{n}|) \right) - P(|N| > \epsilon\phi(|\mathbf{n}|)) \right|$$

$$+ \lim_{\epsilon \to \epsilon_0^+} \alpha_1(\epsilon) \sum_{|\mathbf{n}| > K} \frac{g[\phi(|\mathbf{n}|)]\phi'(|\mathbf{n}|)}{(\log |\mathbf{n}|)^{d-1}} \left| P\left(\left| \frac{S_\mathbf{n}}{W_\mathbf{n}} \right| > \epsilon\phi(|\mathbf{n}|) \right) - P(|N| > \epsilon\phi(|\mathbf{n}|)) \right|$$

$$:= (I) + (II)$$

Since $S_\mathbf{n}/W_\mathbf{n} \to^D N(0, 1)$ as $|\mathbf{n}| \to \infty$ and $P(|N| > x)$ is continuous for $x \geq 0$, it is obvious that

$$\delta_{|\mathbf{n}|} := \sup_x \left| P\left(\left| \frac{S_\mathbf{n}}{W_\mathbf{n}} \right| > x \right) - P(|N| > x) \right| \to 0 \text{ as } |\mathbf{n}| \to \infty.$$

As to (I), we have that

$$(I) = \lim_{\epsilon \to \epsilon_0^+} \alpha_1(\epsilon) \sum_{|\mathbf{n}| \leq K} \frac{g[\phi(|\mathbf{n}|)]\phi'(|\mathbf{n}|)}{(\log |\mathbf{n}|)^{d-1}} \left| P\left(\left| \frac{S_\mathbf{n}}{W_\mathbf{n}} \right| > \epsilon\phi(|\mathbf{n}|) \right) - P(|N| > \epsilon\phi(|\mathbf{n}|)) \right|$$

$$= \lim_{\epsilon \to \epsilon_0^+} \alpha_1(\epsilon) \sum_{|\mathbf{n}| \leq K} g[\phi(|\mathbf{n}|)]\phi'(|\mathbf{n}|)(\log |\mathbf{n}|)^{1-d}\delta_{|\mathbf{n}|}.$$

If $\epsilon_0 > 0$, then $K = K(M, \epsilon)$ is bounded and the summation $\sum_{|\mathbf{n}| \leq K} g[\phi(|\mathbf{n}|)] \times \phi'(|\mathbf{n}|)(\log |\mathbf{n}|)^{1-d}\delta_{|\mathbf{n}|}$ has finite terms for a fixed M. Thus, from condition (A3),

$$\lim_{\epsilon \to \epsilon_0^+} \alpha_1(\epsilon) \sum_{|\mathbf{n}| \leq K} g[\phi(|\mathbf{n}|)]\phi'(|\mathbf{n}|)(\log |\mathbf{n}|)^{1-d}\delta_{|\mathbf{n}|} = 0.$$

If $\epsilon_0 = 0$, we have that

$$\frac{1}{G(\frac{M}{\epsilon})} \sum_{|\mathbf{n}| \leq K} g[\phi(|\mathbf{n}|)]\phi'(|\mathbf{n}|)(\log |\mathbf{n}|)^{1-d}$$

$$= \frac{1}{G(\frac{M}{\epsilon})} \sum_{j \leq K} \sum_{|\mathbf{n}|=j} g[\phi(|\mathbf{n}|)]\phi'(|\mathbf{n}|)(\log |\mathbf{n}|)^{1-d}$$

$$= \frac{1}{G(\frac{M}{\epsilon})} \sum_{j \leq K} d(j)g[\phi(j)]\phi'(j)(\log j)^{1-d}$$

$$= \frac{1}{(d-1)!} \frac{1}{G(\frac{M}{\epsilon})} \int_{\phi(1)}^{\phi(K)} g(u)du = \frac{1}{(d-1)!} \frac{1}{G(\frac{M}{\epsilon})} \int_{\phi(1)}^{\frac{M}{\epsilon}} g(u)du$$

$$\leq \frac{1}{(d-1)!} \frac{1}{G(\frac{M}{\epsilon})} G(\frac{M}{\epsilon}) = \frac{1}{(d-1)!}.$$

Therefore by Toeplitz's lemma [see, e.g., Stout [26], pp. 120–121], one can obtain that

$$\lim_{\epsilon \to \epsilon_0^+} \frac{1}{G(\frac{M}{\epsilon})} \sum_{|\mathbf{n}| \leq K} g[\phi(|\mathbf{n}|)]\phi'(|\mathbf{n}|)(\log |\mathbf{n}|)^{1-d}\delta_{|\mathbf{n}|} = 0$$

and thus by noting that (7) implies that $\alpha(\epsilon)G(\frac{M}{\epsilon})$ are bounded uniformly for $\epsilon \in [a, b]$, we have

$$\lim_{\epsilon \to \epsilon_0^+} \alpha_1(\epsilon) \sum_{|\mathbf{n}| \leq K} g[\phi(|\mathbf{n}|)]\phi'(|\mathbf{n}|)|P(|S_\mathbf{n}| > \epsilon W_\mathbf{n}\phi(|\mathbf{n}|)) - P(|N| > \epsilon\phi(|\mathbf{n}|))|$$

$$= \lim_{\epsilon \to \epsilon_0^+} \alpha_1(\epsilon)G(\frac{M}{\epsilon})\frac{1}{G(\frac{M}{\epsilon})} \sum_{|\mathbf{n}| \leq K} g[\phi(|\mathbf{n}|)]\phi'(|\mathbf{n}|)\delta_{|\mathbf{n}|} = 0.$$

We turn to (II). Firstly note that

$$(II) = \lim_{\epsilon \to \epsilon_0^+} \alpha_1(\epsilon) \sum_{|\mathbf{n}| > K} \frac{g[\phi(|\mathbf{n}|)]\phi'(|\mathbf{n}|)}{(\log|\mathbf{n}|)^{d-1}} \left| P\left(\left|\frac{S_\mathbf{n}}{W_\mathbf{n}}\right| > \epsilon\phi(|\mathbf{n}|)\right) - P(|N| > \epsilon\phi(|\mathbf{n}|)) \right|$$

$$\leq \lim_{\epsilon \to \epsilon_0^+} \alpha_1(\epsilon) \sum_{|\mathbf{n}| > K} g[\phi(|\mathbf{n}|)]\phi'(|\mathbf{n}|)(\log|\mathbf{n}|)^{1-d}P(|S_\mathbf{n}| > \epsilon W_\mathbf{n}\phi(|\mathbf{n}|))$$

$$+ \lim_{\epsilon \to \epsilon_0^+} \alpha_1(\epsilon) \sum_{|\mathbf{n}| > K} g[\phi(|\mathbf{n}|)]\phi'(|\mathbf{n}|)(\log|\mathbf{n}|)^{1-d}P(|N| > \epsilon\phi(|\mathbf{n}|))$$

$$:= \lim_{\epsilon \to \epsilon_0^+} \alpha_1(\epsilon)A_1(\epsilon) + \lim_{\epsilon \to \epsilon_0^+} \alpha_1(\epsilon)A_2(\epsilon).$$

Now, (7) implies that for any give $\epsilon \in [a, b]$, $\alpha_1(\epsilon)E\left[G\left(\frac{|N|}{\epsilon}\right)\right] < \infty$ and thus,

$$(d-1)!\alpha_1(\epsilon)A_2(\epsilon)$$

$$= (d-1)!\alpha_1(\epsilon) \sum_{|\mathbf{n}| > K} g[\phi(|\mathbf{n}|)]\phi'(|\mathbf{n}|)(\log|\mathbf{n}|)^{1-d}P(|N| > \epsilon\phi(|\mathbf{n}|))$$

$$= (d-1)!\alpha_1(\epsilon) \sum_{j>K}\sum_{|\mathbf{n}|=j} g[\phi(|\mathbf{n}|)]\phi'(|\mathbf{n}|)(\log|\mathbf{n}|)^{1-d}P(|N| > \epsilon\phi(|\mathbf{n}|))$$

$$= (d-1)!\alpha_1(\epsilon) \sum_{j>K} d(j)g[\phi(j)]\phi'(j)(\log j)^{1-d}P(|N| > \epsilon\phi(j))$$

$$= \alpha_1(\epsilon)\int_K^\infty g(\phi(x))\phi'(x)P(|N| > \epsilon\phi(x))dx = \alpha_1(\epsilon)\int_{M/\epsilon}^\infty g(u)P(|N| > \epsilon u)du$$

$$= \alpha_1(\epsilon)\int_{\frac{\epsilon M}{\epsilon}}^\infty \int_{\frac{M}{\epsilon}}^{\frac{y}{\epsilon}} g(u)dudF(y) \leq \alpha_1(\epsilon)\int_M^\infty G(\frac{y}{\epsilon})dF(y)$$

$$= \alpha_1(\epsilon)E\left[G\left(\frac{|N|}{\epsilon}\right)I_{\{|N|>M\}}\right] \to 0 \text{ as } M \to \infty.$$

Next, from Lemma 2, as $M \to \infty$,

$$(d-1)!\alpha_1(\epsilon)A_1(\epsilon)$$

$$= (d-1)!\alpha_1(\epsilon) \sum_{|\mathbf{n}|>K} g[\phi(|\mathbf{n}|)]\phi'(|\mathbf{n}|)(\log|\mathbf{n}|)^{1-d}P(|S_\mathbf{n}| > \epsilon W_\mathbf{n}\phi(|\mathbf{n}|))$$

$$= (d-1)!\alpha_1(\epsilon) \sum_{|\mathbf{n}|>K} g[\phi(|\mathbf{n}|)]\phi'(|\mathbf{n}|)(\log|\mathbf{n}|)^{1-d}\exp\left(\frac{\gamma\epsilon^2\phi^2(|\mathbf{n}|)}{2}\right)$$

$$= (d-1)!\alpha_1(\epsilon) \sum_{j>K}\sum_{|\mathbf{n}|=j} g[\phi(|\mathbf{n}|)]\phi'(|\mathbf{n}|)(\log|\mathbf{n}|)^{1-d}\exp\left(\frac{\gamma\epsilon^2\phi^2(|\mathbf{n}|)}{2}\right)$$

$$= (d-1)!\alpha_1(\epsilon) \sum_{j>K} d(j)g[\phi(j)]\phi'(j)(\log j)^{1-d}\exp\left(\frac{\gamma\epsilon^2\phi^2(j)}{2}\right)$$

$$\leq \alpha_1(\epsilon)\int_K^\infty g[\phi(x)]\phi'(x)\exp\left(-\frac{\gamma\epsilon^2\phi^2(x)}{2}\right)dx$$

$$= \alpha_1(\epsilon)G(\frac{y}{\epsilon})\exp\left(-\frac{\gamma y^2}{2}\right)\Big|_M^\infty + \int_M^\infty \gamma\alpha_1(\epsilon)yG(\frac{y}{\epsilon})\exp\left(-\frac{\gamma y^2}{2}\right)dy \to 0.$$

$$(22)$$

Thus, (21) holds. Now it follows from (19) in Lemma 5 and (21) that,

$$\lim_{\epsilon\to\epsilon_0^+}\alpha_1(\epsilon)\sum_\mathbf{n} g[\phi(|\mathbf{n}|)]\phi'(|\mathbf{n}|)(\log|\mathbf{n}|)^{1-d}P(|S_\mathbf{n}| > \epsilon W_\mathbf{n}\phi(|\mathbf{n}|))$$

$$= \lim_{\epsilon\to\epsilon_0^+}\alpha_1(\epsilon)\sum_\mathbf{n} g[\phi(|\mathbf{n}|)]\phi'(|\mathbf{n}|)(\log|\mathbf{n}|)^{1-d}P(|N| > \epsilon\phi(|\mathbf{n}|))$$

$$= \frac{1}{(d-1)!}\lim_{\epsilon\to\epsilon_0^+}\alpha_1(\epsilon)EG\left(\frac{|N|}{\epsilon}\right)$$

This completes the proof of Theorem 5.

The proof of Theorem 6 can be obtained by using Lemma 4, instead of using Lemma 2, in (22) of the proof of Theorem 5, and thus are omitted. Now we are in the position to prove Theorem 7.

4.2   The Proof of Theorem 7

In order to complete the proof of the theorem, we first note that

$$\lim_{\epsilon\to\epsilon_0^+}\alpha_2(\epsilon)\sum_\mathbf{n} g(\phi(|\mathbf{n}|))\phi'(|\mathbf{n}|)(\log|\mathbf{n}|)^{1-d}E(S_\mathbf{n}/W_\mathbf{n})^2I(|S_\mathbf{n}| \geq \epsilon W_\mathbf{n}\phi(|\mathbf{n}|))$$

$$= \lim_{\epsilon\to\epsilon_0^+}\alpha_2(\epsilon)\sum_\mathbf{n} g(\phi(|\mathbf{n}|))\phi'(|\mathbf{n}|)(\log|\mathbf{n}|)^{1-d}\int_{\epsilon\phi(|\mathbf{n}|)}(-x^2)dP\left(\left|\frac{S_\mathbf{n}}{W_\mathbf{n}}\right| \geq x\right)$$

$$= \lim_{\epsilon \to \epsilon_0^+} \alpha_2(\epsilon) \sum_{\mathbf{n}} g(\phi(|\mathbf{n}|))\phi'(|\mathbf{n}|)(\log|\mathbf{n}|)^{1-d} \left\{ \epsilon^2 \phi^2(|\mathbf{n}|) P\left(\left| \frac{S_\mathbf{n}}{W_\mathbf{n}} \right| \geq \epsilon\phi(|\mathbf{n}|) \right) \right.$$

$$\left. + \int_{\epsilon\phi(|\mathbf{n}|)} 2xP\left(\left| \frac{S_\mathbf{n}}{W_\mathbf{n}} \right| \geq x \right) dx \right\}$$

$$= \lim_{\epsilon \to \epsilon_0^+} \epsilon^2 \alpha_2(\epsilon) \sum_{\mathbf{n}} g(\phi(|\mathbf{n}|))\phi'(|\mathbf{n}|)\phi^2(|\mathbf{n}|)(\log|\mathbf{n}|)^{1-d} P\left(\left| \frac{S_\mathbf{n}}{W_\mathbf{n}} \right| \geq \epsilon\phi(|\mathbf{n}|) \right)$$

$$+ \lim_{\epsilon \to \epsilon_0^+} \alpha_2(\epsilon) \sum_{\mathbf{n}} g(\phi(|\mathbf{n}|))\phi'(|\mathbf{n}|)(\log|\mathbf{n}|)^{1-d} \int_{\epsilon\phi(|\mathbf{n}|)} 2xP\left(\left| \frac{S_\mathbf{n}}{W_\mathbf{n}} \right| \geq x \right) dx$$

$$:= (III) + (IV)$$

Similar to the proof of (7), one can prove that

$$(III) = \lim_{\epsilon \to \epsilon_0^+} \epsilon^2 \alpha_2(\epsilon) \sum_{\mathbf{n}} g(\phi(|\mathbf{n}|))\phi'(|\mathbf{n}|)\phi^2(|\mathbf{n}|)(\log|\mathbf{n}|)^{1-d} P\left(\left| \frac{S_\mathbf{n}}{W_\mathbf{n}} \right| \geq \epsilon\phi(|\mathbf{n}|) \right)$$

$$= \lim_{\epsilon \to \epsilon_0^+} \epsilon^2 \alpha_2(\epsilon) \sum_{\mathbf{n}} g(\phi(|\mathbf{n}|))\phi'(|\mathbf{n}|)\phi^2(|\mathbf{n}|)(\log|\mathbf{n}|)^{1-d} P\left(|N| \geq \epsilon\phi(|\mathbf{n}|) \right)$$

$$= \frac{1}{(d-1)!} \lim_{\epsilon \to \epsilon_0^+} \epsilon^2 \alpha_2(\epsilon) EG_2\left(\frac{|N|}{\epsilon} \right)$$

where $G_2(x) = \int_{\phi(1)}^x u^2 g(u) du$.

Next, if

$$\lim_{\epsilon \to \epsilon_0^+} \alpha_2(\epsilon) \sum_{\mathbf{n}} g(\phi(|\mathbf{n}|))\phi'(|\mathbf{n}|)(\log|\mathbf{n}|)^{1-d} \int_{\epsilon\phi(|\mathbf{n}|)} 2xP\left(\left| \frac{S_\mathbf{n}}{W_\mathbf{n}} \right| \geq x \right) dx$$

$$= \lim_{\epsilon \to \epsilon_0^+} \alpha_2(\epsilon) \sum_{\mathbf{n}} g(\phi(|\mathbf{n}|))\phi'(|\mathbf{n}|)(\log|\mathbf{n}|)^{1-d} \int_{\epsilon\phi(|\mathbf{n}|)} 2xP\left(|N| \geq x \right) dx, \quad (23)$$

from (20) in Lemma 5, we have

$$(IV) = \lim_{\epsilon \to \epsilon_0^+} \alpha_2(\epsilon) \sum_{\mathbf{n}} g(\phi(|\mathbf{n}|))\phi'(|\mathbf{n}|)(\log|\mathbf{n}|)^{1-d} \int_{\epsilon\phi(|\mathbf{n}|)} 2xP\left(\left| \frac{S_\mathbf{n}}{W_\mathbf{n}} \right| \geq x \right) dx$$

$$= \frac{1}{(d-1)!} \lim_{\epsilon \to \epsilon_0^+} \epsilon^2 \alpha_2(\epsilon) EG_1\left(\frac{|N|}{\epsilon} \right)$$

and thus by the fact that $G_1(x) + G_2(x) = x^2 G(x) - [\phi(1)]^2 G[\phi(1)]$,

$$\lim_{\epsilon \to \epsilon_0^+} \alpha_2(\epsilon) \sum_{\mathbf{n}} g(\phi(|\mathbf{n}|))\phi'(|\mathbf{n}|)(\log|\mathbf{n}|)^{1-d} E(S_\mathbf{n}/W_\mathbf{n})^2 I(|S_\mathbf{n}| \geq \epsilon W_\mathbf{n}\phi(|\mathbf{n}|))$$

$$= \frac{1}{(d-1)!} \lim_{\epsilon \to \epsilon_0^+} \epsilon^2 \alpha_2(\epsilon) EG_2\left(\frac{|N|}{\epsilon} \right) + \frac{1}{(d-1)!} \lim_{\epsilon \to \epsilon_0^+} \epsilon^2 \alpha_2(\epsilon) EG_1\left(\frac{|N|}{\epsilon} \right)$$

$$= \frac{1}{(d-1)!} \lim_{\epsilon \to \epsilon_0^+} \epsilon^2 \alpha_2(\epsilon) E\left[G_2\left(\frac{|N|}{\epsilon}\right) + G_1\left(\frac{|N|}{\epsilon}\right)\right]$$

$$= \frac{1}{(d-1)!} \lim_{\epsilon \to \epsilon_0^+} \alpha_2(\epsilon) E\left[N^2 G\left(\frac{|N|}{\epsilon}\right) - \epsilon^2 [\phi(1)]^2 G[\phi(1)]\right]$$

$$= \frac{1}{(d-1)!} \lim_{\epsilon \to \epsilon_0^+} \alpha_2(\epsilon) E\left[N^2 G\left(\frac{|N|}{\epsilon}\right)\right].$$

Now, it remains to prove (23). To this end, it suffices to prove that

$$\lim_{\epsilon \to \epsilon_0^+} \alpha_2(\epsilon) \sum_{\mathbf{n}} \frac{g(\phi(|\mathbf{n}|))\phi'(|\mathbf{n}|)}{(\log |\mathbf{n}|)^{1-d}} \times$$

$$\left| \int_{\epsilon\phi(|\mathbf{n}|)} 2xP\left(\left|\frac{S_{\mathbf{n}}}{W_{\mathbf{n}}}\right| \geq x\right) dx - \int_{\epsilon\phi(|\mathbf{n}|)} 2xP\left(|N| \geq x\right) dx \right| = 0 \quad (24)$$

Moreover, (24) can be obtained by proving

$$\lim_{\epsilon \to \epsilon_0^+} \alpha_2(\epsilon) \sum_{|\mathbf{n}| \leq K(\epsilon, M)} \frac{g(\phi(|\mathbf{n}|))\phi'(|\mathbf{n}|)}{(\log |\mathbf{n}|)^{d-1}} \times$$

$$\left| \int_{\epsilon\phi(|\mathbf{n}|)} 2xP\left(\left|\frac{S_{\mathbf{n}}}{W_{\mathbf{n}}}\right| \geq x\right) dx - \int_{\epsilon\phi(|\mathbf{n}|)} 2xP\left(|N| \geq x\right) dx \right| = 0, \quad (25)$$

and

$$\lim_{\epsilon \to \epsilon_0^+} \alpha_2(\epsilon) \sum_{|\mathbf{n}| > K(\epsilon, M)} \frac{g(\phi(|\mathbf{n}|))\phi'(|\mathbf{n}|)}{(\log |\mathbf{n}|)^{d-1}}$$

$$\left| \int_{\epsilon\phi(|\mathbf{n}|)} 2xP\left(\left|\frac{S_{\mathbf{n}}}{W_{\mathbf{n}}}\right| \geq x\right) dx - \int_{\epsilon\phi(|\mathbf{n}|)} 2xP\left(|N| \geq x\right) dx \right| = 0 \quad (26)$$

where $K = K(\epsilon, M) = \phi^{-1}(M/\epsilon)$.

To prove (25), by setting $\delta_{|\mathbf{n}|} = \sup_x |P(|S_{\mathbf{n}}/W_{\mathbf{n}}| \geq x) - P(|N| \geq x)|$, we have that

$$\alpha_2(\epsilon) \sum_{|\mathbf{n}| \leq K} \frac{g(\phi(|\mathbf{n}|))\phi'(|\mathbf{n}|)}{(\log |\mathbf{n}|)^{d-1}} \left| \int_{\epsilon\phi(|\mathbf{n}|)}^{\infty} 2xP\left(\left|\frac{S_{\mathbf{n}}}{W_{\mathbf{n}}}\right| \geq x\right) dx - \int_{\epsilon\phi(|\mathbf{n}|)}^{\infty} 2xP(|N| \geq x) dx \right|$$

$$\leq \alpha_2(\epsilon) \sum_{|\mathbf{n}| \leq K} \frac{g(\phi(|\mathbf{n}|))\phi'(|\mathbf{n}|)}{(\log |\mathbf{n}|)^{d-1}} \int_{\epsilon\phi(|\mathbf{n}|)+\delta_{|\mathbf{n}|}^{-1/4}}^{\infty} 2x \left| P\left(\left|\frac{S_{\mathbf{n}}}{W_{\mathbf{n}}}\right| \geq x\right) dx - P(|N| \geq x) \right| dx$$

$$+ \alpha_2(\epsilon) \sum_{|\mathbf{n}| \leq K} \frac{g(\phi(|\mathbf{n}|))\phi'(|\mathbf{n}|)}{(\log |\mathbf{n}|)^{d-1}} \int_{\epsilon\phi(|\mathbf{n}|)}^{\epsilon\phi(|\mathbf{n}|)+\delta_{|\mathbf{n}|}^{-1/4}} 2x \left| P\left(\left|\frac{S_{\mathbf{n}}}{W_{\mathbf{n}}}\right| \geq x\right) dx - P(|N| \geq x) \right| dx$$

$$\equiv (V) + (VI).$$

Now, from Lemma 3 and (18) in Theorem 7, we have

$(V) \le \alpha_2(\epsilon) \times$

$$\sum_{|\mathbf{n}| \le K} \frac{g(\phi(|\mathbf{n}|))\phi'(|\mathbf{n}|)}{(\log|\mathbf{n}|)^{d-1}} \int_{\epsilon\phi(|\mathbf{n}|)+\delta_{|\mathbf{n}|}^{-1/4}}^{\infty} 2x \left[P\left(\left| \frac{S_{\mathbf{n}}}{W_{\mathbf{n}}} \right| \ge x \right) dx + P(|N| \ge x) \right] dx$$

$$\le \alpha_2(\epsilon) \sum_{|\mathbf{n}| \le K} \frac{g(\phi(|\mathbf{n}|))\phi'(|\mathbf{n}|)}{(\log|\mathbf{n}|)^{d-1}} \int_{\epsilon\phi(|\mathbf{n}|)+\delta_{|\mathbf{n}|}^{-1/4}}^{\infty} 2x \left[A \exp\left\{ -\frac{x^2}{2} \right\} + C \exp\left\{ -\frac{x^2}{2} \right\} \right] dx$$

$$\le \frac{C\alpha_2(\epsilon)}{(d-1)!} \sum_{j \le K} g(\phi(j))\phi'(j) \int_{\epsilon\phi(j)+\delta_j^{-1/4}}^{\infty} x \exp\left\{ -\frac{x^2}{2} \right\} dx$$

$$\le \frac{C\alpha_2(\epsilon)}{(d-1)!} \sum_{j \le K} g(\phi(j))\phi'(j) \exp\left\{ -\frac{\delta_j^{-1/2}}{2} \right\}$$

$$= \frac{C\alpha_2(\epsilon)}{(d-1)!} \sum_{j \le K(\epsilon,M)} g(\phi(j))\phi'(j)\delta_j', \tag{27}$$

where $\delta_j' = \exp\{-\delta_j^{-1/2}/2\} \to 0$ as $j \to \infty$. Next, we have

$(VI) \le \alpha_2(\epsilon) \times$

$$\sum_{|\mathbf{n}| \le K(\epsilon,M)} \frac{g(\phi(|\mathbf{n}|))\phi'(|\mathbf{n}|)}{(\log|\mathbf{n}|)^{d-1}} \int_{\epsilon\phi(|\mathbf{n}|)}^{\epsilon\phi(|\mathbf{n}|)+\delta_{|\mathbf{n}|}^{-1/4}} 2x \left| P\left(\left| \frac{S_{\mathbf{n}}}{W_{\mathbf{n}}} \right| \ge x \right) dx - P(|N| \ge x) \right| dx$$

$$\le \alpha_2(\epsilon) \sum_{|\mathbf{n}| \le K(\epsilon,M)} \frac{g(\phi(|\mathbf{n}|))\phi'(|\mathbf{n}|)}{(\log|\mathbf{n}|)^{d-1}} \int_{\epsilon\phi(|\mathbf{n}|)}^{\epsilon\phi(|\mathbf{n}|)+\delta_{|\mathbf{n}|}^{-1/4}} 2x\delta_{|\mathbf{n}|} dx$$

$$\le \frac{\alpha_2(\epsilon)}{(d-1)!} \sum_{j \le K(\epsilon,M)} g(\phi(j))\phi'(j) \int_{\epsilon\phi(j)}^{\epsilon\phi(j)+\delta_j^{-1/4}} 2x\delta_j dx$$

$$\le \frac{C\alpha_2(\epsilon)}{(d-1)!} \sum_{j \le K(\epsilon,M)} g(\phi(j))\phi'(j)\delta_j^{1/2}$$

where $\delta_j^{1/2} \to 0$. Now again from Toeplitz's lemma, we have that

$$\lim_{\epsilon \to \epsilon_0+} \sum_{|\mathbf{n}| \le K(\epsilon,M)} \frac{g(\phi(|\mathbf{n}|))\phi'(|\mathbf{n}|)}{(\log|\mathbf{n}|)^{d-1}} \times$$

$$\left| \int_{\epsilon\phi(|\mathbf{n}|)}^{\infty} 2xP\left(\left| \frac{S_{\mathbf{n}}}{W_{\mathbf{n}}} \right| \ge x \right) dx - \int_{\epsilon\phi(|\mathbf{n}|)}^{\infty} 2xP(|N| \ge x)dx \right|$$

$$\le \frac{C}{(d-1)!} \lim_{\epsilon \to \epsilon_0+} \alpha_2(\epsilon) \sum_{j \le K(\epsilon,M)} g(\phi(j))\phi'(j)[\delta_j' + \delta_j^{1/2}] = 0.$$

Next, we show (26). Note that from the conditions in Theorem 7 and Lemma 3, we have

$$\alpha_2(\epsilon) \sum_{|\mathbf{n}|>K(\epsilon,M)} \frac{g(\phi(|\mathbf{n}|))\phi'(|\mathbf{n}|)}{(\log|\mathbf{n}|)^{d-1}} \times$$

$$\left| \int_{\epsilon\phi(|\mathbf{n}|)}^{\infty} 2xP\left(\left|\frac{S_\mathbf{n}}{W_\mathbf{n}}\right| \geq x\right)dx - \int_{\epsilon\phi(|\mathbf{n}|)} 2xP\left(|N| \geq x\right)dx \right|$$

$$\leq \alpha_2(\epsilon) \sum_{|\mathbf{n}|>K(\epsilon,M)} \frac{g(\phi(|\mathbf{n}|))\phi'(|\mathbf{n}|)}{(\log|\mathbf{n}|)^{d-1}} \int_{\epsilon\phi(|\mathbf{n}|)}^{\infty} 2x\left[P\left(\left|\frac{S_\mathbf{n}}{W_\mathbf{n}}\right| \geq x\right) + P\left(|N| \geq x\right)\right]dx$$

$$\leq C\alpha_2(\epsilon) \sum_{j>K(\epsilon,M)} \sum_{|\mathbf{n}|=j} \frac{g(\phi(|\mathbf{n}|))\phi'(|\mathbf{n}|)}{(\log|\mathbf{n}|)^{d-1}} \int_{\epsilon\phi(|\mathbf{n}|)}^{\infty} x\exp\left(-\frac{x^2}{2}\right)dx$$

$$\leq C\alpha_2(\epsilon) \sum_{j>K(\epsilon,M)} d(j)\frac{g(\phi(j))\phi'(j)}{(\log j)^{d-1}} \int_{\epsilon\phi(j)}^{\infty} x\exp\left(-\frac{x^2}{2}\right)dx$$

$$= \frac{C\alpha_2(\epsilon)}{(d-1)!} \int_{K(\epsilon,M)}^{\infty} g(\phi(u))\phi'(u)\left(\int_{\epsilon\phi(u)}^{\infty} x\exp\left(-\frac{x^2}{2}\right)dx\right)du$$

$$\leq \frac{C\alpha_2(\epsilon)}{(d-1)!} \int_{M}^{\infty} xG\left(\frac{x}{\epsilon}\right)\exp\left(-\frac{x^2}{2}\right)dx$$

$$\leq \frac{C\alpha_2(\epsilon)}{(d-1)!} E\left[N^2 G\left(\frac{|N|}{\epsilon}\right)I\{|N| \geq M\}\right] \tag{28}$$

which converges to 0 uniformly with respect to ϵ as $M \to \infty$. Therefore (26) follows. This completes the proof of Theorem 7.

Lemma 6. *Suppose that $g(x)$, $\phi(x)$ and $\alpha_i(x)(i = 1,\ldots,4)$ satisfy the conditions (A1)–(A3), respectively.*

(i) Assume that $\kappa(x)$ is a function of x such that

$$\kappa(x)\phi(x) = O(1), \quad as \quad x \to \infty.$$

Then

$$\lim_{\epsilon\to 0^+} \alpha_1(\epsilon) \sum_{\mathbf{n}} g[\phi(|\mathbf{n}|)]\phi'(|\mathbf{n}|)(\log|\mathbf{n}|)^{1-d}P(|N| > \epsilon\phi(|\mathbf{n}|) + \kappa(|\mathbf{n}|))$$

$$= \frac{1}{(d-1)!} \lim_{\epsilon\to 0^+} \alpha_1(\epsilon)EG\left(\frac{|N|}{\epsilon}\right), \tag{29}$$

and

$$\lim_{\epsilon \to 0^+} \alpha_2(\epsilon) \sum_n g[\phi(|\boldsymbol{n}|)]\phi'(|\boldsymbol{n}|)(\log |\boldsymbol{n}|)^{1-d} E[N^2 I(|N| > \epsilon\phi(j) + \kappa(j))]$$

$$= \frac{1}{(d-1)!} \lim_{\epsilon \to 0^+} \alpha_2(\epsilon) E\left[N^2 G\left(\frac{|N|}{\epsilon}\right)\right]. \tag{30}$$

(ii) Assume that $\kappa(\epsilon, x)$ is a nonnegative function such that

$$\phi(x)\kappa(\epsilon, x) \to \rho \quad as \ x \to \infty \quad and \ \epsilon \to \epsilon_0^+ > 0.$$

Then

$$\lim_{\epsilon \to \epsilon_0^+} \alpha_3(\epsilon) \sum_n g[\phi(|\boldsymbol{n}|)]\phi'(|\boldsymbol{n}|)(\log |\boldsymbol{n}|)^{1-d} P(|N| > \epsilon\phi(|\boldsymbol{n}|) + \kappa(\epsilon, |\boldsymbol{n}|))$$

$$= \frac{\exp(-\epsilon_0\rho)}{(d-1)!} \lim_{\epsilon \to \epsilon_0^+} \alpha_3(\epsilon) EG\left(\frac{|N|}{\epsilon}\right), \tag{31}$$

and

$$\lim_{\epsilon \to \epsilon_0^+} \alpha_4(\epsilon) \sum_n g[\phi(|\boldsymbol{n}|)]\phi'(|\boldsymbol{n}|)(\log |\boldsymbol{n}|)^{1-d} E[N^2 I(|N| > \epsilon\phi(|\boldsymbol{n}|) + \kappa(\epsilon, |\boldsymbol{n}|))]$$

$$= \frac{\exp(-\epsilon_0\rho)}{(d-1)!} \lim_{\epsilon \to \epsilon_0^+} \alpha_4(\epsilon) E\left[N^2 G\left(\frac{|N|}{\epsilon}\right)\right]. \tag{32}$$

Proof. (i) From Lemma 5, to prove (29) and (30) it suffices to prove that

$$\lim_{\epsilon \to 0} \alpha_1(\epsilon) \sum_{\mathbf{n}} g[\phi(|\mathbf{n}|)]\phi'(|\mathbf{n}|)(\log |\mathbf{n}|)^{1-d} \times$$

$$|P(|N| > \epsilon\phi(|\mathbf{n}|) + \kappa(|\mathbf{n}|)) - P(|N| > \epsilon\phi(|\mathbf{n}|))| = 0 \tag{33}$$

and

$$\lim_{\epsilon \to 0} \alpha_2(\epsilon) \sum_{\mathbf{n}} g[\phi(|\mathbf{n}|)]\phi'(|\mathbf{n}|)(\log |\mathbf{n}|)^{1-d} \times$$

$$|E[N^2 I(|N| > \epsilon\phi(|\mathbf{n}|) + \kappa(|\mathbf{n}|))] - E[N^2 I(|N| > \epsilon\phi(|\mathbf{n}|))]| = 0. \tag{34}$$

Now note that

$$|P(|N| > \epsilon\phi(|\mathbf{n}|) + \kappa(|\mathbf{n}|)) - P(|N| > \epsilon\phi(|\mathbf{n}|))|$$

$$= \left| \int_{\epsilon\phi(|\mathbf{n}|)}^{\epsilon\phi(|\mathbf{n}|)+\kappa(|\mathbf{n}|)} \sqrt{\frac{2}{\pi}} \exp\left\{-\frac{x^2}{2}\right\} dx \right| \le |\kappa(|\mathbf{n}|)| \sqrt{\frac{2}{\pi}} \exp\left\{-\frac{(\epsilon\phi(|\mathbf{n}|) - |\kappa(|\mathbf{n}|)|)^2}{2}\right\}$$

$$\le \frac{C\exp(\epsilon C)}{\phi(|\mathbf{n}|)} \sqrt{\frac{2}{\pi}} \exp\left\{-\frac{\epsilon^2\phi^2(n)}{2}\right\}, \tag{35}$$

and

$$|E[N^2 I(|N| > \epsilon\phi(|\mathbf{n}|) + \kappa(|\mathbf{n}|))] - E[N^2 I(|N| > \epsilon\phi(|\mathbf{n}|))]|$$

$$= \left| \int_{\epsilon\phi(|\mathbf{n}|)}^{\epsilon\phi(|\mathbf{n}|)+\kappa(|\mathbf{n}|)} x^2 \sqrt{\frac{2}{\pi}} \exp\{-\frac{x^2}{2}\} dx \right|$$

$$\leq \frac{1}{3} |(\epsilon\phi(|\mathbf{n}|) + \kappa(|\mathbf{n}|))^3 - (\epsilon\phi(|\mathbf{n}|))^3| \sqrt{\frac{2}{\pi}} \exp\left\{-\frac{(\epsilon\phi(|\mathbf{n}|) - |\kappa(|\mathbf{n}|)|)^2}{2}\right\}$$

$$\leq C\epsilon^2 \phi(|\mathbf{n}|) \exp(\epsilon C) \sqrt{\frac{2}{\pi}} \exp\left\{-\frac{\epsilon^2\phi^2(|\mathbf{n}|)}{2}\right\}, \tag{36}$$

where C denote a constant which varies from line to line. Therefore from (35) and (36),

$$\alpha_1(\epsilon) \sum_{\mathbf{n}} g[\phi(|\mathbf{n}|)]\phi'(|\mathbf{n}|)(\log|\mathbf{n}|)^{1-d} |P(|N| > \epsilon\phi(|\mathbf{n}|) + \kappa(|\mathbf{n}|)) - P(|N| > \epsilon\phi(|\mathbf{n}|))|$$

$$\leq \alpha_1(\epsilon) \sum_{\mathbf{n}} \frac{g[\phi(|\mathbf{n}|)]\phi'(|\mathbf{n}|)}{(\log|\mathbf{n}|)^{1-d}} \frac{C\exp(\epsilon C)}{\phi(|\mathbf{n}|)} \sqrt{\frac{2}{\pi}} \exp\left\{-\frac{\epsilon^2\phi^2(|\mathbf{n}|)}{2}\right\}$$

$$\leq \alpha_1(\epsilon) \sum_{j=1}^{\infty} \sum_{|\mathbf{n}|=j} \frac{g[\phi(j)]\phi'(j)}{(\log j)^{1-d}} \frac{C\exp(\epsilon C)}{\phi(j)} \sqrt{\frac{2}{\pi}} \exp\left\{-\frac{\epsilon^2\phi^2(j)}{2}\right\}$$

$$\leq \frac{C\exp(\epsilon C)\alpha_1(\epsilon)}{(d-1)!} \sum_{j=1}^{\infty} \frac{g[\phi(j)]}{\phi(j)} \phi'(j) \sqrt{\frac{2}{\pi}} \exp\left\{-\frac{\epsilon^2\phi^2(j)}{2}\right\}$$

$$\leq \frac{C\alpha_1(\epsilon)}{(d-1)!} E\left[\frac{1}{|N|} g\left(\frac{|N|}{\epsilon}\right)\right], \tag{37}$$

and

$$\alpha_2(\epsilon) \sum_{\mathbf{n}} g[\phi(|\mathbf{n}|)]\phi'(|\mathbf{n}|)(\log|\mathbf{n}|)^{1-d} \times$$

$$|E[N^2 I(|N| > \epsilon\phi(|\mathbf{n}|) + \kappa(|\mathbf{n}|))] - E[N^2 I(|N| > \epsilon\phi(|\mathbf{n}|))]|$$

$$\leq \alpha_2(\epsilon) \sum_{\mathbf{n}} \frac{g[\phi(|\mathbf{n}|)]\phi'(|\mathbf{n}|)}{(\log|\mathbf{n}|)^{d-1}} C\epsilon^2 \phi(|\mathbf{n}|) \exp(\epsilon C) \sqrt{\frac{2}{\pi}} \exp\left\{-\frac{\epsilon^2\phi^2(n)}{2}\right\}$$

$$\leq C\exp(\epsilon C)\epsilon^2 \alpha_2(\epsilon) \sum_{j=1}^{\infty} \sum_{|\mathbf{n}|=j} \frac{g[\phi(|\mathbf{n}|)]\phi'(|\mathbf{n}|)\phi(|\mathbf{n}|)}{(\log|\mathbf{n}|)^{d-1}} \sqrt{\frac{2}{\pi}} \exp\left\{-\frac{\epsilon^2\phi^2(|\mathbf{n}|)}{2}\right\}$$

$$\leq \frac{C\exp(\epsilon C)\epsilon^2 \alpha_2(\epsilon)}{(d-1)!} \sum_{j=1}^{\infty} \phi(j)g[\phi(j)]\phi'(j) \sqrt{\frac{2}{\pi}} \exp\left\{-\frac{\epsilon^2\phi^2(j)}{2}\right\}$$

$$\leq C\alpha_2(\epsilon) E\left[|N|g\left(\frac{|N|}{\epsilon}\right)\right]. \tag{38}$$

Next from integral by part, we have that

$$
E\left[|N|g\left(\frac{|N|}{\epsilon}\right)\right] = \int_0^\infty xg\left(\frac{x}{\epsilon}\right) dF(x)
$$

$$
= \int_0^\infty xg\left(\frac{x}{\epsilon}\right)\sqrt{\frac{2}{\pi}}\exp\left(-\frac{x^2}{2}\right) dx = \int_0^\infty \epsilon x \exp\left(-\frac{x^2}{2}\right) dG\left(\frac{x}{\epsilon}\right)
$$

$$
= \epsilon\left[\int_0^\infty x^2 G\left(\frac{x}{\epsilon}\right)\sqrt{\frac{2}{\pi}}\exp\left(-\frac{x^2}{2}\right) dx - \int_0^\infty G\left(\frac{x}{\epsilon}\right)\sqrt{\frac{2}{\pi}}\exp\left(-\frac{x^2}{2}\right) dx\right]
$$

$$
= \epsilon\left[EN^2 G\left(\frac{|N|}{\epsilon}\right) - EG\left(\frac{|N|}{\epsilon}\right)\right].
$$

Therefore

$$
lim_{\epsilon\to 0}\alpha_2(\epsilon)E\left[|N|g\left(\frac{|N|}{\epsilon}\right)\right] = \lim_{\epsilon\to 0}\alpha_2(\epsilon)\epsilon\left[EN^2 G\left(\frac{|N|}{\epsilon}\right) - EG\left(\frac{|N|}{\epsilon}\right)\right] = 0
$$

$$(39)$$

provided that $\lim_{\epsilon\to 0}\alpha_2(\epsilon)E\left[N^2 G\left(\frac{|N|}{\epsilon}\right)\right] < +\infty$. Similarly one can prove that

$$
\lim_{\epsilon\to 0}\alpha_1(\epsilon)E\left[\frac{1}{|N|}g\left(\frac{|N|}{\epsilon}\right)\right] = 0 \qquad (40)
$$

provided that $\lim_{\epsilon\to 0}\alpha_1(\epsilon)E\left[G\left(\frac{|N|}{\epsilon}\right)\right] < +\infty$. Hence (33) and (34) follow from (37) to (40).

(ii) Note that as $x \to \infty$ and $\epsilon \to \epsilon_0 + > 0$, $\epsilon\phi(x) \to \infty$. By the asymptotic result for normal tail probability $P(|N| > x) \sim \frac{1}{x}\sqrt{\frac{2}{\pi}}\exp(-\frac{x^2}{2})(x \to \infty)$, we have that

$$
P(|N| > \epsilon\phi(|\mathbf{n}|) + \kappa(\epsilon, |\mathbf{n}|))
$$

$$
\sim \sqrt{\frac{2}{\pi}}\frac{1}{\epsilon\phi(|\mathbf{n}|) + \kappa(\epsilon, |\mathbf{n}|)}\exp\left[-\frac{1}{2}(\epsilon\phi(|\mathbf{n}|) + \kappa(\epsilon, |\mathbf{n}|))^2\right]
$$

$$
\sim \sqrt{\frac{2}{\pi}}\frac{1}{\epsilon\phi(|\mathbf{n}|)}\exp[-\epsilon\phi(|\mathbf{n}|)\kappa(\epsilon, |\mathbf{n}|)]\exp\left[-\frac{1}{2}\epsilon^2\phi^2(|\mathbf{n}|)\right].
$$

Therefore for any $0 < \theta < 1$, there exist $\epsilon' > 0$ and an integer N_0 such that for all $|\mathbf{n}| \geq N_0$ and $\epsilon_0 < \epsilon < \epsilon_0 + \epsilon'$,

$$
\exp(-\epsilon_0\rho - \theta)P(|N| > \epsilon\phi(|\mathbf{n}|))
$$

$$
\leq P(|N| > \epsilon\phi(|\mathbf{n}|) + \kappa(\epsilon, |\mathbf{n}|)) \leq \exp(-\epsilon_0\rho + \theta)P(|N| > \epsilon\phi(|\mathbf{n}|))
$$

and thus (31) follows from (19) in Lemma 5 and the arbitrariness of θ. Next we have that

$$EN^2I\{|N| > \epsilon\phi(|\mathbf{n}|) + \kappa(\epsilon, |\mathbf{n}|)\}$$

$$= [\epsilon\phi(|\mathbf{n}|) + \kappa(\epsilon, |\mathbf{n}|)]^2 P(|N| > \epsilon\phi(|\mathbf{n}|) + \kappa(\epsilon, |\mathbf{n}|))$$

$$+ \int_{\epsilon\phi(|\mathbf{n}|)+\kappa(\epsilon,|\mathbf{n}|)} 2xP(|N| > x)dx$$

$$\sim [\epsilon\phi(|\mathbf{n}|) + \kappa(\epsilon, |\mathbf{n}|)]^2 \exp\{-\epsilon\phi(|\mathbf{n}|)\kappa(\epsilon, |\mathbf{n}|)\}P(|N| > \epsilon\phi(|\mathbf{n}|)$$

$$+ \int_{\epsilon\phi(|\mathbf{n}|)+\kappa(\epsilon,|\mathbf{n}|)} 2x\frac{1}{x}\sqrt{\frac{2}{\pi}}\exp\{-\frac{x^2}{2}\}dx$$

$$\sim [\epsilon^2\phi^2(|\mathbf{n}|) + 2\epsilon_0\rho + 2]\exp\{-\epsilon_0\rho\}P(|N| > \epsilon\phi(|\mathbf{n}|))$$

$$\sim \epsilon^2\phi^2(|\mathbf{n}|)\exp\{-\epsilon_0\rho\}P(|N| > \epsilon\phi(|\mathbf{n}|)).$$

Therefore

$$\lim_{\epsilon \to \epsilon^+0} \alpha_4(\epsilon)\sum_{\mathbf{n}} g[\phi(|\mathbf{n}|)]\phi'(|\mathbf{n}|)(\log|\mathbf{n}|)^{1-d}E[N^2I(|N| > \epsilon\phi(|\mathbf{n}|) + \kappa(\epsilon, |\mathbf{n}|))]$$

$$= \lim_{\epsilon \to \epsilon_0^+} \alpha_4(\epsilon)\sum_{\mathbf{n}} g[\phi(|\mathbf{n}|)]\phi'(|\mathbf{n}|)(\log|\mathbf{n}|)^{1-d}\epsilon^2\phi^2(|\mathbf{n}|)\exp\{-\epsilon_0\rho\}P(|N| > \epsilon\phi(|\mathbf{n}|)$$

$$= \frac{\exp\{-\epsilon_0\rho\}}{(d-1)!}\lim_{\epsilon \to \epsilon_0^+} \alpha_4(\epsilon)\epsilon^2 E\left[G_1\left(\frac{|N|}{\epsilon}\right)\right]$$

$$= \frac{\exp\{-\epsilon_0\rho\}}{(d-1)!}\lim_{\epsilon \to \epsilon_0^+} \alpha_4(\epsilon)E\left[N^2G\left(\frac{|N|}{\epsilon}\right)\right]$$

$$- \frac{\exp\{-\epsilon_0\rho\}}{(d-1)!}\lim_{\epsilon \to \epsilon_0^+} \alpha_4(\epsilon)\epsilon^2 E\left[G_2\left(\frac{|N|}{\epsilon}\right)\right]. \tag{41}$$

Now note that

$$\alpha_4(\epsilon)\epsilon^2 E\left[G_2\left(\frac{|N|}{\epsilon}\right)\right] = \alpha_4(\epsilon)\epsilon^2\int_0^\infty\int_0^{\frac{x}{\epsilon}} 2ug(u)dudF(x)$$

$$\leq \alpha_4(\epsilon)\epsilon^2\int_0^\infty 2\frac{x}{\epsilon}G\left(\frac{x}{\epsilon}\right)dF(x)$$

$$= \alpha_4(\epsilon)\epsilon\int_0^{x_0} 2xG\left(\frac{x}{\epsilon}\right)dF(x) + \alpha_4(\epsilon)\epsilon^2\int_{x_0}^\infty 2\frac{x^2}{x_0}G\left(\frac{x}{\epsilon}\right)dF(x)$$

$$\leq 2\epsilon_0 G\left(\frac{x_0}{\epsilon_0}\right)\alpha_4(\epsilon) + \frac{2}{x_0}(\epsilon_0 + 1)^2\alpha_4(\epsilon)E\left[N^2G\left(\frac{|N|}{\epsilon}\right)I\{|N| > x_0\}\right] \tag{42}$$

Now if $\lim_{\epsilon \to \epsilon_0^+} \alpha_4(\epsilon) E\left[N^2 G\left(\frac{|N|}{\epsilon}\right)\right] < +\infty$, for any $\theta > 0$ by choosing a large x_0 such that the second term of (42) is less than $\theta/2$ and then choosing a small $\epsilon' > 0$ such that the first term of (42) is less than $\theta/2$ for $\epsilon_0 < \epsilon < \epsilon_0 + \epsilon'$, we have

$$\lim_{\epsilon \to \epsilon_0^+} \alpha_4(\epsilon)\epsilon^2 E\left[G_2\left(\frac{|N|}{\epsilon}\right)\right] = 0. \tag{43}$$

Hence (32) follows from (42) and (43).

4.3   The Proof of Theorem 8

Based on the same procedures in the proofs of Theorem 5, it is easy to show that

$$\lim_{\epsilon \to 0^+} \alpha_1(\epsilon) \sum_{\mathbf{n}} \frac{g[\phi(|\mathbf{n}|)]\phi'(|\mathbf{n}|)}{(\log |\mathbf{n}|)^{d-1}} \times$$

$$\left| P\left(\left|\frac{S_{\mathbf{n}}}{W_{\mathbf{n}}}\right| > \epsilon\phi(|\mathbf{n}|) + \kappa(|\mathbf{n}|)\right) - P(|N| > \epsilon\phi(|\mathbf{n}|) + \kappa(|\mathbf{n}|)) \right| = 0$$

and

$$\lim_{\epsilon \to \epsilon_0^+} \alpha_1(\epsilon) \sum_{\mathbf{n}} \frac{g[\phi(|\mathbf{n}|)]\phi'(|\mathbf{n}|)}{(\log |\mathbf{n}|)^{d-1}} \times$$

$$\left| P\left(\left|\frac{S_{\mathbf{n}}}{W_{\mathbf{n}}}\right| > \epsilon\phi(|\mathbf{n}|) + \kappa(\epsilon, |\mathbf{n}|)\right) - P\left(\left|\frac{S_{\mathbf{n}}}{W_{\mathbf{n}}}\right| > \epsilon\phi(|\mathbf{n}|) + \kappa(\epsilon, |\mathbf{n}|)\right) \right| = 0.$$

Therefore Theorem 8 follows from Lemma 6.

Also, we omit the proof of Theorem 9 due to the same reason.

Acknowledgements The first author's research is partly supported by the Natural Sciences and Engineering Research Council of Canada.

References

1. Csörgő, M., Martsynyuk, Y.V.: Functional central limit theorems for self-normalized least squares processes in regression with possibly infinite variance data. Stoch. Process. Appl. **121**, 2925–2953 (2011)
2. Csörgő, M., Szyszkowicz, B., Wang, Q.: Darling-Eredős theorem for self-normalized sums. Ann. Probab. **31**, 676–692 (2003)
3. Csörgő, M., Szyszkowicz, B., Wang, Q.: Donsker's theorem for self-normalized partial sums processes. Ann. Probab. **31**, 1228–1240 (2003)

4. Dembo, A., Shao, Q.M.: Self-normalized large deviations in vector spaces. In: Eberlein, E., Hahn, M., Talagrand, M. (eds.) Progress in Probability, vol. 43, pp. 27–32. Springer-Verlag (1998). ISBN: 978-3-0348-9790-7
5. Deng, D.: Self-normalized Wittmann;s laws of iterated logarithm in Banach space. Stat. Probab. Lett. **77**, 632–643 (2007)
6. Deng, D.: Precise asymptotics in the deviation probability series of self-normalized sums. J. Math. Anal. Appl. **376**, 136–153 (2010)
7. Giné, E., Götze, F., Mason, D.M.: When is the Student t-statistics asymptotically standard normal? Ann. Probab. **25**, 1514–1531 (1997)
8. Griffin, P.S., Kuelbs, J.: Self-normalized laws of the iterated logarithm. Ann. Probab. **17**, 1571–1601 (1989)
9. Griffin, P.S., Kuelbs, J.: Some extensions of the LIL via Self-normalized sums. Ann. Probab. **19**, 380–395 (1991)
10. Griffin, P.S., Mason, D.M.: On the asymptotic normality of self-normalized sums. Math. Proc. Camb. Philos. Soc. **109**, 597–610 (1991)
11. Gut, A.: Marcinkiewicz laws and convergence rates in the law of large numbers for random variables with multidimemsional indices. Ann. Probab. **6**, 469–482 (1978)
12. Gut, A.: Convergence rates for probabilities of moderate deviations for sums of random variables with multidimensional indices. Ann. Probab. **8**, 298–313 (1980)
13. Gut, A., Spătaru, A.: Precise asymptotics in some strong limit theorems for multidimensionally indexed random variables. J. Multivar. Anal. **86**, 398–422 (2003)
14. Hall, P.: On the effect of random norming on the rate of convergence in the central limit theorem. Ann. Probab. **16**, 1265–1280 (1988)
15. Hardy, G.H., Wright, E.M.: An Introduction to the Theory of Numbers, 3rd edn. Oxford University Press, Oxford (1954)
16. Jiang, C., Yang, X.: Precise asymptotics in self-normalized sums iterated logarithm for multidimensionally indexed random variables. Appl. Math. J. Chin. Univ. Ser. B **22**, 87–94 (2007)
17. Jing, B.Y., Shao, Q.M., Wang, Q.Y.: Self-normalized Cramér type large deviations for independent random variables. Ann. Probab. **31**, 2167–2215 (2003)
18. Liu, W.D., Lin, Z.Y.: Precise asymptotics for a new kind of complete moment convergence. Stat. Probab. Lett. **76**, 1787–1799 (2006)
19. Mason, D.M., Shao, Q.M.: Bootstrapping the student t-statistic. Ann. Probab. **29**, 1435–1450 (2001)
20. Nagaev, S.V.: The Berry-Esseen bound for self-normalized sums. Sib. Adv. Math. **12**, 79–125 (2002)
21. Pang, P., Zhang, L., Wang, W.: Precise asymptotics in the self-normalized law of iterated logarithm. J. Math. Anal. Appl. **340**, 1249–1261 (2008)
22. de la Pena, V., Klass, M., Lai, T.Z.: Self-normalized processes: exponential inequalities, moment bounds and iterated logarithm laws. Ann. Probab. **32**, 1902–1933 (2004)
23. Shao, Q.M.: Self-normalized large deviations. Ann. Probab. **25**, 285–328 (1997)
24. Shao, Q.M.: Cramér-type large deviation for Student's t statistic. J. Theor. Probab. **12**, 387–398 (1999)
25. Slavova, V.V.: On the Berry-Esseen bound for Students statistic. In: Kalashnikov, V.V., Zolotarev, V.M. (eds.) Stability Problems for Stochastic Models. Lecture Notes in Mathematics, vol. 1155, pp. 355–390. Spinger, Berlin (1985)
26. Stout, W.F.: Almost Sure Convergence. Academic, New York/San Francisco/London (1974)
27. Spătaru, A.: Exact asymptotics in log log lowas for random fields. J. Theor. Probab. **17**, 943–965 (2004)
28. Wang, Q.: Limit theorems for self-normalized large deviation. Electron. J. Probab. **10**, 1260–1285 (2005)
29. Wang, Q., Jing, B.Y.: An exponential nonuniform Berry-Esseen bound for self-normalized sums. Ann. Probab. **27**, 2068–2088 (1999)
30. Zhao, Y., Tao, J.: Precise asymptotics in complete moment convergence for self-normalized sums. Comput. Math. Appl. **56**, 1779–1786 (2008)

The Self-normalized Asymptotic Results
for Linear Processes

Magda Peligrad and Hailin Sang

1 Introduction

Let (ξ_i) be a sequence of independent and identically distributed (i.i.d.) centered random variables with $\xi_i \in \mathscr{L}^p, p > 0$; let (a_i) be real coefficients such that $\sum_{i=0}^{\infty} |a_i|^{min(2,p)} < \infty$. Then the linear process

$$X_t = \sum_{i=0}^{\infty} a_i \xi_{t-i} \qquad (1)$$

exists and is well-defined. It is also interesting to replace the i.i.d. innovation process in (1) by white noise processes, martingale difference processes or other dependence structures.

We can study many time series via the research on linear processes. For example, for a causal ARMA(p,q) process defined by the equation

$$X_t - \sum_{i=1}^{p} \phi_i X_{t-i} = \xi_t + \sum_{i=1}^{q} \theta_i \xi_{t-i},$$

M. Peligrad
Department of Mathematical Sciences, University of Cincinnati,
PO Box 210025, Cincinnati, OH 45221-0025, USA
e-mail: peligrm@ucmail.uc.edu

H. Sang (✉)
Department of Mathematics, University of Mississippi,
University, MS 38677-1848, USA
e-mail: sang@olemiss.edu

© Springer Science+Business Media New York 2015
D. Dawson et al. (eds.), *Asymptotic Laws and Methods in Stochastics*,
Fields Institute Communications 76, DOI 10.1007/978-1-4939-3076-0_3

there exists a sequence of constants (φ_j) such that $\sum_{j=0}^{\infty} |\varphi_j| < \infty$ and $X_t = \sum_{j=0}^{\infty} \varphi_j \xi_{t-j}$. In fact, as early as in 1938, Wold proved the Wold decomposition for weakly stationary processes: any mean zero weakly stationary process can be decomposed into a sum of a linear part $\sum_{j=0}^{\infty} a_j Z_{i-j}$ and a singular process. Here (Z_i) is a white noise process. The singular process could be zero under some regularity condition. Traditionally, linear process decomposition plays a key role in the development of time series asymptotics; see Hannan [10], Anderson [1] and Fuller [7].

For stationary time series, it is commonly accepted that the term "short memory" or "short range dependence" describes a time series with summable covariances. We refer to the review work or books on long memory time series by Baillie [3], Robinson [19] and Doukhan, Oppenheim and Taqqu [5] for references to both theory and applications. In terms of a linear process, if the innovations have a second moment, one commonly uses $\sum |a_i|$ as the standard for memory. A linear process with a second moment has short memory if $\sum a_i \neq 0$ and $\sum |a_i| < \infty$. Otherwise it is called a long memory linear process. If the innovations do not have a second moment, there is no completely commonly accepted definition of short memory or long memory. Nevertheless, in the regularly varying tail case with $\alpha < 2$, usually we say that the linear process has short memory if $\sum |a_i|^{\alpha/2} < \infty$ and long memory if $\sum |a_i|^{\alpha/2} = \infty$ but $\sum |a_i|^{\alpha} < \infty$. The case $\alpha = 2$ needs special treatment which is handled in the next section. Recall that, a distribution function $F(x)$ has regularly varying tail with parameter $\alpha > 0$ if $1 - F(x) = x^{-\alpha} L(x)$ for $x > 0$ and some slowly varying function $L(x)$.

This paper is aimed to review some recent developments on the linear process asymptotics including central limit theorem, functional central limit theorem and their self-normalized form. For classical asymptotics of linear processes, see the papers Giraitis and Surgailis [9], Phillips and Solo [17] and Wu and Woodroofe [20] and the references therein.

2   The Central Limit Theorem

Let (ξ_i) be a sequence of i.i.d. centered random variables and

$$H(x) = \mathbb{E}(\xi_0^2 I(|\xi_0| \leq x)) \text{ is a slowly varying function at } \infty. \tag{2}$$

This tail condition (2) is highly relevant to the central limit theory. For i.i.d. centered variables this condition is equivalent to the fact that the variables are in the domain of attraction of the normal law. This means: there is a sequence of constants $\eta_n \to \infty$ such that $\sum_{i=1}^{n} \xi_i / \eta_n$ is convergent in distribution to a standard normal variable (see for instance Fuller, [6]; Ibragimov and Linnik, [11]; Araujo and Giné, [2]). More precisely, if we put $b = \inf\{x > 1 : H(x) > 0\}$ then η_n is defined as

$$\eta_n = \inf\left\{s : s \geq b+1, \frac{H(s)}{s^2} \leq \frac{1}{n}\right\}. \tag{3}$$

To simplify the exposition we shall assume that the sequence of constants is indexed by integers, $(a_i)_{i\in\mathbb{Z}}$, and construct the linear process

$$X_k = \sum_{j=-\infty}^{\infty} a_{k+j}\xi_j. \tag{4}$$

In what follows we shall also make the following conventions:

Convention 1 By convention, for $x = 0$, $|x|H(|x|^{-1}) = 0$. For instance we can write instead $\sum_{j\in\mathbb{Z}, a_j\neq 0} a_j^2 H(|a_j|^{-1}) < \infty$, simply $\sum_{j\in\mathbb{Z}} a_j^2 H(|a_j|^{-1}) < \infty$.

Convention 2 The second convention refers to the function $H(x)$ defined in (2). Since the case $\mathbb{E}(\xi_0^2) < \infty$ is known, we shall consider the case $\mathbb{E}(\xi_0^2) = \infty$. Let $b = \inf\{x \geq 0 : H(x) > 1\}$ and $H_b(x) = H(x \vee (b+1))$. Then clearly $b < \infty$, $H_b(x) \geq 1$ and $H_b(x) = H(x)$ for $x > b+1$. We shall redenote $H_b(x)$ by $H(x)$. Therefore, since our results are asymptotic, without restricting the generality we shall assume that $H(x) \geq 1$ for all $x \geq 0$.

In a recent paper, Peligrad and Sang [16] addressed the question of the central limit theorem for partial sums of a linear process. For independent and identically distributed random variables they showed that the central limit theorem for the linear process is equivalent to the fact that the variables are in the domain of attraction of a normal law, answering in this way an open problem in the literature.

When the variables satisfy (2) a natural question is to point out necessary and sufficient conditions to be imposed to the coefficients which assures the existence of the linear process.

Proposition 1 (Peligrad and Sang [16], Proposition 2.2). *Let* $(\xi_k)_{k\in\mathbb{Z}}$ *be a sequence of i.i.d. centered random variables satisfying (2). The linear process* (X_k) *in (4) is well defined in the almost sure sense if and only if*

$$\sum_{j\in\mathbb{Z}} a_j^2 H(|a_j|^{-1}) < \infty. \tag{5}$$

As an example in this class we mention the particular linear process with regularly varying weights with exponent α where $1/2 < \alpha < 1$. This means that the coefficients are of the form $a_n = n^{-\alpha}L(n)$ for $n \geq 1$ and $a_n = 0$ for $n \leq 0$, where $L(n)$ is a slowly varying function at ∞. It incorporates the fractionally integrated processes that play an important role in financial econometrics, climatology and so on and they are widely studied. Such processes are defined for $0 < d < 1/2$ by

$$X_k = (1-B)^{-d}\xi_k = \sum_{i\geq 0} a_i\xi_{k-i} \text{ where } a_i = \frac{\Gamma(i+d)}{\Gamma(d)\Gamma(i+1)}$$

and B is the backward shift operator, $B\varepsilon_k = \varepsilon_{k-1}$. For this example, by the well known fact that for any real x, $\lim_{n\to\infty} \Gamma(n+x)/n^x\Gamma(n) = 1$, we have

$$\lim_{n\to\infty} a_n/n^{d-1} = 1/\Gamma(d).$$

Notice that these processes have long memory because $\sum_{j\geq1}|a_j| = \infty$.

The partial sums of a linear process can be expressed as an infinite series of independent random variables with double indexed sequence of real coefficients. Denote:

$$b_{nj} = a_{j+1} + \cdots + a_{j+n}$$

and with this notation

$$S_n(X) = \sum_{k=1}^{n} X_k = \sum_{j\in\mathbb{Z}} b_{nj}\xi_j. \tag{6}$$

A key element in establishing the central limit theorem for the partial sums of a linear process is to defined a suitable normalizing sequence

$$D_n = \inf\left\{ s \geq 1 : \sum_{k\geq1} \frac{b_{nk}^2}{s^2} H\left(\frac{s}{b_{nk}}\right) \leq 1 \right\}.$$

Peligrad and Sang ([16], Theorem 2.5), added the point (4) to the well-known results (1), (2) and (3) from the next theorem.

Theorem 1. *Let $(\xi_k)_{k\in\mathbb{Z}}$ be a sequence of independent and identically distributed centered random variables. Then the following four statements are equivalent:*

(1) ξ_0 satisfies condition (2).

(2) The sequence $(\xi_n)_{n\in\mathbb{Z}}$ satisfies the central limit theorem

$$\frac{S_n(\xi)}{\eta_n} \Rightarrow N(0, 1),$$

where $S_n(\xi)$ denotes the partial sums for the sequence $(\xi_n)_{n\in\mathbb{Z}}$.

(3) The sequence $(\xi_n)_{n\in\mathbb{Z}}$ satisfies the functional CLT

$$\frac{S_{[nt]}(\xi)}{\eta_n} \Rightarrow W(t)$$

on the space $D[0, 1]$ of all functions on $[0, 1]$ which have left-hand limits and are continuous from the right, where $W(t)$ is the standard Brownian motion and $[x]$ denotes the integer part of x.

(4) For any sequence of constants $(a_n)_{n \in \mathbb{Z}}$ satisfying (5) and $\sum_k b_{nk}^2 \to \infty$ the central limit theorem holds

$$\frac{S_n(X)}{D_n} \Rightarrow N(0, 1).$$

The point (4) of this theorem extends the Theorem 18.6.5 in Ibragimov and Linnik [11] from i.i.d. innovations with finite second moment to innovations in the domain of attraction of a normal law. It positively answers the question on the stability of the central limit theorem for i.i.d. variables under formation of linear sums.

This theorem has rather theoretical importance. For applying it the normalizing sequence D_n should be known. This sequence depends on the function $H(s)$ which is often unknown. A way to avoid the use of $H(s)$ is via the self-normalization that will be discussed in the next section.

3   Self-normalization

In this section we shall review self-normalized central limit theorem and self-normalized functional central limit theorem for linear processes (1). For a sequence of i.i.d. centered random variables $(\xi_k)_{k \in \mathbb{Z}}$ define

$$V_n^2 = V_n^2(\xi) = \sum_{k=1}^{n} \xi_k^2.$$

Recall that for a sequence of non degenerate i.i.d. centered random variables $(\xi_k)_{k \in \mathbb{Z}}$, the self-normalized sum $S_n(\xi)/V_n \Rightarrow N(0, 1)$ if and only if ξ is in the domain of attraction of a normal law (Giné, Götze and Mason [8]).

The following theorem, due to Csörgő, Szyszkowicz and Wang [4] gives the self-normalized functional central limit theorem.

Theorem 2 (Csörgő, Szyszkowicz and Wang [4]). *The following statements are equivalent:*

(1) The i.i.d. sequence $(\xi_k)_{k \in \mathbb{Z}}$ is centered and in the domain of attraction of a normal law.
(2) $S_{[nt]}(\xi)/V_n \Rightarrow W(t)$ on the space $D[0, 1]$.
(3) On an appropriate probability space,

$$\sup_{0 \le t \le 1} |\frac{1}{V_n} S_{[nt]}(\xi) - \frac{1}{\sqrt{n}} W(nt)| = o_P(1).$$

This result extended the classical weak invariance principle which states that, on an appropriate probability space, as $n \to \infty$,

$$\sup_{0 \le t \le 1} |\frac{1}{\sqrt{n}\sigma} \sum_{j=1}^{[nt]} \xi_j - \frac{1}{\sqrt{n}} W(nt)| = o_P(1)$$

if and only if $Var(\xi_0) = \sigma^2 < \infty$.

One natural question is, can we have the weak invariance principle for the self-normalized partial sums of the linear process (X_k), when the innovation is in the domain of attraction of a normal law?

Define

$$V_n^2(X) = \sum_{k=1}^{n} X_k^2. \tag{7}$$

One of the first self-normalized central limit theorems for linear processes is due to Juodis and Račkauskas [12], who considered a specific form of dependence. Precisely, they assumed that $X_i, i \in \mathbb{Z}$, is $AR(1)$ process obtained as a solution of the equation $X_t = \rho X_{t-1} + \xi_t$. They proved that $S_n(X)/V_n(X) \Rightarrow N(0, (1 + \rho)/(1 - \rho))$ under the condition $|\rho| < 1$ and ξ_t has mean 0 and satisfies (2). They further apply blocking technique to remove the parameter ρ in the self-normalized central limit theorem. More precisely they proved the following theorem:

Theorem 3 (Juodis and Račkauskas [12]). *For the AR(1) process $X_t = \rho X_{t-1} + \xi_t$, assume that $|\rho| < 1$ and ξ_0 has mean 0 and satisfies (2). Further assume that $n = mN$. Let $Y_j = \sum_{(j-1)m < i \le jm} X_i$, $j = 1, 2, \cdots, N$ and define $U_n^2 = Y_1^2 + \cdots + Y_N^2$. Then*

$$\frac{1}{U_n} S_n(X) \Rightarrow N(0, 1) \tag{8}$$

under the conditions $m \to \infty$ and $m/n \to 0$ as $n \to \infty$.

It is well known that, in this case, $(X_i; i \in \mathbb{Z})$ is a stationary sequence which can be expressed as the infinite time series $X_t = \sum_{j=1}^{\infty} \rho^j \xi_{t-j}$. So the problem Juodis and Račkauskas [12] have solved is for a specific short memory linear process.

For general short memory linear processes, Kulik [14] provides a nice self-normalized functional central limit theorem.

Theorem 4 (Kulik [14]). *For the class of linear processes (1) with $\sum_{k=1}^{\infty} |a_k| < \infty$, assume the innovations are centered and satisfy (2). Then*

$$\frac{S_n(X)}{V_n(X)} \Rightarrow N(0, \beta^2)$$

where $\beta^2 = (\sum_{k=0}^{\infty} a_k)^2/(\sum_{k=0}^{\infty} a_k^2)$. Also, the following invariance principle holds

$$\sup_{t \in (0,1]} \left| \frac{S_{[nt]}(X)}{V_n(X)} - \frac{|\beta| W(nt)}{\sqrt{n}} \right| = o_P(1).$$

Juodis and Računkas [13] also considered the linear process (1) satisfying a stronger condition than Kulik [14], namely, $\sum_{i=0}^{\infty} i|a_i| < \infty$. Under this condition, they establish (8), avoiding in this way to use the sequence of constants (a_i) in the limit. Later, Računkas and Suquet [18] give the self-normalized central limit theorem (8) under the same conditions as in Theorem 4. Furthermore, they provide the invariance principle $S_{[nt]}(X)/U_n \Rightarrow W(t)$ in the space $D[0, 1]$ or the Banach space $C[0, 1]$ of continuous functions on $[0, 1]$.

The results we discussed so far are for short memory linear processes only. Peligrad and Sang [15] provide self-normalized central limit theorem and self-normalized functional central limit theorem for long memory linear processes with regularly varying coefficients

$$a_n = n^{-\alpha} L(n), \text{ where } 1/2 < \alpha < 1, n \geq 1. \tag{9}$$

Recall (2) and (3) and set

$$B_n^2 := c_\alpha l_n n^{3-2\alpha} L^2(n) \text{ with } l_n = l(\eta_n) \tag{10}$$

where

$$c_\alpha = \{\int_0^{\infty} [x^{1-\alpha} - \max(x-1,0)^{1-\alpha}]^2 dx\}/(1-\alpha)^2. \tag{11}$$

Theorem 5 (Peligrad and Sang [15]). *Define $(X_n; n \geq 1)$ by (1) and assume that the innovations are centered and conditions (2) and (9) are satisfied. Then, $S_{[nt]}(X)/B_n$ converges weakly on the space $D[0, 1]$ endowed with Skorohod topology to the fractional Brownian motion W_H with Hurst index $H = 3/2 - \alpha$; i.e. to a Gaussian process with covariance structure $\frac{1}{2}(t^{3-2\alpha} + s^{3-2\alpha} - (t-s)^{3-2\alpha})$ for $s < t$. In particular, for $t = 1$, $S_n(X)/B_n \to N(0, 1)$.*

Denote $\sum_{i=0}^{\infty} a_i^2 = A^2$ and recall the definition (7). The corresponding self-normalized result is:

Theorem 6 (Peligrad and Sang [15]). *Under the same conditions as in Theorem 5,*

$$\frac{1}{nl_n} V_n^2(X) \xrightarrow{P} A^2 \tag{12}$$

and therefore

$$\frac{S_{[nt]}(X)}{na_n V_n(X)} \Rightarrow \frac{\sqrt{c_\alpha}}{A} W_H(t).$$

In particular

$$\frac{S_n(X)}{na_n V_n(X)} \Rightarrow N(0, \frac{c_\alpha}{A^2}).$$

The question of selfnormalized central limit theorem for a general linear process when the sequence of constants $(a_n)_{n \in \mathbb{Z}}$ satisfies (5) and the innovation is in the domain of attraction of a normal law is an interesting and useful problem. We can say that by combining the result on self-normalized CLT in Giné, Götze and Mason [8] with Theorem 1, we have:

Theorem 7 (Theorem 2.5, Peligrad and Sang [15]). *Let $(\xi_k)_{k \in \mathbb{Z}}$ be a sequence of i.i.d. centered random variables such that ξ_0 is centered and satisfies condition (2). Assume $(a_n)_{n \in \mathbb{Z}}$ satisfies (5). Then (1)–(4) in Theorem 1 are equivalent to (5) For any sequence of constants $(a_n)_{n \in \mathbb{Z}}$ satisfying (5) and $\sum_k b_{nk}^2 \to \infty$ the self-normalized CLT*

$$\frac{S_n(X)}{(\sum_k b_{nk}^2 \xi_k^2)^{1/2}} \Rightarrow N(0, 1) \quad as \quad n \to \infty$$

holds.

However, the selfnormalizer in this result is based on innovations and the coefficients rather than on observable variables (X_k). Further study is needed to determine a suitable normalizer for the general case, based on the linear process itself or partial sums in blocks of variables.

Acknowledgements We thank Rafal Kulik and the referees for helpful comments. Magda Peligrad was supported in part by a Charles Phelps Taft Memorial Fund grant, the NSA grant H98230-11-1-0135, and the NSF grant DMS-1208237.

References

1. Anderson, T.W.: The Statistical Analysis of Time Series. Wiley, New York (1971)
2. Araujo, A., Giné, E.: The Central Limit Theorem for Real and Banach Valued Random Variables. Wiley Series in Probability and Mathematical Statistics. Wiley, New York/Chichester/Brisbane (1980)
3. Baillie, R.T.: Long memory processes and fractional integration in econometrics. J. Econom. **73**, 5–59 (1996)
4. Csörgő, M., Szyszkowicz, B., Wang, Q.: Donsker's theorem for self-normalized partial sums processes. Ann. Probab. **31**, 1228–1240 (2003)

5. Doukhan, P., Oppenheim, G., Taqqu, M.S. (eds.): Theory and Applications of Long-Range Dependence. Birkhäuser, Boston (2003)
6. Fuller, W.A.: An Introduction to Probability Theory and Its Applications, vol. 2. Willey, New York (1966)
7. Fuller, W.A.: Introduction to Statistical Time Series. Wiley, New York (1976)
8. Giné, E., Götze, F., Mason, D.: When is the student t-statistic asymptotically standard normal? Ann. Probab. **25**, 1514–1531 (1997)
9. Giraitis, L., Surgailis, D.: A central limit theorem for quadratic forms in strongly dependent linear variables and its application to asymptotic normality of Whittle's estimate. Probab. Theory Relat. Fields **86**, 87–104 (1990)
10. Hannan, E.J.: Multiple Time Series. Wiley, New York (1970)
11. Ibragimov, I.A., Linnik, Y.V.: Independent and Stationary Sequences of Random Variables. Wolters, Groningen (1971)
12. Juodis, M., Račkauskas, A.: A remark on self-normalization for dependent random variables. Lith. Math. J. **45**, 142–151 (2005)
13. Juodis, M., Račkauskas, A.: A central limit theorem for self-normalized sums of linear process. Stat. Probab. Lett. **77**, 1535–1541 (2007)
14. Kulik, R.: Limit theorems for self-normalized linear processes. Stat. Probab. Lett. **76**, 1947–1953 (2006)
15. Peligrad, M., Sang, H.: Asymptotic properties of self-normalized linear processes with long memory. Econom. Theory **28**, 548–569 (2012)
16. Peligrad, M., Sang, H.: Central limit theorem for linear processes with infinite variance. J. Theor. Probab. **26**, 222–239 (2013)
17. Phillips, P.C.B., Solo, V.: Asymptotics for linear processes. Ann. Stat. **20**, 971–1001 (1992)
18. Račkauskas, A., Suquet, C.: Functional central limit theorems for self-normalized partial sums of linear processes. Lith. Math. J. **51**, 251–259 (2011)
19. Robinson, P.M. (ed.): Time Series with Long Memory. Oxford University Press, Oxford/New York (2003)
20. Wu, W.B., Woodroofe, M.: Martingale approximations for sums of stationary processes. Ann. Probab. **32**, 1674–1690 (2004)

Part II
Planar Processes

Some Results and Problems for Anisotropic Random Walks on the Plane

Endre Csáki, Antónia Földes, and Pál Révész

1 Introduction

We consider random walks on the square lattice $\mathbb{Z}^2$ with possibly unequal symmetric horizontal and vertical step probabilities, so that these probabilities can only depend on the value of the vertical coordinate. In particular, if such a random walk is situated at a site on the horizontal line $y = j \in \mathbb{Z}$, then at the next step it moves with probability p_j to either vertical neighbor, and with probability $1/2 - p_j$ to either horizontal neighbor. More formally, consider the random walk $\{\mathbf{C}(N) = (C_1(N), C_2(N)) ; N = 0, 1, 2, \ldots\}$ on $\mathbb{Z}^2$ with the transition probabilities

$$\mathbf{P}(\mathbf{C}(N + 1) = (k + 1, j) | \mathbf{C}(N) = (k, j))$$

$$= \mathbf{P}(\mathbf{C}(N + 1) = (k - 1, j) | \mathbf{C}(N) = (k, j)) = \frac{1}{2} - p_j,$$

$$\mathbf{P}(\mathbf{C}(N + 1) = (k, j + 1) | \mathbf{C}(N) = (k, j))$$

$$= \mathbf{P}(\mathbf{C}(N + 1) = (k, j - 1) | \mathbf{C}(N) = (k, j)) = p_j, \tag{1}$$

E. Csáki (✉)
Alfréd Rényi Institute of Mathematics, Hungarian Academy of Sciences,
Budapest, P.O.B. 127, H-1364, Hungary
e-mail: csaki.endre@renyi.mta.hu

A. Földes
Department of Mathematics, College of Staten Island, CUNY,
2800 Victory Blvd., Staten Island, New York, NY 10314, USA
e-mail: antonia.foldes@csi.cuny.edu

P. Révész
Institut für Statistik und Wahrscheinlichkeitstheorie, Technische Universität Wien,
Wiedner Hauptstrasse 8-10/107, A-1040 Vienna, Austria
e-mail: revesz.paul@renyi.mta.hu

© Springer Science+Business Media New York 2015
D. Dawson et al. (eds.), *Asymptotic Laws and Methods in Stochastics*,
Fields Institute Communications 76, DOI 10.1007/978-1-4939-3076-0_4

for $(k, j) \in \mathbb{Z}^2, N = 0, 1, 2, \ldots$ We assume throughout the paper that $0 < p_j \leq 1/2$ and $\min_{j \in \mathbb{Z}} p_j < 1/2$. Unless otherwise stated we assume also that $\mathbf{C}(0) = (0, 0)$. This model has a number of physical applications and the topic has a broad literature. We refer to Silver et al. [28], Seshadri et al. [26], Shuler [27], Westcott [30], where certain properties of this random walk were studied under various conditions. Heyde [14] proved an almost sure approximation for $C_2(\cdot)$ under the condition

$$n^{-1} \sum_{j=1}^{n} p_j^{-1} = 2\gamma + o(n^{-\eta}), \qquad n^{-1} \sum_{j=1}^{n} p_{-j}^{-1} = 2\gamma + o(n^{-\eta}) \qquad (2)$$

as $n \to \infty$ for some constants γ, $1 < \gamma < \infty$ and $1/2 < \eta < \infty$.

Heyde et al. [16] treated the case when conditions similar to (2) are assumed but γ can be different for the two parts of (2) and obtained almost sure convergence to the so-called oscillating Brownian motion. In Heyde [15] limiting distributions were given for $\mathbf{C}(\cdot)$ under the condition (2) but without remainder. Den Hollander [12] proved strong approximations for $\mathbf{C}(\cdot)$ in the case when p_j-s are random variables with values $1/4$ and $1/2$. Roerdink and Shuler [25] proved some asymptotic properties, including local limit theorems, under certain conditions. For more detailed history see [12].

First we give a general construction and discuss the issue of recurrence and transience of this random walk. In Sect. 2 we discuss strong approximations of the random walk $\{\mathbf{C}(N), N = 0, 1, \ldots\}$. Section 3 treats the local time and in Sect. 4 some results on the range will be given.

1.1 General Construction

Suppose that in a probability space we have two independent simple symmetric random walks, i.e.,

$$S_1(n), n = 0, 1, 2, \ldots, \qquad S_2(n), n = 0, 1, 2, \ldots,$$

where $S_1(0) = S_2(0) = 0$, $S_i(\cdot)$ are sums of i.i.d. random variables each taking the values 1 and -1 with probability $1/2$. The local times of S_i are defined by

$$\xi_i(j, n) = \#\{0 \leq k \leq n : S_i(k) = j\}, \quad j \in \mathbb{Z}, \quad n = 0, 1, 2, \ldots$$

Moreover, on the same probability space we have a double array of independent geometric random variables

$$G_i^{(j)}, i = 1, 2, \ldots, j \in \mathbb{Z}$$

with distribution

$$\mathbf{P}(G_i^{(j)} = k) = 2p_j(1 - 2p_j)^k, \quad k = 0, 1, 2, \ldots$$

We now construct our walk $\mathbf{C}(N)$ as follows. We will take all the horizontal steps consecutively from $S_1(\cdot)$ and all the vertical steps consecutively from $S_2(\cdot)$. First we will take some horizontal steps from $S_1(\cdot)$, then exactly one vertical step from $S_2(\cdot)$, then again some horizontal steps from $S_1(\cdot)$ and exactly one vertical step from $S_2(\cdot)$, and so on. Now we explain how to get the number of horizontal steps on each occasion. Consider our walk starting from the origin proceeding first horizontally $G_1^{(0)}$ steps (note that $G_1^{(0)} = 0$ is possible with probability $2p_0$), after which it takes exactly one vertical step, arriving either to the level 1 or -1, where it takes $G_1^{(1)}$ or $G_1^{(-1)}$ horizontal steps (which might be no steps at all) before proceeding with another vertical step. If this step carries the walk to the level j, then it will take $G_1^{(j)}$ horizontal steps, if this is the first visit to level j, otherwise it takes $G_2^{(j)}$ horizontal steps. In general, if we finished the k-th vertical step and arrived to the level j for the i-th time, then it will take $G_i^{(j)}$ horizontal steps.

In this paper N will denote the number of steps of the walk out of which H_N denotes the number of horizontal steps and $V_N = n$ the number of vertical steps, i.e., $H_N + V_N = N$. Then clearly

$$\mathbf{C}(N) = (C_1(N), C_2(N)) = (S_1(H_N), S_2(V_N)).$$

1.2 Recurrence, Transience

Our result on recurrence is a simple application of the celebrated Nash-Williams theorem [21]. To state this result we need some definitions. Consider a Markov chain $(\mathbf{X}, \mathbf{Y}, p)$ with countable state space $\mathbf{X}$, process $\mathbf{Y}$ and transition probabilities $p(\mathbf{u}, \mathbf{v})$. The chain is reversible if there exist strictly positive weights $\pi_{\mathbf{u}}$ for all $\mathbf{u} \in \mathbf{X}$ such that

$$\pi_{\mathbf{u}} p(\mathbf{u}, \mathbf{v}) = \pi_{\mathbf{v}} p(\mathbf{v}, \mathbf{u}). \tag{3}$$

If the chain is reversible we will use

$$a(\mathbf{u}, \mathbf{v}) := \pi_{\mathbf{u}} p(\mathbf{u}, \mathbf{v}).$$

Obviously the above defined anisotropic walk is a Markov chain on the state space $\mathbf{X} = \mathbb{Z}^2$, with the transition probabilities defined in (1). Furthermore, this Markov chain is reversible with the strictly positive weights

$$\pi(k, j) = \frac{1}{p_j}$$

and

$$a\left((k,j),(k,j+1)\right) = a\left((k,j),(k,j-1)\right) = 1$$

$$a\left((k,j),(k+1,j)\right) = a\left((k,j),(k-1,j)\right) = \frac{1}{2p_j} - 1 \qquad (4)$$

(and for non nearest neighbor sites $a(.,.) = 0$). This Markov chain is also time homogeneous, irreducible, i.e. it is possible to get to any state from any state with positive probability. The invariant measure is given by

$$\mu(k,j) = \pi(k,j) = \frac{1}{p_j}, \quad (k,j) \in \mathbb{Z}^2, \qquad (5)$$

i.e.,

$$\mu(\mathbf{u}) = \sum_{\mathbf{v}} \mu(\mathbf{v}) p(\mathbf{v}, \mathbf{u}),$$

where the summation goes for the four nearest neighbors of $\mathbf{u}$.

Theorem A (Nash-Williams [21]). *Suppose that* $(\mathbf{X}, \mathbf{Y}, p)$ *is a reversible Markov chain and that* $\mathbf{X} = \bigcup_{k=0}^{\infty} \Lambda^k$ *where* Λ^k *are disjoint. Suppose further that* $\mathbf{u} \in \Lambda^k$ *and* $a(\mathbf{u}, \mathbf{v}) > 0$ *together imply that* $\mathbf{v} \in \Lambda^{k-1} \bigcup \Lambda^k \bigcup \Lambda^{k+1}$, *and that for each* k *the sum* $\displaystyle\sum_{\mathbf{u} \in \Lambda^k, \mathbf{v} \in \mathbf{X}} a(\mathbf{u}, \mathbf{v}) < \infty$. *Let* $[\Lambda^k, \Lambda^{k+1}]$ *denote the set of pairs* $(\mathbf{u}, \mathbf{v})$ *such that* $\mathbf{u} \in \Lambda^k$ *and* $\mathbf{v} \in \Lambda^{k+1}$. *The Markov chain is recurrent if*

$$\sum_{k=0}^{\infty} \left(\sum_{(\mathbf{u},\mathbf{v}) \in [\Lambda^k, \Lambda^{k+1}]} a(\mathbf{u}, \mathbf{v}) \right)^{-1} = \infty. \qquad (6)$$

To apply this theorem, let Λ^k be the set of $8k$ lattice points on the square of width $2k$, centered at the origin. Furthermore, let $[\Lambda^k, \Lambda^{k+1}]$ be the set of $8k + 4$ nearest neighbor pairs (edges) between Λ^k and Λ^{k+1}.

It is easy to see by (4) that the sum in (6) is equal to

$$\sum_{k=0}^{\infty} \left(2 \left(\sum_{j=-k}^{k} \left(\frac{1}{2p_j} - 1 \right) + \sum_{j=-k}^{k} 1 \right) \right)^{-1} = \sum_{k=0}^{\infty} \left(\sum_{j=-k}^{k} \frac{1}{p_j} \right)^{-1}.$$

So we got the following result.

Theorem 1. *The anisotropic walk is recurrent if*

$$\sum_{k=0}^{\infty} \left(\sum_{j=-k}^{k} \frac{1}{p_j} \right)^{-1} = \infty. \tag{7}$$

As a simple consequence, if $\min_{j \in \mathbb{Z}} p_j > 0$, then the anisotropic walk is recurrent.

It is an intriguing question whether the converse of this statement is true as well. That is to say, is it true that

Conjecture 1. If

$$\sum_{k=0}^{\infty} \left(\sum_{j=-k}^{k} \frac{1}{p_j} \right)^{-1} < \infty, \tag{8}$$

then the anisotropic walk is transient.

We can't prove this conjecture, but a somewhat weaker result is true.

Theorem 2. *Assume that*

$$\sum_{j=-k}^{k} \frac{1}{p_j} = Ck^{1+A} + O(k^{1+A-\delta}), \quad k \to \infty \tag{9}$$

for some $C > 0$, $A > 0$ and $0 < \delta \leq 1$. Then the anisotropic random walk is transient.

Proof. Consider the simple symmetric random walk $S_2(\cdot)$ of the vertical steps and let $\xi_2(\cdot, \cdot)$ be its local time, and $\rho_2(\cdot)$ be its return time to zero. Consider the anisotropic random walk of N steps, where $N = N(m)$ is the time of m-th return of $S_2(\cdot)$ to zero, i.e., let $V_N = \rho_2(m)$.

First we give a lower bound for the number of the horizontal steps H_N.

Lemma 1. *For small enough $\varepsilon > 0$ we have almost surely for large enough m*

$$H_N = H_{N(m)} \geq m^{1+(1-\varepsilon)(A+1)}. \tag{10}$$

Proof. For simplicity in the proof, we denote ξ_2 by ξ and ρ_2 by ρ. From the construction in Sect. 1.1 it can be seen that the number of horizontal steps up to the m-th return to zero by the vertical component is given by

$$H_N = \sum_{j=-\infty}^{\infty} \sum_{i=1}^{\xi(j,\rho(m))} G_i^{(j)},$$

where G_i^j are as in Sect. 1.1. Since $\rho(m) \geq m^{2-\varepsilon}$ for small $\varepsilon > 0$ and large enough m almost surely, it follows from the stability of local time (see [23], Theorem 11.22, p. 127), that for any $\varepsilon > 0$, $|j| \leq m^{1-\varepsilon}$ we have

$$(1 - \varepsilon)m \leq \xi(j, \rho(m)) \leq (1 + \varepsilon)m$$

almost surely for large enough m. Hence

$$H_N \geq \sum_{|j| \leq m^{1-\varepsilon}} \sum_{i=1}^{(1-\varepsilon)m} G_i^{(j)} =: U_m.$$

We consider the expectation of U_m and show that the other terms are negligible. We have

$$\mathbf{E}G_i^{(j)} = \frac{1 - 2p_j}{2p_j},$$

$$Var G_i^{(j)} = \frac{1 - 2p_j}{4p_j^2}.$$

Hence by (9) we get

$$\mathbf{E}U_m = m(1 - \varepsilon) \sum_{|j| \leq m^{1-\varepsilon}} \frac{1 - 2p_j}{2p_j} \geq cm^{1+(1-\varepsilon)(A+1)},$$

where $c > 0$ is a constant. In what follows the value of such a c might change from line to line. We have

$$Var U_m = m(1 - \varepsilon) \sum_{|j| \leq m^{1-\varepsilon}} \frac{1 - 2p_j}{4p_j^2}.$$

It follows from (9) that $\frac{1}{2p_j} \leq c|j|^{1+A-\delta}$, hence

$$Var U_m \leq cm(1 - \varepsilon) \sum_{|j| \leq m^{1-\varepsilon}} \frac{|j|^{1+A-\delta}}{p_j} \leq cm^{1+(1-\varepsilon)(2+2A-\delta)}.$$

By Chebyshev inequality

$$\mathbf{P}(|U_m - \mathbf{E}U_m| \geq m^{(1-\varepsilon)(A+2)}) \leq c\frac{m^{3+2A-2\varepsilon(1+A)-(1-\varepsilon)\delta}}{m^{2(1-\varepsilon)(A+2)}} = cm^{-1-(1-\varepsilon)\delta+2\varepsilon}$$

which, by choosing ε small enough, is summable. Hence, as $m \to \infty$,

$$U_m = \mathbf{E}U_m + O(m^{(1-\varepsilon)(A+2)}) \quad a.s.,$$

consequently

$$H_N \geq U_m \geq cm^{1+(1-\varepsilon)(1+A)}$$

almost surely for large m. □

Lemma 2. *Let $S(\cdot)$ be a simple symmetric random walk and let $r(m)$ be a sequence of integer valued random variables independent of $S(\cdot)$ and such that $r(m) \geq m^\beta$ almost surely for large m with $\beta > 2$. Then with small enough $\varepsilon > 0$ we have*

$$|S(r(m))| \geq m^{\beta/2-1-\varepsilon}$$

almost surely for large m.

Proof. From the local central limit theorem

$$\mathbf{P}(S(k) = j) \leq \frac{c}{\sqrt{k}}$$

for all $k \geq 0$ and $j \in \mathbb{Z}$ with an absolute constant $c > 0$. Hence

$$\mathbf{P}\left(\frac{|S(k)|}{\sqrt{k}} \leq x\right) = \sum_{|j| \leq x\sqrt{k}} \mathbf{P}(S(k) = j) \leq cx,$$

This remains true if k is replaced by a random variable, independent of $S(\cdot)$, e.g. $k = r(m)$, i.e. we have

$$\mathbf{P}\left(\frac{|S(r(m))|}{\sqrt{r(m)}} \leq \frac{1}{m^{1+\varepsilon}}\right) \leq c\frac{1}{m^{1+\varepsilon}},$$

consequently by Borel-Cantelli Lemma

$$|S(r(m))| \geq \frac{\sqrt{r(m)}}{m^{1+\varepsilon}} \geq m^{\beta/2-1-\varepsilon},$$

almost surely for all large enough m. This completes the proof of the Lemma. □

Applying the two lemmas with $r(m) = H_{N(m)}$, we get

$$|S_1(H_N)| \geq m^{A/2-\varepsilon A/2-3\varepsilon/2} = m^\gamma$$

with $\gamma > 0$ by choosing $\varepsilon > 0$ small enough. It follows that for large N, $S_1(H_N)$ almost surely can't be equal to zero.

Let

$$A_m := \{\exists j, \rho_2(m) < j < \rho_2(m+1) \quad \text{such that} \quad \mathbf{C}(j) = (0,0)\}.$$

Clearly A_m could only happen if from $\rho_2(m)$ to $\rho_2(m+1)$ the walk only steps horizontally (if it makes one vertical step the return to the origin could only happen after or at $\rho_2(m+1)$). Thus by our lemmas in order that A_m could happen, the walk needs to have at least m^γ consecutive steps on the x-axis, thus

$$\sum_m^\infty \mathbf{P}(A_m) \leq \sum_m^\infty (1/2 - p_0)^{m^\gamma} < \infty.$$

So the anisotropic random walk cannot return to zero infinitely often with probability 1, it is transient. This proves the theorem.   □

2 Strong Approximations

In this section we present results concerning strong approximations of the two-dimensional anisotropic random walks. Of course, the results are different in the various cases, and in some cases the problem is open. We also mention weak convergence results available in the literature. First we describe the general method how to obtain these strong approximations.

Assume that our anisotropic random walk is constructed on a probability space as described in Sect. 1.1, and in accordance with Theorems 6.1 and 10.1 of Révész [23] we may assume that on the same probability space there are also two independent standard Wiener processes (Brownian motions) $W_i(\cdot)$, $i = 1, 2$ with local times $\eta_i(\cdot, \cdot)$ such that for all $\varepsilon > 0$, as $n \to \infty$, we have

$$S_i(n) = W_i(n) + O(n^{1/4+\varepsilon}) \quad a.s.$$

and

$$\xi_i(j, n) = \eta_i(j, n) + O(n^{1/4+\varepsilon}) \quad a.s.$$

Then

$$C_1(N) = S_1(H_N) = W_1(H_N) + O(H_N^{1/4+\varepsilon}) \quad a.s.,$$

and

$$C_2(N) = S_2(V_N) = W_2(V_N) + O(V_N^{1/4+\varepsilon}) \quad a.s.,$$

if $H_N \to \infty$ and $V_N \to \infty$ as $N \to \infty$, almost surely.

So we have to give reasonable approximations to H_N and V_N, or at least to one of them, and use $V_N + H_N = N$.

It turned out that in many cases treated, the following is a good approximation of H_N.

$$H_N \sim \sum_j \sum_{i=1}^{\xi_2(j,n)} G_i^{(j)} \sim \sum_j \xi_2(j,n) \frac{1-2p_j}{2p_j},$$

with $n = V_N$. H_N and the double sum above are not necessarily equal, since the last geometric variable might be truncated in H_N. So we have to investigate the additive functional

$$A(n) = \sum_j f(j)\xi_2(j,n) = \sum_{k=0}^{n} f(S_2(k)), \quad f(j) = \frac{1-2p_j}{2p_j}$$

of the vertical component and approximate it by the additive functional of $W_2(\cdot)$

$$B(t) = \int_{-\infty}^{\infty} f(x)\eta_2(x,t)\,dx = \int_0^t f(W_2(s))\,ds,$$

where between integers we define $f(x) = f(j), j \le x < j+1$.

In certain cases the following Lemma, equivalent to Lemma 2.3 of Horváth [17], giving strong approximation of additive functionals, may be useful.

Lemma A. *Let $g(t)$ be a non-negative function such that $g(t) = g(j), j \le t < j+1$, for $j \in \mathbb{Z}$ and assume that $g(t) \le c(|t|^\beta + 1)$ for some $0 < c$ and $\beta \ge 0$. Then*

$$\left| \sum_{j=0}^{n} g(S_2(j)) - \int_0^n g(W_2(s))\,ds \right| = o(n^{\beta/2+3/4+\varepsilon}) \quad a.s.$$

as $n \to \infty$.

Now let us introduce the notations

$$\sum_{j=1}^{k} f(j) = b_k, \quad \sum_{j=1}^{k} f(-j) = c_k$$

The next assumption is a reasonable one used in the literature: as $k \to \infty$

$$b_k = (\gamma - 1)k^\alpha + o(k^{\alpha-\delta}) \tag{11}$$

$$c_k = (\gamma - 1)k^\alpha + o(k^{\alpha-\delta}) \tag{12}$$

with some $\gamma \ge 1, \alpha \ge 0$ and $\delta > 0$. Observe that (2) is a particular case with $\alpha = 1$.

We consider the following cases:

(1) $\alpha = 0$
(2) $0 < \alpha < 1$
(3) $\alpha = 1$
(4) $\alpha > 1$
(5) nonsymmetric case, i.e. the constants γ in (11) and (12) are different.

2.1 The Case $\alpha = 0$

The most interesting and well established case is the so-called comb structure, i.e., $p_0 = 1/4, p_j = 1/2,\ j = \pm1, \pm2, \dots$. It follows from Theorem 1 that the random walk in this case is recurrent. We note in passing the interesting result of Krishnapur-Peres [18]: two independent random walks on the comb meet only finitely often with probability 1.

For random walk on comb we refer to Weiss and Havlin [29], Bertacchi and Zucca [2] and references given there. The following result on weak convergence was established by Bertacchi [1].

Theorem B.

$$\left(\frac{C_1(Nt)}{N^{1/4}}, \frac{C_2(Nt)}{N^{1/2}}; t \geq 0 \right) \xrightarrow{\text{Law}} (W_1(\eta_2(0,t)), W_2(t); t \geq 0), \quad N \to \infty.$$

Strong approximation was given in Csáki et al. [5].

Theorem 3. *On an appropriate probability space we have*

$$N^{-1/4}|C_1(N) - W_1(\eta_2(0,N))| + N^{-1/2}|C_2(N) - W_2(N)| = O(N^{-1/8+\varepsilon}) \quad a.s.,$$

as $N \to \infty$.

We have the following consequences.

Corollary 1.

$$\limsup_{N \to \infty} \frac{C_1(N)}{2^{5/4}3^{-3/4}N^{1/4}(\log \log N)^{3/4}} = 1 \quad a.s.$$

$$\limsup_{N \to \infty} \frac{C_2(N)}{(2N \log \log N)^{1/2}} = 1 \quad a.s.$$

For general results in the case $\alpha = 0$ we just remark that in this case $\bar{f} = \sum_j f(j)$ is convergent, then by our assumptions its terms are non-negative and at least one of them is strictly positive, hence $\bar{f} > 0$. By the ratio ergodic theorem (cf., e.g., Theorem 3.6 in Revuz [24])

$$A(n) \sim \bar{f}\xi_2(0,n), \quad \bar{f} = \sum_j f(j) = 2(\gamma - 1) + f(0),$$

almost surely, as $n \to \infty$, hence

$$A(n) = O((n \log \log n)^{1/2}) \quad \text{a.s., } n \to \infty.$$

Let

$$H_N^+ = \sum_j \sum_{i=1}^{\xi_2(j,n)} G_i^{(j)}, \quad H_N^- = \sum_j \sum_{i=1}^{\xi_2(j,n)-1} G_i^{(j)}.$$

Obviously, $H_N^- \leq H_N \leq H_N^+$. Consider H_N^+, which is a (random) sum of independent random variables. Under the condition $\mathscr{F} = \{S_2(k), \, k \geq 0\}$ we have

$$E(H_N^+ | \mathscr{F}) = \sum_j f(j)\xi_2(j,n) = A(n)$$

$$Var(H_N^+ | \mathscr{F}) = \sum_j \frac{f(j)}{2p_j}\xi_2(j,n).$$

It is easy to see that the sum $\sum_j f(j)/2p_j$ is also convergent, hence

$$Var(H_N^+ | \mathscr{F}) \sim c\xi(0,n)$$

with some $c > 0$. Now apply Theorem 6.17 in Petrov [22] saying that for sums of independent (not necessary identically distributed) random variables we have

$$\sum_i X_i = \sum_i EX_i + o\left(\left(\sum_i VarX_i\right)^{1/2+\varepsilon}\right)$$

almost surely. Thus

$$H_N^+ = \bar{f}\xi_2(0,n)(1 + o(1)) = \bar{f}\xi_2(0, V_N)(1 + o(1))$$

almost surely as $N \to \infty$. Similar results are true for H_N^-, hence also for H_N, i.e.

$$H_N = \bar{f}\xi_2(0,n)(1 + o(1)) = \bar{f}\xi_2(0, V_N)(1 + o(1)).$$

Since $C_1(N) = S_1(H_N)$, using that $H_N = O((N \log \log N)^{1/2})$ and the strong approximations of $S_1(\cdot), S_2(\cdot)$ by $W_1(\cdot), W_2(\cdot)$ and $\xi_2(0, \cdot)$ by $\eta_2(0, \cdot)$, we can obtain the following limit distribution: as $N \to \infty$,

$$\left(\frac{C_1(N)}{N^{1/4}}, \frac{C_2(N)}{N^{1/2}}\right) \xrightarrow{d} \left(W_1(\bar{f}\eta_2(0,1)), W_2(1)\right).$$

Further results, like strong approximations, remain to be established in this case.

2.2 The Case $0 < \alpha < 1$

This is also a recurrent case, but approximations, limit theorems, etc. remain to be worked out in detail. We just note that from the law of the iterated logarithm for the local time we have

$$A(n) = \sum_j f(j)\xi_2(j,n) \le c(n\log\log n)^{(1+\alpha)/2},$$

a.s., $n \to \infty$, hence the vertical part dominates, i.e., as $N \to \infty$ we should have

$$H_N = O((N\log\log N)^{(1+\alpha)/2}) << N \quad a.s.,$$

and we expect that

$$C_1(N) = W_1(Z(N)) + O(N^{(1+\alpha)/4+\varepsilon}) \quad a.s.,$$

where

$$Z(N) = \int_0^N f(W_2(s))\, ds,$$

and for the vertical component

$$C_2(N) = W_2(N) + O(N^{1/2-\varepsilon}), \quad a.s.$$

as $N \to \infty$.

2.3 The Case $\alpha = 1$

Assume also that $\delta > 1/2, \gamma > 1$.

It can be seen from Theorem 1 that the anisotropic random walk in this case is recurrent.

Heyde [14] gave the following strong approximation:

Theorem C. *On an appropriate probability space we have*

$$\gamma^{1/2} C_2(N) = W_2(N(1 + \varepsilon_N)) + O(N^{1/4} (\log N)^{1/2} (\log \log N)^{1/2}) \quad a.s.$$

as $N \to \infty$, where $W(\cdot)$ is a standard Wiener process and $\lim_{N \to \infty} \varepsilon_N = 0$ a.s.

In another paper Heyde [15] gave weak convergence result for both coordinates.

Theorem D.

$$\left(\frac{C_1(N)}{N^{1/2}}, \frac{C_2(N)}{N^{1/2}} \right) \xrightarrow{d} \left(W_1(1 - \gamma^{-1}), W_2(\gamma^{-1}) \right).$$

Strong approximation result for both coordinates was given in Csáki et al. [9]:

Theorem 4. *On an appropriate probability space we have for any $\varepsilon > 0$*

$$\left| C_1(N) - W_1 \left(\frac{(\gamma - 1)N}{\gamma} \right) \right| + \left| C_2(N) - W_2 \left(\frac{N}{\gamma} \right) \right| = O(N^{5/8 - \delta/4 + \varepsilon}) \quad a.s.,$$

as $N \to \infty$.

Moreover, in the periodic case, when $p_j = p_{j+L}$ for each $j \in \mathbb{Z}$ and a fixed integer $L \geq 1$ we have

$$\left| C_1(N) - W_1 \left(\frac{(\gamma - 1)N}{\gamma} \right) \right| + \left| C_2(N) - W_2 \left(\frac{N}{\gamma} \right) \right| = O(N^{1/4 + \varepsilon}) \quad a.s.,$$

as $N \to \infty$, where

$$\gamma = \frac{\sum_{j=0}^{L-1} p_j^{-1}}{2L}.$$

Some consequences are the following laws of the iterated logarithm.

$$\limsup_{N \to \infty} \frac{C_1(N)}{(N \log \log N)^{1/2}} = \left(\frac{2(\gamma - 1)}{\gamma} \right)^{1/2} \quad a.s.$$

$$\limsup_{N \to \infty} \frac{C_2(N)}{(N \log \log N)^{1/2}} = \left(\frac{2}{\gamma} \right)^{1/2} \quad a.s.$$

2.4 The Case $\alpha > 1$

In this case the random walk is transient by Theorem 2.

It is an open problem to give strong approximations in this case. Horváth [17] established weak convergence of $C_2(\cdot)$ to some time changed Wiener process. We mention a particular case of his results, valid for all $\alpha > 1$.

Let

$$I_\alpha(t) = \int_0^t |W_2(s)|^{\alpha-1} \, ds.$$

I_α is strictly increasing, so we can define its inverse, denoted by J_α. Then we have

$$\frac{C_2(Nt)}{N^{1/(1+\alpha)}} \xrightarrow{\text{Law}} c_0 W_2(J_\alpha(t))$$

with some constant c_0.

In this case the number of horizontal steps dominates the number of vertical steps, therefore $C_1(N)$ might be approximated by $W_1(N)$.

2.5 Unsymmetric Case

Some weak convergence in this case was treated in Heyde et al. [16] and Horváth [17]. Strong approximation in a particular case, the so-called half-plane half-comb structure (HPHC) was given in Csáki et al. [8].

Let $p_j = 1/4, j = 0, 1, 2, \ldots$ and $p_j = 1/2, j = -1, -2, \ldots$, i.e., a square lattice on the upper half-plane, and a comb structure on the lower half plane. Let furthermore

$$\alpha_2(t) := \int_0^t I\{W_2(s) \geq 0\} \, ds,$$

i.e., the time spent by W_2 on the non-negative side during the interval $[0, t]$. The process $\gamma_2(t) := \alpha_2(t) + t$ is strictly increasing, hence we can define its inverse: $\beta_2(t) := (\gamma_2(t))^{-1}$.

Theorem 5. *On an appropriate probability space we have*

$$|C_1(N) - W_1(N - \beta_2(N))| + |C_2(N) - W_2(\beta_2(N))| = O(N^{3/8+\varepsilon}) \quad a.s.,$$

as $N \to \infty$.

The following laws of the iterated logarithm hold:

Corollary 2.

$$\limsup_{t\to\infty} \frac{W_1(t - \beta_2(t))}{\sqrt{t \log\log t}} = \limsup_{N\to\infty} \frac{C_1(N)}{\sqrt{N \log\log N}} = 1 \quad a.s.,$$

$$\liminf_{t\to\infty} \frac{W_1(t - \beta_2(t))}{\sqrt{t \log\log t}} = \liminf_{N\to\infty} \frac{C_1(N)}{\sqrt{N \log\log N}} = -1 \quad a.s.,$$

$$\limsup_{t \to \infty} \frac{W_2(\beta_2(t))}{\sqrt{t \log \log t}} = \limsup_{N \to \infty} \frac{C_2(N)}{\sqrt{N \log \log N}} = 1 \quad a.s.,$$

$$\liminf_{t \to \infty} \frac{W_2(\beta_2(t))}{\sqrt{t \log \log t}} = \liminf_{N \to \infty} \frac{C_2(N)}{\sqrt{N \log \log N}} = -\sqrt{2} \quad a.s.$$

3 Local Time

We don't know any general result about the local time of the anisotropic walk. It would require to determine asymptotic results or at least good estimations for the return probabilities, i.e., we would need local limit theorems. In fact, we know such results in two cases: the periodic and the comb structure case.

3.1 Periodic Anisotropic Walk

In case of the periodic anisotropic walk, i.e., when $p_j = p_{j+L}$, for some fixed integer $L \geq 1$ and $j = 0, \pm 1, \pm 2, \ldots$ we know the following local limit theorem for the random walk denoted by $\mathbf{C^P}(\cdot)$.

Lemma 3. *As $N \to \infty$, we have*

$$\mathbf{P}(\mathbf{C^P}(2N) = (0,0)) \sim \frac{1}{4\pi N p_0 \sqrt{\gamma - 1}} \tag{13}$$

with $\gamma = \sum_{j=0}^{L-1} p_j^{-1} / (2L)$.

The proof of this lemma is based on the work of Roerdink and Shuler [25]. It follows from this lemma, that the truncated Green function $g(\cdot)$ is given by

$$g(N) = \sum_{k=0}^{N} \mathbf{P}(\mathbf{C^P}(k) = (0,0)) \sim \frac{\log N}{4 p_0 \pi \sqrt{\gamma - 1}}, \qquad N \to \infty,$$

which implies that our anisotropic random walk in this case is recurrent and also Harris recurrent.

First, we define the local time by

$$\Xi((k,j), N) = \sum_{r=1}^{N} I\{\mathbf{C^P}(r) = (k,j)\}, \quad (k,j) \in \mathbb{Z}^2. \tag{14}$$

In the case when the random walk is (Harris) recurrent, then we have (cf. e.g. Chen [4])

$$\lim_{N\to\infty} \frac{\Xi((k_1,j_1),N)}{\Xi((k_2,j_2),N)} = \frac{\mu(k_1,j_1)}{\mu(k_2,j_2)} \quad a.s.,$$

where $\mu(\cdot)$ is an invariant measure. Hence by (5)

$$\lim_{N\to\infty} \frac{\Xi((0,0),N)}{\Xi((k,j),N)} = \frac{p_j}{p_0} \quad a.s.$$

for $(k,j) \in \mathbb{Z}^2$ fixed.

Thus, using now $g(N)$, it follows from Darling and Kac [11] that we have

Corollary 3.

$$\lim_{N\to\infty} \mathbf{P}\left(\frac{\Xi((0,0),N)}{g(N)} \geq x\right) = \lim_{N\to\infty} \mathbf{P}\left(\frac{4p_0\pi\sqrt{\gamma-1}\,\Xi((0,0),N)}{\log N} \geq x\right) = e^{-x}$$

for $x \geq 0$.

For a limsup result, via Chen [4] we conclude

Corollary 4.

$$\limsup_{N\to\infty} \frac{\Xi((0,0),N)}{\log N \log\log\log N} = \frac{1}{4p_0\pi\sqrt{\gamma-1}} \quad a.s.$$

3.2 Comb

Now we consider the case of the two-dimensional comb structure $\mathbb{C}^2$, i.e., when $p_0 = 1/4$ and $p_j = 1/2$ for $j = \pm 1, \pm 2, \ldots$

First we give the return probability from Woess [31], p. 197:

$$\mathbf{P}(\mathbf{C}(2N) = (0,0)) \sim \frac{2^{1/2}}{\Gamma(1/4)N^{3/4}}, \quad N \to \infty.$$

This result indicates that the local time tipically is of order $N^{1/4}$. In Csáki et al. [6] and [7] we have shown the following results.

Theorem 6. *The limiting distribution of the local time is given by*

$$\lim_{N\to\infty} \mathbf{P}(\Xi((0,0),N)/N^{1/4} < x) = \mathbf{P}(2\eta_1(0,\eta_2(0,1)) < x) = \mathbf{P}(2|U|\sqrt{|V|} < x),$$

where U and V are two independent standard normal random variables.

Concerning strong approximation, in Csáki et al. [7] we proved the following results.

Theorem 7. *On an appropriate probability space for the random walk* $\{C(N) = (C_1(N), C_2(N)); N = 0, 1, 2, \ldots\}$ *on* $\mathbb{C}^2$, *one can construct two independent standard Wiener processes* $\{W_1(t); t \geq 0\}$, $\{W_2(t); t \geq 0\}$ *with their corresponding local time processes* $\eta_1(\cdot, \cdot), \eta_2(\cdot, \cdot)$ *such that, as* $N \to \infty$, *we have for any* $\delta > 0$

$$\sup_{x \in \mathbb{Z}} |\Xi((x, 0), N) - 2\eta_1(x, \eta_2(0, N))| = O(N^{1/8+\delta}) \quad a.s.$$

The next result shows that on the backbone up to $|x| \leq N^{1/4-\epsilon}$ we have uniformity, all the sites have approximately the same local time as the origin. Furthermore if we consider a site on a tooth of the comb its local time is roughly half of the local time of the origin. This is pretty natural, as it turns out from the proof that on the backbone the number of horizontal and vertical visits to any particular site are approximately equal.

Theorem 8. *On the probability space of* Theorem 7, *as* $N \to \infty$, *we have for any* $0 < \epsilon < 1/4$

$$\max_{|x| \leq N^{1/4-\epsilon}} |\Xi((x, 0), N) - \Xi((0, 0), N)| = O(N^{1/4-\delta}) \quad a.s.$$

and

$$\max_{0 < |y| \leq N^{1/4-\epsilon}} \max_{|x| \leq N^{1/4-\epsilon}} |\Xi((x, y), N) - \frac{1}{2}\Xi((0, 0), N)| = O(N^{1/4-\delta}) \quad a.s.,$$

for any $0 < \delta < \epsilon/2$, *where the maximum is taken on the integers.*

It would be an interesting problem to investigate the local time for $|y| > N^{1/4}$. We believe e.g. that the maximal local time taken for all $(x, y) \in \mathbb{Z}^2$ is of order $N^{1/2}$. Such results however remain to be established.

One of our old results [10] describes the Strassen class of $\eta_1(0, \eta_2(0, zt))$ as follows. This, combined with Theorems 7 and 8, allows us to conclude the corresponding Strassen class result for the local times of the walk.

Theorem 9. *The net*

$$\left\{ \frac{\eta_1(0, \eta_2(0, zt))}{2^{5/4}3^{-3/4}t^{1/4}(\log \log t)^{3/4}}; 0 \leq z \leq 1 \right\}_{t \geq 3},$$

as $t \to \infty$, *is almost surely relatively compact in the space* $C([0, 1], \mathbb{R})$ *of continuous functions from* $[0, 1]$ *to* $\mathbb{R}$, *and the set of its limit points is the class of nondecreasing absolutely continuous functions (with respect to the Lebesgue measure) on* $[0, 1]$ *for which*

$$\mathscr{S}^* : \left\{ f(0) = 0 \text{ and } \int_0^1 |\dot{f}(x)|^{4/3}\, dx \le 1 \right\}.$$

Some obvious consequences of these results are the following

- $\displaystyle \limsup_{t \to \infty} \frac{\eta_1(0, \eta_2(0,t))}{t^{1/4}(\log\log t)^{3/4}} = \frac{2^{5/4}}{3^{3/4}} \quad \text{a.s.}$

- $\displaystyle \limsup_{N \to \infty} \frac{\Xi((x,0),N)}{N^{1/4}(\log\log N)^{3/4}} = \frac{2^{9/4}}{3^{3/4}} \quad \text{a.s.,}$

- $\displaystyle \limsup_{N \to \infty} \frac{\Xi((x,y),N)}{N^{1/4}(\log\log N)^{3/4}} = \frac{2^{5/4}}{3^{3/4}} \quad \text{a.s. } y \ne 0.$

A beautiful classical result of Lévy, P. [19] reads as follows

Theorem E. *Let $W(\cdot)$ be a standard Wiener process with local time process $\eta(\cdot, \cdot)$. The following equality in distribution holds:*

$$\{\eta(0,t),\ t \ge 0\} \stackrel{d}{=} \{\sup_{0 \le s \le t} W(s),\ t \ge 0\}.$$

Consequently using a Hirsch type result of Bertoin [3], we get

Corollary 5. *Let $\beta(t) > 0,\ t \ge 0$, be a non-increasing function. Then we have almost surely that*

$$\liminf_{t \to \infty} \frac{\eta_1(0, \eta_2(0,t))}{t^{1/4}\beta(t)} = 0 \quad \text{or} \quad \infty$$

according as the integral $\int_1^\infty \beta(t)/t\, dt$ diverges or converges.

So we also have

Corollary 6. *Let $\beta(n), n = 1, 2, \ldots$ be a non-increasing sequence of positive numbers. Then, for any fixed $(x,y) \in \mathbb{Z}^2$, we have almost surely that*

$$\liminf_{n \to \infty} \frac{\Xi((x,y),n)}{n^{1/4}\beta(n)} = 0 \quad \text{or} \quad \infty$$

according as the series $\sum_1^\infty \beta(n)/n$ diverges or converges.

Now we also might consider the behavior of the supremum of the local time over the backbone. To this end we first had to prove the following pair of integral tests for the $\sup_{x \in \mathbb{R}} \eta_1(x, \eta_2(0,t))$ process.

Theorem 10. *Let $f(t) > 0$ be a non-decreasing function and put*

$$I(f) := \int_1^\infty \frac{f^2(t)}{t} \exp\left(-\frac{3}{2^{5/3}} f^{4/3}(t)\right) dt.$$

Then, as $t \to \infty$,

$$\mathbf{P}(\sup_{x \in \mathbb{R}} \eta_1(x, \eta_2(0, t)) > t^{1/4} f(t) \ i.o.) = 0 \ or \ 1$$

according as $I(f)$ *converges or diverges.*

Theorem 11. *Let* $g(t) > 0$ *be a non-increasing function and*

$$J(g) := \int_1^\infty \frac{g^2(t)}{t} \, dt.$$

Then, as $t \to \infty$,

$$\mathbf{P}(\sup_{x \in \mathbb{R}} \eta_1(x, \eta_2(0, t)) < t^{1/4} g(t) \ i.o.) = 0 \ or \ 1$$

according as whether $J(g)$ *converges or diverges.*

The above theorems imply the following integral tests for $\sup_{x \in \mathbb{Z}} \Xi((x, 0), n)$;

Theorem 12. *Let* $a(n)$ *be a non-decreasing sequence. Then, as* $n \to \infty$,

$$\mathbf{P}(\sup_{x \in \mathbb{Z}} \Xi((x, 0), n) > 2n^{1/4} a(n) \ i.o.) = 0 \ or \ 1$$

according as

$$\sum_{n=1}^\infty \frac{a^2(n)}{n} \exp\left(-\frac{3a^{4/3}(n)}{2^{5/3}}\right) < \infty \ or \ = \infty.$$

Theorem 13. *Let* $b(n)$ *be a non-increasing sequence. Then, as* $n \to \infty$,

$$\mathbf{P}(\sup_{x \in \mathbb{Z}} \Xi((x, 0), n) < n^{1/4} b(n) \ i.o.) = 0 \ or \ 1$$

according as

$$\sum_{n=1}^\infty \frac{b^2(n)}{n} < \infty \ or \ = \infty.$$

4 Range

The range of the anisotropic walk is defined in the usual way as

$$R(N) = \sum_{(k,j) \in \mathbb{Z}^2} I(\Xi((k, j), N) > 0)$$

i.e., the number of distinct sites visited by the random walk up to time N, where $\Xi((k,j),N)$ is the local time of the point (k,j) at time N.

We are not aware of any all embracing result about the range of the anisotropic walk in general. However the case of the periodic walk is completely understood.

Recall that the walk is periodic if $p_j = p_{j+L}$ for each $j \in \mathbb{Z}$, where $L \geq 1$ is a positive integer. In this case it is easy to see that

$$\gamma = \frac{\sum_{j=0}^{L-1} p_j^{-1}}{2L}.$$

Roerdink and Shuler [25] gives the asymptotic expected value of the range:

$$\mathbf{E}(R(N)) \sim \frac{2\pi \sqrt{\gamma - 1}}{\gamma} \frac{N}{\log N}, \quad N \to \infty.$$

Moreover, it can be seen that our walk in this case is equivalent to the so-called random walk with internal states, consequently, a law of large numbers follows from Nándori [20]

$$\lim_{N\to\infty} \frac{R(N)}{\mathbf{E}(R(N))} = \lim_{N\to\infty} \frac{\gamma R(N) \log N}{2\pi \sqrt{\gamma - 1} N} = 1 \qquad a.s.$$

As a special case from these results we recover the well-known Dvoretzky-Erdős [13] results for the simple random walk on the plane (without the remainder term), as for the plane $L = 1$ and $\gamma = 2$. Thus we get

$$\mathbf{E}(R(N)) \sim \frac{\pi N}{\log N}, \quad N \to \infty.$$

and

$$\lim_{N\to\infty} \frac{R(N)}{\mathbf{E}(R(N))} = \lim_{N\to\infty} \frac{R(N) \log N}{\pi N} = 1 \qquad a.s.$$

Acknowledgements The authors thank the referees for valuable comments and suggestions. Research supported by PSC CUNY Grant, No. 68080-0043 and by the Hungarian National Foundation for Scientific Research, No. K108615.

References

1. Bertacchi, D.: Asymptotic behaviour of the simple random walk on the 2-dimensional comb. Electron. J. Probab. **11**, 1184–1203 (2006)
2. Bertacchi, D., Zucca, F.: Uniform asymptotic estimates of transition probabilities on combs. J. Aust. Math. Soc. **75**, 325–353 (2003)

3. Bertoin, J.: Iterated Brownian Motion and stable (1/4) subordinator. Stat. Probab. Lett. **27**, 111–114 (1996)
4. Chen, X.: How often does a Harris recurrent Markov chain recur? Ann. Probab. **27**, 1324–1346 (1999)
5. Csáki, E., Csörgő, M., Földes, A., Révész, P.: Strong limit theorems for a simple random walk on the 2-dimensional comb. Electron. J. Probab. **14**, 2371–2390 (2009)
6. Csáki, E., Csörgő, M., Földes, A., Révész, P.: On the supremum of iterated local time. Publ. Math. Debr. **76**, 255–270 (2010)
7. Csáki, E., Csörgő, M., Földes, A., Révész, P.: On the local time of random walk on the 2-dimensional comb. Stoch. Process. Appl. **121**, 1290–1314 (2011)
8. Csáki, E., Csörgő, M., Földes, A., Révész, P.: Random walk on half-plane half-comb structure. Annales Mathematicae et Informaticae. **39**, 29–44 (2012)
9. Csáki, E., Csörgő, M., Földes, A., Révész, P.: Strong limit theorems for anisotropic random walks on $\mathbb{Z}^2$. Periodica Math. Hungar. **67**, 71–94 (2013)
10. Csáki, E., Földes, A., Révész, P.: Strassen theorems for a class of iterated processes. Trans. Am. Math. Soc. **349**, 1153–1167 (1997)
11. Darling, D.A., Kac, M.: On occupation times for Markoff processes. Trans. Am. Math. Soc. **84**, 444–458 (1957)
12. den Hollander, F.: On three conjectures by K. Shuler. J. Stat. Phys. **75**, 891–918 (1994)
13. Dvoretzky, A., Erdős, P.: Some problems on random walk in space. In: Proceedings of Second Berkeley Symposium, Berkeley, pp. 353–367 (1951)
14. Heyde, C.C.: On the asymptotic behaviour of random walks on an anisotropic lattice. J. Stat. Phys. **27**, 721–730 (1982)
15. Heyde, C.C.: Asymptotics for two-dimensional anisotropic random walks. In: Stochastic Processes, pp. 125–130. Springer, New York (1993)
16. Heyde, C.C., Westcott, M., Williams, E.R.: The asymptotic behavior of a random walk on a dual-medium lattice. J. Stat. Phys. **28**, 375–380 (1982)
17. Horváth, L.: Diffusion approximation for random walks on anisotropic lattices. J. Appl. Probab. **35**, 206–212 (1998)
18. Krishnapur, M., Peres, Y.: Recurrent graphs where two independent random walks collide finitely often. Electron. Commun. Probab. **9**, 72–81 (2004)
19. Lévy, P.: Processus Stochastiques et Mouvement Brownien. Gauthier Villars, Paris (1948)
20. Nándori, P.: Number of distinct sites visited by a random walk with internal states. Probab. Theory Relat. Fields **150**, 373–403 (2011)
21. Nash-Williams, C. St. J.A.: Random walks and electric currents in networks. Proc. Camb. Philos. Soc. **55**, 181–194 (1959)
22. Petrov, V.V.: Limit Theorems of Probability Theory. Sequences of Independent Random Variables. Clarendon Press, Oxford (1995)
23. Révész, P.: Random Walk in Random and Non-random Environments, 2nd edn. World Scientific, Singapore (2005)
24. Revuz, D.: Markov Chain. North-Holland, Amsterdam (1975)
25. Roerdink, J., Shuler, K.E.: Asymptotic properties of multistate random walks. I. Theory. J. Stat. Phys. **40**, 205–240 (1985)
26. Seshadri, V., Lindenberg, K., Shuler, K.E.: Random walks on periodic and random lattices. II. Random walk properties via generating function techniques. J. Stat. Phys. **21**, 517–548 (1979)
27. Shuler, K.E.: Random walks on sparsely periodic and random lattices I. Physica A **95**, 12–34 (1979)
28. Silver, H., Shuler, K.E., Lindenberg, K.: Two-dimensional anisotropic random walks. In: Statistical Mechanics and Statistical Methods in Theory and Application (Proc. Sympos., Univ. Rochester, Rochester, 1976). Plenum, New York, pp. 463–505 (1977)
29. Weiss, G.H., Havlin, S.: Some properties of a random walk on a comb structure. Physica A **134**, 474–482 (1986)
30. Westcott, M.: Random walks on a lattice. J. Stat. Phys. **27**, 75–82 (1982)
31. Woess, W.: Random Walks on Infinite Graphs and Groups. Cambridge Tracts in Mathematics, vol. 138. Cambridge University Press, Cambridge (2000)

On the Area of the Largest Square Covered by a Comb-Random-Walk

Pal Révész

1 Introduction

In recent years Miklós and some of his friends (Bandi, Toncsi and myself) have investigated some asymptotic properties of the comb-random-walk. In the present paper I continue this project. In particular, I am interested in the area of the largest disc around the origin completely covered by a comb-random-walk, and also by a random walk on a half-plane half-comb (cf. Theorems 1 and 2 respectively).

Let $\mathbf{C}(n) = (C_1(n), C_2(n))$ be a comb-random-walk, i.e., $\mathbf{C}(n)$ is a Markov chain on $\mathbb{Z}^2$ with $\mathbf{C}(0) = (0,0)$ and

$$\mathbf{P}\{\mathbf{C}(n+1) = (x, y \pm 1) \mid \mathbf{C}(n) = (x,y)\} = \frac{1}{2} \quad \text{if} \quad y \neq 0,$$

$$\mathbf{P}\{\mathbf{C}(n+1) = (x \pm 1, 0) \mid \mathbf{C}(n) = (x,0)\} =$$

$$= \mathbf{P}\{\mathbf{C}(n+1) = (x, \pm 1) \mid \mathbf{C}(n) = (x,0)\} = \frac{1}{4}.$$

Various properties of $\mathbf{C}$ were studied in many papers (cf., e.g.,[2–5]).

We say that a lattice point (x, y) of $\mathbb{Z}^2$ is covered by $\mathbf{C}$ at time n if there is a $k \leq n$ for which $\mathbf{C}(k) = (x, y)$. A set A is covered if each $(x, y) \in A$ is covered. Let R_n be the largest integer for which $[-R_n, R_n]^2$ is covered at time n.

Our main result in this regard reads as follows.

P. Révész (✉)
Technische Universität Wien, Institut für Statistik,
Wiedner Hauptstr. 8–10/107, A–1040 Wien, Austria
e-mail: revesz@ci.tuwien.ac.at

© Springer Science+Business Media New York 2015

D. Dawson et al. (eds.), *Asymptotic Laws and Methods in Stochastics*,
Fields Institute Communications 76, DOI 10.1007/978-1-4939-3076-0_5

Theorem 1. *For any $\varepsilon > 0$ we have*

$$\frac{n^{1/4}}{(\log n)^{5/2+\varepsilon}} \le R_n \le (1+\varepsilon)\Delta_n := (1+\varepsilon)\frac{2^{5/4}}{3^{3/4}}n^{1/4}(\log\log n)^{3/4} \ a.s. \qquad (1)$$

if n is large enough.

The above mentioned problem is due to Erdős and Chen [9] who studied the largest square covered by a simple random walk on $\mathbb{Z}^2$.

Let $S_2(n)$ be the simple random walk on $\mathbb{Z}^2$, i.e., $S_2(n)$ is a Markov chain with $S_2(0) = (0,0)$ and

$$\mathbf{P}\{S_2(n+1) = (x \pm 1, y \pm 1) \mid S_2(n) = (x, y)\} = \frac{1}{4}.$$

Let U_n be the largest integer for which the square $[-U_n, U_n]^2$ is covered by S_2 at time n. The properties of U_n were investigated in many papers (cf., e.g., [1, 8–11]). The corresponding results conclude that

$$U_n \sim \exp(C(\log n)^{1/2})$$

with some constant $C > 0$.

2   Proof of Theorem 1

We recall the following result.

Lemma 1 ([4] (1.23)).

$$\limsup_{n\to\infty} \frac{C_1(n)}{n^{1/4}(\log\log n)^{3/4}} = \frac{2^{5/4}}{3^{3/4}} \quad a.s. \qquad (2)$$

Clearly (2) implies

$$R_n \le (1+\varepsilon)\Delta_n \quad a.s. \qquad (3)$$

if n is large enough which, in turn, implies the upper part of (1).

In order to prove the lower part of (1), we recall some known results on **C** and present their simple consequences.

Let

$$\varXi((x,y),n) = \#\{k : \ k \le n, \mathbf{C}(k) = (x,y)\},$$

$$Z_n = \min_{|x| \le n^{1/4}(\log n)^{-3/2}} \varXi((x,0),n).$$

Then we have

$$\limsup_{n\to\infty} \frac{\Xi((0,0),n)}{n^{1/4}(\log\log n)^{3/4}} = \frac{2^{9/4}}{3^{3/4}} \quad \text{a.s.} \quad ([5],(1.21)), \tag{4}$$

and, if n is large enough,

$$\Xi((0,0),n) \geq \frac{n^{1/4}}{(\log n)^2} \quad \text{a.s.} \quad ([5], \text{Corollary } 1.5). \tag{5}$$

Before we continue the proof of the lower part of (1) we recall some known results on the simple random walk S_1 on $\mathbb{Z}^1$. Let

$$\xi_1(x,n) = \#\{k : k \leq n, S_1(k) = x\},$$

$$p(0,i,k) =$$

$$= \mathbf{P}\{\min\{j : j \geq m, S_1(j) = 0\} < \min\{j : j > k, S_1(j) = k\} \mid S_1(m) = i\}.$$

Then we have

$$p(0,i,k) = \frac{k-i}{k} \quad ([12],(3.1)). \tag{6}$$

Now (6) implies

$$\mathbf{P}\{S_1 \text{ hits } G > 0 \text{ during its first excursion }\} = \frac{1}{2G} \tag{7}$$

and (7) implies

$$\mathbf{P}\{S_1 \text{ does not hit } G = n^{1/4}(\log n)^{-a} \text{ during its first} \tag{8}$$

$$n^{1/4}(\log n)^{-3/2} \text{ excursions }\} = \left(1 - \frac{(\log n)^a}{2n^{1/4}}\right)^{n^{1/4}(\log n)^{-3/2}}$$

$$\leq \exp(-(\log n)^{a-3/2}/2)$$

if $a > 3/2$. We also recall that, if n is large enough, then

$$\frac{n^{1/2}}{(\log n)^{1+\varepsilon}} \leq \xi_1(0,n) \leq n^{1/2}(\log n)^\varepsilon \quad ([12] \text{ Theorem } 11.1), \tag{9}$$

and, in turn, the following result.

Lemma 2 ([7]). *Let*

$$g(n) = \frac{n^{1/2}}{(\log n)^{1+\varepsilon}}.$$

Then

$$\lim_{n \to \infty} \sup_{|x| \le g(n)} \left| \frac{\xi_1(x, n)}{\xi_1(0, n)} - 1 \right| = 0 \quad a.s. \tag{10}$$

A trivial consequence of (9) and Lemma B is that for large n we have

$$\frac{n^{1/2}}{(\log n)^{1+\varepsilon}} \le (1 - \varepsilon)\xi_1(0, n) \le \min_{|x| \le g(n)} \xi_1(x, n)$$

$$\le \max_{|x| < g(n)} \xi_1(x, n) \le (1 + \varepsilon)\xi_1(0, n) \le n^{1/2}(\log n)^{\varepsilon}. \text{ a.s.} \tag{11}$$

Now we present a simple generalization of the Borel–Cantelli lemma.

Lemma 3. *Let $\{A_n\}$ and $\{B_n\}$ be two sequences of events for which*

$$\sum_{n=1}^{\infty} \mathbf{P}\{\overline{A}_n \mid B_n\} < \infty, \tag{i}$$

and

$$B_n \text{ occurs a.s. if } n \text{ is large enough.} \tag{ii}$$

Then $\overline{A}_n$ occurs a.s. only finitely many times.

Proof. Since

$$\mathbf{P}\{\overline{A}_n B_n\} = \mathbf{P}\{\overline{A}_n \mid B_n\}\mathbf{P}\{B_n\} \le \mathbf{P}\{\overline{A}_n \mid B_n\},$$

$\overline{A}_n B_n$ occurs finitely many times. Since $\overline{A}_n \overline{B}_n \subset \overline{B}_n$, $\overline{A}_n \overline{B}_n$ also occurs a.s. only finitely many times.   □

Going back to study of the properties of $\mathbf{C}$, we let

1. $V(n) = \#\{k : k \le n, C_2(k) \ne C_2(k + 1)\}$,
2. $H(n) = \#\{k : k \le n, C_1(k) \ne C_1(k + 1)\}$,
3. $v(n) = \max_{k \le n}|C_2(k)|$,
4. $h(n) = \max_{k \le n}|C_1(k)|$,
5. $\Xi((x, y), n) = \#\{k : k \le n, \mathbf{C}(k) = (x, y)\}$,
6. $\zeta(n) = \#\{k : k < n, C_2(k) \ne 0, C_2(k + 1) = 0\}$,
7. $M_n(x) = \max_{k \le n, \, C_1(k)=x}|C_2(k)|$,
8. $m_n = \min_{|x| \le n^{1/4}(\log n)^{-3/2}} M_n(x)$.

Let $X_i(i = 1, 2, \dots)$ be $+1$ resp. -1 if the i-th horizontal step of $\mathbf{C}$ is $+1$ resp. -1. Similarly let $Y_i(i = 1, 2, \dots)$ be $+1$ resp. -1 if the i-th vertical step of $\mathbf{C}$ is $+1$

resp. -1. For example if $\mathbf{C}(0) = (0,0), \mathbf{C}(1) = (1,0), \mathbf{C}(2) = (1,1), \mathbf{C}(3) = (1,0), \mathbf{C}(4) = (0,0)$ then $X_1 = 1, Y_1 = 1, Y_2 = -1, X_2 = -1$.

Let $S_1(n) = X_1 + X_2 + \cdots + X_n$, $S_2(n) = Y_1 + Y_2 + \cdots + Y_n$.

Clearly $S_1(n)$ and $S_2(n)$ are independent, simple random walks on $\mathbb{Z}^1$ and

$$C_1(n) = S_1(H(n)),$$
$$C_2(n) = S_2(V(n)).$$

Now we present a two lemmas on the above random sequences.

Lemma 4.

$$V(n) \le n, \tag{12}$$

$$v(n) \le (1+\varepsilon)b_n^{-1} \quad a.s \tag{13}$$

$$\zeta(n) \le (1+\varepsilon)b_n^{-1} \quad a.s. \tag{14}$$

$$\zeta(n) - (\zeta(n))^{1/2+\varepsilon} \le H(n) \le \zeta(n) + (\zeta(n))^{1/2+\varepsilon} \quad a.s. \tag{15}$$

$$h(n) \le (1+\varepsilon)(b(\zeta(n)))^{-1} \le (1+\varepsilon)2^{3/4}n^{1/4}(\log\log n)^{3/4} \quad a.s. \tag{16}$$

Proof. Equation (12) is trivial. Equation (13) follows from the law of the iterated logarithm (LIL) and from (12). Equation (14) follows from the LIL. In order to see (15) let $C_2(k) = 0$ and $\ell_k = \min\{j : j \ge 0, \ C_2(k+j) \ne 0\}$.

Then

$$\mathbf{P}\{\ell_k = j\} = \frac{1}{2^{j+1}} \quad (j = 0, 1, \ldots),$$

$$\mathbf{E}\ell_k = 1 \quad \text{and} \quad \text{Var } \ell_k = 2. \tag{17}$$

Hence we have (15). Equation (16) is trivial. □

Lemma 5.

$$V(n) = n - H(n) \ge n - \zeta(n) - (\zeta(n))^{1/2+\varepsilon} \ge n - n^{1/2+\varepsilon} \quad a.s. \tag{18}$$

$$v(n) \ge (1-\varepsilon)(b(V(n)))^{-1} \ge (1-\varepsilon)b_n^{-1} \quad a.s. \ i.o. \tag{19}$$

$$\zeta(n) \ge (1-\varepsilon)b_n^{-1} \quad a.s. \ i.o. \tag{20}$$

Proof. is trivial.

By (9), if n is large enough, we have

$$\frac{V_n^{1/2}}{(\log V_n)^{1+\varepsilon}} \le \zeta_n \le V_n^{1/2}(\log V_n)^\varepsilon \quad a.s. \tag{21}$$

and

$$\frac{V_n^{1/2}}{(\log V_n)^{1+2\varepsilon}} \le \frac{V_n^{1/2}}{(\log V_n)^{1+\varepsilon}} - V_n^{1/4+\varepsilon} \le \zeta_n - \zeta_n^{1/2+\varepsilon} \le H_n \le \qquad (22)$$

$$\le \zeta_n + \zeta_n^{1/2+\varepsilon} \le V_n^{1/2}(\log V_n)^{2\varepsilon} \quad \text{a.s.}$$

Since $n = H_n + V_n$, we have

$$V_n + \frac{V_n^{1/2}}{(\log V_n)^{1+2\varepsilon}} \le n \le V_n + V_n^{1/2}(\log V_n)^{2\varepsilon} \quad \text{a.s.}$$

Consequently, if n is large enough,

$$n - n^{1/2}(\log n)^{3\varepsilon} \le V_n \le n - \frac{n^{1/2}}{(\log n)^{1+3\varepsilon}}$$

and

$$\frac{n^{1/2}}{(\log n)^{1+3\varepsilon}} \le n - V_n = H_n \le n^{1/2}(\log n)^{1+3\varepsilon}. \qquad (23)$$

Clearly,

$$\Xi((x,0),n) \ge \xi_1(x,H_n). \qquad (24)$$

By (11) and (23), for large n we have

$$\frac{n^{1/4}}{(\log n)^{3/2+\varepsilon}} \le \frac{H_n^{1/2}}{(\log H_n)^{1+\varepsilon}} \le \min_{|x| \le g(H_n)} \xi_1(x, H_n) \le$$

$$\le \min_{|x| \le g\left(\frac{n^{1/2}}{(\log n)^{1+\varepsilon}}\right)} \xi_1(x, H_n) \le \min_{|x| \le \frac{n^{1/4}}{(\log n)^{3/2}}} \xi_1(x, H_n) \le$$

$$\le Z_n := \min_{|x| \le \frac{n^{1/4}}{(\log n)^{3/2}}} \Xi((x,0),n) \quad \text{a.s.} \qquad (25)$$

Now, if $a > 3/2$, then by (8) we have

$\mathbf{P}\{$among the at least $n^{1/4}(\log n)^{-3/2}$ excursions going vertically

from $(0,0)$ no one hits $(0, n^{1/4}(\log n)^{-a}\} \le$

$$\le (1 - n^{-1/4}(\log n)^a)^{n^{1/4}(\log n)^{-3/2}} = \exp(-(\log n)^{a-3/2}) \le$$

$$\le 1 - \exp(-\exp(-(\log n)^{a-3/2})).$$

Consequently,

$$\mathbf{P}\left\{m_n \geq \frac{n^{1/4}}{(\log n)^{5/2+\varepsilon}} \mid Z_n > \frac{n^{1/4}}{(\log n)^{3/2+\varepsilon}}\right\} \geq$$

$$\geq \exp\left(-\frac{n^{1/4}}{(\log n)^{3/2}} \exp(-(\log n)^{1+\varepsilon})\right) \geq 1 - \frac{n^{1/4}}{(\log n)^{3/2}} \frac{1}{n^{1+(\log n)^\varepsilon}}.$$

Apply now Lemma 1 with

$$A_n = \left\{m_n \geq \frac{n^{1/4}}{(\log n)^{5/2+\varepsilon}}\right\},$$

$$B_n = \left\{Z_n > \frac{n^{1/4}}{(\log n)^{3/2+\varepsilon}}\right\}.$$

Then we also have the lower part of (1) and, this combined with (3), concludes the proof of Theorem 1. □

3 The Largest Square Covered by a HPHC Random Walk

Quite recently Miklós and his friends [6] investigated the properties of a random walk on a half-plane-half-comb (HPHC). Let $D(n) = (D_1(n), D_2(n))$ be a Markov chain on $\mathbb{Z}^2$ with $D(0) = (0, 0)$ and

$$\mathbf{P}\{D(N+1) = (k+1, j) \mid D(N) = (k, j)\} =$$

$$= \mathbf{P}\{D(N+1) = (k-1, j) \mid D(N) = (k, j)\} = \frac{1}{2} - p_j,$$

$$\mathbf{P}\{D(N+1) = (k, j+1) \mid D(N) = (k, j)\} =$$

$$= \mathbf{P}\{D(N+1) = (k, j-1) \mid D(N) = (k, j)\} = p_j,$$

where

$$p_j = 1/4 \quad \text{if} \quad j = 0, 1, 2, \ldots,$$

$$p_j = 1/2 \quad \text{if} \quad j = -1, -2, \ldots,$$

i.e., we have a square lattice on the upper half-plane and a comb structure on the lower half-plane. Let L_n be the largest integer for which $[-L_n, L_n]^2$ is covered by D at time n. Our main result on an HPHC reads as follows.

Theorem 2. *For any $\varepsilon > 0$, if n is large enough, we have*

$$L_n \leq (\log n)^{1+\varepsilon} \quad a.s., \tag{26}$$

and

$$\mathbf{P}\{L_n \leq (\log n)^{1-\varepsilon}\} \leq \exp(-(\log n)^{\varepsilon/2}). \tag{27}$$

In order to prove Theorem 2, we recall a few known results.

Lemma C ([12] p. 215). *Let*

$$\xi_2((x,y),n) = \#\{k : k \leq n, \ S_2(k) = (x,y)\}.$$

Then

$$(\log n)^{1-\varepsilon} \leq \xi_2((0,0),n) \leq (\log n)^{1+\varepsilon} \quad a.s.$$

if n is large enough.

Lemma D ([12] p. 34, p. 117). *Let*

$$v_n = \min\{k : \xi_1(0,k) = n\}.$$

Then

$$\max_{j \leq v_n} |S_1(j)| \leq v_n^{1/2+\varepsilon} \leq n^{1+2\varepsilon} \quad a.s.$$

if n is large enough.

Lemma E ([12] p. 100).

$$\mathbf{P}\{\max_{j \leq v_n} |S_1(j)| \leq n^{1-\varepsilon}\} \leq \exp(-n^{\varepsilon}).$$

Lemma F ([1]).

$$\lim_{n \to \infty} \sup_{|x| \leq (\log n)^{1+\varepsilon}} \left| \frac{\xi_2((x,0),n)}{\xi_2((0,0),n)} - 1 \right| = 0 \quad a.s.$$

Consequently,

$$\min_{|x| \leq (\log n)^{1+\varepsilon}} \xi_2((x,0),n) \geq (\log n)^{1-\varepsilon} \quad a.s. \tag{28}$$

if n is large enough.

Proof of (26). Lemmas C and D combined imply

$$\max_{k \leq n:\, D_1(k)=0} |D_2(k)| \leq (\log n)^{1+\varepsilon} \quad \text{a.s.,}$$

if n is large enough, which, in turn, implies (26). □

Proof of (27). By Lemmas E and F we conclude (27) as well. □

Acknowledgements Research supported by Hungarian Research Grant OTKA K108615.

References

1. Auer, P.: The circle homogeneously covered by random walk on $\mathbb{Z}^2$. Stat. Probab. Lett. **9**, 403–407 (1990)
2. Bertacchi, D.: Asymptotic behaviour of the simple random walk on the 2-dimensional comb. Electron. J. Probab. **11**, 1184–1203 (2006)
3. Bertacchi, D., Zucca, F.: Uniform asymptotic estimates of transition probabilities on comb. J. Aust. Math. Soc. **75**, 325–353 (2003)
4. Csáki, E., Csörgő, M., Földes, A., Révész, P.: Strong limit theorems for a simple random walk on the 2-dimensional comb. Electron. J. Probab. **14**, 2371–2390 (2009)
5. Csáki, E., Csörgő, M., Földes, A., Révész, P.: On the local time of random walk on the 2-dimensional comb. Stoch. Process. Appl. **121**, 1290–1314 (2011)
6. Csáki, E., Csörgő, M., Földes, A., Révész, P.: Random walk on half-plane-half-comb structure. Annales Mathematicae et Informaticae **39**, 29–44 (2012)
7. Csáki, E., Földes, A.: A note on the stability of the local time of Wiener process. Stoch. Process. Appl. **25**, 203–213 (1987)
8. Dembo, A., Peres, Y., Rosen, J.: How large a disc is covered by a random walk in n steps? Ann. Probab. **35**, 577–601 (2007)
9. Erdős, P., Chen, R.W.: Random walk on $\mathbb{Z}_2^n$. J. Multivar. Anal. **25**, 111–118 (1988)
10. Erdős, P., Révész, P.: On the area of the circles covered by a random walk. J. Multivar. Anal. **27**, 169–180 (1988)
11. Révész, P.: Estimates of the largest disc covered by a random walk. Ann. Probab. **18**, 1784–1789 (1990)
12. Révész, P.: Random Walk in Random and Non-Random Environments, 2nd edn. World Scientific, Singapore (2005)

A Compensator Characterization of Planar Point Processes

B. Gail Ivanoff

*This paper is dedicated to Professor Miklós Csörgő, a wonderful
mentor and friend, on the occasion of his 80th birthday.*

1 Background and Motivation

If N is a point process on $\mathbf{R}_+$ with $E[N(t)] < \infty$ $\forall t \in \mathbf{R}_+$, the compensator of N
is the unique predictable increasing process $\tilde{N}$ such that $N - \tilde{N}$ is a martingale with
respect to the minimal filtration generated by N, possibly augmented by information
at time 0. Why is $\tilde{N}$ so important? Some reasons include:

- The law of N determines and *is determined by* $\tilde{N}$ [11].
- The asymptotic behaviour of a sequence of point processes can be determined
  by the asymptotic behaviour of the corresponding sequence of compensators
  [2, 3, 7].
- Martingale methods provide elegant and powerful nonparametric methods for
  point process inference, state estimation, change point problems, and easily
  incorporate censored data [13].

Can martingale methods be applied to point processes in higher dimensions?
This is an old question, dating back more than 30 years to the 1970s and 1980s
when multiparameter martingale theory was an active area of research. However,
since there are many different definitions of planar martingales, there is no single
definition of "the compensator" of a point process on $\mathbf{R}_+^2$. A discussion of the
various definitions and a more extensive literature review can be found in [10]
and [7].

In this article, we revisit the following question: When can a compensator be
defined for a planar point process in such a way that it exists, it is unique and it
characterizes the distribution of the point process? Since there are many possible

B.G. Ivanoff (✉)
Department of Mathematics & Statistics, University of Ottawa, 585 King Edward,
Ottawa, ON K1N 6N5, Canada
e-mail: givanoff@uottawa.ca

© Springer Science+Business Media New York 2015
D. Dawson et al. (eds.), *Asymptotic Laws and Methods in Stochastics*,
Fields Institute Communications 76, DOI 10.1007/978-1-4939-3076-0_6

definitions of a point process compensator in two dimensions, we focus here on the one that has been the most useful in practice: the so-called *-compensator. Although existence and uniqueness of the *-compensator is well understood [5, 6, 14], in general it does not determine the law of the point process and it must be calculated on a case-by-case basis. However, it will be proven in Theorem 6 that when the point process satisfies a certain property of conditional independence (usually denoted by (F4), see Definition 2), the *-compensator determines the law of the point process and an explicit regenerative formula can be given. Although it seems to be widely conjectured that under (F4) the law must be characterized by the *-compensator, we have been unable to find a proof in the literature and, in particular, the related regenerative formula (14) appears to be completely new.

The basic building block of the planar model is the *single line* process (a point process with incomparable jump points). This approach was first introduced in [15] and further exploited in [10]. In both cases, the planar process is embedded into a point process with totally ordered jumps on a larger partially ordered space. "Compensators" are then defined on the larger space. In the case of [10], this is a family of one-dimensional compensators that, collectively, do in fact characterize the original distribution. Although the results in [10] do not require the assumption (F4) and are of theoretical significance, they seem to be difficult to apply in practice due to the abstract nature of the embedding. So, although in some sense the problem of a compensator characterization has been resolved for general planar point processes, for practical purposes it is important to be able to work on the original space, $\mathbf{R}_+^2$, if possible. We will see here that the assumption (F4) allows us to do so.

Returning to the single line process, when (F4) is satisfied we will see that its law can be characterized by a class of avoidance probabilities that form the two-dimensional counterpart of the survival function of a single jump point on $[0, \infty)$. Conditional avoidance probabilities then play the same role in the construction of the *-compensator as conditional survival distributions do for compensators in one dimension. For clarity and ease of exposition, we will be assuming throughout continuity of the so-called avoidance probabilities; this will automatically ensure the necessary predictability conditions and connects the avoidance probabilities and the *-compensator via a simple logarithmic formula. The more technical issues of discontinuous avoidance probabilities and other related problems will be dealt with in a separate publication. We comment further on these points in the Conclusion.

Our arguments involve careful manipulation of conditional expectations with respect to different σ-fields, making repeated use of the conditional independence assumption (F4). For a good review of conditional independence and its implications, we refer the reader to [12].

We proceed as follows: in Sect. 2, we begin with a brief review of the point process compensator on $\mathbf{R}_+$, including its heuristic interpretation and its regenerative formula. In Sect. 3 we define compensators for planar point processes. We discuss the geometry and decomposition of planar point processes into "single line processes" in Sect. 4, and in Sect. 5 we show how the single line processes can be interpreted via stopping sets, the two-dimensional analogue of a stopping time.

The compensator of the single line process is developed in Sect. 6 and combined with the decomposition of Sect. 4, this leads in Sect. 7 to the main result, Theorem 6, which gives an explicit regenerative formula for the compensator of a planar point process that characterizes its distribution. We conclude with some directions for further research in Sect. 8.

2   A Quick Review of the Compensator on $\mathbf{R}_+$

There are several equivalent characterizations of a point process on $\mathbf{R}_+$, and we refer the reader to [4] or [13] for details. For our purposes, given a complete probability space $(\Omega, \mathscr{F}, P)$, we interpret a simple point process N to be a pure jump stochastic process on $\mathbf{R}_+$ defined by

$$N(t) := \sum_{i=1}^{\infty} I(\tau_i \le t), \tag{1}$$

where $0 < \tau_1 < \tau_2 < \ldots$ is a strictly increasing sequence of random variables (the jump points of N). Assume that $E[N(t)] < \infty$ for every $t \in \mathbf{R}_+$. Let $\mathscr{F}(t) \equiv \mathscr{F}_0 \vee \mathscr{F}^N(t)$, where $\mathscr{F}^N(t) := \sigma\{N(s) : s \le t\}$, suitably completed, and $\mathscr{F}_0$ can be interpreted as information available at time 0. This is a right-continuous filtration on $\mathbf{R}_+$ and without loss of generality we assume $\mathscr{F} = \mathscr{F}(\infty)$. The law of N is determined by its finite dimensional distributions.

Since N is non-decreasing, it is an integrable submartingale and so has a Doob-Meyer decomposition $N - \tilde{N}$ where $\tilde{N}$ is the unique $\mathscr{F}$-predictable increasing process such that $N - \tilde{N}$ is a martingale. Heuristically,

$$\tilde{N}(dt) \approx P(N(dt) = 1 \mid \mathscr{F}(t)).$$

More formally, for each t,

$$\tilde{N}(t) = \lim_{n \to \infty} \sum_{k=0}^{2^n-1} E\left[N\left(\frac{(k+1)t}{2^n} \right) - N\left(\frac{kt}{2^n} \right) \mid \mathscr{F}\left(\frac{kt}{2^n} \right) \right], \tag{2}$$

where convergence is in the weak L^1-topology.

We have the following examples:

1. If N is a Poisson process with mean measure Γ and if $\mathscr{F} \equiv \mathscr{F}^N$, then by independence of the increments of N, it is an immediate consequence of (2) that $\tilde{N} = \Gamma$.

2. Let N be a Cox process (doubly stochastic Poisson process): given a realization Γ of a random measure γ on $\mathbf{R}_+$, N is (conditionally) a Poisson process with mean measure Γ. If $\mathscr{F}_0 = \sigma\{\gamma\}$, then $\tilde{N} = \gamma$. We refer to γ as the driving measure of the Cox process.

3. The single jump process: Suppose that N has a single jump point τ_1, a r.v. with continuous distribution F and let $\mathscr{F} \equiv \mathscr{F}^N$. In this case [4, 13]

$$\tilde{N}(t) = \int_0^t I(u \le \tau_1) \frac{dF(u)}{1 - F(u)} = \Lambda(t \wedge \tau_1), \tag{3}$$

where $\Lambda(t) := -\ln(1 - F(t))$ is the cumulative (or integrated) hazard of F. F is determined by its hazard $\frac{dF(\cdot)}{1-F(\cdot)}$. The relationship $\Lambda(t) = -\ln P(N(t) = 0)$ in Eq. (3) will be seen to have a direct analogue in two dimensions.

4. The general simple point process: We note that the jump points (τ_i) are $\mathscr{F}$-stopping times and so we define $\mathscr{F}(\tau_i) := \{F \in \mathscr{F} : F \cap \{\tau_i \le t\} \in \mathscr{F}(t) \; \forall t\}$. Assume that for every n, there exists a continuous regular version $F_n(\cdot | \mathscr{F}(\tau_{n-1}))$ of the conditional distribution of τ_n given $\mathscr{F}(\tau_{n-1})$ (we define $\tau_0 = 0$). Then if $\Lambda_n \equiv -\ln(1 - F_n)$, we have the following regenerative formula for the compensator (cf. [4], Theorem 14.1.IV):

$$\tilde{N}(t) = \sum_{n=1}^{\infty} \Lambda_n(t \wedge \tau_n) I(\tau_{n-1} < t). \tag{4}$$

Let $Q = P|_{\mathscr{F}_0}$ (the restriction of P to $\mathscr{F}_0$). Since there is a 1–1 correspondence between F_n and Λ_n, together, Q and $\tilde{N}$ characterize the law of N ([11], Theorem 3.4). When $\mathscr{F} \equiv \mathscr{F}^N$ (i.e. $\mathscr{F}_0$ is trivial), $\tilde{N}$ characterizes the law of N.

Comment 1. Note that Λ_n can be regarded as a random measure with support on (τ_{n-1}, ∞). Of course, in general we do not need to assume that F_n is continuous in order to define the compensator (cf. [4]). However, the logarithmic relation above between Λ_n and F_n holds only in the continuous case, and we will be making analogous continuity assumptions for planar point processes.

3   Compensators on $\mathbf{R}_+^2$

We begin with some notation: For $s = (s_1, s_2), t = (t_1, t_2) \in \mathbf{R}_+^2$,

- $s \le t \Leftrightarrow s_1 \le t_1$ and $s_2 \le t_2$
- $s \ll t \Leftrightarrow s_1 < t_1$ and $s_2 < t_2$.

We let $A_t := \{s \in \mathbf{R}_+^2 : s \le t\}$ and $D_t := \{s \in \mathbf{R}_+^2 : s_1 \le t_1 \text{ or } s_2 \le t_2\}$. A set $L \subseteq \mathbf{R}_+^2$ is a lower layer if for every $t \in \mathbf{R}_+^2, t \in L \Leftrightarrow A_t \subseteq L$. In analogy to (1), given a complete probability space $(\Omega, \mathscr{F}, P)$ and distinct $\mathbf{R}_+^2$-valued random variables $\tau_1, \tau_2, \ldots$ (the jump points), the point process N is defined by

$$N(t) := \sum_{i=1}^{\infty} I(\tau_i \le t) = \sum_{i=1}^{\infty} I(\tau_i \in A_t). \tag{5}$$

As pointed out in [13], in $\mathbf{R}_+^2$ there is no unique ordering of the indices of the jump points. Now letting $\tau_i = (\tau_{i,1}, \tau_{i,2})$, we assume that $P(\tau_{i,1} = \tau_{j,1}$ for some $i \neq j) = P(\tau_{i,2} = \tau_{j,2}$ for some $i \neq j) = 0$ and that $P(\tau_{i,1} = 0) = P(\tau_{i,2} = 0) = 0 \; \forall i$. In this case, we say that N is a strictly simple point process on $\mathbf{R}_+^2$ (i.e. there is at most one jump point on each vertical and horizontal line and there are no points on the axes). The law of N is determined by its finite dimensional distributions:

$$P(N(t_1) = k_1, \ldots N(t_i) = k_i), i \geq 1, t_1, \ldots, t_i \in \mathbf{R}_+^2, k_1, \ldots k_i \in \mathbf{Z}_+.$$

For any lower layer L, define

$$\mathscr{F}^N(L) := \sigma(N(t) : t \in L)$$

and

$$\mathscr{F}(L) = \mathscr{F}_0 \vee \mathscr{F}^N(L), \tag{6}$$

where $\mathscr{F}_0$ denotes the sigma-field of events known at time $(0,0)$. In particular, since there are no jumps on the axes, $\mathscr{F}(L) = \mathscr{F}_0$ for L equal to the axes. Furthermore, for any two lower layers L_1, L_2 it is easy to see that

$$\mathscr{F}(L_1) \vee \mathscr{F}(L_2) = \mathscr{F}(L_1 \cup L_2) \text{ and } \mathscr{F}(L_1) \cap \mathscr{F}(L_2) = \mathscr{F}(L_1 \cap L_2).$$

For $t \in \mathbf{R}_+^2$, denote

$$\mathscr{F}(t) := \mathscr{F}(A_t) \text{ and } \mathscr{F}^*(t) := \mathscr{F}(D_t).$$

Both $(\mathscr{F}(t))$ and $(\mathscr{F}^*(t))$ are right continuous filtrations indexed by $\mathbf{R}_+^2$: i.e. $\mathscr{F}^{(*)}(s) \subseteq \mathscr{F}^{(*)}(t)$ for all $s \leq t \in \mathbf{R}_+^2$ and if $t_n \downarrow t$, then $\mathscr{F}^{(*)}(t) = \cap_n \mathscr{F}^{(*)}(t_n)$. More generally, if (L_n) is a decreasing sequence of closed lower layers, $\mathscr{F}(\cap_n L_n) = \cap_n \mathscr{F}(L_n)$ (cf. [8]).

Definition 1. Let $(X(t) : t \in \mathbf{R}_+^2)$ be an integrable stochastic process on $\mathbf{R}_+^2$ and let $(\mathscr{F}(t) : t \in \mathbf{R}_+^2)$ be any filtration to which X is adapted (i.e. $X(t)$ is $\mathscr{F}(t)$-measurable for all $t \in \mathbf{R}_+^2$). X is a weak $\mathscr{F}$-martingale if for any $s \leq t$,

$$E[X(s,t] \mid \mathscr{F}(s)] = 0$$

where $X(s,t] := X(t_1, t_2) - X(s_1, t_2) - X(t_1, s_2) + X(s_1, s_2)$.

We now turn our attention to point process compensators on $\mathbf{R}_+^2$. It will always be assumed that $E[N(t)] < \infty$ for every $t \in \mathbf{R}_+^2$. For $t = (t_1, t_2) \in \mathbf{R}_+^2$ and $0 \leq k, j \leq 2^n - 1$ define

$$\Delta N(k,j) := N\left(\left(\frac{kt_1}{2^n}, \frac{jt_2}{2^n}\right), \left(\frac{(k+1)t_1}{2^n}, \frac{(j+1)t_2}{2^n}\right)\right].$$

In analogy to $\mathbf{R}_+$, the *weak $\mathscr{F}$-compensator* of N is defined by

$$\tilde{N}(t) := \lim_{n\to\infty} \sum_{j=0}^{2^n-1} \sum_{k=0}^{2^n-1} E\left[\Delta N(k,j) \mid \mathscr{F}\left(\frac{kt_1}{2^n}, \frac{jt_2}{2^n}\right)\right],$$

and the *$\mathscr{F}^*$-compensator (strong $\mathscr{F}$-compensator)* of N is defined by

$$\tilde{N}^*(t) := \lim_{n\to\infty} \sum_{j=0}^{2^n-1} \sum_{k=0}^{2^n-1} E\left[\Delta N(k,j) \mid \mathscr{F}^*\left(\frac{kt_1}{2^n}, \frac{jt_2}{2^n}\right)\right],$$

where both limits are in the weak L^1 topology. When there is no ambiguity, reference to $\mathscr{F}$ will be suppressed in the notation. Note that although $\tilde{N}^*$ is $\mathscr{F}^*$-adapted, it is not $\mathscr{F}$-adapted in general.

Comment 2. Under very general conditions, the compensators exist and $N - \tilde{N}$ and $N - \tilde{N}^*$ are weak martingales with respect to $\mathscr{F}$ and $\mathscr{F}^*$, respectively [7, 14]. Furthermore, each has a type of predictability property that ensures uniqueness (cf. [7]). Both compensators have non-negative increments: $\tilde{N}^{(*)}(s, t] \geq 0\ \forall s, t \in \mathbf{R}_+^2$. However, neither compensator determines the distribution of N in general, as can be seen in the following examples.

Examples. 1. The Poisson and Cox processes: Let N be a Poisson process on $\mathbf{R}_+^2$ with mean measure Γ and let $\mathscr{F} = \mathscr{F}^N$. By independence of the increments, both the weak and *-compensators of N ($\tilde{N}$ and $\tilde{N}^*$) are equal to Γ ([7], Theorem 4.5.2). A deterministic *-compensator characterizes the Poisson process, but a deterministic weak compensator does not (see [7] for details). Likewise, if N is a Cox process with driving measure γ on $\mathbf{R}_+^2$ and if $\mathscr{F}_0 = \sigma\{\gamma\}$, then $\tilde{N}^* \equiv \gamma$; this too characterizes the Cox process (cf. [7], Theorem 5.3.1). This discussion can be summarized as follows:

Theorem 1. *Let N be a strictly simple point process on $\mathbf{R}_+^2$ and let γ be a random measure on $\mathbf{R}_+^2$ that puts 0 mass on every vertical and horizontal line. Let $\mathscr{F}_0 = \sigma\{\gamma\}$ and $\mathscr{F}(t) = \mathscr{F}_0 \vee \mathscr{F}^N(t)$, $\forall t \in \mathbf{R}_+^2$. Then N is a Cox process with driving measure γ if and only if $\tilde{N}^* \equiv \gamma$. The law of N is therefore determined by $Q := P|_{\mathscr{F}_0}$ and $\tilde{N}^*$. In the case that γ is deterministic, $\mathscr{F}_0$ is trivial and N is a Poisson process.*

2. The single jump process: Assume that N has a single jump point $\tau \in \mathbf{R}_+^2$, a random variable with continuous distribution F and survival function

$$S(u) = P(\tau \geq u).$$

Then (cf. [7]):

$$\tilde{N}(t) = \int_{[0,t_1] \times [0,t_2]} I(u \leq \tau) \frac{dF(u)}{1 - F(u)}, \text{ and}$$

$$\tilde{N}^*(t) = \int_{[0,t_1] \times [0,t_2]} I(u \leq \tau) \frac{dF(u)}{S(u)}.$$

Although both formulas look very similar to (3), in two dimensions it is well known that neither $dF(u)/(1 - F(u))$ nor $dF(u)/S(u)$ determines F.

So we see that neither $\tilde{N}$ nor $\tilde{N}^*$ determines the law of N in general. The problem is that the filtration $\mathscr{F}$ does not provide enough information about N, and in some sense the filtration $\mathscr{F}^*$ can provide too much. As was observed in [10], the correct amount of information at time t lies between $\mathscr{F}(t)$ and $\mathscr{F}^*(t)$. The solution would be to identify a condition under which the two filtrations provide essentially the same information – this occurs under a type of conditional independence, a condition usually denoted by (F4) in the two-dimensional martingale literature.

To be precise, for $t = (t_1, t_2) \in \mathbf{R}_+^2$ and any filtration $(\mathscr{F}(t))$, define the following σ-fields:

$$\mathscr{F}^1(t) := \vee_{s \in \mathbf{R}_+} \mathscr{F}(t_1, s)$$

$$\mathscr{F}^2(t) := \vee_{s \in \mathbf{R}_+} \mathscr{F}(s, t_2).$$

Definition 2. We say that the filtration $(\mathscr{F}(t))$ satisfies condition (F4) if for all $t \in \mathbf{R}_+^2$, the σ-fields $\mathscr{F}^1(t)$ and $\mathscr{F}^2(t)$ are conditionally independent, given $\mathscr{F}(t)$ $(\mathscr{F}^1(t) \perp \mathscr{F}^2(t) \mid \mathscr{F}(t))$.

For the point process filtration $\mathscr{F}(t) = \mathscr{F}_0 \vee \mathscr{F}^N(t)$, in practical terms (F4) means that the behaviour of the point process is determined only by points in the past (in terms of the partial order): geographically, this means by points from the southwest. N could denote the points of infection in the spread of an air-born disease under prevailing winds from the southwest: since there are no points in $[0, t_1] \times (t_2, \infty)$ southwest of $(t_1, \infty) \times [0, t_2]$ and vice versa, the behaviour of N in either region will not affect the other.

While it appears that (F4) is related to the choice of the axes, it can be expressed in terms of the partial order on $\mathbf{R}_+^2$. In fact, it is equivalent to the requirement that for any $s, t \in \mathbf{R}_+^2$,

$$E[E[\cdot \mid \mathscr{F}(s)] \mid \mathscr{F}(t)] = E[\cdot \mid \mathscr{F}(s \wedge t)]].$$

This concept can be extended in a natural way to other partially ordered spaces; see Definition 1.4.2 of [7], for example.

Condition (F4) has the following important consequence: if $F \in \mathscr{F}(t) = \mathscr{F}_0 \vee \mathscr{F}^N(t)$, then for any lower layer D,

$$P[F \mid \mathscr{F}(D)] = P[F \mid \mathscr{F}(t) \cap \mathscr{F}(D)]. \tag{7}$$

This is proven in [5] for $D = D_s$ for $s \in \mathbf{R}_+^2$, and the result is easily generalized as follows. To avoid trivialities, assume $t \notin D$. Let $s_1 := \sup\{s \in \mathbf{R}_+ : (s, t_2) \in D\}$ and $s_2 := \sup\{s \in \mathbf{R}_+ : (t_1, s) \in D\}$ and define the lower layers D_1 and D_2 as follows:

$$D_1 := \{u = (u_1, u_2) \in D : u_1 \leq s_1\}$$

$$D_2 := \{u = (u_1, u_2) \in D : u_2 \leq s_2\}$$

We have that $D = (D \cap A_t) \cup D_1 \cup D_2$ and $\mathscr{F}(D) = \mathscr{F}(A_t \cap D) \vee \mathscr{F}(D_1) \vee \mathscr{F}(D_2)$. By (F4), $\mathscr{F}(D_2) \perp (\mathscr{F}(t) \vee \mathscr{F}(D_1)) \mid \mathscr{F}((t_1, s_2))$. Now use the chain rule for conditional expectation ([12], Theorem 5.8):

$$\mathscr{F}(D_2) \perp (\mathscr{F}(t) \vee \mathscr{F}(D_1)) \mid \mathscr{F}((t_1, s_2))$$

$$\Rightarrow \mathscr{F}(D_2) \perp \mathscr{F}(t) \mid (\mathscr{F}((t_1, s_2)) \vee \mathscr{F}(D_1))$$

$$\Rightarrow \mathscr{F}(D_2) \perp \mathscr{F}(t) \mid (\mathscr{F}((t_1, s_2)) \vee \mathscr{F}(D_1) \vee \mathscr{F}(A_t \cap D)) \tag{8}$$

$$\Rightarrow \mathscr{F}(D_2) \perp \mathscr{F}(t) \mid (\mathscr{F}(D_1) \vee \mathscr{F}(A_t \cap D)). \tag{9}$$

(8) and (9) follow since $\mathscr{F}((t_1, s_2)) \subseteq \mathscr{F}(A_t \cap D) \subseteq \mathscr{F}(t)$. But once again by (F4) we have $\mathscr{F}(D_1) \perp \mathscr{F}(t) \mid \mathscr{F}((s_1, t_2))$, and since $\mathscr{F}((s_1, t_2)) \subseteq \mathscr{F}(A_t \cap D) \subseteq \mathscr{F}(t)$ we have

$$\mathscr{F}(D_1) \perp \mathscr{F}(t) \mid \mathscr{F}((s_1, t_2)) \Rightarrow \mathscr{F}(D_1) \perp \mathscr{F}(t) \mid (\mathscr{F}((s_1, t_2)) \vee \mathscr{F}(A_t \cap D))$$

$$\Rightarrow \mathscr{F}(D_1) \perp \mathscr{F}(t) \mid \mathscr{F}(A_t \cap D). \tag{10}$$

Finally, if $F \in \mathscr{F}(t)$,

$$P[F \mid \mathscr{F}_D] = P[F \mid \mathscr{F}(A_t \cap D) \vee \mathscr{F}(D_1) \vee \mathscr{F}(D_2)]$$

$$= P[F \mid \mathscr{F}(A_t \cap D) \vee \mathscr{F}(D_1)] \text{ by (9)}$$

$$= P[F \mid \mathscr{F}(A_t \cap D)] \text{ by (10)},$$

and (7) follows since $\mathscr{F}(A_t \cap D) = \mathscr{F}(t) \cap \mathscr{F}(D)$.

We can use (7) to argue heuristically that (F4) ensures that $\mathscr{F}$ and $\mathscr{F}^*$ provide roughly the same information:

$$E\left[\Delta N(k, j) \mid \mathscr{F}^*\left(\frac{kt_1}{2^n}, \frac{jt_2}{2^n}\right)\right]$$

$$= E\left[\Delta N(k, j) \mid \mathscr{F}\left(\frac{(k+1)t_1}{2^n}, \frac{jt_2}{2^n}\right) \vee \mathscr{F}\left(\frac{kt_1}{2^n}, \frac{(j+1)t_2}{2^n}\right)\right] \text{ by (F4) (cf. (7))}$$

$$\approx E\left[\Delta N(k, j) \mid \mathscr{F}\left(\frac{kt_1}{2^n}, \frac{jt_2}{2^n}\right)\right] \text{ as } n \to \infty.$$

Therefore, $\tilde{N} \approx \tilde{N}^*$ and in particular, $\tilde{N}^*$ is $\mathscr{F}$-adapted. In this case, we refer to $N - \tilde{N}^*$ as a *strong* $\mathscr{F}$-martingale:

Definition 3. Let $(X(t) : t \in \mathbf{R}_+^2)$ be an integrable stochastic process on $\mathbf{R}_+^2$ and let $(\mathscr{F}(t) : t \in \mathbf{R}_+^2)$ be any filtration to which X is adapted. X is a strong $\mathscr{F}$-martingale if for any $s \leq t$,

$$E[X(s, t] \mid \mathscr{F}^*(s)] = 0.$$

As mentioned before, to avoid a lengthy discussion of predictability we will deal only with continuous compensators. In this case, we have the following (cf. [5, 6]):

Theorem 2. *Let N be a strictly simple point process and assume that the filtration $\mathscr{F} = \mathscr{F}_0 \vee \mathscr{F}^N$ satisfies (F4). If γ is a continuous increasing $\mathscr{F}$-adapted process such that $N - \gamma$ is a strong martingale, then $\tilde{N}^* \equiv \gamma$. (We say that γ is increasing if $\gamma(s, t] \geq 0 \; \forall \, s \leq t \in \mathbf{R}_+^2$.)*

We now address the following question: if (F4) is satisfied, will the *-compensator characterize the distribution of N? In the case of both the Poisson and Cox processes, (F4) is satisfied for the appropriate filtration ($\mathscr{F}(t) = \mathscr{F}^N(t)$ for the Poisson process and $\mathscr{F}(t) = \sigma\{\gamma\} \vee \mathscr{F}^N(t)$ for the Cox process) and the answer is yes, as noted in Theorem 1. For these two special cases, it is possible to exploit one dimensional techniques since conditioned on $\mathscr{F}_0$, the *-compensator is deterministic (see [7], Theorem 5.3.1). Unfortunately, this one dimensional approach cannot be used for more general point process compensators. Nonetheless, Theorem 1 turns out to be the key to the general construction of the compensator.

Before continuing, we note here that when (F4) is assumed a priori and the point process is strictly simple, there are many other characterizations of the two-dimensional Poisson process – for a thorough discussion see [16]. Assuming (F4), another approach is to project the two-dimensional point process onto a family of increasing paths. Under different sets of conditions, it is shown in [1] and [10] that if the compensators of the corresponding one-dimensional point processes are deterministic, the original point process is Poisson. (For a comparison of these results, see [10].) However, the characterization of the Poisson and Cox processes given in Theorem 1 does *not* require the hypothesis of (F4), and in fact implies it. Furthermore, it can be extended to more general spaces and to point processes that are not strictly simple (cf. [7], Theorem 5.3.1), although (F4) will no longer necessarily be satisfied.

Returning to the general case, the first step is to analyze the geometry of strictly simple point processes from the point of view taken in [10] and [15].

4   The Geometry of Point Processes on $\mathbf{R}_+^2$

Let $d = 1$ or 2. If N is a strictly simple point process on $\mathbf{R}_+^d$, then N can be characterized via the increasing family of random sets

$$\xi_k(N) := \{t \in \mathbf{R}_+^d : N(s) < k \; \forall s \ll t\}, k \geq 1.$$

By convention, in $\mathbf{R}_+$ we define $\xi_0(N)$ to be the origin, and in $\mathbf{R}_+^2$ we define $\xi_0(N)$ to be the axes. We observe that:

- In $\mathbf{R}_+$, $\xi_k(N) = [0, \tau_k]$.
- $N(t) = k \Leftrightarrow t \in \{\xi_{k+1}^o(N) \setminus \xi_k^o(N)\}$. ($\xi_k^o(N)$ denotes the interior of $\xi_k(N)$.)
- In $\mathbf{R}_+^2$, $\xi_k(N)$ is defined by the set of its *exposed points*:

$$\mathcal{E}_k := \min\{t \in \mathbf{R}_+^2 : N(t) \geq k\}$$

where for a nonempty Borel set $B \subseteq \mathbf{R}_+^d$, $\min(B) := \{t \in B : s \not\leq t, \forall s \in B, s \neq t\}$. By convention, $\min(\emptyset) := \infty$. It is easily seen that

$$\xi_k(N) = \cap_{\tau \in \mathcal{E}_k} D_\tau.$$

To illustrate, in Fig. 1 we consider the random sets $\xi_1(N)$ and $\xi_2(N)$ of a point process with five jump points, each indicated by a "$\bullet$". While the exposed points $\tau_1^{(1)}, \tau_2^{(1)}, \tau_3^{(1)}$ of $\xi_1(N)$ are all jump points of N, the exposed points of $\xi_2(N)$ include $\tau_1^{(1)} \vee \tau_2^{(1)}$ and $\tau_2^{(1)} \vee \tau_3^{(1)}$ (each indicated by a "o") , which are not jump points. In fact, if

$$\xi_k^+(N) := \cap_{\epsilon, \epsilon' \in \mathcal{E}_k, \epsilon \neq \epsilon'} D_{\epsilon \vee \epsilon'},$$

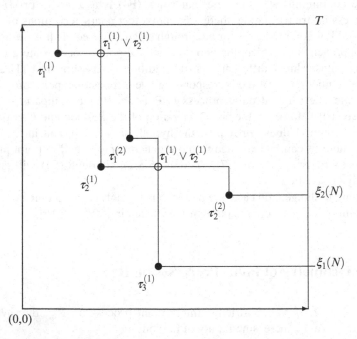

Fig. 1 Upper boundaries of the random sets $\xi_1(N)$ and $\xi_2(N)$. Jump points of N indicated by $\bullet$. Other exposed points indicated by o

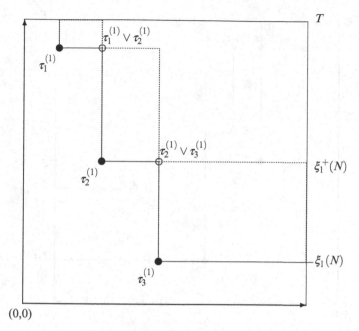

Fig. 2 Upper boundaries of the random sets $\xi_1(N)$ and $\xi_1^+(N)$

then $\xi_k(N) \subseteq \xi_{k+1}(N) \subseteq \xi_k^+(N)$. If $\mathscr{E}_k$ is empty or consists of a single point, then $\xi_k^+(N) := \mathbf{R}_+^2$. For the same example, the upper boundaries of the sets $\xi_1(N)$ and $\xi_1^+(N)$ are illustrated in Fig. 2.

We can now define N in terms of *single line* point processes:

Definition 4. A point process on $\mathbf{R}_+^2$ whose jump points are all incomparable is a single line process. (Points $s, t \in \mathbf{R}_+^2$ are incomparable if $s \not\leq t$ and $t \not\leq s$.)

Definition 5. Let N be a strictly simple point process on $\mathbf{R}_+^2$ and let $J(N)$ denote the set of jump points of N. Then $N(t) = \sum_1^\infty M_k(t)$ where for $k \geq 1$, M_k is the single line process whose set of jump points is

$$J(M_k) := \min\left(J(N) \cap (\xi_{k-1}^+(N) \setminus \xi_{k-1}(N))\right),$$

where $\xi_0 = \{\{0\} \times \mathbf{R}_+\} \cup \{\mathbf{R}_+ \times \{0\}\}$ and $\xi_0^+ = \mathbf{R}_+^2$.

Returning to our example, in Fig. 3 we illustrate each of the jump points of M_1 with ●, and each of jump points of M_2 with ⊗.

Before continuing, we make a few observations:

• $\xi_k(N) = \xi_{k-1}^+(N) \cap \xi_1(M_k)$ (this is illustrated in Fig. 3 for $k = 2$). We note that M_k has no jump points if $J(M_k) = \emptyset$; in this case $\xi_1(M_k) = \mathbf{R}_+^2$ and $\xi_k(N) = \xi_{k-1}^+(N)$.

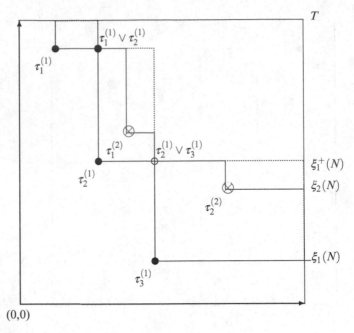

Fig. 3 Jump points of M_1 indicated by ●, jump points of M_2 indicated by ⊗

- Since $\{N(t) = k\} = \{t \in \xi^o_{k+1}(N) \setminus \xi^o_k(N)\}$, in a manner that will be made precise, the law of N (its finite dimensional distributions) is determined by the joint (finite dimensional) distributions of the random sets $\xi^o_k(N)$. We will see that this can be done by successive conditioning, as in one dimension where the joint distribution of the successive jump times is built up through conditioning.
- If M is a single line process, it is completely determined by $\xi_1(M)$ (cf. [10] – the jump points of M are the exposed points of $\xi_1(M)$).
- Since the point process and its related random sets $\xi_k(N)$ are determined by single line processes, we will be able to reduce our problem to the following question: will the *-compensator of the single line process M_k characterize its distribution if (F4) is satisfied?

First, we need to consider the concept of stopping in higher dimensions.

5   Stopping Sets and Their Distributions

We begin with the definition of adapted random sets and stopping sets; in particular, a stopping set is the multidimensional analogue of a stopping time.

Definition 6. Let $d = 1$ or 2. An adapted random set ζ with respect to the filtration $\mathscr{F}$ on $\mathbf{R}_+^d$ is a random Borel subset of $\mathbf{R}_+^d$ such that $\{t \in \zeta\} \in \mathscr{F}(t) \; \forall t \in \mathbf{R}_+^d$. An adapted random set ξ is an $\mathscr{F}$-stopping set if ξ is a closed lower layer.

For $d = 1$, we see that if τ is an $\mathscr{F}$-stopping time, then $\zeta = [0, \tau)$ is an adapted random set and $\xi = [0, \tau]$ is an $\mathscr{F}$-stopping set. Since $\mathscr{F}$ is right-continuous, it is easily seen that $\xi = [0, \tau]$ is an $\mathscr{F}$-stopping set if and only if τ is an $\mathscr{F}$-stopping time. For $d = 2$ and $\mathscr{F}(t) = \mathscr{F}_0 \vee \mathscr{F}^N(t)$ for a point process N, and if $\mathscr{F}(L)$ is defined as in (6) for a lower layer L, then it is shown in [7] that both $\{\xi \subseteq L\} \in \mathscr{F}(L)$ and $\{L \subseteq \xi\} \in \mathscr{F}(L)$.

The law of an adapted random set ζ is determined by its finite dimensional distributions:

$$P(t_1, \ldots, t_n \in \zeta), n \in \mathbf{N}, t_1, \ldots, t_n \in \mathbf{R}_+^d, d = 1 \text{ or } 2.$$

In analogy to the history of a stopping time, the history of a stopping set ξ is

$$\mathscr{F}(\xi) := \{G \in \mathscr{F} : G \cap \{\xi \subseteq L\} \in \mathscr{F}(L) \; \forall \text{ lower layers } L\}.$$

If ξ takes on at most countably many values in the class of lower layers, then equality can be used in the definition above, and it is easy to see that $\mathscr{F}(\xi) = \mathscr{F}(L)$ on $\{\xi = L\}$. For any point process N on $\mathbf{R}_+^2$ and filtration $\mathscr{F} \equiv \mathscr{F}_0 \vee \mathscr{F}^N$, we have the following:

- Since $\{t \in \xi_k^o\} = \{N(t) < k\} \in \mathscr{F}_t \; \forall k, t$, $\xi_k^o(N)$ is an $\mathscr{F}$-adapted random set.
- It is shown in [9] that the sets $\xi_k(N)$ and $\xi_k^+(N)$ are both $\mathscr{F}$-stopping sets. As well, both are $\mathscr{F}(\xi_k(N))$-measurable for every k (i.e. $\{t \in \xi_k^{(+)}(N)\} \in \mathscr{F}(\xi_k) \; \forall t$).
- Since N is strictly simple, a priori there are no jumps on the axes and so $\mathscr{F}(\xi_0) = \mathscr{F}_0$.

Just as the joint distributions of the increasing jump times $\tau_1 < \tau_2 < \ldots$ determine the law of a point process on $\mathbf{R}_+$ and can be built up by successive conditioning on $\mathscr{F}_0 \subseteq \mathscr{F}(\tau_1) \subseteq \mathscr{F}(\tau_2) \subseteq \ldots$, we see that the law (finite dimensional distributions) of a planar point process N can be reconstructed from the joint finite dimensional distributions of the related adapted random sets:

$$P(N(t_1) = k_1, \ldots N(t_n) = k_n) = P(t_i \in \xi_{k_i+1}^o \setminus \xi_{k_i}^o, i = 1, \ldots, n).$$

As well, it is clear that the joint distributions of the increasing random sets $\xi_1^o \subset \xi_2^o \subset \ldots$ can be built up by successive conditioning on $\mathscr{F}_0 = \mathscr{F}(\xi_0) \subseteq \mathscr{F}(\xi_1) \subseteq \mathscr{F}(\xi_2) \subseteq \ldots$.

6   The Compensator of a Single Line Point Process

We are now ready to construct the *-compensator of a single line process M on $\mathbf{R}_+^2$. Of course, we continue to assume that $E[M(t)] < \infty \; \forall t \in \mathbf{R}_+^2$.

Although in principle the law of a point process is determined by the joint laws of the sets $\xi_k^o(M), k \geq 1$, in the case of a single line process, the law of M is completely determined by the law of $\xi_1^o(M)$ ([10], Proposition 5.1). In other words, the set of probabilities

$$P(M(t_1) = 0, \ldots, M(t_n) = 0) = P(t_1, \ldots, t_n \in \xi_1^o(M))$$

for $t_1, \ldots, t_n \in \mathbf{R}_+^2, n \geq 1$, characterize the law of M. (This can be compared with the characterization of the law of a point process on an arbitrary complete measurable metric space via the so-called avoidance function; see [4], Theorem 7.3.II.)

However, when (F4) is satisfied, we have a further simplification. Define the *avoidance probability function* P_0 of a single line process M by

$$P_0(t) := P(M(t) = 0), t \in \mathbf{R}_+^2.$$

Theorem 3 ([10], Lemma 5.3). *Let M be a single line process whose minimal filtration $\mathscr{F} \equiv \mathscr{F}^M$ satisfies (F4). The law (the f.d.d.'s) of $\xi_1^o(M)$ (and hence the law of M) is determined by the avoidance probability function P_0 of M.*

A complete proof is given in [10], but to illustrate, we consider two incomparable points $s, t \in \mathbf{R}_+^2$. If $s_1 < t_1$ and $t_2 < s_2$, recalling that $\mathscr{F} = \mathscr{F}^M$ satisfies (F4) and that M is a single line process, we have:

$$P(M(t) = 0 \mid \mathscr{F}^1(s))$$
$$= P(M(t) = 0 \mid \mathscr{F}(s \wedge t)) \text{ by (F4) (cf. (7))}$$
$$= P(M(t) = 0 \mid M(s \wedge t) = 0)I(M(s \wedge t) = 0)$$
$$= \frac{P(M(t) = 0)}{P(M(s \wedge t) = 0)}I(M(s \wedge t) = 0).$$

Therefore,

$$P(M(s) = 0, M(t) = 0)$$
$$= P(s, t \in \xi_1^o(M))$$
$$= E\left[I(M(s) = 0)P(M(t) = 0 \mid \mathscr{F}^1(s))\right]$$
$$= E\left[I(M(s) = 0)I(M(s \wedge t) = 0)\frac{P(M(t) = 0)}{P(M(s \wedge t) = 0)}\right]$$

$$= E\left[I(M(s) = 0)\frac{P(M(t) = 0)}{P(M(s \wedge t) = 0)}\right]$$

$$= \frac{P(M(s) = 0)P(M(t) = 0)}{P(M(s \wedge t) = 0)} = \frac{P_0(s)P_0(t)}{P_0(s \wedge t)}. \tag{11}$$

Under (F4), the avoidance probability function of a single line process can be regarded as the two-dimensional analogue of the survival function of the jump time τ of a single jump point process on $\mathbf{R}_+$. *Henceforth, we will assume that the avoidance probability function is continuous.* Obviously, the avoidance probability function is non-increasing in the partial order on $\mathbf{R}_+^2$, but when is a continuous function bounded by 0 and 1 and non-increasing in each variable an avoidance probability? The answer lies in its logarithm.

Let $\Lambda(t) := -\ln P_0(t) = -\ln P(M(t) = 0)$. Returning to (11) and taking logarithms on both sides, if $s, t \in \mathbf{R}_+^2$ are incomparable,

$$\Lambda(s \vee t) = -\ln P(M(s \vee t) = 0)$$

$$\geq -\ln P(M(A_s \cup A_t) = 0) \text{ since } A_s \cup A_t \subseteq A_{s \vee t}$$

$$= -\ln P(M(s) = 0, M(t) = 0)$$

$$= \Lambda(s) + \Lambda(t) - \Lambda(s \wedge t) \text{ by (11)}.$$

If P_0 is continuous, then Λ is continuous and increasing on $\mathbf{R}_+^2$: i.e. it has non-negative increments. Therefore, $\Lambda = -\ln P_0$ is the distribution function of a measure on $\mathbf{R}_+^2$. In what follows, we will use the same notation for both the measure and its distribution function; for example, for B a Borel set, $\Lambda(B)$ and $M(B)$ are the measures assigned to B by the distribution functions $\Lambda(t) = \Lambda(A_t)$ and $M(t) = M(A_t)$, respectively. To summarize, when $\mathscr{F} = \mathscr{F}^M$ satisfies (F4):

- If P_0 is continuous, $\Lambda = -\ln P_0$ defines a measure on $\mathbf{R}_+^2$, and it is straightforward that for any lower layer L,

$$P(L \subseteq \xi_1(M)) = e^{-\Lambda(L)} = P(M(L) = 0).$$

- Conversely, a measure Λ that puts mass 0 on each vertical and horizontal line uniquely defines the (continuous) avoidance probability function P_0 (and therefore the law) of a single line point process whose minimal filtration satisfies (F4).
- Heuristically, $d\Lambda$ can be interpreted as the hazard of M:

$$P(M(dt) = 1|\mathscr{F}^*(t)) \stackrel{(F4)}{\approx} I(M(A_t) = 0)d\Lambda(t).$$

We will refer to Λ as the *cumulative hazard* of M.

All of the preceding discussion can be applied to conditional avoidance probability functions and conditional cumulative hazard functions, but first we need to define regularity of conditional avoidance probabilities; this is analogous to the definition of a regular conditional distribution.

Definition 7. Given an arbitrary σ-field $\mathscr{F}' \subseteq \mathscr{F}$, we say that a family $(P_0(t, \omega) : (t, \omega) \in \mathbf{R}_+^2 \times \Omega)$ is a continuous regular version of a conditional avoidance probability function given $\mathscr{F}'$ if for each $t \in \mathbf{R}_+^2$, $P_0(t, \cdot)$ is $\mathscr{F}'$-measurable, and for each $\omega \in \Omega$, $P_0(\cdot, \omega)$ is equal to one on the axes, and $-\ln P_0(\cdot, \omega)$ is continuous and increasing on $\mathbf{R}_+^2$.

We have the following generalization of Theorem 3:

Theorem 4. *Let M be a single line process with filtration $\mathscr{F}(t) = \mathscr{F}_0 \vee \mathscr{F}^M(t)$ that satisfies (F4). If there exists a continuous regular version $P_0^{(0)}(\cdot, \omega)$ of the conditional avoidance probability of M given $\mathscr{F}_0$, then the conditional law of $\xi_1^o(M)$ (and hence M) given $\mathscr{F}_0$ is determined by $P_0^{(0)}$, or equivalently by the conditional cumulative hazard $\Lambda_0 := -\ln P_0^{(0)}$.*

Now we can define the *-compensator of the single line process; to do so, we will make use of Theorem 1. Suppose first that we have the minimal filtration: $\mathscr{F}(t) = \mathscr{F}^M(t)$. Since $P(M(L) = 0) = e^{-\Lambda(L)}$ for any lower layer L, we can identify the single line process M with the single line process M_1 (the first line) in the decomposition of a Poisson process N with continuous mean measure Λ (cf. Definition 5): we have $\xi_1(M) = \xi_1(M_1) = \xi_1(N)$. As shown in Example 7.4 of [9], it is easy to see that the $(\mathscr{F}^N)^*$-compensator of M is $\tilde{M}^*(t) = \Lambda(A_t \cap \xi_1(M))$. However, since $\mathscr{F}^M \subseteq \mathscr{F}^N$ and $\tilde{M}^*$ is $\mathscr{F}^M$-adapted, by Theorem 2 it follows that $\tilde{M}^*$ is also the $(\mathscr{F}^M)^*$-compensator of M. Similarly, if $\mathscr{F}(t) = \mathscr{F}_0 \vee \mathscr{F}^M(t)$, since $P(M(L) = 0 \mid \mathscr{F}_0) = e^{-\Lambda_0(L)}$ for any lower layer L, we make the same identification with a Cox process with driving measure Λ_0 to obtain $\tilde{M}^*(t) = \Lambda_0(A_t \cap \xi_1(M))$ (as above, this is both the $(\mathscr{F}_0 \vee \mathscr{F}^N)^*$ and the $(\mathscr{F}_0 \vee \mathscr{F}^M)^*$-compensator). We summarize this as follows:

Theorem 5. *Let M be a single line process with filtration $\mathscr{F}_0 \vee \mathscr{F}^M$ satisfying (F4). If there exists a continuous regular version $P_0^{(0)}$ of the conditional avoidance probability function of M given $\mathscr{F}_0$, then the $(\mathscr{F}_0 \vee \mathscr{F}^M)^*$-compensator of M is*

$$\tilde{M}^*(t) = \Lambda_0(A_t \cap \xi_1(M)), \tag{12}$$

where $\Lambda_0 = -\ln P_0^{(0)}$. Furthermore, if $Q = P|_{\mathscr{F}_0}$, then the law of M is characterized by Q and $\tilde{M}^*$.

Note: Compare (12) with (3), the formula for the compensator of a single jump process M on $\mathbf{R}_+$ (with $\mathscr{F}_0$ trivial). If the jump point of M has continuous distribution F, then $P_0 = 1 - F$ and from (3), the compensator is $-\ln P_0(t \wedge \tau_1) = \Lambda(A_t \cap \xi_1(M))$. Thus, (12) and (3) are identical and in both cases, $\Lambda = -\ln P_0$ can be interpreted as a cumulative hazard. The same will be true if $\mathscr{F}_0$ is not trivial.

7 The Compensator of a General Point Process

We are now ready to develop a recursive formula for the general point process compensator. Let N be a general strictly simple point process on $\mathbf{R}_+^2$ with filtration $\mathscr{F} = \mathscr{F}_0 \vee \mathscr{F}^N$ satisfying (F4) and let $N = \sum_{k=1}^{\infty} M_k$ be the decomposition into single line point processes of Definition 5. We will proceed as follows, letting $k \geq 1$:

1. We will show that if the filtration $\mathscr{F}(t) = \mathscr{F}_0 \vee \mathscr{F}^N(t)$ satisfies (F4) under P, then so does $\mathscr{G}(t) := \mathscr{F}(\xi_{k-1}(N)) \vee \mathscr{F}^{M_k}(t)$. This is the key point in the development of the general point process compensator.
2. Since $\xi_k(N) = \xi_{k-1}^+(N) \cap \xi_1(M_k)$ and $\xi_{k-1}^+(N)$ is $\mathscr{F}(\xi_{k-1}(N))$-measurable, the conditional law of $\xi_k(N)$ given $\mathscr{F}(\xi_{k-1}(N))$ is determined by the conditional law of $\xi_1(M_k)$. By Theorem 4 and the preceding point, this in turn is characterized by the conditional avoidance probability function

$$P_0^{(k)}(t) := P(M_k(t) = 0 \mid \mathscr{F}(\xi_{k-1}(N))). \tag{13}$$

Therefore, the law of N is determined by $Q = P|_{\mathscr{F}_0}$ and the conditional avoidance probability functions $P_0^{(k)}, k \geq 1$.
3. Define $\Lambda_k(A_t, \omega) := -\ln P_0^{(k)}(t, \omega)$. Letting $\mathscr{F}(\xi_{k-1}(N))$ play the role of $\mathscr{F}_0$ for M_k and defining $\mathscr{G}(t)$ as in point 1 above, it will be shown that

$$\tilde{M}_k^*(t) = \Lambda_k(A_t \cap \xi_k(N))I(t \in \xi_{k-1}^c(N)).$$

is both the $\mathscr{G}^*$- and the $\mathscr{F}^*$-compensator of M_k.
4. Since $\tilde{N}^* = \sum_k \tilde{M}_k^*$, when (F4) holds the law of N is therefore characterized by Q and $\tilde{N}^*$.

Putting the preceding points together, we arrive at our main result:

Theorem 6. *Let N be a strictly simple point process on $\mathbf{R}_+^2$ with filtration $(\mathscr{F}(t) = \mathscr{F}_0 \vee \mathscr{F}^N(t))$ satisfying (F4). Assume that there exists a continuous regular version of $P_0^{(k)} \ \forall k \geq 1$, where $P_0^{(k)}$ is as defined in (13). Then the *-compensator of N has the regenerative form:*

$$\tilde{N}^*(t) = \sum_{k=1}^{\infty} \Lambda_k(A_t \cap \xi_k(N))I(t \in \xi_{k-1}^c(N)) \tag{14}$$

where $\Lambda_k(t) = -\ln P_0^{(k)}(t)$. If $Q = P|_{\mathscr{F}_0}$, then the law of N is characterized by Q and $\tilde{N}^*$.

Comment 3. Theorem 6 is the two-dimensional analogue of the corresponding result for point processes on $\mathbf{R}_+$, and in fact the formulas in one and two dimensions are identical: recalling (4) (the compensator on $\mathbf{R}_+$),

$$\tilde{N}(t) = \sum_{n=1}^{\infty} \Lambda_n(t \wedge \tau_n) I(\tau_{n-1} < t)$$

$$= \sum_{n=1}^{\infty} \Lambda_n(A_t \cap \xi_n(N)) I(t \in \xi_{n-1}^c(N)),$$

which is the same as (14).

Proof of Theorem 6. We must fill in the details of points 1–4, listed above.

1. • We will begin by showing that that for any $\mathscr{F}$-stopping set ξ and incomparable points $s, t \in \mathbf{R}_+^2$,

$$\mathscr{F}(s) \perp \mathscr{F}(t) \mid (\mathscr{F}(\xi) \vee \mathscr{F}(s \wedge t)),$$

or equivalently that for any $F \in \mathscr{F}(t)$,

$$P(F \mid \mathscr{F}(\xi) \vee \mathscr{F}(s)) = P(F \mid \mathscr{F}(\xi) \vee \mathscr{F}(s \wedge t)). \tag{15}$$

This then shows that

$$(\mathscr{F}(s) \vee \mathscr{F}(\xi)) \perp (\mathscr{F}(t) \vee \mathscr{F}(\xi)) \mid (\mathscr{F}(\xi) \vee \mathscr{F}(s \wedge t))$$

and so if $\mathscr{G}(s) := \mathscr{F}(\xi_{k-1}) \vee \mathscr{F}^{M_k}(s)$, then $\mathscr{G}(s) \perp \mathscr{G}(t) \mid (\mathscr{F}(\xi_{k-1}) \vee \mathscr{F}(s \wedge t))$.

   • We will then show that for $G \in \mathscr{G}(t)$,

$$P(G \mid \mathscr{F}(\xi_{k-1}) \vee \mathscr{F}(s \wedge t)) = P(G \mid \mathscr{F}(\xi_{k-1}) \vee \mathscr{F}^{M_k}(s \wedge t))$$

$$= P(G \mid \mathscr{G}(s \wedge t)). \tag{16}$$

Since $\mathscr{G}(s) \subseteq \mathscr{F}(\xi_{k-1}) \vee \mathscr{F}(s)$, (15) and (16) prove that $\mathscr{G}(s) \perp \mathscr{G}(t) \mid \mathscr{G}(s \wedge t)$.

Therefore, the proof of point 1 will be complete provided that (15) and (16) are verified.

To prove (15), we recall (7): if (F4) holds and if $F \in \mathscr{F}(t)$, then for any lower layer D,

$$P[F \mid \mathscr{F}(D)] = P[F \mid \mathscr{F}(D) \cap \mathscr{F}(t)].$$

Next, as is shown in [7], any stopping set ξ can be approximated from above by a decreasing sequence $(g_m(\xi))$ of *discrete* stopping sets (i.e. $g_m(\xi)$ is a stopping set taking on at most countably many values in the set of lower layers and $\xi = \cap_m g_m(\xi)$). Since $\mathscr{F}(\xi) = \cap_m \mathscr{F}(g_m(\xi))$ ([7], Proposition 1.5.12), it is enough to verify (15) for ξ a discrete stopping set. Let $\mathscr{D}$ be a countable class of lower layers such that $\sum_{D \in \mathscr{D}} P(\xi = D) = 1$. As noted before, for ξ discrete,

$$\mathscr{F}(\xi) = \{G \in \mathscr{F} : G \cap \{\xi = D\} \in \mathscr{F}(D) \ \forall D \in \mathscr{D}\}$$

and it is straightforward that $\mathscr{F}(\xi) = \mathscr{F}(D)$ on $\{\xi = D\}$. For $F \in \mathscr{F}(t)$, we consider $F \cap \{t \in \xi\}$ and $F \cap \{t \in \xi^c\}$ separately. First,

$$P(F \cap \{t \in \xi\} \mid \mathscr{F}(\xi) \vee \mathscr{F}(s))$$

$$= \sum_{D \in \mathscr{D}} P(F \cap \{t \in \xi\} \mid \mathscr{F}(\xi) \vee \mathscr{F}(s))I(\xi = D)$$

$$= \sum_{D \in \mathscr{D}} P(F \cap \{t \in D\} \mid \mathscr{F}(D) \vee \mathscr{F}(s))I(\xi = D)$$

$$= \sum_{D \in \mathscr{D}} I(F \cap \{t \in D\})I(\xi = D)$$

$$= \sum_{D \in \mathscr{D}} P(F \cap \{t \in D\} \mid \mathscr{F}(D) \vee \mathscr{F}(s \wedge t))I(\xi = D)$$

$$= \sum_{D \in \mathscr{D}} P(F \cap \{t \in \xi\} \mid \mathscr{F}(\xi) \vee \mathscr{F}(s \wedge t))I(\xi = D)$$

$$= P(F \cap \{t \in \xi\} \mid \mathscr{F}(\xi) \vee \mathscr{F}(s \wedge t)). \tag{17}$$

Next,

$$P(F \cap \{t \in \xi^c\} \mid \mathscr{F}(\xi) \vee \mathscr{F}(s))$$

$$= \sum_{D \in \mathscr{D}} P(F \cap \{t \in \xi^c\} \mid \mathscr{F}(\xi) \vee \mathscr{F}(s))I(\xi = D)$$

$$= \sum_{D \in \mathscr{D}} P(F \cap \{t \in D^c\} \mid \mathscr{F}(D) \vee \mathscr{F}(s))I(\xi = D)$$

$$= \sum_{D \in \mathscr{D}} P(F \cap \{t \in D^c\} \mid \mathscr{F}(D \cup A_s))I(\xi = D)$$

$$= \sum_{D \in \mathscr{D}} P(F \cap \{t \in D^c\} \mid \mathscr{F}(D \cup A_s) \cap \mathscr{F}(t))I(\xi = D) \tag{18}$$

$$= \sum_{D \in \mathscr{D}} P(F \cap \{t \in D^c\} \mid \mathscr{F}((D \cup A_s) \cap A_t)I(\xi = D)$$

$$= \sum_{D \in \mathscr{D}} P(F \cap \{t \in D^c\} \mid \mathscr{F}(D \cup (A_s \cap A_t))I(\xi = D) \tag{19}$$

$$= \sum_{D \in \mathscr{D}} P(F \cap \{t \in D^c\} \mid \mathscr{F}(D) \vee \mathscr{F}(A_s \cap A_t))I(\xi = D)$$

$$= \sum_{D \in \mathscr{D}} P(F \cap \{t \in \xi^c\} \mid \mathscr{F}(\xi) \vee \mathscr{F}(A_s \cap A_t))I(\xi = D)$$

$$= P(F \cap \{t \in \xi^c\} \mid \mathscr{F}(\xi) \vee \mathscr{F}(s \wedge t)). \tag{20}$$

Equations (18) and (19) follow from (7). Putting (17) and (20) together yields (15).

Now we prove (16). Since s and t are incomparable, without loss of generality we will assume that $t_1 < s_1$ and $t_2 > s_2$ and so $s \wedge t = (t_1, s_2)$. We have $\mathscr{F}(\xi_{k-1}) \vee \mathscr{F}(s \wedge t) = \mathscr{F}(\xi_{k-1}) \vee \mathscr{F}^N(s \wedge t)$. Let $\tau := \inf(v \in \mathbf{R}_+ : M_k(v, s_2) > 0) \wedge t_1$; τ is a stopping time with respect to the one-dimensional filtration $\mathscr{F}(\xi_{k-1}) \vee \mathscr{F}^{M_k}(\cdot, s_2)$. Note that $\mathscr{F}(\xi_{k-1}) \vee \mathscr{F}(\tau, s_2) = \mathscr{F}(\xi_{k-1}) \vee \mathscr{F}^N(\tau, s_2) = \mathscr{F}(\xi_{k-1}) \vee \mathscr{F}^{M_k}(\tau, s_2)$ since N has no jumps on $A_{(\tau, s_2)} \setminus \xi_{k-1}$ other than (possibly) a single jump from M_k on the line segment $\{(\tau, u), 0 \le u \le s_2\}$. Approximate τ from above with discrete stopping times $\tau_m \le t_1$, $\tau_m \downarrow \tau$. By right continuity of the filtrations,

$$\mathscr{F}(\xi_{k-1}) \vee \mathscr{F}(\tau_m, s_2) = \mathscr{F}(\xi_{k-1}) \vee \mathscr{F}^N(\tau_m, s_2)$$

$$\downarrow \mathscr{F}(\xi_{k-1}) \vee \mathscr{F}^N(\tau, s_2)$$

$$= \mathscr{F}(\xi_{k-1}) \vee \mathscr{F}^{M_k}(\tau, s_2). \tag{21}$$

Without loss of generality, let $G = \{M_k(t) = j\}$ in (16). Observe that on $\{M_k(s \wedge t) > 0\}$, $M_k(t) = M_k(\tau_m, t_2) + M_k(((\tau_m, 0), (t_1, s_2)])$ for every m, since $\{M_k(\tau_m) > 0\}$ and the jumps of M_k are incomparable. On $\{M_k(s \wedge t) = 0\}$, $\tau_m = t_1 \, \forall m$ and $M_k(t) = M_k(\tau_m, t_2)$. For ease of notation in what follows, let $X(\tau_m) := M_k(\tau_m, t_2)$ and $Y(\tau_m) := M_k((\tau_m, 0), (t_1, s_2)])$. Recall that $s \wedge t = (t_1, s_2)$ and let R_m denote the (countable) set of possible values of τ_m.

$$P(M_k = j \mid \mathscr{F}(\xi_{k-1}) \vee \mathscr{F}(s \wedge t))$$

$$= \sum_{r \in R_m} P(M_k = j \mid \mathscr{F}(\xi_{k-1}) \vee \mathscr{F}(s \wedge t)) I(\tau_m = r)$$

$$= \sum_h \sum_{r \in R_m} P(X(\tau_m) = h, Y(\tau_m) = j - h \mid \mathscr{F}(\xi_{k-1}) \vee \mathscr{F}(s \wedge t)) I(\tau_m = r)$$

$$= \sum_h \sum_{r \in R_m} P(X(r) = h \mid \mathscr{F}(\xi_{k-1}) \vee \mathscr{F}(s \wedge t)) I(Y(r) = j - h) I(\tau_m = r)$$

$$= \sum_h \sum_{r \in R_m} P(X(r) = h \mid \mathscr{F}(\xi_{k-1}) \vee \mathscr{F}(r, s_2))$$

$$\times I(Y(r) = j - h) I(\tau_m = r) \tag{22}$$

$$= \sum_h \sum_{r \in R_m} P(X(r) = h, Y(r) = j - h \mid \mathscr{F}(\xi_{k-1}) \vee \mathscr{F}(r, s_2) \vee \mathscr{F}^{M_k}(s \wedge t))$$

$$\times I(\tau_m = r) \tag{23}$$

$$= P(M_k = j \mid \mathscr{F}(\xi_{k-1}) \vee \mathscr{F}(\tau_m, s_2) \vee \mathscr{F}^{M_k}(s \wedge t))$$

$$\overset{m \to \infty}{\longrightarrow} P(M_k = j \mid \mathscr{F}(\xi_{k-1}) \vee \mathscr{F}^{M_k}(\tau, s_2) \vee \mathscr{F}^{M_k}(s \wedge t)) \tag{24}$$

$$= P(M_k = j \mid \mathscr{F}(\xi_{k-1}) \vee \mathscr{F}^{M_k}(s \wedge t)).$$

(22) and (23) follow from (7) and the fact that $X(r)$ is $\mathcal{F}(r, t_2)$-measurable, and (24) follows from (21). This proves (16) and completes the proof of point 1.

2. This follows immediately from point 1 and Theorem 4.

3. Begin by recalling that M_k has its support on $\xi_{k-1}^+(N) \setminus \xi_{k-1}(N)$ and so $P_0^{(k)}(t) = P(M_k(t) = 0|\mathcal{F}(\xi_{k-1}(N))) = P(M_k(A_t \cap \xi_{k-1}^+(N)) = 0|\mathcal{F}(\xi_{k-1}(N)))$. Therefore, we will identify M_k with the first line of a Cox process whose driving measure $\Lambda_k(t) = -\ln P_0^{(k)}(t)$ has support $\xi_{k-1}^+(N) \setminus \xi_{k-1}(N)$. Now, identifying $\mathcal{G}_0 = \mathcal{F}(\xi_{k-1}(N))$ and $\mathcal{G}(t) = \mathcal{F}(\xi_{k-1}(N)) \vee \mathcal{F}^{M_k}(t)$, as in Theorem 5 we have that the $\mathcal{G}^*$-compensator of M_k is:

$$
\begin{aligned}
\tilde{M}_k^*(t) &= \Lambda_k(A_t \cap \xi_1(M_k)) \\
&= \Lambda_k(A_t \cap \xi_1(M_k))I(t \in \xi_{k-1}^c(N)) \\
&= \Lambda_k(A_t \cap \xi_{k-1}^+(N) \cap \xi_1(M_k))I(t \in \xi_{k-1}^c(N)) \\
&= \Lambda_k(A_t \cap \xi_k(N))I(t \in \xi_{k-1}^c(N)).
\end{aligned}
$$

The last two equalities follow since $\{t \in \xi_{k-1}^+(N)\}$ is $\mathcal{F}(\xi_{k-1}(N))$-measurable and $\xi_k(N) = \xi_{k-1}^+(N) \cap \xi_1(M_k)$.

We must now show that $\tilde{M}_k^*$ is also the $\mathcal{F}^*$-compensator of M_k. First we show that $\tilde{M}_k^*$ is $\mathcal{F}$-adapted. On $\{t \in \xi_{k-1}\} \in \mathcal{F}(t)$, $P_0^{(k)}(t) = 0$. On $\{t \in \xi_{k-1}^c\}$, by (7) and taking discrete approximations of ξ_{k-1}, arguing as in the proof of (20) we have

$$
P_0^{(k)}(t)I(t \in \xi_{k-1}^c) = P(M_k(t) = 0 \mid \mathcal{F}(\xi_{k-1}) \cap \mathcal{F}(t))I(t \in \xi_{k-1}^c).
$$

Therefore, $-\ln P_0^{(k)}$ is $\mathcal{F}$-adapted. Since $\tilde{M}_k^*$ is $\mathcal{F}$-adapted and continuous, by Theorem 2 it remains only to prove that

$$
E\left[(M_k - \tilde{M}_k^*)(s, t] \mid \mathcal{F}^*(s)\right] = 0.
$$

First, if $t \in \xi_{k-1}$ or if $s \in (\xi_{k-1}^+)^c$, then $(M_k - \tilde{M}_k^*)(s, t] = 0$ and so trivially

$$
E\left[(M_k - \tilde{M}_k^*)(s, t]I(t \in \xi_{k-1}) \mid \mathcal{F}^*(s)\right] = 0, \tag{25}
$$

and

$$
E\left[(M_k - \tilde{M}_k^*)(s, t]I(s \in (\xi_{k-1}^+)^c) \mid \mathcal{F}^*(s)\right] = 0. \tag{26}
$$

For $s < t$, $t \in \xi_{k-1}^c$ and $s \in \xi_{k-1}^+$, it is enough to show that

$$
\begin{aligned}
&E\left[M_k(s, t]I(s \in \xi_{k-1}^+, t \in \xi_{k-1}^c) \mid \mathcal{G}^*(s)\right] \\
&= E\left[M_k(s, t]I(s \in \xi_{k-1}^+, t \in \xi_{k-1}^c) \mid \mathcal{F}(\xi_{k-1}) \vee (\mathcal{F}^{M_k})^*(s)\right] \\
&= E\left[M_k(s, t]I(s \in \xi_{k-1}^+, t \in \xi_{k-1}^c) \mid \mathcal{F}(\xi_{k-1}) \vee (\mathcal{F}^N)^*(s)\right]. \tag{27}
\end{aligned}
$$

If (27) is true, then since $\mathscr{F}^*(s) = \mathscr{F}_0 \vee (\mathscr{F}^N)^*(s) \subseteq \mathscr{F}(\xi_{k-1}) \vee (\mathscr{F}^N)^*(s)$,

$$
\begin{aligned}
0 &= E\left[(M_k - \tilde{M}_k^*)(s, t]I(s \in \xi_{k-1}^+, t \in \xi_{k-1}^c) \mid \mathscr{G}^*(s)\right] \\
&= E\left[(M_k - \tilde{M}_k^*)(s, t]I(s \in \xi_{k-1}^+, t \in \xi_{k-1}^c) \mid \mathscr{F}(\xi_{k-1}) \vee (\mathscr{F}^N)^*(s)\right] \\
&= E\left[(M_k - \tilde{M}_k^*)(s, t]I(s \in \xi_{k-1}^+, t \in \xi_{k-1}^c) \mid \mathscr{F}^*(s)\right].
\end{aligned}
\tag{28}
$$

To prove (27), let $\tau_1 = \inf\{v : M_k(v, s_2) > 0\} \wedge t_1$. Similar to the argument used to prove (16), we have

$$
\mathscr{F}(\xi_{k-1}) \vee \mathscr{F}^N(\tau_1, s_2) = \mathscr{F}(\xi_{k-1}) \vee \mathscr{F}^{M_k}(\tau_1, s_2)
$$

and using (F4) (cf. (7)) and discrete approximations for τ_1, it follows that

$$
\begin{aligned}
&E\left[M_k(s, t]I(s \in \xi_{k-1}^+, t \in \xi_{k-1}^c) \mid \mathscr{F}(\xi_{k-1}) \vee (\mathscr{F}^N)^*(s)\right] \\
&= E\left[M_k(s, t]I(s \in \xi_{k-1}^+, t \in \xi_{k-1}^c) \mid \mathscr{F}(\xi_{k-1}) \vee (\mathscr{F}^N)^1(s) \vee \mathscr{F}^{M_k}(t_1, s_2)\right].
\end{aligned}
$$

Next, letting $\tau_2 = \inf\{u : M_k(s_1, u) > 0\} \wedge t_2$, we argue as above and apply (F4) (cf. (7)) twice to obtain

$$
\begin{aligned}
&E\left[M_k(s, t]I(s \in \xi_{k-1}^+, t \in \xi_{k-1}^c) \mid \mathscr{F}(\xi_{k-1}) \vee (\mathscr{F}^N)^1(s) \vee \mathscr{F}^{M_k}(t_1, s_2)\right] \\
&= E\left[M_k(s, t]I(s \in \xi_{k-1}^+, t \in \xi_{k-1}^c) \mid \mathscr{F}(\xi_{k-1}) \vee \mathscr{F}^{M_k}(s_1, t_2) \vee \mathscr{F}^{M_k}(t_1, s_2)\right] \\
&= E\left[M_k(s, t]I(s \in \xi_{k-1}^+, t \in \xi_{k-1}^c) \mid \mathscr{G}^*(s)\right].
\end{aligned}
$$

This completes the proof of (27) and (28). Combining (25), (26) and (28), it follows that $\tilde{M}_k^*$ is the $\mathscr{F}^*$-compensator of M_k.

4. This is immediate because of the decomposition $N = \sum_{k=1}^{\infty} M_k$.

This completes the proof of Theorem 6. $\square$

8   Conclusion

In this paper we have proven a two-dimensional analogue of Jacod's characterization of the law of a point process via a regenerative formula for its compensator. For clarity we have restricted our attention to continuous avoidance probabilities. There remain many open questions that merit further investigation, for example:

- Extend the regenerative formula to discontinuous avoidance probability functions. In this case, the logarithmic relation between the avoidance probability and the cumulative hazard will be replaced by a product limit formula.
- Extend the regenerative formula to marked point processes.

- Find a complete characterization of the class of predictable increasing functions that are *-compensators for planar point processes satisfying (F4), in analogy to Theorem 3.6 of [11].
- Generalize the results of this paper to point processes on $\mathbf{R}_+^d$, $d > 2$. The main challenge will be to find an appropriate d-dimensional analogue of (F4).

Acknowledgements Research supported by a grant from the Natural Sciences and Engineering Research Council of Canada.

References

1. Aletti, G., Capasso, V.: Characterization of spatial Poisson along optional increasing paths – a problem of dimension's reduction. Stat. Probab. Lett. **43**, 343–347 (1999)
2. Brown, T.: A martingale approach to the Poisson convergence of simple point processes. Ann. Probab. **6**, 615–628 (1978)
3. Brown, T., Ivanoff, B.G., Weber, N.C.: Poisson convergence in two dimensions with applications to row and column exchangeable arrays. Stoch. Proc. Appl. **23**, 307–318 (1986)
4. Daley, D.J., Vere-Jones, D.: An Introduction to the Theory of Point Processes, vol. II, 2nd edn. Springer, New York (2008)
5. Dozzi, M.: On the decomposition and integration of two-parameter stochastic processes. In: Colloque ENST-CNET. Lecture Notes in Mathematics, vol. 863, pp. 162–171. Springer, Berlin/Heidelberg (1981)
6. Gushchin, A.A.: On the general theory of random fields on the plane. Russ. Math. Surv. **37**(6), 55–80 (1982)
7. Ivanoff, B.G., Merzbach, E.: Set-Indexed Martingales. Chapman & Hall/CRC, Boca Raton (2000)
8. Ivanoff, B.G., Merzbach, E.: Set-indexed Markov processes. Can. Math. Soc. Conf. Proc. **26**, 217–232 (2000)
9. Ivanoff, B.G., Merzbach, E.: What is a multi-parameter renewal process? Stochastics **78**, 411–441 (2006)
10. Ivanoff, B.G., Merzbach, E., Plante, M.: A compensator characterization of point processes on topological lattices. Electron. J. Prob. **12**, 47–74 (2007)
11. Jacod, J.: Multivariate point processes: Predictable projection, Radon-Nikodym derivatives, representation of martingales. Z. Wahr. **31**, 235–253 (1975)
12. Kallenberg, O.: Foundations of Modern Probability. Springer, New York (1997)
13. Karr, A.F.: Point Processes and Their Statistical Inference, 2nd edn. Marcel Dekker, New York (1991)
14. Last, G.: Predictable projections for point process filtrations. Probab. Theory Relat. Fields **99**, 361–388 (1994)
15. Mazziotto, G., Merzbach, E.: Point processes indexed by directed sets. Stoch. Proc. Appl. **30**, 105–119 (1988)
16. Merzbach, E.: Point processes in the plane. Acta Appl. Math. **12**, 79–101 (1988)

Part III
Central Limit Theorems and Laws of Large Numbers

Central Limit Theorem Related to MDR-Method

Alexander Bulinski

1 Introduction

High dimensional data arise naturally in a number of experiments. Very often such data are viewed as the values of some factors $X_1, \dots, X_n$ and the corresponding response variable Y. For example, in medical studies such response variable Y can describe the health state (e.g., $Y = 1$ or $Y = -1$ mean "sick" or "healthy") and $X_1, \dots, X_m$ and $X_{m+1}, \dots, X_n$ are genetic and non-genetic factors, respectively. Usually X_i $(1 \leq i \leq m)$ characterizes a single nucleotide polymorphism (SNP), i.e. a certain change of nucleotide bases adenine, cytosine, thymine and guanine (these genetic notions can be found, e.g., in [2]) in a specified segment of DNA molecule. In this case one considers X_i with three values, for instance, 0, 1 and 2 (see, e.g., [4]). It is convenient to suppose that other X_i $(m + 1 \leq i \leq n)$ take values in $\{0, 1, 2\}$ as well. For example, the range of blood pressure can be partitioned into zones of low, normal and high values. However, further we will suppose that all factors take values in arbitrary finite set. The binary response variable can also appear in pharmacological experiments where $Y = 1$ means that the medicament is efficient and $Y = -1$ otherwise.

A challenging problem is to find the genetic and non-genetic (or environmental) factors which could increase the risk of complex diseases such as diabetes, myocardial infarction and others. Now the most part of specialists share the paradigm that in contrast to simple disease (such as sickle anemia) certain combinations of the "damages" of the DNA molecule could be responsible for provoking the complex disease whereas the single mutations need not have dangerous effects (see, e.g., [15]). The important research domain called the *genome-wide association*

A. Bulinski (✉)
Faculty of Mathematics and Mechanics, Lomonosov Moscow State University,
Moscow 119991, Russia
e-mail: bulinski@mech.math.msu.su

© Springer Science+Business Media New York 2015

D. Dawson et al. (eds.), *Asymptotic Laws and Methods in Stochastics*,
Fields Institute Communications 76, DOI 10.1007/978-1-4939-3076-0_7

studies (GWAS) inspires development of new methods for handling large massives of biostatistical data. Here we will continue our treatment of the *multifactor dimensionality reduction* (MDR) method introduced by M. Ritchie et al. [13]. The idea of this method goes back to the Michalski algorithm. A comprehensive survey concerning the MDR method is provided in [14], on subsequent modifications and applications see, e.g., [5, 7–12, 17] and [18]. Other complementary methods applied in GWAS are discussed, e.g., in [4], there one can find further references.

In [3] the basis for application of the MDR-method was proposed when one uses an arbitrary penalty function to describe the prediction error of the binary response variable by means of a function in factors. The goal of the present paper is to establish the new multidimensional central limit theorem (CLT) for statistics which permit to justify the optimal choice of a subcollection of the explanatory variables.

2 Auxiliary Results

Let $X = (X_1, \ldots, X_n)$ be a random vector with components $X_i : \Omega \to \{0, 1, \ldots, q\}$, $i = 1, \ldots, n$ (q, n are positive integers). Thus, X takes values in $\mathbb{X} = \{0, 1, \ldots, q\}^n$. Introduce a random (response) variable $Y : \Omega \to \{-1, 1\}$, non-random function $f : \mathbb{X} \to \{-1, 1\}$ and a penalty function $\psi : \{-1, 1\} \to \mathbb{R}_+$ (the trivial case $\psi \equiv 0$ is excluded). The quality of approximation of Y by $f(X)$ is defined as follows

$$Err(f) := \mathbf{E}|Y - f(X)|\psi(Y). \tag{1}$$

Set $M = \{x \in \mathbb{X} : \mathbf{P}(X = x) > 0\}$ and

$$F(x) = \psi(-1)\mathbf{P}(Y = -1|X = x) - \psi(1)\mathbf{P}(Y = 1|X = x), \quad x \in M.$$

It is not difficult to show (see [3]) that the collection of *optimal functions*, i.e. all functions $f : \mathbb{X} \to \{-1, 1\}$ which are solutions of the problem $Err(f) \to inf$, has the form

$$f = \mathbb{I}\{A\} - \mathbb{I}\{\overline{A}\}, \quad A \in \mathscr{A}, \tag{2}$$

$\mathbb{I}\{A\}$ stands for an indicator of A ($\mathbb{I}\{\varnothing\} := 0$) and $\mathscr{A}$ consists of sets

$$A = \{x \in M : F(x) < 0\} \cup B \cup C.$$

Here B is an arbitrary subset of $\{x \in M : F(x) = 0\}$ and C is any subset of $\overline{M} := \mathbb{X} \setminus M$. If we take $A^* = \{x \in M : F(x) < 0\}$, then A^* has the minimal cardinality among all subsets of $\mathscr{A}$. In view of the relation $\psi(-1) + \psi(1) \neq 0$ we have

$$A^* = \{x \in M : \mathbf{P}(Y = 1|X = x) > \gamma(\psi)\}, \quad \gamma(\psi) := \psi(-1)/(\psi(-1) + \psi(1)). \tag{3}$$

If $\psi(1) = 0$ then $A^* = \varnothing$. If $\psi(1) \neq 0$ and $\psi(-1)/\psi(1) = a$ where $a \in \mathbb{R}_+$ then $A^* = \{x \in M : \mathbf{P}(Y = 1 | X = x) > a/(1 + a)\}$. Note that we can rewrite (1) as follows

$$Err(f) = 2 \sum_{y \in \{-1,1\}} \psi(y)\mathbf{P}(Y = y, f(X) \neq y).$$

The value $Err(f)$ is unknown as we do not know the law of a random vector (X, Y). Thus, statistical inference on the quality of approximation of Y by means of $f(X)$ is based on the estimate of $Err(f)$.

Let $\xi^1, \xi^2, \dots$ be i.i.d. random vectors with the same law as a vector (X, Y). For $N \in \mathbb{N}$ set $\xi_N = \{\xi^1, \dots, \xi^N\}$. To approximate $Err(f)$, as $N \to \infty$, we will use a *prediction algorithm*. It involves a function $f_{PA} = f_{PA}(x, \xi_N)$ with values $\{-1, 1\}$ which is defined for $x \in \mathbb{X}$ and ξ_N. In fact we use a *family* of functions $f_{PA}(x, v_m)$ defined for $x \in \mathbb{X}$ and $v_m \in \mathbb{V}_m$ where $\mathbb{V}_m := (\mathbb{X} \times \{-1, 1\})^m$, $m \in \mathbb{N}$, $m \leq N$. To simplify the notation we write $f_{PA}(x, v_m)$ instead of $f_{PA}^m(x, v_m)$. For $S \subset \{1, \dots, N\}$ ("$\subset$" means non-strict inclusion "$\subseteq$") put $\xi_N(S) = \{\xi^j, j \in S\}$ and $\overline{S} := \{1, \dots, N\} \setminus S$. For $K \in \mathbb{N}$ $(K > 1)$ introduce a partition of $\{1, \dots, N\}$ formed by subsets

$$S_k(N) = \{(k - 1)[N/K] + 1, \dots, k[N/K]\mathbb{I}\{k < K\} + N\mathbb{I}\{k = K\}\}, \quad k = 1, \dots, K,$$

here $[b]$ is the integer part of a number $b \in \mathbb{R}$. Generalizing [4] we can construct an estimate of $Err(f)$ using a sample ξ_N, a prediction algorithm with f_{PA} and K-fold cross-validation where $K \in \mathbb{N}$, $K > 1$ (on cross-validation see, e.g., [1]). Namely, let

$$\hat{Err}_K(f_{PA}, \xi_N) := 2 \sum_{y \in \{-1,1\}} \frac{1}{K} \sum_{k=1}^{K} \sum_{j \in S_k(N)} \frac{\hat{\psi}(y, S_k(N))\mathbb{I}\{Y^j = y, f_{PA}(X^j, \xi_N(\overline{S_k(N)})) \neq y\}}{\sharp S_k(N)}.$$

$$(4)$$

For each $k = 1, \dots, K$, random variables $\hat{\psi}(y, S_k(N))$ denote strongly consistent estimates (as $N \to \infty$) of $\psi(y)$, $y \in \{-1, 1\}$, constructed from data $\{Y^j, j \in S_k(N)\}$, and $\sharp S$ stands for a finite set S cardinality. We call $\hat{Err}_K(f_{PA}, \xi_N)$ an *estimated prediction error*.

The following theorem giving a criterion of validity of the relation

$$\hat{Err}_K(f_{PA}, \xi_N) \to Err(f) \quad \text{a.s.,} \quad N \to \infty,$$

$$(5)$$

was established in [3] (further on a sum over empty set is equal to 0 as usual).

Theorem 1. *Let f_{PA} define a prediction algorithm for a function $f : \mathbb{X} \to \{-1, 1\}$. Assume that there exists such set $U \subset \mathbb{X}$ that for each $x \in U$ and any $k = 1, \dots, K$ one has*

$$f_{PA}(x, \xi_N(\overline{S_k(N)})) \to f(x) \quad \text{a.s.,} \quad N \to \infty.$$

$$(6)$$

Then (5) *is valid if and only if, for* $N \to \infty$,

$$\sum_{k=1}^{K} \Big(\sum_{x \in \mathbb{X}^+} \mathbb{I}\{f_{PA}(x, \xi_N(\overline{S_k(N)})) = -1\}L(x) - \sum_{x \in \mathbb{X}^-} \mathbb{I}\{f_{PA}(x, \xi_N(\overline{S_k(N)})) = 1\}L(x) \Big) \to 0 \; a.s.$$

(7)

Here $\mathbb{X}^+ := (\mathbb{X} \setminus U) \cap \{x \in M : f(x) = 1\}$, $\mathbb{X}^- := (\mathbb{X} \setminus U) \cap \{x \in M : f(x) = -1\}$
and

$$L(x) = \psi(1)\mathbf{P}(X = x, Y = 1) - \psi(-1)\mathbf{P}(X = x, Y = -1), \quad x \in \mathbb{X}.$$

The sense of this result is the following. It shows that one has to demand condition (7) outside the set U (i.e. outside the set where f_{PA} provides the a.s. approximation of f) to obtain (5).

Corollary 1 ([3]). *Let, for a function* $f : \mathbb{X} \to \{-1, 1\}$, *a prediction algorithm be defined by* f_{PA}. *Suppose that there exists a set* $U \subset \mathbb{X}$ *such that for each* $x \in U$ *and any* $k = 1, \ldots, K$ *relation* (6) *is true. If*

$$L(x) = 0 \; for \; x \in (\mathbb{X} \setminus U) \cap M$$

then (5) *is satisfied.*

Note also that Remark 4 from [3] explains why the choice of a penalty function proposed by Velez et al. [17]:

$$\psi(y) = c(\mathbf{P}(Y = y))^{-1}, \quad y \in \{-1, 1\}, \quad c > 0,$$

(8)

is natural. Further discussion and examples can be found in [3].

3 Main Results and Proofs

In many situations it is reasonable to suppose that the response variable Y depends only on subcollection $X_{k_1}, \ldots, X_{k_r}$ of the explanatory variables, $\{k_1, \ldots, k_r\}$ being a subset of $\{1, \ldots, n\}$. It means that for any $x \in M$

$$\mathbf{P}(Y = 1 | X_1 = x_1, \ldots, X_n = x_n) = \mathbf{P}(Y = 1 | X_{k_1} = x_{k_1}, \ldots, X_{k_r} = x_{k_r}).$$

(9)

In the framework of the complex disease analysis it is natural to assume that only part of the risk factors could provoke this disease and the impact of others can be neglected. Any collection $\{k_1, \ldots, k_r\}$ implying (9) is called *significant*. Evidently if $\{k_1, \ldots, k_r\}$ is significant then any collection $\{m_1, \ldots, m_i\}$ such that $\{k_1, \ldots, k_r\} \subset \{m_1, \ldots, m_i\}$ is significant as well. For a set $D \subset \mathbb{X}$ let $\pi_{k_1, \ldots, k_r} D :=$

$\{u = (x_{k_1}, \ldots, x_{k_r}) : x = (x_1, \ldots, x_n) \in D\}$. For $B \in \mathbb{X}_r$ where $\mathbb{X}_r := \{0, 1, \ldots, q\}^r$ define in $\mathbb{X} = \mathbb{X}_n$ a cylinder

$$C_{k_1, \ldots, k_r}(B) := \{x = (x_1, \ldots, x_n) \in \mathbb{X} : (x_{k_1}, \ldots, x_{k_r}) \in B\}.$$

For $B = \{u\}$ where $u = (u_1, \ldots, u_r) \in \mathbb{X}_r$ we write $C_{k_1, \ldots, k_r}(u)$ instead of $C_{k_1, \ldots, k_r}(\{u\})$. Obviously

$$\mathbf{P}(Y = 1 | X_{k_1} = x_{k_1}, \ldots, X_{k_r} = x_{k_r}) \equiv \mathbf{P}(Y = 1 | X \in C_{k_1, \ldots, k_r}(u)),$$

here

$$u = \pi_{k_1, \ldots, k_r}\{x\}, \quad \text{i.e.} \quad u_i = x_{k_i}, \quad i = 1, \ldots, r. \tag{10}$$

For $C \subset \mathbb{X}$, $N \in \mathbb{N}$ and $W_N \subset \{1, \ldots, N\}$ set

$$\hat{\mathbf{P}}_{W_N}(Y = 1 | X \in C) := \frac{\sum_{j \in W_N} \mathbb{I}\{Y^j = 1, X^j \in C\}}{\sum_{j \in W_N} \mathbb{I}\{X^j \in C\}}. \tag{11}$$

When $C = \mathbb{X}$ we write simply $\hat{\mathbf{P}}_{W_N}(Y = 1)$ in (11). According to the *strong law of large numbers for arrays* (SLLNA), see, e.g., [16], for any $C \subset \mathbb{X}$ with $\mathbf{P}(X \in C) > 0$

$$\hat{\mathbf{P}}_{W_N}(Y = 1 | X \in C) \to \mathbf{P}(Y = 1 | X \in C) \quad \text{a.s.,} \quad \sharp W_N \to \infty, \quad N \to \infty.$$

If (9) is valid then the optimal function f^* defined by (2) with $A = A^*$ introduced in (3) has the form

$$f^{k_1, \ldots, k_r}(x) = \begin{cases} 1, & \text{if} \quad \mathbf{P}(Y = 1 | X \in C_{k_1, \ldots, k_r}(u)) > \gamma(\psi) \text{ and } x \in M, \\ -1, & \text{otherwise}, \end{cases} \tag{12}$$

here u and x satisfy (10) ($\mathbf{P}(X \in C_{k_1, \ldots, k_r}(u)) \geq \mathbf{P}(X = x) > 0$ as $x \in M$). Hence, for each significant $\{k_1, \ldots, k_r\} \subset \{1, \ldots, n\}$ and any $\{m_1, \ldots, m_r\} \subset \{1, \ldots, n\}$ one has

$$Err(f^{k_1, \ldots, k_r}) \leq Err(f^{m_1, \ldots, m_r}). \tag{13}$$

For arbitrary $\{m_1, \ldots, m_r\} \subset \{1, \ldots, n\}$, $x \in \mathbb{X}$, $u = \pi_{m_1, \ldots, m_r}\{x\}$ and a penalty function ψ we consider the prediction algorithm with a function $\hat{f}_{PA}^{m_1, \ldots, m_r}$ such that

$$\hat{f}_{PA}^{m_1, \ldots, m_r}(x, \xi_N(W_N)) = \begin{cases} 1, & \hat{\mathbf{P}}_{W_N}(Y = 1 | X \in C_{m_1, \ldots, m_r}(u)) > \hat{\gamma}_{W_N}(\psi), \quad x \in M, \\ -1, & \text{otherwise}, \end{cases}$$

$$\tag{14}$$

here $\hat{\gamma}_{W_N}(\psi)$ is a strongly consistent estimate of $\gamma(\psi)$ constructed by means of $\xi_N(W_N)$. Introduce

$$U := \{x \in M : \mathbf{P}(Y = 1 | X_{m_1} = x_{m_1}, \dots, X_{m_r} = x_{m_r}) \neq \gamma(\psi)\}. \tag{15}$$

Using Corollary 1 (and in view of Examples 1 and 2 of [3]) we conclude that for any $\{m_1, \dots, m_r\} \subset \{1, \dots, n\}$

$$\hat{Err}_K(\hat{f}_{PA}^{m_1, \dots, m_r}, \xi_N) \to Err(f^{m_1, \dots, m_r}) \quad \text{a.s.,} \quad N \to \infty. \tag{16}$$

Relations (13) and (16) show that for each $\varepsilon > 0$, any significant collection $\{k_1, \dots, k_r\} \subset \{1, \dots, n\}$ and arbitrary set $\{m_1, \dots, m_r\} \subset \{1, \dots, n\}$ one has

$$\hat{Err}_K(\hat{f}_{PA}^{k_1, \dots, k_r}, \xi_N) \leq \hat{Err}_K(\hat{f}_{PA}^{m_1, \dots, m_r}, \xi_N) + \varepsilon \quad \text{a.s.} \tag{17}$$

when N is large enough.

Thus, for a given $r = 1, \dots, n - 1$, according to (17) we come to the following conclusion. It is natural to choose among factors $X_1, \dots, X_n$ a collection $X_{k_1}, \dots, X_{k_r}$ leading to the smallest estimated prediction error $\hat{Err}_K(\hat{f}_{PA}^{k_1, \dots, k_r}, \xi_N)$. After that it is desirable to apply the permutation tests (see, e.g., [4] and [6]) for validation of the prediction power of selected factors. We do not tackle here the choice of r, some recommendations can be found in [14]. Note also in passing that a nontrivial problem is to estimate the importance of various collections of factors, see, e.g., [15].

Remark 1. It is essential that for each $\{m_1, \dots, m_r\} \subset \{1, \dots, n\}$ we have strongly consistent estimates of $Err(f^{m_1, \dots, m_r})$. So to compare these estimates we can use the subset of Ω having probability one. If we had only the convergence in probability instead of a.s. convergence in (16) then to compare different $\hat{Err}_K(\hat{f}_{PA}^{m_1, \dots, m_r}, \xi_N)$ one should take into account the Bonferroni corrections for all subsets $\{m_1, \dots, m_r\}$ of $\{1, \dots, n\}$.

Further on we consider a function ψ having the form (8). In view of (3) w.l.g. we can assume that $c = 1$ in (8). In this case $\gamma(\psi) = \mathbf{P}(Y = 1)$. Introduce events

$$A_{N,k}(y) = \{Y^j = -y, \; j \in S_k(N)\}, \quad N \in \mathbb{N}, \quad k = 1, \dots, K, \quad y \in \{-1, 1\},$$

and random variables

$$\hat{\psi}_{N,k}(y) := \frac{\mathbb{I}\{\overline{A_{N,k}(y)}\}}{\hat{\mathbf{P}}_{S_k(N)}(Y = y)},$$

trivial cases $\mathbf{P}(Y = y) \in \{0, 1\}$ are excluded. Here we formally set $0/0 := 0$. Then

$$\hat{\psi}_{N,k}(y) - \psi(y) = \frac{\mathbf{P}(Y = y) - \hat{\mathbf{P}}_{S_k(N)}(Y = y)}{\hat{\mathbf{P}}_{S_k(N)}(Y = y)\mathbf{P}(Y = y)}\mathbb{I}\{\overline{A_{N,k}(y)}\} - \frac{1}{\mathbf{P}(Y = y)}\mathbb{I}\{A_{N,k}(y)\}.$$

(18)

Clearly

$$\mathbb{I}\{A_{N,k}(y)\} \to 0 \text{ a.s., } N \to \infty,$$

(19)

and the following relation is true

$$\frac{\mathbb{I}\{\overline{A_{N,k}(y)}\}}{\hat{\mathbf{P}}_{S_k(N)}(Y = y)} \to \frac{1}{\mathbf{P}(Y = y)} \text{ a.s., } N \to \infty.$$

(20)

Therefore, by virtue of (18)–(20) we have that for $y \in \{-1, 1\}$ and $k = 1, \ldots, K$

$$\hat{\psi}_{N,k}(y) - \psi(y) \to 0 \text{ a.s., } N \to \infty.$$

(21)

Let $\{m_1, \ldots, m_r\} \subset \{1, \ldots, n\}$. We define the functions which can be viewed as the *regularized versions* of the estimates $\hat{f}_{PA}^{m_1,\ldots,m_r}$ of $f^{m_1,\ldots,m_r}$ (see (14) and (12)). Namely, for $W_N \subset \{1, \ldots, N\}$, $N \in \mathbb{N}$, and $\varepsilon = (\varepsilon_N)_{N \in \mathbb{N}}$ where non-random positive $\varepsilon_N \to 0$, as $N \to \infty$, put

$$\hat{f}_{PA,\varepsilon}^{m_1,\ldots,m_r}(x, \xi_N(W_N)) = \begin{cases} 1, & \hat{\mathbf{P}}_{W_N}(Y = 1 | X \in C_{m_1,\ldots,m_r}(u)) > \hat{\gamma}_{W_N}(\psi) + \varepsilon_N, \; x \in M, \\ -1, & \text{otherwise,} \end{cases}$$

where $u = \pi_{m_1,\ldots,m_r}\{x\}$. Regularization of $\hat{f}_{PA}^{m_1,\ldots,m_r}$ means that instead of the threshold $\hat{\gamma}_{W_N}(\psi)$ we use $\hat{\gamma}_{W_N}(\psi) + \varepsilon_N$.

Take now U appearing in (15). Applying Corollary 1 once again (and in view of Examples 1 and 2 of [3]) we can claim that the statements which are analogous to (16) and (17) are valid for the regularized versions of the estimates introduced above. Now we turn to the principle results, namely, central limit theorems.

Theorem 2. *Let $\varepsilon_N \to 0$ and $N^{1/2}\varepsilon_N \to \infty$ as $N \to \infty$. Then, for each $K \in \mathbb{N}$, any subset $\{m_1, \ldots m_r\}$ of $\{1, \ldots, n\}$, the corresponding function $f = f^{m_1,\ldots,m_r}$ and prediction algorithm defined by $f_{PA} = \hat{f}_{PA,\varepsilon}^{m_1,\ldots,m_r}$, the following relation holds:*

$$\sqrt{N}(\hat{Err}_K(f_{PA}, \xi_N) - Err(f)) \xrightarrow{law} Z \sim N(0, \sigma^2), \; N \to \infty,$$

(22)

where σ^2 is variance of the random variable

$$V = 2 \sum_{y \in \{-1, 1\}} \frac{\mathbb{I}\{Y = y\}}{\mathbf{P}(Y = y)}\left(\mathbb{I}\{f(X) \neq y\} - \mathbf{P}(f(X) \neq y | Y = y)\right).$$

(23)

Proof. For a fixed $K \in \mathbb{N}$ and any $N \in \mathbb{N}$ set

$$T_N(f) := \frac{2}{K} \sum_{k=1}^{K} \frac{1}{\sharp S_k(N)} \sum_{y \in \{-1,1\}} \psi(y) \sum_{j \in S_k(N)} \mathbb{I}\{Y^j = y, f(X^j) \neq y\},$$

$$\hat{T}_N(f) := \frac{2}{K} \sum_{k=1}^{K} \frac{1}{\sharp S_k(N)} \sum_{y \in \{-1,1\}} \hat{\psi}_{N,k}(y) \sum_{j \in S_k(N)} \mathbb{I}\{Y^j = y, f(X^j) \neq y\}.$$

One has

$$\hat{Err}_K(f_{PA}, \xi_N) - Err(f) = (\hat{Err}_K(f_{PA}, \xi_N) - \hat{T}_N(f))$$

$$+ (\hat{T}_N(f) - T_N(f)) + (T_N(f) - Err(f)). \tag{24}$$

First of all we show that

$$\sqrt{N}(\hat{Err}_K(f_{PA}, \xi_N) - \hat{T}_N(f)) \xrightarrow{\mathbf{P}} 0, \quad N \to \infty. \tag{25}$$

For $x \in \mathbb{X}, y \in \{-1, 1\}, k = 1, \ldots, K$ and $N \in \mathbb{N}$ introduce

$$F_{N,k}(x, y) := \mathbb{I}\{f_{PA}(x, \xi_N(\overline{S_k(N)})) \neq y\} - \mathbb{I}\{f(x) \neq y\}.$$

Then

$$\hat{Err}_K(f_{PA}, \xi_N) - \hat{T}_N(f) = \frac{2}{K} \sum_{k=1}^{K} \frac{1}{\sharp S_k(N)} \sum_{y \in \{-1,1\}} \hat{\psi}_{N,k}(y) \sum_{j \in S_k(N)} \mathbb{I}\{Y^j = y\} F_{N,k}(X^j, y).$$

$$\tag{26}$$

We define the random variables

$$B_{N,k}(y) := \frac{1}{\sqrt{\sharp S_k(N)}} \sum_{j \in S_k(N)} \mathbb{I}\{Y^j = y\} F_{N,k}(X^j, y)$$

and verify that for each $k = 1, \ldots, K$

$$\sum_{y \in \{-1,1\}} \hat{\psi}_{N,k}(y) B_{N,k}(y) \xrightarrow{\mathbf{P}} 0, \quad N \to \infty. \tag{27}$$

Clearly (27) implies (25) in view of (26) as $\sharp S_k(N) = [N/K]$ for $k = 1, \ldots, K-1$ and $[N/K] \leq \sharp S_K(N) < [N/K] + K$. Write $B_{N,k}(y) = B_{N,k}^{(1)}(y) + B_{N,k}^{(2)}(y)$ where

$$B_{N,k}^{(1)}(y) = \frac{1}{\sqrt{\sharp S_k(N)}} \sum_{j \in S_k(N)} \mathbb{I}\{X^j \in U)\}\mathbb{I}\{Y^j = y\}F_{N,k}(X^j, y),$$

$$B_{N,k}^{(2)}(y) = \frac{1}{\sqrt{\sharp S_k(N)}} \sum_{j \in S_k(N)} \mathbb{I}\{X^j \notin U\}\mathbb{I}\{Y^j = y\}F_{N,k}(X^j, y).$$

Obviously

$$|B_{N,k}^{(1)}(y)| \leq \sum_{x \in U} \frac{1}{\sqrt{\sharp S_k(N)}} \sum_{j \in S_k(N)} |\mathbb{I}\{f_{PA}(x, \xi_N(\overline{S_k(N)})) \neq y\} - \mathbb{I}\{f(x) \neq y\}|.$$

Functions f_{PA} and f take values in the set $\{-1, 1\}$. Thus, for any $x \in U$ (where U is defined in (15)), $k = 1, \ldots, K$ and almost all $\omega \in \Omega$ relation (6) ensures the existence of an integer $N_0(x, k, \omega)$ such that $f_{PA}(x, \xi_N(\overline{S_k(N)})) = f(x)$ for $N \geq N_0(x, k, \omega)$. Hence $B_{N,k}^{(1)}(y) = 0$ for any y belonging to $\{-1, 1\}$, each $k = 1, \ldots, K$ and almost all $\omega \in \Omega$ when $N \geq N_{0,k}(\omega) = \max_{x \in U} N_0(x, k, \omega)$. Evidently by Clearly to avoid the interruption of the formula $N_{0,k} < \infty$ a.s., because $U < \infty$. We obtain that

$$\sum_{y \in \{-1,1\}} \hat{\psi}_{N,k}(y)B_{N,k}^{(1)}(y) \to 0 \quad \text{a.s.,} \quad N \to \infty. \tag{28}$$

If $U = \mathbb{X}$ then $B_{N,k}^{(2)}(y) = 0$ for all N, k and y under consideration. Consequently, (27) is valid and thus, for $U = \mathbb{X}$, relation (25) holds. Let now $U \neq \mathbb{X}$. Then for $k = 1, \ldots, K$ and $N \in \mathbb{N}$ one has

$$\sum_{y \in \{-1,1\}} \hat{\psi}_{N,k}(y)B_{N,k}^{(2)}(y) = \sum_{x \in \mathbb{X}_+} \sum_{y \in \{-1,1\}} H_{N,k}(x, y) + \sum_{x \in \mathbb{X}_-} \sum_{y \in \{-1,1\}} H_{N,k}(x, y),$$

here $\mathbb{X}_+ = (\mathbb{X} \setminus U) \cap \{x \in \mathbb{X} : f(x) = 1\}$, $\mathbb{X}_- = (\mathbb{X} \setminus U) \cap \{x \in \mathbb{X} : f(x) = -1\}$ and

$$H_{N,k}(x, y) := \frac{\hat{\psi}_{N,k}(y)}{\sqrt{\sharp S_k(N)}} \sum_{j \in S_k(N)} \mathbb{I}\{A^j(x, y)\}(\mathbb{I}\{f_{PA}(x, \xi_N(\overline{S_k(N)})) \neq y\} - \mathbb{I}\{f(x) \neq y\})$$

where $A^j(x, y) = \{X^j = x, Y^j = y\}$. The definition of U yields that $\mathbb{X}_+ = \emptyset$ and

$$\mathbb{X}_- = \overline{M} \cup \{x \in M : \mathbf{P}(Y = 1 | X_{m_1} = x_{m_1}, \ldots, X_{m_r} = x_{m_r}) = \gamma(\psi)\}.$$

Set

$$\hat{R}_{N,k}^j(x) = \mathbb{I}\{X^j = x\}(\hat{\psi}_{N,k}(1)\mathbb{I}\{Y^j = 1\} - \hat{\psi}_{N,k}(-1)\mathbb{I}\{Y^j = -1\}).$$

It is easily seen that

$$\sum_{x \in \mathbb{X}_-} \sum_{y \in \{-1,1\}} H_{N,k}(x,y) = - \sum_{x \in \mathbb{X}_-} \mathbb{I}\{f_{PA}(x, \xi_N(\overline{S_k(N)})) = 1)\} \sum_{j \in S_k(N)} \frac{\hat{R}_{N,k}^j(x)}{\sqrt{\sharp S_k(N)}}.$$

Note that $\hat{R}_{N,k}^j(x) = 0$ a.s. for all $x \in \overline{M}$, $k = 1, \ldots, K$, $j = 1, \ldots, N$ and $N \in \mathbb{N}$. Let us prove that, for any $x \in M \cap \mathbb{X}_-$ and $k = 1, \ldots, K$,

$$\mathbb{I}\{f_{PA}(x, \xi_N(\overline{S_k(N)})) = 1\} \xrightarrow{P} 0, \quad N \to \infty. \tag{29}$$

For any $\nu > 0$ and $x \in M \cap \mathbb{X}_-$ we have

$$\mathbf{P}(\mathbb{I}\{f_{PA}(x, \xi_N(\overline{S_k(N)})) = 1\} > \nu)$$

$$= \mathbf{P}\left(\hat{\mathbf{P}}_{\overline{S_k(N)}}(Y = 1 | X_{m_1} = x_{m_1}, \ldots, X_{m_r} = x_{m_r}) > \hat{\gamma}_{\overline{S_k(N)}}(\psi) + \varepsilon_N\right).$$

Now we show that, for $k = 1, \ldots, K$, this probability tends to 0 as $N \to \infty$. For $W_N \subset \{1, \ldots, N\}$ and $x \in M \cap \mathbb{X}_-$, put

$$\Delta_N(W_N, x) := \mathbf{P}\left(\frac{\frac{1}{\sharp W_N} \sum_{j \in W_N} \eta^j}{\frac{1}{\sharp W_N} \sum_{j \in W_N} \zeta^j} > \hat{\gamma}_{W_N}(\psi) + \varepsilon_N\right)$$

where $\eta^j = \mathbb{I}\{Y^j = 1, X_{m_1}^j = x_{m_1}, \ldots, X_{m_r}^j = x_{m_r}\}$, $\zeta^j = \mathbb{I}\{X_{m_1}^j = x_{m_1}, \ldots, X_{m_r}^j = x_{m_r}\}$, $j = 1, \ldots, N$. Set $p = \mathbf{P}(X_{m_1} = x_{m_1}, \ldots, X_{m_r} = x_{m_r})$. It follows that, for any $\alpha_N > 0$,

$$\Delta_N(W_N, x)$$

$$\leq \mathbf{P}\left(\frac{\sum_{j \in W_N} \eta^j}{\sum_{j \in W_N} \zeta^j} > \hat{\gamma}_{W_N}(\psi) + \varepsilon_N, \left|\frac{1}{\sharp W_N} \sum_{j \in W_N} \zeta^j - p\right| < \alpha_N, \left|\hat{\gamma}_{W_N}(\psi) - \gamma(\psi)\right| < \alpha_N\right)$$

$$+ \mathbf{P}\left(\left|\frac{1}{\sharp W_N} \sum_{j \in W_N} \zeta^j - p\right| \geq \alpha_N\right) + \mathbf{P}\left(\left|\frac{1}{\sharp W_N} \sum_{j \in W_N} \mathbb{I}\{Y^j = 1\} - \mathbf{P}(Y = 1)\right| \geq \alpha_N\right). \tag{30}$$

Due to the Hoeffding inequality

$$\mathbf{P}\left(\left|\frac{1}{\sharp W_N} \sum_{j \in W_N} \zeta^j - p\right| \geq \alpha_N\right) \leq 2 \exp\{-2\sharp W_N \alpha_N^2\} =: \delta_N(W_N, \alpha_N).$$

We have an analogous estimate for the last summand in (30). Consequently, taking into account that $p > 0$ we see that for all N large enough

$$\Delta_N(W_N, x) \le \mathbf{P}\Big(\frac{1}{\sharp W_N} \sum_{j \in W_N} \eta^j > (p - \alpha_N)(\gamma(\psi) - \alpha_N + \varepsilon_N)\Big) + 2\delta_N(W_N, \alpha_N).$$

Whenever $x \in M \cap \mathbb{X}_-$ one has

$$\mathbf{P}(Y = 1, X_{m_1} = x_{m_1}, \ldots, X_{m_r} = x_{m_r}) = \mathbf{P}(Y = 1)\mathbf{P}(X_{m_1} = x_{m_1}, \ldots, X_{m_r} = x_{m_r}),$$

therefore

$$\Delta_N(W_N, x) \le \mathbf{P}\Big(\sum_{j \in W_N} \frac{\eta^j - \mathbf{E}\eta^j}{\sqrt{\sharp W_N}} > \sqrt{\sharp W_N}\big(p\varepsilon_N - \alpha_N(\gamma(\psi) + p - \alpha_N + \varepsilon_N)\big)\Big)$$
$$+ 2\delta_N(W_N, \alpha_N).$$

The CLT holds for an array $\{\eta^j, j \in W_N, N \in \mathbb{N}\}$ consisting of i.i.d. random variables, thus

$$\frac{1}{\sqrt{\sharp W_N}} \sum_{j \in W_N} (\eta^j - \mathbf{E}\eta^j) \xrightarrow{law} Z \sim N(0, \sigma_0^2),$$

here $\sigma_0^2 = var\mathbb{I}\{Y = 1, X_{m_1} = x_{m_1}, \ldots, X_{m_r} = x_{m_r}\}$. Hence $\Delta_N(W_N, x) \to 0$ if, for some $\alpha_N > 0$,

$$\alpha_N \sqrt{\sharp W_N} \to \infty, \quad \varepsilon_N \sqrt{\sharp W_N} \to \infty, \quad \alpha_N/\varepsilon_N \to 0 \text{ as } N \to \infty. \tag{31}$$

Take $W_N = \overline{S_k(N)}$ with $k = 1, \ldots, K$. Then $\sharp \overline{S_k(N)} \ge (K - 1)[N/K]$ for $k = 1, \ldots, K$ and we conclude that (31) is satisfied when $\varepsilon_N N^{1/2} \to \infty$ as $N \to \infty$ if we choose a sequence $(\alpha_N)_{N \in \mathbb{N}}$ in appropriate way. So, relation (29) is established.

Let

$$R_j(x) = \mathbb{I}\{X^j = x\}(\psi(1)\mathbb{I}\{Y^j = 1\} - \psi(-1)\mathbb{I}\{Y^j = -1\}), \quad x \in \mathbb{X}, \ j \in \mathbb{N}.$$

For all $x \in M \cap \mathbb{X}_-$ one has

$$\frac{1}{\sqrt{\sharp S_k(N)}} \sum_{j \in S_k(N)} \hat{R}_{N,k}^j(x) = \frac{1}{\sqrt{\sharp S_k(N)}} \sum_{j \in S_k(N)} R_j(x)$$

$$+ \sum_{j \in S_k(N)} \mathbb{I}\{X^j = x\} \frac{(\hat{\psi}_{N,k}(1) - \psi(1))\mathbb{I}\{Y^j = 1\} - (\hat{\psi}_{N,k}(-1) - \psi(-1))\mathbb{I}\{Y^j = -1\}}{\sqrt{\sharp S_k(N)}}.$$

Note that $\mathbf{E}R_j(x) = 0$ for all $j \in \mathbb{N}$ and $x \in \mathbb{X}_-$. The CLT for an array of i.i.d. random variables $\{R_j(x), j \in S_k(N), N \in \mathbb{N}\}$ provides that

$$\frac{1}{\sqrt{\sharp S_k(N)}} \sum_{j \in S_k(N)} R_j(x) \xrightarrow{law} Z_1 \sim N(0, \sigma_1^2(x)), \quad N \to \infty,$$

where $\sigma_1^2(x) = var(\mathbb{I}\{X = x\}(\psi(1)\mathbb{I}\{Y = 1\} - \psi(-1)\mathbb{I}\{Y = -1\}))$, $x \in \mathbb{X}_-$. For each $y \in \{-1, 1\}$,

$$(\hat{\psi}_{N,k}(y) - \psi(y)) \frac{1}{\sqrt{\sharp S_k(N)}} \sum_{j \in S_k(N)} \mathbb{I}\{X^j = x\}\mathbb{I}\{Y^j = y\}$$

$$= (\hat{\psi}_{N,k}(y) - \psi(y)) \frac{1}{\sqrt{\sharp S_k(N)}} \sum_{j \in S_k(N)} (\mathbb{I}\{X^j = x\}\mathbb{I}\{Y^j = y\} - \mathbf{E}\mathbb{I}\{X^j = x\}\mathbb{I}\{Y^j = y\})$$

$$+ (\hat{\psi}_{N,k}(y) - \psi(y)) \sqrt{\sharp S_k(N)} \mathbf{P}(X = x, Y = y).$$

Due to the CLT

$$\sum_{j \in S_k(N)} \frac{\mathbb{I}\{X^j = x\}\mathbb{I}\{Y^j = y\} - \mathbf{E}\mathbb{I}\{X^j = x\}\mathbb{I}\{Y^j = y\}}{\sqrt{\sharp S_k(N)}} \xrightarrow{law} Z_2 \sim N(0, \sigma_2^2(x, y))$$

as $N \to \infty$, where $\sigma_2^2(x, y) = var\mathbb{I}\{X^j = x, Y^j = y\}$. In view of (21) we have

$$\frac{\hat{\psi}_{N,k}(y) - \psi(y)}{\sqrt{\sharp S_k(N)}} \sum_{j \in S_k(N)} (\mathbb{I}\{X^j = x\}\mathbb{I}\{Y^j = y\} - \mathbf{E}\mathbb{I}\{X^j = x\}\mathbb{I}\{Y^j = y\}) \xrightarrow{P} 0$$

as $N \to \infty$. Now we apply (18)–(20) once again to conclude that

$$(\hat{\psi}_{N,k}(y) - \psi(y)) \sqrt{\sharp S_k(N)} \xrightarrow{law} Z_3 \sim N(0, \sigma_3^2(y)), \quad N \to \infty,$$

with $\sigma_3^2(y) = \mathbf{P}(Y = -y)(\mathbf{P}(Y = y))^{-3}$. Thus,

$$\sum_{y \in \{-1, 1\}} \hat{\psi}_{N,k}(y) B_{N,k}^{(2)}(y) \xrightarrow{P} 0, \quad N \to \infty. \tag{32}$$

Taking into account (28) and (32) we come to (27) and consequently to (25).

Now we turn to the study of $\hat{T}_N(f) - T_N(f)$ appearing in (24). One has

$$\sqrt{N}(\hat{T}_N(f) - T_N(f))$$

$$= \frac{2\sqrt{N}}{K} \sum_{k=1}^{K} \frac{1}{\sharp S_k(N)} \sum_{y \in \{-1, 1\}} (\hat{\psi}_{N,k}(y) - \psi(y)) \sum_{j \in S_k(N)} \mathbb{I}\{Y^j = y, f(X^j) \neq y\}.$$

Put $Z^j = \mathbb{I}\{Y^j = y, f(X^j) \neq y\}, j = 1, \ldots, N$. For each $k = 1, \ldots, K$

$$\sum_{y \in \{-1,1\}} (\hat{\psi}_{N,k}(y) - \psi(y)) \frac{1}{\sqrt{\sharp S_k(N)}} \sum_{j \in S_k(N)} \mathbb{I}\{Y^j = y, f(X^j) \neq y\}$$

$$= \sum_{y \in \{-1,1\}} (\hat{\psi}_{N,k}(y) - \psi(y)) \frac{1}{\sqrt{\sharp S_k(N)}} \sum_{j \in S_k(N)} (Z^j - \mathbf{E}Z^j)$$

$$+ \sqrt{\sharp S_k(N)} \sum_{y \in \{-1,1\}} (\hat{\psi}_{N,k}(y) - \psi(y)) \mathbf{P}(Y = y, f(X) \neq y).$$

Due to (21) and CLT for an array of $\{Z^j, j \in S_k(N), N \in \mathbb{N}\}$ we have

$$\sum_{y \in \{-1,1\}} (\hat{\psi}_{N,k}(y) - \psi(y)) \frac{1}{\sqrt{\sharp S_k(N)}} \sum_{j \in S_k(N)} (Z^j - \mathbf{E}Z^j) \xrightarrow{\mathbf{P}} 0$$

as $N \to \infty$. Consequently the limit distribution of

$$\sqrt{N}[(\hat{T}_N(f) - T_N(f)) + (T_N(f) - Err(f))]$$

will be the same as for random variables

$$\sqrt{N}[(T_N(f) - Err(f)) + \frac{2}{K} \sum_{k=1}^{K} \sum_{y \in \{-1,1\}} (\hat{\psi}_{N,k}(y) - \psi(y)) \mathbf{P}(Y = y, f(X) \neq y)]. \qquad (33)$$

Note that for each $y \in \{-1, 1\}$ and $k = 1, \ldots, K$

$$\hat{\mathbf{P}}_{S_k(N)}(Y = y) - \mathbf{P}(Y = y) \xrightarrow{\mathbf{P}} 0,$$

$$\sqrt{\sharp S_k(N)}(\hat{\mathbf{P}}_{S_k(N)}(Y = y) - \mathbf{P}(Y = y)) \xrightarrow{law} Z_4 \sim N(0, \sigma_4^2),$$

as $N \to \infty$, where $\sigma_4^2 = \mathbf{P}(Y = -1)\mathbf{P}(Y = 1)$.

Now the Slutsky lemma shows that the limit behavior of the random variables introduced in (33) will be the same as for random variables

$$\sqrt{N}(T_N(f) - Err(f))$$

$$- \frac{2\sqrt{N}}{K} \sum_{k=1}^{K} \sum_{y \in \{-1,1\}} \frac{(\hat{\mathbf{P}}_{S_k(N)}(Y = y) - \mathbf{P}(Y = y))\mathbf{P}(Y = y, f(X) \neq y)}{\mathbf{P}(Y = y)^2}$$

$$= \frac{2\sqrt{N}}{K} \sum_{k=1}^{K} \sum_{y \in \{-1,1\}} \frac{1}{\sharp S_k(N)} \sum_{j \in S_k(N)} \left(\frac{\mathbb{I}\{Y^j = y, f(X^j) \neq y\} - \mathbf{P}(Y = y, f(X) \neq y)}{\mathbf{P}(Y = y)} \right)$$

$$
\begin{aligned}
&-\frac{\mathbb{I}\{Y^j = y\} - \mathbf{P}(Y = y)\mathbf{P}(Y = y, f(X) \neq y)}{\mathbf{P}(Y = y)^2}\Bigg) \\
&= \frac{\sqrt{N}}{K} \sum_{k=1}^{K} \frac{1}{\sharp S_k(N)} \sum_{j \in S_k(N)} (V^j - \mathsf{E}V^j)
\end{aligned}
$$

where

$$
V^j = \sum_{y \in \{-1,1\}} \frac{2\mathbb{I}\{Y^j = y\}}{\mathbf{P}(Y = y)} \left(\mathbb{I}\{f(X^j) \neq y\} - \frac{\mathbf{P}(Y = y, f(X) \neq y)}{\mathbf{P}(Y = y)} \right).
$$

For each $k = 1, \ldots, K$, the CLT for an array $\{V^j, j \in S_k(N), N \in \mathbb{N}\}$ of i.i.d. random variables yields the relation

$$
Z_{N,k} := \frac{1}{\sqrt{\sharp S_k(N)}} \sum_{j \in S_k(N)} (V^j - \mathsf{E}V^j) \xrightarrow{law} Z \sim N(0, \sigma^2), \quad N \to \infty,
$$

where $\sigma^2 = var\, V$ and V was introduced in (23). Since $Z_{N,1}, \ldots, Z_{N,K}$ are independent and $\sqrt{N}/\sqrt{\sharp S_k(N)} \to \sqrt{K}$ for $k = 1, \ldots, K$, as $N \to \infty$, we come to (22). The proof is complete. $\qquad \square$

Recall that for a sequence of random variables $(\eta_N)_{N \in \mathbb{N}}$ and a sequence of positive numbers $(a_N)_{N \in \mathbb{N}}$ one writes $\eta_N = o_P(a_N)$ if $\eta_N/a_N \xrightarrow{\mathbf{P}} 0, N \to \infty$.

Remark 2. As usual one can view the CLT as a result describing the exact rate of approximation for random variables under consideration. Theorem 2 implies that

$$
\hat{Err}_K(f_{PA}, \xi_N) - Err(f) = o_P(a_N), \quad N \to \infty, \tag{34}
$$

where $a_N = o(N^{-1/2})$. The last relation is optimal in a sense whenever $\sigma^2 > 0$, i.e. one cannot take $a_N = O(N^{-1/2})$ in (34).

Remark 3. In view of (11) it is not difficult to construct the consistent estimates $\hat{\sigma}_N$ of unknown σ appearing in (22). Therefore (if $\sigma^2 \neq 0$) we can claim that under conditions of Theorem 1

$$
\frac{\sqrt{N}}{\hat{\sigma}_N}(\hat{Err}_K(f_{PA}, \xi_N) - Err(f)) \xrightarrow{law} \frac{Z}{\sigma} \sim N(0, 1), \quad N \to \infty.
$$

Now we consider the multidimensional version of Theorem 2. To simplify notation set $\alpha = (m_1, \ldots, m_r)$. We write $\hat{f}_{PA,\varepsilon}^{\alpha}$ and f^{α} instead of $\hat{f}_{PA,\varepsilon}^{m_1,\ldots,m_r}$ and $f^{m_1,\ldots,m_r}$, respectively. Employing the Cramér–Wold device and the proof of Theorem 2 we come to the following statement (as usual we use the column vectors and write $\top$ for transposition).

Theorem 3. *Let $\varepsilon_N \to 0$ and $N^{1/2}\varepsilon_N \to \infty$ as $N \to \infty$. Then, for each $K \in \mathbb{N}$, any $\alpha(i) = \{m_1^{(i)}, \ldots, m_r^{(i)}\} \subset \{1, \ldots, n\}$ where $i = 1, \ldots, s$, one has*

$$\sqrt{N}(Z_N^{(1)}, \ldots, Z_N^{(s)})^\top \xrightarrow{law} Z \sim N(0, C), \quad N \to \infty.$$

Here $Z_N^{(i)} = \hat{Err}_K(\hat{f}_{PA,\varepsilon}^{\alpha(i)}, \xi_N) - Err(f^{\alpha(i)})$, $i = 1, \ldots, s$, and the elements of covariance matrix $C = (c_{i,j})$ have the form

$$c_{i,j} = cov(V(\alpha(i)), V(\alpha(j))), \quad i, j = 1, \ldots, s,$$

the random variables $V(\alpha(i))$ being defined in the same way as V in (23) with $f^{m_1, \ldots, m_r}$ replaced by $f^{\alpha(i)}$.

To conclude we note (see also Remark 3) that one can construct the consistent estimates $\hat{C}_N$ of the unknown (nondegenerate) covariance matrix C to obtain the statistical version of the last theorem. Namely, under conditions of Theorem 3 the following relation is valid

$$(\hat{C}_N)^{-1/2}(Z_N^{(1)}, \ldots, Z_N^{(s)})^\top \xrightarrow{law} C^{-1/2}Z \sim N(0, I), \quad N \to \infty,$$

where I stands for the unit matrix of order s.

Acknowledgements The author is grateful to Organizing Committee for invitation to participate in the Fields Institute International Symposium on Asymptotic Methods in Stochastics, in Honour of Miklós Csörgő's Work. Special thanks are due to Professor Csörgő and his colleagues for hospitality.

The work is partially supported by RFBR grant 13-01-00612.

References

1. Arlot, S., Celisse, A.: A survey of cross-validation procedures for model selection. Stat. Surv. **4**, 40–79 (2010)
2. Bradley-Smith, G., Hope, S., Firth, H.V., Hurst, J.A.: Oxford Handbook of Genetics. Oxford University Press, New York (2009)
3. Bulinski, A.V.: On foundation of the dimesionality reduction method for explanatory variables. Zapiski Nauchnyh Seminarov POMI. **408**, 84–101 (2012) (In Russian; English translation in Journal of Mathematical Sciences **199**(2), 113–122 (2014))
4. Bulinski, A., Butkovsky, O., Sadovnichy, V., Shashkin, A., Yaskov, P., Balatskiy, A., Samokhodskaya, L., Tkachuk, V.: Statistical methods of SNP data analysis and applications. Open J. Stat. **2**(1), 73–87 (2012)
5. Edwards, T.L., Torstenson, E.S., Martin, E.M., Ritchie, M.D.: A cross-validation procedure for general pedigrees and matched odds ratio fitness metric implemented for the multifactor dimensionality reduction pedigree disequilibrium test MDR-PDT and cross-validation: power studies. Genet. Epidemiol. **34**(2), 194–199 (2010)

6. Golland, P., Liang, F., Mukherjee, S., Panchenko, D.: Permutation tests for classification. In: COLT'05 Proceedings of the 18th Annual Conference on Learning Theory, pp. 501–515. Springer, Berlin (2005)
7. Gui, J., Andrew, A.S., Andrews, P., Nelson, H.M., Kelsey, K.T., Karagas, M.R., Moore, J.H.: A robust multifactor dimensionality reduction method for detecting gene-gene interactions with application to the genetic analysis of bladder cancer susceptibility. Ann. Hum. Genet. **75**(1), 20–28 (2011)
8. He, H., Oetting, W.S., Brott, M.J., Basu, S.: Power of multifactor dimensionality reduction and penalized logistic regression for detecting gene-gene interaction in a case-control study. BMC Med. Genet. **10**, 127 (2009)
9. Mei, H., Cuccaro, M.L., Martin, E.R.: Multifactor dimensionality reduction-phenomics: a novel method to capture genetic heterogeneity with use of phenotypic variables. Am. J. Hum. Genet. **81**(6), 1251–1261 (2007)
10. Namkung, J., Elston, R.C., Yang, J.M., Park, T.: Identification of gene-gene interactions in the presence of missing data using the multifactor dimensionality reduction method. Genet. Epidemiol. **33**(7), 646–656 (2009)
11. Niu, A., Zhang, S., Sha, Q.: A novel method to detect gene-gene interactions in structured populations: MDR-SP. Ann. Hum. Genet. **75**(6), 742–754 (2011)
12. Oh, S., Lee, J., Kwon, M-S., Weir, B., Ha, K., Park, T.: A novel method to identify high order gene-gene interactions in genome-wide association studies: gene-based MDR. BMC Bioinformatics **13**(Suppl 9), S5 (2012)
13. Ritchie, M.D., Hahn, L.W., Roodi, N., Bailey, R.L., Dupont, W.D., Parl, F.F., Moore, J.H.: Multifactor-dimensionality reduction reveals high-order interactions among estrogen-metabolism genes in sporadic breast cancer. Am. J. Hum. Genet. **69**(1), 138–147 (2001)
14. Ritchie, M.D., Motsinger, A.A.: Multifactor dimensionality reduction for detecting gene-gene and gene-environment interactions in pharmacogenomics studies. Pharmacogenomics **6**(8), 823–834 (2005)
15. Schwender, H., Ruczinski, I., Ickstadt K.: Testing SNPs and sets of SNPs for importance in association studies. Biostatistics **12**(1), 18–32 (2011)
16. Taylor, R.L., Hu, T.-C.: Strong laws of large numbers for arrays of row-wise independent random elements. Int. J. Math. Math. Sci. **10**(4), 805–814 (1987)
17. Velez, D.R., White, B.C., Motsinger, A.A., Bush, W.S., Ritchie, M.D., Williams, S.M., Moore, J.H.: A balanced accuracy function for epistasis modeling in imbalanced datasets using multifactor dimensionality reduction. Genet. Epidemiol. **31**(4), 306–315 (2007)
18. Winham, S.J., Slater, A.J., Motsinger-Reif, A.A.: A comparison of internal validation techniques for multifactor dimensionality reduction. BMC Bioinformatics **11**, 394 (2010)

An Extension of Theorems of Hechner and Heinkel

Deli Li, Yongcheng Qi, and Andrew Rosalsky

1 Introduction and Main Result

It is a great pleasure for us to contribute this paper in honour of Professor Miklós Csörgő's work on the occasion of his 80th birthday.

Throughout, let $(\mathbf{B}, \| \cdot \|)$ be a real separable Banach space equipped with its Borel σ-algebra $\mathscr{B}$ (= the σ-algebra generated by the class of open subsets of $\mathbf{B}$ determined by $\| \cdot \|$) and let $\{X_n;\ n \geq 1\}$ be a sequence of independent copies of a $\mathbf{B}$-valued random variable X defined on a probability space $(\Omega, \mathscr{F}, \mathbb{P})$. As usual, let $S_n = \sum_{k=1}^{n} X_k$, $n \geq 1$ denote their partial sums. If $0 < p < 2$ and if X is a real-valued random variable (that is, if $\mathbf{B} = \mathbb{R}$), then

$$\lim_{n \to \infty} \frac{S_n}{n^{1/p}} = 0 \text{ almost surely (a.s.)}$$

if and only if

$$\mathbb{E}|X|^p < \infty \text{ where } \mathbb{E}X = 0 \text{ whenever } p \geq 1.$$

D. Li (✉)
Department of Mathematical Sciences, Lakehead University,
Thunder Bay, ON P7B 5E1, Canada
e-mail: dli@lakeheadu.ca

Y. Qi
Department of Mathematics and Statistics, University of Minnesota Duluth,
Duluth, MN 55812, USA
e-mail: yqi@d.umn.edu

A. Rosalsky
Department of Statistics, University of Florida, Gainesville, FL 32611, USA
e-mail: rosalsky@stat.ufl.edu

© Springer Science+Business Media New York 2015
D. Dawson et al. (eds.), *Asymptotic Laws and Methods in Stochastics*,
Fields Institute Communications 76, DOI 10.1007/978-1-4939-3076-0_8

This is the celebrated Kolmogoroff-Marcinkiewicz-Zygmund strong law of large numbers (SLLN); see Kolmogoroff [9] for $p = 1$ and Marcinkiewicz and Zygmund [14] for $p \neq 1$.

The classical Kolmogoroff SLLN in real separable Banach spaces was established by Mourier [15]. The extension of the Kolmogoroff-Marcinkiewicz-Zygmund SLLN to **B**-valued random variables is independently due to Azlarov and Volodin [1] and de Acosta [4].

Theorem 1 (Azlarov and Volodin [1] and de Acosta [4]). *Let* $0 < p < 2$ *and let* $\{X_n; \ n \geq 1\}$ *be a sequence of independent copies of a* **B**-*valued random variable* X. *Then*

$$\lim_{n \to \infty} \frac{S_n}{n^{1/p}} = 0 \ \ a.s.$$

if and only if

$$\mathbb{E}\|X\|^p < \infty \ \ and \ \ \frac{S_n}{n^{1/p}} \to_{\mathbb{P}} 0.$$

Let $0 < p \leq 2$ and let $\{\Theta_n; \ n \geq 1\}$ be a sequence of i.i.d. stable random variables each with characteristic function $\psi(t) = \exp\{-|t|^p\}$, $-\infty < t < \infty$. Then **B** is said to be of *stable type* p if $\sum_{n=1}^{\infty} \Theta_n v_n$ converges a.s. whenever $\{v_n : \ n \geq 1\} \subseteq \mathbf{B}$ with $\sum_{n=1}^{\infty} \|v_n\|^p < \infty$. Equivalent characterizations of a Banach space being of stable type p and properties of stable type p Banach spaces may be found in Ledoux and Talagrand [10]. Some of these properties are summarized in Li, Qi, and Rosalsky [12].

At the origin of the current investigation is the following recent and striking result by Hechner [6] for $p = 1$ and Hechner and Heinkel [7, Theorem 5] for $1 < p < 2$ which are new even in the case where the Banach space **B** is the real line. The earliest investigation that we are aware of concerning the convergence of the series $\sum_{n=1}^{\infty} \frac{1}{n} \left(\frac{\mathbb{E}|S_n|}{n} \right)$ was carried out by Hechner [5] for the case where $\{X_n; \ n \geq 1\}$ is a sequence of i.i.d. mean zero real-valued random variables.

Theorem 2 (Hechner [6, Theorem 2.4.1] for $p = 1$ and Hechner and Heinkel [7, Theorem 5] for $1 < p < 2$). *Suppose that* **B** *is of stable type* p *for some* $p \in [1, 2)$ *and let* $\{X_n; \ n \geq 1\}$ *be a sequence of independent copies of a* **B**-*valued variable* X *with* $\mathbb{E}X = 0$. *Then*

$$\sum_{n=1}^{\infty} \frac{1}{n} \left(\frac{\mathbb{E}\|S_n\|}{n^{1/p}} \right) < \infty$$

if and only if

$$\begin{cases} \mathbb{E}\|X\| \ln(1 + \|X\|) < \infty & if \ p = 1, \\ \\ \int_0^{\infty} \mathbb{P}^{1/p} \left(\|X\| > t \right) dt < \infty & if \ 1 < p < 2. \end{cases}$$

Inspired by the above discovery by Hechner [6] and Hechner and Heinkel [7], Li, Qi, and Rosalsky [12] obtained sets of necessary and sufficient conditions for

$$\sum_{n=1}^{\infty} \frac{1}{n} \left(\frac{\|S_n\|}{n^{1/p}} \right) < \infty \text{ a.s.}$$

for the three cases: $0 < p < 1, p = 1, 1 < p < 2$ (see Theorem 2.4, Theorem 2.3, and Corollary 2.1, respectively of Li, Qi, and Rosalsky [12]). Again, these results are new when $\mathbf{B} = \mathbb{R}$; see Theorem 2.5 of Li, Qi, and Rosalsky [12]. Moreover for $1 \leq p < 2$, Li, Qi, and Rosalsky [12, Theorems 2.1 and 2.2] obtained necessary and sufficient conditions for

$$\sum_{n=1}^{\infty} \frac{1}{n} \left(\frac{\mathbb{E}\|S_n\|}{n^{1/p}} \right) < \infty$$

for general separable Banach spaces.

This paper is devoted to an extension of Theorem 2 above and Theorems 2.1 and 2.2 of Li, Qi, and Rosalsky [12]. More specifically, the main result of this paper is the following theorem. We note that no conditions are being imposed on the Banach space $\mathbf{B}$.

Theorem 3. *Let* $0 < p < 2$ *and* $0 < q < \infty$. *Let* $\{X_n; n \geq 1\}$ *be a sequence of independent copies of a* $\mathbf{B}$*-valued random variable* X. *Then*

$$\sum_{n=1}^{\infty} \frac{1}{n} \mathbb{E} \left(\frac{\|S_n\|}{n^{1/p}} \right)^q < \infty \tag{1}$$

if and only if

$$\sum_{n=1}^{\infty} \frac{1}{n} \left(\frac{\|S_n\|}{n^{1/p}} \right)^q < \infty \text{ a.s.} \tag{2}$$

and

$$\begin{cases} \displaystyle\int_0^{\infty} \mathbb{P}^{q/p} \left(\|X\|^q > t \right) dt < \infty & \text{if } 0 < q < p, \\[2ex] \mathbb{E}\|X\|^p \ln(1 + \|X\|) < \infty & \text{if } q = p, \\[2ex] \mathbb{E}\|X\|^q < \infty & \text{if } q > p. \end{cases} \tag{3}$$

Furthermore, each of (1) *and* (2) *implies that*

$$\lim_{n \to \infty} \frac{S_n}{n^{1/p}} = 0 \text{ a.s.} \tag{4}$$

For $0 < q < p$, (1) *and* (2) *are equivalent so that each of them implies that* (3) *and* (4) *hold.*

Remark 1. Let $q = 1$. Then one can easily see that Theorems 2.1 and 2.2 of Li, Qi, and Rosalsky [12] follow from Theorem 3 above.

Remark 2. It follows from the conclusion (4) of Theorem 3 that, if (2) holds for some $q = q_1 > 0$ then (2) holds for all $q > q_1$.

The proof of Theorem 3 will be given in Sect. 3. For proving Theorem 3, we employ new versions of the classical Lévy [11], Ottaviani [3, p. 75], and Hoffmann-Jørgensen [8] inequalities which have recently been obtained by Li and Rosalsky [13] (stated in Sect. 2). As an application of the new versions of the classical Lévy [11] and Hoffmann-Jørgensen [8] inequalities, in Theorem 7 some general results concerning sums of the form $\sum_{n=1}^{\infty} a_n \| \sum_{k=1}^{n} V_k \|^q$ (where the $a_n \geq 0$ and $\{V_k;\ k \geq 1\}$ is a sequence of independent symmetric **B**-valued random variables and $q > 0$) are established; these results are key components in the proof of Theorem 3.

2 New Versions of Some Classical Stochastic Inequalities

Li and Rosalsky [13] have recently obtained new versions of the classical Lévy [11], Ottaviani [3, p. 75], and Hoffmann-Jørgensen [8] inequalities. In this section we state the results obtained by Li and Rosalsky [13] which we use for proving the main result in this paper. Then, as an application of the new versions of the classical Lévy and Hoffmann-Jørgensen [8] inequalities, we establish some general results for sums of the form $\sum_{n=1}^{\infty} a_n \| \sum_{k=1}^{n} V_k \|^q$, where the a_n are nonnegative and where $\{V_k;\ k \geq 1\}$ is a sequence of independent symmetric **B**-valued random variables and $q > 0$.

Let $\{V_n;\ n \geq 1\}$ be a sequence of independent **B**-valued random variables defined on a probability space $(\Omega, \mathscr{F}, \mathbb{P})$. Let $\mathbf{B}^{\infty} = \mathbf{B} \times \mathbf{B} \times \mathbf{B} \times \cdots$ and $g : \mathbf{B}^{\infty} \to \overline{\mathbb{R}}_+ = [0, \infty]$ be a measurable function. Let

$$T_n = g(V_1, \ldots, V_n, 0, \ldots), \quad Y_n = g(0, \ldots, 0, V_n, 0, \ldots), \quad M_n = \max_{1 \leq j \leq n} T_j, \quad N_n = \max_{1 \leq j \leq n} Y_j$$

for $n \geq 1$, and

$$M = \sup_{n \geq 1} T_n, \quad N = \sup_{n \geq 1} Y_n.$$

The following result, which is a new general version of Lévy's inequality, is Theorem 2.1 of Li and Rosalsky [13].

Theorem 4 (Li and Rosalsky [13]). *Let $\{V_n;\ n \geq 1\}$ be a sequence of independent symmetric **B**-valued random variables. Let $g : \mathbf{B}^{\infty} \to \overline{\mathbb{R}}_+ = [0, \infty]$ be a measurable function such that for all $\mathbf{x}, \mathbf{y} \in \mathbf{B}^{\infty}$,*

$$g\left(\frac{\mathbf{x} + \mathbf{y}}{2}\right) \leq \alpha \max\left(g(\mathbf{x}), g(\mathbf{y})\right), \tag{5}$$

where $1 \leq \alpha < \infty$ is a constant, depending only on the function g. Then for all $t \geq 0$, we have

$$\mathbb{P}(M_n > t) \leq 2\mathbb{P}\left(T_n > \frac{t}{\alpha}\right)$$

and

$$\mathbb{P}(N_n > t) \leq 2\mathbb{P}\left(T_n > \frac{t}{\alpha}\right).$$

Moreover if $T_n \to T$ in law, then for all $t \geq 0$, we have

$$\mathbb{P}(M > t) \leq 2\mathbb{P}\left(T > \frac{t}{\alpha}\right)$$

and

$$\mathbb{P}(N > t) \leq 2\mathbb{P}\left(T > \frac{t}{\alpha}\right).$$

Remark 3. Theorem 4 includes the classical Lévy inequality [11] as a special case if $\mathbf{B} = \mathbb{R}$ and $g(x_1, x_2, \ldots, x_n, \ldots) = \left|\sum_{i=1}^{n} x_i\right|$, $(x_1, x_2, \ldots, x_n, \ldots) \in \mathbb{R}^{\infty}$. Theorem 4 is due to Hoffmann-Jørgensen [8] for the special case of $\alpha = 1$.

The following result, which is Theorem 2.2 of Li and Rosalsky [13], is a new general version of the classical Ottaviani [3, p. 75] inequality.

Theorem 5 (Li and Rosalsky [13]). Let $\{V_n; n \geq 1\}$ be a sequence of independent B-valued random variables. Let $g : \mathbf{B}^{\infty} \to \overline{\mathbb{R}}_+ = [0, \infty]$ be a measurable function such that for all $\mathbf{x}, \mathbf{y} \in \mathbf{B}^{\infty}$,

$$g(\mathbf{x} + \mathbf{y}) \leq \beta(g(\mathbf{x}) + g(\mathbf{y})), \tag{6}$$

where $1 \leq \beta < \infty$ is a constant, depending only on the function g. Then for all $n \geq 1$ and all nonnegative real numbers t and u, we have

$$\mathbb{P}(M_n > t + u) \leq \frac{\mathbb{P}\left(T_n > \frac{t}{\beta}\right)}{1 - \max_{1 \leq k \leq n-1} \mathbb{P}\left(D_{n,k} > \frac{u}{\beta}\right)},$$

where

$$D_{n,j} = g\left(0, \cdots, 0, -V_{j+1}, \cdots, -V_n, 0, \cdots\right), \ j = 1, 2, \ldots, n-1.$$

In particular, if for some $\delta \geq 0$,

$$\max_{1 \leq k \leq n-1} \mathbb{P}\left(D_{n,k} > \frac{\delta}{\beta}\right) \leq \frac{1}{2},$$

then for every $t \geq \delta$, we have

$$\mathbb{P}\left(M_n > 2t\right) \leq 2\mathbb{P}\left(T_n > \frac{t}{\beta}\right).$$

Remark 4. The classical Ottaviani inequality follows from Theorem 5 if $\mathbf{B} = \mathbb{R}$ and

$$g\left(x_1, x_2, \ldots, x_n, \ldots\right) = \left|\sum_{k=1}^{n} x_i\right|, \quad \left(x_1, x_2, \ldots, x_n, \ldots\right) \in \mathbb{R}^{\infty}.$$

The following result, which is Theorem 2.3 of Li and Rosalsky [13], is a new general version of the classical Hoffmann-Jørgensen inequality [8].

Theorem 6 (Li and Rosalsky [13]). *Let $\{V_n; \ n \geq 1\}$ be a sequence of independent symmetric $\mathbf{B}$-valued random variables. Let $g : \mathbf{B}^{\infty} \to \overline{\mathbb{R}}_+ = [0, \infty]$ be a measurable function satisfying conditions (5) and (6). Then for all nonnegative real numbers $s, t,$ and u, we have*

$$\mathbb{P}\left(T_n > s + t + u\right) \leq \mathbb{P}\left(N_n > \frac{s}{\beta^2}\right) + 2\mathbb{P}\left(T_n > \frac{u}{\alpha\beta}\right)\mathbb{P}\left(M_n > \frac{t}{\beta^2}\right)$$

$$\leq \mathbb{P}\left(N_n > \frac{s}{\beta^2}\right) + 4\mathbb{P}\left(T_n > \frac{u}{\alpha\beta}\right)\mathbb{P}\left(T_n > \frac{t}{\alpha\beta^2}\right),$$

$$\mathbb{P}\left(M_n > s + t + u\right) \leq 2\mathbb{P}\left(N_n > \frac{s}{\alpha\beta^2}\right) + 8\mathbb{P}\left(T_n > \frac{u}{\alpha^2\beta}\right)\mathbb{P}\left(T_n > \frac{t}{\alpha^2\beta^2}\right),$$

and

$$\mathbb{P}\left(M > s + t + u\right) \leq 2\mathbb{P}\left(N > \frac{s}{\alpha\beta^2}\right) + 4\mathbb{P}\left(M > \frac{u}{\alpha^2\beta}\right)\mathbb{P}\left(M > \frac{t}{\alpha\beta^2}\right).$$

Remark 5. The classical Hoffmann-Jørgensen inequality [8] follows from Theorem 6 if $\alpha = 1$ and $\beta = 1$.

For illustrating the new versions of the classical Lévy [11] and Hoffmann-Jørgensen [8] inequalities, i.e., Theorems 4 and 6 above, we now establish the following general result.

Theorem 7. *Let $q > 0$ and let $\{a_n; \ n \geq 1\}$ be a sequence of nonnegative real numbers such that $\sum_{n=1}^{\infty} a_n < \infty$. Let $\{V_k; \ k \geq 1\}$ be a sequence of independent symmetric $\mathbf{B}$-valued random variables. Write*

$$b_n = \sum_{k=n}^{\infty} a_k, \quad n \geq 1$$

and

$$\alpha = \begin{cases} 2^{1-q}, \text{ if } 0 < q \le 1 \\ 1, \quad \text{ if } q > 1. \end{cases} \quad \text{and } \beta = \begin{cases} 1, \quad \text{ if } 0 < q \le 1 \\ 2^{q-1}, \text{ if } q > 1. \end{cases} \tag{7}$$

Then, for all nonnegative real numbers s, t, and u, we have that

$$\mathbb{P}\left(\sup_{n \ge 1} b_n \|V_n\|^q > t\right) \le 2\mathbb{P}\left(\sum_{n=1}^{\infty} a_n \left\|\sum_{i=1}^{n} V_i\right\|^q > \frac{t}{\alpha}\right) \tag{8}$$

and

$$\mathbb{P}\left(\sum_{n=1}^{\infty} a_n \left\|\sum_{i=1}^{n} V_i\right\|^q > s + t + u\right) \le \mathbb{P}\left(\sup_{n \ge 1} b_n \|V_n\|^q > \frac{s}{\beta^2}\right)$$
$$+ 4\mathbb{P}\left(\sum_{n=1}^{\infty} a_n \left\|\sum_{i=1}^{n} V_i\right\|^q > \frac{u}{\alpha\beta}\right) \mathbb{P}\left(\sum_{n=1}^{\infty} a_n \left\|\sum_{i=1}^{n} V_i\right\|^q > \frac{t}{\alpha\beta^2}\right). \tag{9}$$

Furthermore, we have that

$$\mathbb{E}\left(\sup_{n \ge 1} b_n \|V_n\|^q\right) \le 2\alpha \mathbb{E}\left(\sum_{n=1}^{\infty} a_n \left\|\sum_{i=1}^{n} V_i\right\|^q\right) \tag{10}$$

and

$$\mathbb{E}\left(\sum_{n=1}^{\infty} a_n \left\|\sum_{i=1}^{n} V_i\right\|^q\right) \le 6(\alpha + \beta)^3 \mathbb{E}\left(\sup_{n \ge 1} b_n \|V_n\|^q\right) + 6(\alpha + \beta)^3 t_0, \tag{11}$$

where

$$t_0 = \inf\left\{t > 0; \ \mathbb{P}\left(\sum_{n=1}^{\infty} a_n \left\|\sum_{i=1}^{n} V_i\right\|^q > t\right) \le 24^{-1}(\alpha + \beta)^{-3}\right\}.$$

Proof. For $m \ge 1$ and $(x_1, x_2, \ldots, x_m) \in \mathbf{B}^m$, write

$$g_m(x_1, x_2, \ldots, x_m) = \sum_{n=1}^{m} a_n \left\|\sum_{i=1}^{m} x_i\right\|^q.$$

One can easily check that, for each $m \ge 1$, the function g_m satisfies conditions (5) and (6) with α and β given by (7). Let

$$T_{m,n} = g_m(V_1, \ldots, V_n, 0, \ldots, 0), \quad Y_{m,n} = g_m(0, \ldots, 0, V_n, 0, \ldots, 0), \quad 1 \le n \le m.$$

Clearly,

$$T_{m,m} = \sum_{n=1}^{m} a_n \left\| \sum_{i=1}^{n} V_i \right\|^q$$

and

$$\max_{1 \le n \le m} Y_{m,n} = \max_{1 \le n \le m} \left(\sum_{i=n}^{m} a_i \right) \|V_n\|^q = \max_{1 \le n \le m} (b_n - b_{m+1}) \|V_n\|^q.$$

Then by Theorem 4 we have for all nonnegative real numbers t,

$$\mathbb{P}\left(\max_{1 \le n \le m} (b_n - b_{m+1}) \|V_n\|^q > t \right) = \mathbb{P}\left(\max_{1 \le n \le m} Y_{m,n} > t \right)$$

$$\le 2\mathbb{P}\left(T_{m,m} > \frac{t}{\alpha} \right) \qquad (12)$$

$$= 2\mathbb{P}\left(\sum_{n=1}^{m} a_n \left\| \sum_{i=1}^{n} V_i \right\|^q > \frac{t}{\alpha} \right),$$

and by Theorem 6 we have for all nonnegative real numbers s, t, and u,

$$\mathbb{P}(T_{m,m} > s + t + u) \le \mathbb{P}\left(\max_{1 \le n \le m} (b_n - b_{m+1}) \|V_n\|^q > \frac{s}{\beta^2} \right)$$

$$+ 4\mathbb{P}\left(T_{m,m} > \frac{u}{\alpha\beta} \right) \mathbb{P}\left(T_{m,m} > \frac{t}{\alpha\beta^2} \right). \qquad (13)$$

Note that with probability 1,

$$T_{m,m} = \sum_{n=1}^{m} a_n \left\| \sum_{i=1}^{n} V_i \right\|^q \nearrow \sum_{n=1}^{\infty} a_n \left\| \sum_{i=1}^{n} V_i \right\|^q$$

and

$$\max_{1 \le n \le m} (b_n - b_{m+1}) \|V_n\|^q \nearrow \sup_{n \ge 1} b_n \|V_n\|^q \quad \text{as } m \to \infty.$$

Thus, letting $m \to \infty$, (8) and (9) follow from (12) and (13) respectively. We only need to verify (11) since (10) follows from (8). Set

$$\gamma = \alpha + \beta \quad \text{and} \quad T = \sum_{n=1}^{\infty} a_n \left\| \sum_{i=1}^{n} V_i \right\|^q.$$

Let $c > t_0$. Noting $\gamma > 1$, $\gamma/\alpha > 1$, and $\gamma/\beta > 1$, by (9) with $s = t = u = \gamma^3 x$, we have that

$$\mathbb{E}(T) = 3\gamma^3 \int_0^\infty \mathbb{P}\left(T > 3\gamma^3 x\right) dx$$

$$= 3\gamma^3 \left(\int_0^c + \int_c^\infty\right) \mathbb{P}\left(T > 3\gamma^3 x\right) dx$$

$$\leq 3\gamma^3 \left(c + \int_c^\infty \mathbb{P}\left(\sup_{n\geq 1} b_n \|V_n\|^q > x\right) dx + 4\int_c^\infty \mathbb{P}^2(T > x) dx\right)$$

$$\leq 3\gamma^3 \left(c + \mathbb{E}\left(\sup_{n\geq 1} b_n \|V_n\|^q\right) + 4\mathbb{P}(T > c)\int_0^\infty \mathbb{P}(T > x) dx\right)$$

$$\leq 3\gamma^3 c + 3\gamma^3 \mathbb{E}\left(\sup_{n\geq 1} b_n \|V_n\|^q\right) + \frac{1}{2}\mathbb{E}(T)$$

since $12\gamma^3 \mathbb{P}(T > c) \leq 1/2$ by the choice of c. We thus conclude that

$$\mathbb{E}(T) \leq 6(\alpha + \beta)^3 \mathbb{E}\left(\sup_{n\geq 1} b_n \|V_n\|^q\right) + 6(\alpha + \beta)^3 c \;\; \forall \, c > t_0$$

and hence (11) is established. $\qquad\qquad\qquad\qquad\qquad\qquad\qquad\qquad\qquad\qquad\qquad$ $\square$

3 Proof of Theorem 3

For the proof of Theorem 3, we need the following five preliminary lemmas.

Lemma 1. *Let $\{c_k; \; k \geq 1\}$ be a sequence of real numbers such that*

$$\sum_{k=1}^\infty |c_k| < \infty$$

and let $\{a_{n,k}; \; k \geq 1, n \geq 1\}$ be an array of real numbers such that

$$\sup_{n\geq 1, k\geq 1} |a_{n,k}| < \infty \;\; and \;\; \lim_{n\to\infty} a_{n,k} = 0 \; \forall \, k \geq 1.$$

Then

$$\lim_{n\to\infty} \sum_{k=1}^\infty a_{n,k} c_k = 0.$$

Proof. This follows immediately from the Lebesgue dominated convergence theorem with counting measure on the positive integers. $\qquad\square$

The proofs of Lemmas 2 and 3 and Theorem 3 involve a symmetrization argument. For the sequence $\{X_n;\ n \geq 1\}$ of independent copies of the **B**-valued random variable X with partial sums $S_n = \sum_{k=1}^{n} X_k$, $n \geq 1$, let $\{X',X'_n;\ n \geq 1\}$ be an independent copy of $\{X,X_n;\ n \geq 1\}$. The symmetrized random variables are defined by $\hat{X} = X - X'$, $\hat{X}_n = X_n - X'_n$, $n \geq 1$. Set $S'_n = \sum_{k=1}^{n} X'_k$, $\hat{S}_n = \sum_{k=1}^{n} \hat{X}_k$, $n \geq 1$.

Lemma 2. *Let $0 < p < 2$ and let $\{X_n;\ n \geq 1\}$ be a sequence of independent copies of a **B**-valued random variable X. Then*

$$\lim_{n \to \infty} \frac{S_n}{n^{1/p}} = 0 \ \ a.s. \tag{14}$$

if and only if

$$\mathbb{E}\|X\|^p < \infty \ \ and \ \ \frac{S_{2^n}}{2^{n/p}} \to_{\mathbb{P}} 0. \tag{15}$$

Proof. By Theorem 1, we see that (15) immediately follows from (14). We now show that (15) implies (14). For $0 < p < 1$, (14) follows from (15) since

$$\lim_{n \to \infty} \frac{\sum_{k=1}^{n} \|X_k\|}{n^{1/p}} = 0 \ \text{ a.s. if and only if } \ \mathbb{E}\|X\|^p < \infty.$$

Clearly, for $1 \leq p < 2$, (15) implies that $\mathbb{E}\|X\| < \infty$ and hence by the SLLN of Mourier [15]

$$\frac{S_n}{n} \to \mathbb{E}X \ \text{ a.s.}$$

Then

$$\frac{S_{2^n}}{2^n} \to_{\mathbb{P}} \mathbb{E}X$$

and so $\mathbb{E}X = 0$ in view of the second half of (15). We thus conclude that when $1 \leq p < 2$, (15) entails $\mathbb{E}X = 0$.

Next, it follows from the second half of (15) that

$$\frac{\hat{S}_{2^n}}{2^{n/p}} \to_{\mathbb{P}} 0.$$

Hence for any given $\epsilon > 0$, there exists a positive integer n_ϵ such that

$$\mathbb{P}\left(\left\|\hat{S}_{2^n}\right\| > 2^{n/p}\epsilon\right) \leq 1/24, \ \ \forall\, n \geq n_\epsilon.$$

Note that $\{\hat{X}_n; n \geq 1\}$ is a sequence of i.i.d. **B**-valued random variables. Thus, by the second part of Proposition 6.8 of Ledoux and Talagrand [10, p. 156], we have

$$\mathbb{E}\left\|\hat{S}_{2^n}\right\| \leq 6\mathbb{E}\max_{1\leq i\leq 2^n}\left\|\hat{X}_i\right\| + 6 \times 2^{n/p}\epsilon \leq 12\mathbb{E}\max_{1\leq i\leq 2^n}\|X_i\| + 6 \times 2^{n/p}\epsilon, \quad \forall\, n \geq n_\epsilon$$

and hence

$$\frac{\mathbb{E}\left\|\hat{S}_{2^n}\right\|}{2^{n/p}} \leq 12\left(\frac{\mathbb{E}\max_{1\leq i\leq 2^n}\|X_i\|}{2^{n/p}}\right) + 6\epsilon, \quad \forall\, n \geq n_\epsilon.$$

It is easy to show that, for $1 \leq p < 2$, the first half of (15) implies that

$$\lim_{n\to\infty}\frac{\mathbb{E}\max_{1\leq i\leq 2^n}\|X_i\|}{2^{n/p}} = 0.$$

We thus have that

$$\lim_{n\to\infty}\frac{\mathbb{E}\left\|\hat{S}_{2^n}\right\|}{2^{n/p}} = 0. \tag{16}$$

Since $\mathbb{E}X = 0$, applying (2.5) of Ledoux and Talagrand [10, p. 46], we have that

$$\max_{2^{n-1}\leq m<2^n}\frac{\mathbb{E}\|S_m\|}{m^{1/p}} \leq 2^{1/p}\max_{2^{n-1}\leq m<2^n}\frac{\mathbb{E}\|S_m\|}{2^{n/p}} \leq 2^{1/p} \times \frac{\mathbb{E}\|S_{2^n}\|}{2^{n/p}} \leq 2^{1/p} \times \frac{\mathbb{E}\left\|\hat{S}_{2^n}\right\|}{2^{n/p}}$$

for $n \geq 1$. It now follows from (16) that

$$\lim_{n\to\infty}\frac{\mathbb{E}\|S_n\|}{n^{1/p}} = 0$$

and hence that

$$\frac{S_n}{n^{1/p}} \to_{\mathbb{P}} 0.$$

By Theorem 1 again, we see that (14) follows. $\qquad\square$

Lemma 3. *Let $0 < p < 2$ and $0 < q < \infty$. Let $\{X_n; n \geq 1\}$ be a sequence of independent copies of a* **B**-*valued random variable X. If (2) holds, i.e., if*

$$\sum_{n=1}^{\infty}\frac{1}{n}\left(\frac{\|S_n\|}{n^{1/p}}\right)^q < \infty \quad a.s.,$$

then (14) holds, i.e.,

$$\lim_{n\to\infty}\frac{S_n}{n^{1/p}} = 0 \quad a.s.$$

Proof. We first show that (2) implies that

$$\frac{S_{2^n}}{2^{n/p}} \to_{\mathbb{P}} 0. \tag{17}$$

To see this, for $n \geq 1$ and $\mathbf{x} = (x_1, x_2, \ldots, x_{2^n}) \in \mathbf{B}^{2^n}$ write

$$g_n(\mathbf{x}) = g_n(x_1, x_2, \ldots, x_{2^n}) = \sum_{k=2^n}^{2^{n+1}-1} \frac{1}{k} \left(\frac{\left\| \sum_{i=1}^{k+1-2^n} x_i \right\|}{k^{1/p}} \right)^q.$$

Clearly, $g_n : \mathbf{B}^{2^n} \to [0, \infty]$ is a measurable function satisfying condition (6) with β given by (7). Set

$$V_1 = S_{2^n}, \; V_j = X_{2^n+j-1}, \; 2 \leq j \leq 2^n,$$

$$M_{n,j} = g_n(V_1, \ldots, V_j, 0, \ldots, 0), \; D_{n,j} = g_n(0, \ldots, 0, -V_j, \ldots, -V_{2^n}), \; 1 \leq j \leq 2^n.$$

By Theorem 5 (i.e., Theorem 2.2 of of Li and Rosalsky [13]), we have that

$$\mathbb{P}\left(\max_{1 \leq j \leq 2^n} M_{n,j} > t + u \right) \leq \frac{\mathbb{P}(M_{n,2^n} > t/\beta)}{1 - \max_{2 \leq j \leq 2^n} \mathbb{P}(D_{n,j} > u/\beta)}, \; \forall s \geq 0, u \geq 0. \tag{18}$$

It is easy to see that

$$M_{n,1} = g_n(S_{2^n}, 0, \ldots, 0) = \left(\sum_{k=2^n}^{2^{n+1}-1} \frac{1}{k^{1+q/p}} \right) (\|S_{2^n}\|)^q \geq 2^{-1-q/p} \left(\frac{\|S_{2^n}\|}{2^{n/p}} \right)^q \tag{19}$$

and it follows from (2) that

$$M_{n,2^n} = g_n(S_{2^n}, X_{2^n+1}, \ldots, X_{2^{n+1}-1}) = \sum_{k=2^n}^{2^{n+1}-1} \frac{1}{k} \left(\frac{\|S_k\|}{k^{1/p}} \right)^q \to 0 \; \text{a.s.} \tag{20}$$

Since $\{X_n; \; n \geq 1\}$ is a sequence of independent copies of X, we have that for all $u \geq 0$,

$$\mathbb{P}(D_{n,j} > u) = \mathbb{P}\left(g_n(0, \ldots, 0, X_1, \ldots, X_{2^n-j+1}) > u \right), \; 2 \leq j \leq 2^n.$$

Note that

$$g_n(0, \ldots, 0, X_1, X_2, \ldots, X_{2^n-j+1}) \leq g_n(0, \ldots, 0, X_1, X_2, \ldots, X_{2^n-j+2}), \; 2 \leq j \leq 2^n.$$

We thus conclude that for all $u \geq 0$,

$$\max_{2 \leq j \leq 2^n} \mathbb{P}(D_{n,j} > u/\beta) \leq \mathbb{P}(g_n(X_1, X_2, \ldots, X_{2^n}) > u/\beta). \tag{21}$$

Set

$$a_{n,k} = \begin{cases} \left(\dfrac{k}{2^n}\right)^{1+q/p} & \text{if } 1 \le k \le 2^n \\[2ex] 0 & \text{if } k > 2^n. \end{cases}$$

Then clearly $\{a_{n,k};\ k \ge 1, n \ge 1\}$ is an array of nonnegative real numbers such that

$$\sup_{n \ge 1, k \ge 1} a_{n,k} \le 1 < \infty \quad \text{and} \quad \lim_{n \to \infty} a_{n,k} = 0 \ \forall\, k \ge 1.$$

Note that, for $n \ge 1$,

$$g_n(X_1, X_2, \ldots, X_{2^n}) = \sum_{k=2^n}^{2^{n+1}-1} \frac{1}{k} \left(\frac{\left\| \sum_{i=1}^{k+1-2^n} X_i \right\|}{k^{1/p}} \right)^q$$

$$\le \sum_{j=1}^{2^n} \frac{1}{2^n} \left(\frac{\|S_j\|}{2^{n/p}} \right)^q$$

$$= \sum_{j=1}^{2^n} \left(\frac{j}{2^n} \right)^{1+q/p} \left(\frac{1}{j} \left(\frac{\|S_j\|}{j^{1/p}} \right)^q \right)$$

$$= \sum_{k=1}^{\infty} a_{n,k} \left(\frac{1}{k} \left(\frac{\|S_k\|}{k^{1/p}} \right)^q \right).$$

Then, by Lemma 1, (2) implies that

$$\lim_{n \to \infty} g_n(X_1, X_2, \ldots, X_{2^n}) = 0 \quad \text{a.s.} \tag{22}$$

It now follows from (18) and (20)–(22) that

$$\lim_{n \to \infty} \mathbb{P}\left(M_{n,1} > \epsilon\right) \le \lim_{n \to \infty} \frac{\mathbb{P}\left(M_{n,2^n} > \frac{\epsilon}{2\beta}\right)}{1 - \mathbb{P}\left(g_n(X_1, X_2, \ldots, X_{2^n}) > \frac{\epsilon}{2\beta}\right)} = 0 \ \forall\, \epsilon > 0;$$

that is,

$$M_{n,1} \to_{\mathbb{P}} 0$$

and hence (17) follows from (19).

We now show that (2) implies that

$$\mathbb{E}\|X\|^p < \infty. \tag{23}$$

To see this, (2) clearly ensures that

$$\sum_{n=1}^{\infty} a_n \left\|\hat{S}_n\right\|^q = \sum_{n=1}^{\infty} \frac{1}{n} \left(\frac{\|\hat{S}_n\|}{n^{1/p}}\right)^q \le \beta \left(\sum_{n=1}^{\infty} \frac{1}{n} \left(\frac{\|S_n\|}{n^{1/p}}\right)^q + \sum_{n=1}^{\infty} \frac{1}{n} \left(\frac{\|S_n'\|}{n^{1/p}}\right)^q\right) < \infty \quad \text{a.s.,} \tag{24}$$

where $a_n = n^{-1-q/p}$, $n \ge 1$. Since $\{\hat{X}_n; \ n \ge 1\}$ is a sequence of independent copies of the **B**-valued random variable $\hat{X}$, it follows from (8) of Theorem 7 that

$$\mathbb{P}\left(\sup_{n\ge1} b_n \left\|\hat{X}_n\right\|^q > t\right) \le 2\mathbb{P}\left(\sum_{n=1}^{\infty} \frac{1}{n} \left(\frac{\|\hat{S}_n\|}{n^{1/p}}\right)^q > \frac{t}{\alpha}\right) \quad \forall \, t \ge 0,$$

where

$$b_n = \sum_{k=n}^{\infty} n^{-1-q/p}, \ n \ge 1$$

which, together with (24), ensures that

$$\sup_{n\ge1} b_n \left\|\hat{X}_n\right\|^q < \infty \quad \text{a.s.} \tag{25}$$

It is easy to check that

$$\lim_{n\to\infty} \frac{b_n}{n^{-q/p}} = \frac{p}{q},$$

and so we have by (25) that

$$\left(\sup_{n\ge1} \frac{\left\|\hat{X}_n\right\|}{n^{1/p}}\right)^q = \sup_{n\ge1} n^{-q/p} \left\|\hat{X}_n\right\|^q < \infty \quad \text{a.s.}$$

Since the $\hat{X}_n, n \ge 1$ are i.i.d., it follows from the Borel-Cantelli lemma that for some finite $\lambda > 0$,

$$\sum_{n=1}^{\infty} \mathbb{P}\left(\|\hat{X}\| > \lambda n^{1/p}\right) < \infty$$

and hence

$$\mathbb{E}\|X - X'\|^p < \infty$$

which is equivalent to (23) . By Lemma 2, (14) now follows from (17) and (23) . The proof of Lemma 3 is complete. □

Lemma 4. *Let* $(\mathbf{E}, \mathscr{G})$ *be a measurable linear space and* $g : \mathbf{E} \to [0, \infty]$ *be a measurable even function such that for all* $\mathbf{x}, \mathbf{y} \in \mathbf{E}$,

$$g(\mathbf{x} + \mathbf{y}) \leq \beta \left(g(\mathbf{x}) + g(\mathbf{y}) \right),$$

where $1 \leq \beta < \infty$ *is a constant, depending only on the function* g. *If* $\mathbf{V}$ *is an* $\mathbf{E}$*-valued random variable and* $\hat{\mathbf{V}}$ *is a symmetrized version of* $\mathbf{V}$ *(i.e.,* $\hat{\mathbf{V}} = \mathbf{V} - \mathbf{V}'$ *where* $\mathbf{V}'$ *is an independent copy of* $\mathbf{V}$*), then for all* $t \geq 0$, *we have that*

$$\mathbb{P}(g(\mathbf{V}) \leq t)\mathbb{E}g(\mathbf{V}) \leq \beta\mathbb{E}g(\hat{\mathbf{V}}) + \beta t \tag{26}$$

and

$$\mathbb{E}g(\hat{\mathbf{V}}) \leq 2\beta\mathbb{E}g(\mathbf{V}). \tag{27}$$

Moreover, if

$$g(\mathbf{V}) < \infty \ \ a.s., \tag{28}$$

then

$$\mathbb{E}g(\mathbf{V}) < \infty \ \ if \ and \ only \ if \ \mathbb{E}g(\hat{\mathbf{V}}) < \infty. \tag{29}$$

Proof. We only give the proof of the second part of this lemma since the first part of this lemma is a special case of Lemma 3.2 of Li and Rosalsky [13]. Note that, by (28), there exists a finite positive number τ such that

$$\mathbb{P}(g(\mathbf{V}) \leq \tau) \geq 1/2.$$

It thus follows from (26) and (27) that

$$\frac{1}{2\beta}\mathbb{E}g(\hat{\mathbf{V}}) \leq \mathbb{E}g(\mathbf{V}) \leq 2\beta\mathbb{E}g(\hat{\mathbf{V}}) + 2\beta\tau$$

which ensures that (29) holds. □

The following nice result is Proposition 3 of Hechner and Heinkel [7].

Lemma 5 (Hechner and Heinkel [7]). *Let* $p > 1$ *and let* $\{X_n;\ n \geq 1\}$ *be a sequence of independent copies of a* **B**-*valued random variable X. Write*

$$u_n = \inf\left\{t:\ \mathbb{P}(\|X\| > t) < \frac{1}{n}\right\},\ n \geq 1.$$

Then the following three statements are equivalent:

(i) $\displaystyle\int_0^\infty \mathbb{P}^{1/p}(\|X\| > t)dt < \infty;$

(ii) $\displaystyle\sum_{n=1}^\infty \frac{u_n}{n^{1+1/p}} < \infty;$

(iii) $\displaystyle\sum_{n=1}^\infty \frac{1}{n^{1+1/p}}\mathbb{E}\left(\max_{1\leq k\leq n}\|X_k\|\right) < \infty.$

Proof of Theorem 3. Firstly, we see that (1) immediately implies that (2) holds. Thus, by Lemma 3, for $0 < q < \infty$, each of (1) and (2) implies that (4) holds.

Secondly, we show that (1) follows from (2) and (3). To see this, by Lemma 4, we conclude that (1) is equivalent to

$$\sum_{n=1}^\infty \frac{1}{n}\mathbb{E}\left(\frac{\|\hat{S}_n\|}{n^{1/p}}\right)^q < \infty. \tag{30}$$

Since (2) ensures that (24) holds, by (10) and (11) of Theorem 7, we see that (30) holds if and only if

$$\mathbb{E}\left(\sup_{n\geq 1} b_n \left\|\hat{X}_n\right\|^q\right) < \infty, \tag{31}$$

where $b_n = \sum_{k=n}^\infty n^{-1-q/p}, n \geq 1$. Since $\lim_{n\to\infty} b_n/n^{-q/p} = p/q$, we conclude that (31) is equivalent to

$$\mathbb{E}\left(\sup_{n\geq 1}\frac{\left\|\hat{X}_n\right\|^p}{n}\right)^{q/p} = \mathbb{E}\left(\sup_{n\geq 1}\frac{\left\|\hat{X}_n\right\|^q}{n^{q/p}}\right) < \infty. \tag{32}$$

Note that we have from (3) that

$$\begin{cases} \mathbb{E}\|X\|^p < \infty & \text{if } 0 < q < p, \\ \\ \mathbb{E}\|X\|^p \ln(1 + \|X\|) < \infty & \text{if } q = p, \\ \\ \mathbb{E}\|X\|^q < \infty & \text{if } q > p \end{cases}$$

which is equivalent to

$$
\begin{cases}
\mathbb{E}\|\hat{X}\|^p < \infty & \text{if } 0 < q < p, \\[2mm]
\mathbb{E}\|\hat{X}\|^p \log(1 + \|\hat{X}\|) < \infty & \text{if } q = p, \\[2mm]
\mathbb{E}\|\hat{X}\|^q < \infty & \text{if } q > p.
\end{cases}
\tag{33}
$$

Burkholder [2] proved that (33) and (32) are equivalent. We thus conclude that (1) follows from (2) and (3).

Since (1) and (30) are equivalent, (30) implies that (32) holds, and (32) and (33) are equivalent, we conclude that (3) follows from (1) if $q \geq p$.

We now show that (1) implies that (3) holds if $0 < q < p$. By the Lévy inequality, we have that, for every $n \geq 1$ and all $t \geq 0$,

$$
\mathbb{P}\left(\max_{1 \leq k \leq n} \left\|\hat{X}_k\right\|^q > t\right) = \mathbb{P}\left(\max_{1 \leq k \leq n} \left\|\hat{X}_k\right\| > t^{1/q}\right)
$$
$$
\leq 2\mathbb{P}\left(\left\|\hat{S}_n\right\| > t^{1/q}\right) = 2\mathbb{P}\left(\left\|\hat{S}_n\right\|^q > t\right),
$$

which ensures that, for every $n \geq 1$,

$$
\mathbb{E}\left(\max_{1 \leq k \leq n} \left\|\hat{X}_k\right\|^q\right) \leq 2\mathbb{E}\left\|\hat{S}_n\right\|^q.
\tag{34}
$$

Since (1) and (30) are equivalent, it now follows from (1) and (34) that

$$
\sum_{n=1}^{\infty} \frac{1}{n^{1+1/p_1}} \mathbb{E}\left(\max_{1 \leq k \leq n} Y_k\right) = \sum_{n=1}^{\infty} \frac{1}{n^{1+q/p}} \mathbb{E}\left(\max_{1 \leq k \leq n} \left\|\hat{X}_k\right\|^q\right) < \infty,
\tag{35}
$$

where $p_1 = p/q > 1$ (since $0 < q < p$) and $Y = \|\hat{X}\|^q$, $Y_n = \left\|\hat{X}_n\right\|^q$, $n \geq 1$. By Lemma 5, (35) is equivalent to

$$
\int_0^{\infty} \mathbb{P}^{1/p_1}(Y > t) dt < \infty,
$$

i.e.,

$$
\int_0^{\infty} \mathbb{P}^{q/p}\left(\|X - X'\|^q > t\right) dt < \infty.
\tag{36}
$$

Let $m(\|X\|)$ denote a median of $\|X\|$. Since, by the weak symmetrization inequality, we have that

$$\mathbb{P}(|\,\|X\| - m(\|X\|)| > t) \leq 2\mathbb{P}\left(|\,\|X\| - \|X'\|\,| > t\right)$$

$$\leq 2\mathbb{P}\left(\|X - X'\| > t\right) \leq 4\mathbb{P}\left(\|X\| > \frac{t}{2}\right) \ \forall\, t \geq 0,$$

we conclude that (36) is equivalent to

$$\int_0^\infty \mathbb{P}^{q/p}\left(\|X\|^q > t\right) dt < \infty,$$

i.e., (3) holds if $0 < q < p$.

Finally, by Lemma 3, (2) implies that $\mathbb{E}\|X\|^p < \infty$. Then (32) holds and hence (30) holds if $0 < q < p$. Since, under (2) , (1) and (30) are equivalent, we see that (1) follows from (2) if $0 < q < p$. □

Acknowledgements The authors are grateful to the referees for carefully reading the manuscript and for providing many constructive comments and suggestions which enabled them to improve the paper. In particular, one of the referees so kindly pointed out to us the relationship of our Theorem 3 to Theorem 2.4.1 of the Doctoral Thesis of Florian Hechner [6] prepared for l'Université de Strasbourg, France. The research of Deli Li was partially supported by a grant from the Natural Sciences and Engineering Research Council of Canada and the research of Yongcheng Qi was partially supported by NSF Grant DMS-1005345.

References

1. Azlarov, T.A., Volodin, N.A.: Laws of large numbers for identically distributed Banach-space valued random variables. Teor. Veroyatnost. i Primenen. **26**, 584–590 (1981). In Russian. English translation in Theory Probab. Appl. **26**, 573–580 (1981)
2. Burkholder, D.L.: Successive conditional expectations of an integrable function. Ann. Math. Statist. **33**, 887–893 (1962)
3. Chow, Y.S., Teicher, H.: Probability Theory: Independence, Interchangeability, Martingales, 3rd edn. Springer, New York (1997)
4. de Acosta, A.: Inequalities for B-valued random vectors with applications to the strong law of large numbers. Ann. Probab. **9**, 157–161 (1981)
5. Hechner, F.: Comportement asymptotique de sommes de Cesàro aléatoires. C. R. Math. Acad. Sci. Paris **345**, 705–708 (2007)
6. Hechner, F.: Lois Fortes des Grands Nombres et Martingales Asymptotiques. Doctoral thesis, l'Université de Strasbourg, France (2009)
7. Hechner, F., Heinkel, B.: The Marcinkiewicz-Zygmund LLN in Banach spaces: a generalized martingale approach. J. Theor. Probab. **23**, 509–522 (2010)
8. Hoffmann-Jørgensen, J.: Sums of independent Banach space valued random variables. Studia Math. **52**, 159–186 (1974)
9. Kolmogoroff, A.: Sur la loi forte des grands nombres. C. R. Acad. Sci. Paris Sér. Math. **191**, 910–912 (1930)
10. Ledoux, M., Talagrand, M.: Probability in Banach Spaces: Isoperimetry and Processes. Springer, Berlin (1991)
11. Lévy, P.: Théorie de L'addition des Variables Aléatoires. Gauthier-Villars, Paris (1937)
12. Li, D., Qi, Y., Rosalsky, A.: A refinement of the Kolmogorov-Marcinkiewicz-Zygmund strong law of large numbers. J. Theoret. Probab. **24**, 1130–1156 (2011)

13. Li, D., Rosalsky, A.: New versions of some classical stochastic inequalities. Stoch. Anal. Appl. **31**, 62–79 (2013)
14. Marcinkiewicz, J., Zygmund, A.: Sur les fonctions indépendantes. Fund. Math. **29**, 60–90 (1937)
15. Mourier, E.: Eléments aléatoires dans un espace de Banach. Ann. Inst. H. Poincaré **13**, 161–244 (1953)

Quenched Invariance Principles via Martingale Approximation

Magda Peligrad

1 Introduction and General Considerations

In recent years there has been an intense effort towards a better understanding of the structure and asymptotic behavior of stochastic processes. For dependent sequences there are two basic techniques: approximation with independent random variables or with martingales. Each of these methods have its own strength. On one hand the processes that can be treated by coupling with an independent sequence exhibit faster rates of convergence in various limit theorems; on the other hand the class of processes that can be treated by a martingale approximation is larger. There are plenty of processes that benefit from approximation with a martingale. Examples are: linear processes with martingale innovations, functions of linear processes, reversible Markov chains, normal Markov chains, various dynamical systems and the discrete Fourier transform of general stationary sequences. A martingale approximation provides important information about these structures because of their rich properties. They satisfy a broad range of inequalities, they can be embedded into Brownian motion and they satisfy various asymptotic results such as the functional conditional central limit theorem and the law of the iterated logarithm. Moreover, martingale approximation provides a simple and unified approach to asymptotic results for many dependence structures. For all these reasons, in recent years martingale approximation, "coupling with a martingale", has gained a prominent role in analyzing dependent data. This is also due to important developments by Liverani [30], Maxwell-Woodroofe [31], Derriennic-Lin [15–17] Wu-Woodroofe [51] and developments by Peligrad-Utev [35], Zhao-Woodroofe [49, 50], Volný [46],

M. Peligrad (✉)
Department of Mathematical Sciences, University of Cincinnati,
PO Box 210025, Cincinnati, OH 45221-0025, USA
e-mail: peligrm@ucmail.uc.edu

© Springer Science+Business Media New York 2015
D. Dawson et al. (eds.), *Asymptotic Laws and Methods in Stochastics*,
Fields Institute Communications 76, DOI 10.1007/978-1-4939-3076-0_9

Peligrad-Wu [37] among others. Many of these new results, originally designed for Markov operators, (see Kipnis-Varadhan [29] and Derriennic-Lin [16]) have made their way into limit theorems for stochastic processes.

This method has been shown to be well suited to transport from the martingale to the stationary process either the conditional central limit theorem or conditional invariance principle in probability. As a matter of fact, papers by Dedecker-Merlevède-Volný [13], Zhao and Woodroofe [50], Gordin and Peligrad [24], point out characterizations of stochastic processes that can be approximated by martingales in quadratic mean. These results are useful to treat evolutions in "annealed" media.

In this survey we address the question of limit theorems started at a point for almost all points. These types of results are also known under the name of quenched limit theorems or almost sure conditional invariance principles. Limit theorems for stochastic processes that do not start from equilibrium is timely and motivated by recent development in evolutions in quenched random environment, random walks in random media, for instance as in Rassoul-Agha and Seppäläinen [40]. Moreover recent discoveries by Volný and Woodroofe [47] show that many of the central limit theorems satisfied by classes of stochastic processes in equilibrium, fail to hold when the processes are started from a point. Special attention will be devoted to normal and reversible Markov chains and several results and open problems will be pointed out. These results are very important since reversible Markov chains have applications to statistical mechanics and to Metropolis Hastings algorithms used in Monte Carlo simulations. The method of proof of this type of limiting results are approximations with martingale in an almost sure sense.

The field of limit theorems for stochastic processes is closely related to ergodic theory and dynamical systems. All the results for stationary sequences can be translated in the language of Markov operators.

2   Limit Theorems Started at a Point via Martingale Approximation

In this section we shall use the framework of strictly stationary sequences adapted to a stationary filtrations that can be introduced in several equivalent ways, either by using a measure preserving transformation or as a functional of a Markov chain with a general state space. It is just a difference of language to present the theory in terms of stationary processes or functionals of Markov chains.

Let $(\Omega, \mathscr{A}, \mathbb{P})$ be a probability space, and $T : \Omega \mapsto \Omega$ be a bijective bimeasurable transformation preserving the probability $\mathbb{P}$. A set $A \in \mathscr{A}$ is said to be invariant if $T(A) = A$. We denote by $\mathscr{I}$ the σ-algebra of all invariant sets. The transformation T is ergodic with respect to probability $\mathbb{P}$ if each element of $\mathscr{I}$ has measure 0 or 1. Let $\mathscr{F}_0$ be a σ-algebra of $\mathscr{A}$ satisfying $\mathscr{F}_0 \subseteq T^{-1}(\mathscr{F}_0)$ and define the nondecreasing filtration $(\mathscr{F}_i)_{i \in \mathbb{Z}}$ by $\mathscr{F}_i = T^{-i}(\mathscr{F}_0)$. Let X_0 be a $\mathscr{F}_0$-measurable, square integrable and centered random variable. Define the sequence $(X_i)_{i\in\mathbb{Z}}$ by $X_i = X_0 \circ T^i$. Let $S_n = X_1 + \cdots + X_n$. For $p \geq 1$, $\|.\|_p$ denotes the norm in $\mathbb{L}_p(\Omega, \mathscr{A}, \mathbb{P})$. In the sequel we shall denote by $\mathbb{E}_0(X) = \mathbb{E}(X|\mathscr{F}_0)$.

The conditional central limit theorem plays an essential role in probability theory and statistics. It asserts that the central limit theorem holds in probability under the measure conditioned by the past of the process. More precisely this means that for any function f which is continuous and bounded we have

$$\mathbb{E}_0(f(S_n/\sqrt{n})) \to \mathbb{E}(f(\sigma N)) \text{ in probability,} \tag{1}$$

where N is a standard normal variable and σ is a positive constant. Usually we shall have the interpretation $\sigma^2 = \lim_{n\to\infty} \text{var}(S_n)/n$.

This conditional form of the CLT is a stable type of convergence that makes possible the change of measure with a majorizing measure, as discussed in Billingsley [1], Rootzén [44], and Hall and Heyde [25]. Furthermore, if we consider the associated stochastic process

$$W_n(t) = \frac{1}{\sqrt{n}} S_{[nt]},$$

where $[x]$ denotes the integer part of x, then the conditional CLT implies the convergence of the finite dimensional distributions of $W_n(t)$ to those of $\sigma W(t)$ where $W(t)$ is the standard Brownian Motion; this constitutes an important step in establishing the functional CLT (FCLT). Note that $W_n(t)$ belongs to the space $D[0, 1]$, the set of functions on $[0, 1]$ which are right continuous and have left hands limits. We endow this space with the uniform topology.

By the conditional functional central limit theorem we understand that for any function f continuous and bounded on $D[0, 1]$ we have

$$\mathbb{E}_0(f(W_n)) \to \mathbb{E}(f(\sigma W)) \text{ in probability.} \tag{2}$$

There is a considerable amount of research concerning this problem. We mention papers by Dedecker and Merlevède [10], Wu and Woodroofe [51] and Zhao and Woodroofe [50] among others.

The quenched versions of these theorems are obtained by replacing the convergence in probability by convergence almost sure. In other words the almost sure conditional theorem states that, on a set of probability one, for any function f which is continuous and bounded we have

$$\mathbb{E}_0(f(S_n/\sqrt{n})) \to \mathbb{E}(f(\sigma N)), \tag{3}$$

while by almost sure conditional functional central limit theorem we understand that, on a set of probability one, for any function f continuous and bounded on $D[0, 1]$ we have

$$\mathbb{E}_0(f(W_n)) \to \mathbb{E}(f(\sigma W)). \tag{4}$$

We introduce now the stationary process as a functional of a Markov chain.

We assume that $(\xi_n)_{n\in\mathbb{Z}}$ is a stationary ergodic Markov chain defined on a probability space $(\Omega, \mathscr{F}, \mathbb{P})$ with values in a Polish space $(S, \mathscr{S})$. The marginal distribution is denoted by $\pi(A) = \mathbb{P}(\xi_0 \in A)$, $A \in \mathscr{S}$. Next, let $\mathbb{L}_2^0(\pi)$ be the set of functions h such that $||h||_{2,\pi}^2 = \int h^2 d\pi < \infty$ and $\int h d\pi = 0$. Denote by $\mathscr{F}_k$ the σ-field generated by ξ_j with $j \leq k$, $X_j = h(\xi_j)$. Notice that any stationary sequence $(Y_k)_{k\in\mathbb{Z}}$ can be viewed as a function of a Markov process $\xi_k = (Y_j; j \leq k)$ with the function $g(\xi_k) = Y_k$. Therefore the theory of stationary processes can be imbedded in the theory of Markov chains.

In this context by the central limit theorem started at a point (quenched) we understand the following fact: let $\mathbb{P}^x$ be the probability associated with the process started from x and let $\mathbb{E}^x$ be the corresponding expectation. Then, for π-almost every x, for every continuous and bounded function f,

$$\mathbb{E}^x(f(S_n/\sqrt{n})) \to \mathbb{E}(f(\sigma N)). \tag{5}$$

By the functional CLT started at a point we understand that, for π-almost every x, for every function f continuous and bounded on $D[0, 1]$,

$$\mathbb{E}^x(f(W_n)) \to \mathbb{E}(f(\sigma W)). \tag{6}$$

where, as before W is the standard Brownian motion on $[0, 1]$.

It is remarkable that a martingale with square integrable stationary and ergodic differences satisfies the quenched CLT in its functional form. For a complete and careful proof of this last fact we direct to Derriennic and Lin ([15], page 520). This is the reason why a fruitful approach to find classes of processes for which quenched limit theorems hold is to approximate partial sums by a martingale.

The martingale approximation as a tool for studying the asymptotic behavior of the partial sums S_n of stationary stochastic processes goes back to Gordin [22] who proposed decomposing the original stationary sequence into a square integrable stationary martingale $M_n = \sum_{i=1}^n D_i$ adapted to $(\mathscr{F}_n)$, such that $S_n = M_n + R_n$ where R_n is a telescoping sum of random variables, with the basic property that $\sup_n ||R_n||_2 < \infty$.

For proving conditional CLT for stationary sequences, a weaker form of martingale approximation was pointed out by many authors (see for instance Merlevède-Peligrad-Utev [32], for a survey).

An important step forward was the result by Heyde [28] who found sufficient conditions for the decomposition

$$S_n = M_n + R_n \text{ with } R_n/\sqrt{n} \to 0 \text{ in } \mathbb{L}_2. \tag{7}$$

Recently, papers by Dedecker-Merlevède-Volný [13] and by Zhao-Woodroofe [50] deal with necessary and sufficient conditions for martingale approximation with an error term as in (7).

The approximation of type (7) is important since it makes possible to transfer from martingale the conditional CLT defined in (1), where $\sigma = ||D_0||_2$.

The theory was extended recently in Gordin-Peligrad [24] who developed necessary and sufficient conditions for a martingale decomposition with the error term satisfying

$$\max_{1 \leq j \leq n} |S_j - M_j|/\sqrt{n} \to 0 \text{ in } \mathbb{L}_2. \tag{8}$$

This approximation makes possible the transport from the martingale to the stationary process the conditional functional central limit theorem stated in (2). These results were surveyed in Peligrad [38].

The martingale approximation of the type (8) brings together many disparate examples in probability theory. For instance, it is satisfied under Hannan [26, 27] and Heyde [28] projective condition.

$$\mathbb{E}(X_0|\mathscr{F}_{-\infty}) = 0 \quad \text{almost surely and} \quad \sum_{i=1}^{\infty} ||\mathbb{E}_{-i}(X_0) - \mathbb{E}_{-i-1}(X_0)||_2 < \infty; \tag{9}$$

It is also satisfied for classes of mixing processes; additive functionals of Markov chains with normal or symmetric Markov operators.

A very important question is to establish quenched version of conditional CLT and conditional FCLT, i.e. the invariance principles as in (3) and also in (4) (or equivalently as in (5) and also in (6)). There are many examples of stochastic processes satisfying (8) for which the conditional CLT does not hold in the almost sure sense. For instance condition (9) is not sufficient for (3) as pointed out by Volný and Woodroofe [47]. In order to transport from the martingale to the stationary process the almost sure invariance principles the task is to investigate the approximations of types (7) or (8) with an error term well adjusted to handle this type of transport. These approximations should be of the type, for every $\varepsilon > 0$

$$\mathbb{P}_0[|S_n - M_n|/\sqrt{n} > \varepsilon] \to 0 \text{ } a.s. \text{ or } \mathbb{P}_0[\max_{1 \leq i \leq n} |S_i - M_i|/\sqrt{n} > \varepsilon] \to 0 \text{ } a.s. \tag{10}$$

where $(M_n)_n$ is a martingale with stationary and ergodic differences and we used the notation $\mathbb{P}_0(A) = \mathbb{P}(A|\mathscr{F}_0)$. They are implied in particular by stronger approximations such that

$$|S_n - M_n|/\sqrt{n} \to 0 \text{ } a.s. \text{ or } \max_{1 \leq i \leq n} |S_i - M_i|/\sqrt{n} \to 0 \text{ } a.s.$$

Approximations of these types have been considered in papers by Zhao-Woodroofe [49], Cuny [5], Merlevède-Peligrad M.-Peligrad C. [34] among others.

In the next subsection we survey resent results and point out several classes of stochastic processes for which approximations of the type (10) hold.

For cases where a stationary martingale approximation does not exist or cannot be pointed out, a nonstationary martingale approximation is a powerful tool. This method was occasionally used to analyze a stochastic process. Many ideas are helpful in this situation ranging from simple projective decomposition of sums as in Gordin and Lifshitz [23] to more sophisticated tools. One idea is to divide the variables into blocks and then to approximate the sums of variables in each block by a martingale difference, usually introducing a new parameter, the block size, and changing the filtration. This method was successfully used in the literature by Philipp-Stout [39], Shao [45], Merlevède-Peligrad [33], among others. Alternatively, one can proceed as in Wu-Woodroofe [51], who constructed a nonstationary martingale approximation for a class of stationary processes without partitioning the variables into blocks.

Recently Dedecker-Merlevède-Peligrad [14] used a combination of blocking technique and a row-wise stationary martingale decomposition in order to enlarge the class of random variables known to satisfy the quenched invariance principles. To describe this approach, roughly speaking, one considers an integer $m = m(n)$ large but such that $n/m \to \infty$. Then one forms the partial sums in consecutive blocks of size m, $Y_j^n = X_{m(j-1)+1} + \cdots + X_{mj}$, $1 \le j \le k$, $k = [n/m]$. Finally, one considers the decomposition

$$S_n = M_n^n + R_n^n, \tag{11}$$

where $M_n^n = \sum_{j=1}^{n} D_j^n$, with $D_j^n = Y_j^n - E(Y_j^n | \mathscr{F}_{m(j-1)})$ a triangular array of row-wise stationary martingale differences.

2.1 Functional Central Limit Theorem Started at a Point Under Projective Criteria

We have commented that condition (9) is not sufficient for the validity of the almost sure CLT started from a point. Here is a short history of the quenched CLT under projective criteria. A result in Borodin and Ibragimov ([2], ch.4, section 8) states that if $||\mathbb{E}_0(S_n)||_2$ is bounded, then the CLT in its functional form started at a point (4) holds. Later, Derriennic-Lin [15–17] improved on this result imposing the condition $||\mathbb{E}_0(S_n)||_2 = O(n^{1/2-\epsilon})$ with $\epsilon > 0$ (see also Rassoul-Agha and Seppäläinen [41]). A step forward was made by Cuny [5] who improved the condition to $||\mathbb{E}_0(S_n)||_2 = O(n^{1/2}(\log n)^{-2}(\log\log n)^{-1-\delta})$ with $\delta > 0$, by using sharp results on ergodic transforms in Gaposhkin [21].

We shall describe now the recent progress made on the functional central limit theorem started at a point under projective criteria. We give here below three classes of stationary sequences of centered square integrable random variables for which both quenched central limit theorem and its quenched functional form given in (3) and (4) hold with $\sigma^2 = \lim_{n \to \infty} \text{var}(S_n)/n$, provided the sequences are ergodic. If the sequences are not ergodic then then the results still hold but with σ^2 replaced by the random variable η described as $\eta = \lim_{n \to \infty} \mathbb{E}(S_n^2|\mathscr{I})/n$ and $\mathbb{E}(\eta) = \sigma^2$. For simplicity we shall formulate the results below only for ergodic sequences.

1. **Hannan-Heyde projective criterion.** Cuny-Peligrad [6] (see also Volný-Woodroofe [48]) showed that (3) holds under the condition

$$\frac{\mathbb{E}(S_n|\mathscr{F}_0)}{\sqrt{n}} \to 0 \quad \text{almost surely and} \quad \sum_{i=1}^{\infty} \|\mathbb{E}_{-i}(X_0) - \mathbb{E}_{-i-1}(X_0)\|_2 < \infty.$$

$$(12)$$

The functional form of this result was established in Cuny-Volný [8].

2. **Maxwell and Woodroofe condition.** The convergence in (4) holds under Maxwell-Woodroofe [31] condition,

$$\sum_{k=1}^{\infty} \frac{\|\mathbb{E}_0(S_k)\|_2}{k^{3/2}} < \infty,$$

$$(13)$$

as recently shown in Cuny-Merlevède [7]. In particular both conditions (12) and (13) and is satisfied if

$$\sum_{k=1}^{\infty} \frac{\|\mathbb{E}_0(X_k)\|_2}{k^{1/2}} < \infty.$$

$$(14)$$

3. **Dedecker-Rio condition.** In a recent paper Dedecker-Merlevède-Peligrad [14] proved (4) under the condition

$$\sum_{k \geq 0} \|X_0 \mathbb{E}_0(X_k)\|_1 < \infty.$$

$$(15)$$

The first two results were proved using almost sure martingale approximation of type (10). The third one was obtained using the large block method described in (11).

Papers by Durieu-Volný [19] and Durieu [20] suggest that conditions (12), (13) and (15) are independent. They have different areas of applications and they lead to optimal results in all these applications. Condition (12) is well adjusted for linear processes. It was shown in Peligrad and Utev [35] that the Maxwell-Woodroofe condition (13) is satisfied by ρ-mixing sequences with logarithmic rate of convergence to 0. Dedecker-Rio [9] have shown that condition (15) is verified for strongly mixing processes under a certain condition combining the tail probabilities of the individual summands with the size of the mixing coefficients. For example, one needs a polynomial rate on the strong mixing coefficients when moments higher than two are available. However, the classes described by projection conditions have a much larger area of applications than mixing sequences. They can be verified by linear processes and dynamical systems that satisfy only weak mixing conditions (Dedecker-Prieur [11, 12], Dedecker-Merlevède-Peligrad [14] among others). More details about the applications are given in Sect. 3.

Certainly, these projective conditions can easily be formulated in the language of Markov operators by using the fact that $\mathbb{E}_0(X_k) = Q(f)(\xi_0)$. In this language $\mathbb{E}_0(S_k) = (Q + Q^2 + \cdots + Q^k)(f)(\xi_0)$.

2.2   Functional Central Limit Theorem Started at a Point for Normal and Reversible Markov Chains

In 1986 Kipnis and Varadhan proved the functional form of the central limit theorem as in (2) for square integrable mean zero additive functionals $f \in \mathbb{L}_2^0(\pi)$ of stationary reversible ergodic Markov chains $(\xi_n)_{n\in\mathbb{Z}}$ with transition function $Q(\xi_0, A) = P(\xi_1 \in A|\xi_0)$ under the natural assumption $var(S_n)/n$ is convergent to a positive constant. This condition has a simple formulation in terms of spectral measure ρ_f of the function f with respect to self-adjoint operator Q associated to the reversible Markov chain, namely

$$\int_{-1}^{1} \frac{1}{1-t} \rho_f(dt) < \infty. \tag{16}$$

This result was established with respect to the stationary probability law of the chain. (Self-adjoint means $Q = Q^*$, where Q also denotes the operator $Qf(\xi) = \int f(x)Q(\xi, dx)$; Q^* is the adjoint operator defined by $< Qf, g >=< f, Q^*g >$, for every f and g in $\mathbb{L}_2(\pi)$).

The central limit theorem (1) for stationary and ergodic Markov chains with normal operator Q ($QQ^* = Q^*Q$), holds under a similar spectral assumption, as discovered by Gordin-Lifshitz [23] (see also and Borodin-Ibragimov [2], ch. 4 sections 7–8). A sharp sufficient condition in this case in terms of spectral measure is

$$\int_D \frac{1}{|1-z|} \rho_f(dz) < \infty. \tag{17}$$

where D is the unit disk.

Examples of reversible Markov chains frequently appear in the study of infinite systems of particles, random walks or processes in random media. A simple example of a normal Markov chain is a random walk on a compact group. Other important example of reversible Markov chain is the extremely versatile (independent) Metropolis Hastings Algorithm which is the modern base of Monte Carlo simulations.

An important problem is to investigate the validity of the almost sure central limit theorem started at a point for stationary ergodic normal or reversible Markov chains. As a matter of fact, in their remark (1.7), Kipnis-Varadhan [29] raised the question if their result also holds with respect to the law of the Markov chain started from x, for almost all x, as in (6).

Conjecture. For any square integrable mean 0 function of reversible Markov chains satisfying condition (16) the functional central limit theorem started from a point holds for almost all points. The same question is raised for continuous time reversible Markov chains.

The answer to this question for reversible Markov chains with continuous state space is still unknown and has generated a large amount of research. The problem of quenched CLT for normal stationary and ergodic Markov chains was considered by Derriennic-Lin [15] and Cuny [5], among others, under some reinforced assumptions on the spectral condition. Concerning normal Markov chains, Derriennic-Lin [15] pointed out that the central limit theorem started at a point does not hold for almost all points under condition (17). Furthermore, Cuny-Peligrad [6] proved that there is a stationary and ergodic normal Markov chain and a function $f \in \mathbb{L}_2^0(\pi)$ such that

$$\int_D \frac{|\log(|1-z|)\log\log(|1-z|)|}{|1-z|} \rho_f(dz) < \infty$$

and such that the central limit theorem started at a point fails, for π-almost all starting points.

However the condition

$$\int_{-1}^1 \frac{(\log^+|\log(1-t)|)^2}{1-t} \rho_f(dt) < \infty, \tag{18}$$

is sufficient to imply central limit theorem started at a point (5) for reversible Markov chains for π-almost all starting points. Note that this condition is a slight reinforcement of condition (17).

It is interesting to note that by Cuny ([5], Lemma 2.1), condition (18) is equivalent to the following projective criterion

$$\sum_n \frac{(\log\log n)^2 \|\mathbb{E}_0(S_n)\|_2^2}{n^2} < \infty. \tag{19}$$

Similarly, condition (17) in the case where Q is symmetric, is equivalent to

$$\sum_n \frac{\|\mathbb{E}_0(S_n)\|_2^2}{n^2} < \infty. \tag{20}$$

3 Applications

Here we list several classes of stochastic processes satisfying quenched CLT and quenched invariance principles. They are applications of the results given in Sect. 2.

3.1 Mixing Processes

In this subsection we discuss two classes of mixing sequences which are extremely relevant for the study of Markov chains, Gaussian processes and dynamical systems.

We shall introduce the following mixing coefficients: For any two σ-algebras $\mathscr{A}$ and $\mathscr{B}$ define the strong mixing coefficient $\alpha(\mathscr{A}, \mathscr{B})$:

$$\alpha(\mathscr{A}, \mathscr{B}) = \sup\{|\mathbb{P}(A \cap B) - \mathbb{P}(A)\mathbb{P}(B)|; A \in \mathscr{A}, B \in \mathscr{B}\}.$$

The ρ-mixing coefficient, known also under the name of maximal coefficient of correlation $\rho(\mathscr{A}, \mathscr{B})$ is defined as:

$$\rho(\mathscr{A}, \mathscr{B}) = \sup\{\operatorname{Cov}(X, Y)/\|X\|_2\|Y\|_2 \, : \, X \in \mathbb{L}_2(\mathscr{A}), \, Y \in \mathbb{L}_2(\mathscr{B})\}.$$

For the stationary sequence of random variables $(X_k)_{k \in \mathbb{Z}}$, we also define $\mathscr{F}_m^n$ the σ-field generated by X_i with indices $m \leq i \leq n$, $\mathscr{F}^n$ denotes the σ-field generated by X_i with indices $i \geq n$, and $\mathscr{F}_m$ denotes the σ-field generated by X_i with indices $i \leq m$. The sequences of coefficients $\alpha(n)$ and $\rho(n)$ are then defined by

$$\alpha(n) = \alpha(\mathscr{F}_0, \mathscr{F}^n), \; \rho(n) = \rho(\mathscr{F}_0, \mathscr{F}^n).$$

An equivalent definition for $\rho(n)$ is

$$\rho(n) = \sup\{\|\mathbb{E}(Y|\mathscr{F}_0)\|_2/\|Y\|_2 : \, Y \in \mathbb{L}_2(\mathscr{F}^n), \mathbb{E}(Y) = 0\}. \tag{21}$$

Finally we say that the stationary sequence is strongly mixing if $\alpha(n) \to 0$ as $n \to \infty$, and ρ-mixing if $\rho(n) \to 0$ as $n \to \infty$. It should be mentioned that a ρ-mixing sequence is strongly mixing. Furthermore, a stationary strongly mixing sequence is ergodic. For an introduction to the theory of mixing sequences we direct the reader to the books by Bradley [3].

In some situations weaker forms of strong and ρ-mixing coefficients can be useful, when $\mathscr{F}^n$ is replaced by the sigma algebra generated by only one variable, X_n, denoted by $\mathscr{F}_n^n$. We shall use the notations $\tilde{\alpha}(n) = \alpha(\mathscr{F}_0, \mathscr{F}_n^n)$ and $\tilde{\rho}(n) = \rho(\mathscr{F}_0, \mathscr{F}_n^n)$.

By verifying the conditions in Sect. 3, we can formulate:

Theorem 1. *Let* $(X_n)_{n\in\mathbb{Z}}$ *be a stationary and ergodic sequence of centered square integrable random variables. The quenched CLT and its quenched functional form as in (3) and (4) hold with* $\sigma^2 = \lim_{n\to\infty} var(S_n)/n$ *under one of the following three conditions:*

$$\sum_{k=1}^{\infty} \frac{\tilde{\rho}(k)}{\sqrt{k}} < \infty. \tag{22}$$

$$\sum_{k=1}^{\infty} \frac{\rho(k)}{k} < \infty. \tag{23}$$

$$\sum_{k=1}^{\infty} \int_0^{\tilde{\alpha}(k)} Q^2(u)du < \infty, \tag{24}$$

where Q *denotes the generalized inverse of the function* $t \to \mathbb{P}(|X_0| > t)$.

We mention that under condition (23) the condition of ergodicity is redundant. Also if (24) holds with $\tilde{\alpha}(k)$ replaced by $\alpha(k)$, then the sequence is again ergodic.

In order to prove this theorem under (22) one verifies condition (14) via the estimate

$$\mathbb{E}(\mathbb{E}_0(X_k))^2 = \mathbb{E}(X_k\mathbb{E}_0(X_k)) \le \tilde{\rho}(k)||X_0||_2^2,$$

which follows easily from the definition of $\tilde{\rho}$.

Condition (23) is used to verify condition (13). This was verified in the Peligrad-Utev-Wu [36] via the inequalities

$$||\mathbb{E}(S_{2^r+1}|\mathscr{F}_0)||_2 \le c \sum_{j=0}^{r} 2^{j/2}\rho(2^j)$$

and

$$\sum_{r=0}^{\infty} \frac{||\mathbb{E}(S_{2^r}|\mathscr{F}_0)||_2}{2^{r/2}} \le c \sum_{j=0}^{\infty} \rho(2^j) < \infty. \tag{25}$$

Furthermore (25) easily implies (13). For more details on this computation we also direct the reader to the survey paper by Merlevède-Peligrad-Utev [32].

To get the quenched results under condition (24) the condition (15) is verified via the following identity taken from Dedecker-Rio ([9], (6.1))

$$\mathbb{E}|X_0\mathbb{E}(X_k|\mathscr{F}_0)| = \text{Cov}(|X_0|(I_{\{\mathbb{E}(X_k|\mathscr{F}_0)>0\}} - I_{\{\mathbb{E}(X_k|\mathscr{F}_0)\le 0\}}), X_k). \tag{26}$$

By applying now Rio's [42] covariance inequality we obtain

$$\mathbb{E}|X_0\mathbb{E}(X_k|\mathscr{F}_0)| \leq c \int_0^{\tilde{\alpha}(k)} Q^2(u)du.$$

It is obvious that condition (22) requires a polynomial rate of convergence to 0 of $\tilde{\rho}(k)$; condition (23) requires only a logarithmic rate for $\rho(n)$. To comment about condition (24) it is usually used in the following two forms:

- either the variables are almost sure bounded by a constant, and then the requirement is $\sum_{k=1}^{\infty} \tilde{\alpha}(k) < \infty$.
- the variables have finite moments of order $2 + \delta$ for some $\delta > 0$, and then the condition on mixing coefficients is $\sum_{k=1}^{\infty} k^{2/\delta}\tilde{\alpha}(k) < \infty$.

3.2 Shift Processes

In this sub-section we apply condition (13) to linear processes which are not mixing in the sense of previous subsection. This class is known under the name of one-sided shift processes, also known under the name of Raikov sums.

Let us consider a Bernoulli shift. Let $\{\varepsilon_k; k \in \mathbb{Z}\}$ be an i.i.d. sequence of random variables with $\mathbb{P}(\varepsilon_1 = 0) = \mathbb{P}(\varepsilon_1 = 1) = 1/2$ and let

$$Y_n = \sum_{k=0}^{\infty} 2^{-k-1}\varepsilon_{n-k} \quad and \quad X_n = g(Y_n) - \int_0^1 g(x)dx,$$

where $g \in \mathbb{L}_2(0, 1)$, $(0, 1)$ being equipped with the Lebesgue measure.

By applying Proposition 3 in Maxwell and Woodroofe [31] for verifying condition (13), we see that if $g \in \mathbb{L}_2(0, 1)$ satisfies

$$\int_0^1 \int_0^1 [g(x) - g(y)]^2 \frac{1}{|x - y|} \left(\log \left[\log \frac{1}{|x - y|} \right] \right)^t dxdy < \infty \qquad (27)$$

for some $t > 1$, then (13) is satisfied and therefore (3) and (4) hold with $\sigma^2 = \lim_{n \to \infty} \text{var}(S_n)/n$. A concrete example of a map satisfying (27), pointed out in Merlevède-Peligrad-Utev [32] is

$$g(x) = \frac{1}{\sqrt{x}} \frac{1}{[1 + \log(2/x)]^4} \sin\left(\frac{1}{x}\right), \quad 0 < x < 1.$$

3.3   Random Walks on Orbits of Probability Preserving Transformation

The following example was considered in Derriennic-Lin [18] and also in Cuny-Peligrad [6]. Let us recall the construction.

Let τ be an invertible ergodic measure preserving transformation on $(S, \mathscr{A}, \pi)$, and denote by U, the unitary operator induced by τ on $\mathbb{L}_2(\pi)$. Given a probability $\nu = (p_k)_{k \in \mathbb{Z}}$ on $\mathbb{Z}$, we consider the Markov operator Q with invariant measure π, defined by

$$Qf = \sum_{k \in \mathbb{Z}} p_k f \circ \tau^k, \quad \text{for every } f \in \mathbb{L}_1(\pi).$$

This operator is associated to the transition probability

$$Q(x, A) = \sum_{k \in \mathbb{Z}} p_k \mathbf{1}_A(\tau^k s), \qquad s \in S, A \in \mathscr{A}.$$

We assume that ν is ergodic, i.e. the the group generated by $\{k \in \mathbb{Z} : p_k > 0\}$ is $\mathbb{Z}$. As shown by Derriennic-Lin [18], since τ is ergodic, Q is ergodic too. We assume ν is symmetric implying that the operator Q is symmetric.

Denote by Γ the unit circle. Define the Fourier transform of ν by $\varphi(\lambda) = \sum_{k \in \mathbb{Z}} p_k \lambda^k$, for every $\lambda \in \Gamma$. Since ν is symmetric, $\varphi(\lambda) \in [-1, 1]$, and if μ_f denotes the spectral measure (on Γ) of $f \in \mathbb{L}_2(\pi)$, relative to the unitary operator U, then, the spectral measure ρ_f (on $[-1, 1]$) of f, relative to the symmetric operator Q is given by

$$\int_{-1}^{1} \psi(s) \rho_f(ds) = \int_{\Gamma} \psi(\varphi(\lambda)) \mu_f(d\lambda),$$

for every positive Borel function ψ on $[-1, 1]$. Condition (19) is verified under the assumption

$$\int_{\Gamma} \frac{(\log^+ |\log(1 - \varphi(\lambda))|)^2}{1 - \varphi(\lambda)} \mu_f(d\lambda) < \infty.$$

and therefore (5) holds.

When ν is centered and admits a moment of order 2 (i.e. $\sum_{k \in \mathbb{Z}} k^2 p_k < \infty$), Derriennic and Lin [18] proved that the condition $\int_{\Gamma} \frac{1}{|1 - \varphi(\lambda)|} \mu_f(d\lambda) < \infty$, is sufficient for (5).

Let $a \in \mathbb{R} - \mathbb{Q}$, and let τ be the rotation by a on $\mathbb{R}/\mathbb{Z}$. Define a measure σ on $\mathbb{R}/\mathbb{Z}$ by $\sigma = \sum_{k \in \mathbb{Z}} p_k \delta_{ka}$. For that τ, the canonical Markov chain associated to Q is the random walk on $\mathbb{R}/\mathbb{Z}$ of law σ. In this setting, if $(c_n(f))$ denotes the Fourier coefficients of a function $f \in \mathbb{L}_2(\mathbb{R}/\mathbb{Z})$, condition (19) reads

$$\sum_{n \in \mathbb{Z}} \frac{(\log^+ |\log(1 - \varphi(e^{2i\pi na}))|)^2 |c_n(f)|^2}{1 - \varphi(e^{2i\pi na})} < \infty.$$

3.4 CLT Started from a Point for a Metropolis Hastings Algorithm

In this subsection we mention a standardized example of a stationary irreducible and aperiodic Metropolis-Hastings algorithm with uniform marginal distribution. This type of Markov chain is interesting since it can easily be transformed into Markov chains with different marginal distributions. Markov chains of this type are often studied in the literature from different points of view. See, for instance Rio [43].

Let $E = [-1, 1]$ and let υ be a symmetric atomless law on E. The transition probabilities are defined by

$$Q(x, A) = (1 - |x|)\delta_x(A) + |x|\upsilon(A),$$

where δ_x denotes the Dirac measure. Assume that $\theta = \int_E |x|^{-1}\upsilon(dx) < \infty$. Then there is a unique invariant measure

$$\pi(dx) = \theta^{-1}|x|^{-1}\upsilon\,(dx)$$

and the stationary Markov chain (γ_k) generated by $Q(x, A)$ and π is reversible and positively recurrent, therefore ergodic.

Theorem 2. *Let f be a function in $\mathbb{L}_2^0(\pi)$ satisfying $f(-x) = -f(x)$ for any $x \in E$. Assume that for some positive t, $|f| \leq g$ on $[-t, t]$ where g is an even positive function on E such that g is nondecreasing on $[0, 1]$, $x^{-1}g(x)$ is nonincreasing on $[0, 1]$ and*

$$\int_{[0,1]} [x^{-1}g(x)]^2 dx < \infty. \tag{28}$$

Define $X_k = f(\gamma_k)$. Then (5) holds.

Proof. Because the chain is Harris recurrent if the annealed CLT holds, then the CLT also holds for any initial distribution (see Chen [4]), in particular started at a point. Therefore it is enough to verify condition (20). Denote, as before, by $\mathbb{E}^x$ the expected value for the process started from $x \in E$. We mention first relation (4.6) in Rio [43]. For any $n \geq 1/t$

$$|\mathbb{E}^x(S_n(g))| \leq ng(1/n) + t^{-1}|f(x)| \text{ for any } x \in [-1, 1].$$

Then

$$|\mathbb{E}^x(S_n(g))|^2 \leq 2[ng(1/n)]^2 + 2t^{-2}|f(x)|^2 \text{ for any } x \in [-1, 1],$$

and so, for any $n \geq 1/t$

$$||\mathbb{E}^x(S_n)||_{2,\pi}^2 \leq 2[ng(1/n)]^2 + 2t^{-2}||f(x)||_{2,\pi}^2.$$

Now we impose condition (20) involving $||\mathbb{E}^x(S_n)||_{2,\pi}^2$, and note that

$$\sum_n \frac{[ng(1/n)]^2}{n^2} < \infty \text{ if and only if (28) holds.}$$

Acknowledgements The author would like to thank the referees for carefully reading the manuscript and for many useful suggestions that improved the presentation of this paper. This paper was Ssupported in part by a Charles Phelps Taft Memorial Fund grant, the NSA grant H98230-11-1-0135 and the NSF grant DMS-1208237.

References

1. Billingsley, P.: Convergence of Probability Measures. Wiley, New York (1968)
2. Borodin, A.N., Ibragimov, I.A.: Limit theorems for functionals of random walks. Trudy Mat. Inst. Steklov. **195** (1994). Transl. into English: Proc. Steklov Inst. Math. , **195**(2) (1995)
3. Bradley, R.C.: Introduction to Strong Mixing Conditions, vol. I, II, III. Kendrick Press, Heber City (2007)
4. Chen, X.: Limit theorems for functionals of ergodic Markov chains in general state space. Mem. Am. Math. Soc. **139**(664) (1999)
5. Cuny, C.: Pointwise ergodic theorems with rate and application to limit theorems for stationary processes. Stoch. Dyn. **11**, 135–155 (2011)
6. Cuny, C., Peligrad, M.: Central limit theorem started at a point for stationary processes and additive functional of reversible Markov Chains. J. Theoret. Probab. **25**, 171–188 (2012)
7. Cuny, C., Merlevède, F. On martingale approximations and the quenched weak invariance principle. Ann. Probab. **42**, 760–793 (2014)
8. Cuny, C., Volný, D.: A quenched invariance principle for stationary processes. ALEA **10**, 107–115 (2013)
9. Dedecker, J., Rio, E.: On the functional central limit theorem for stationary processes. Ann. Inst. H. Poincaré Probab. Statist. **36**, 1–34 (2000)
10. Dedecker, J., Merlevède, F.: Necessary and sufficient conditions for the conditional central limit theorem. Ann. Probab. **30**, 1044–1081 (2002)
11. Dedecker, J., Prieur, C.: Coupling for tau-dependent sequences and applications. J. Theoret. Probab. **17**, 861–885 (2004)
12. Dedecker, J., Prieur, C.: New dependence coefficients. Examples and applications to statistics. Probab. Theory Relat. Fields. **132**, 203–236 (2005)
13. Dedecker, J., Merlevède, F., Volný, D.: On the weak invariance principle for non-adapted stationary sequences under projective criteria. J. Theoret. Probab. **20**, 971–1004 (2007)

14. Dedecker, J., Merlevède, F., Peligrad, M.: A quenched weak invariance principle. Ann. Inst. H. Poincaré Probab. Statist. **50**, 872–898 (2014)
15. Derriennic, Y., Lin, M.: The central limit theorem for Markov chains with normal transition operators, started at a point. Probab. Theory Relat. Fields. **119**, 508–528 (2001)
16. Derriennic, Y., Lin, M.: Fractional Poisson equations and ergodic theorems for fractional coboundaries. Isr. J. Math. **123**, 93–130 (2001)
17. Derriennic, Y., Lin, M.: The central limit theorem for Markov chains started at a point. Probab. Theory Relat. Fields **125**, 73–76 (2003)
18. Derriennic, Y., Lin, M.: The central limit theorem for random walks on orbits of probability preserving transformation. Contemp. Math. **444**, 31–51 (2007)
19. Durieu, O., Volný, D.: Comparison between criteria leading to the weak invariance principle. Ann. Inst. H. Poincaré Probab. Statist. **44**, 324–340 (2008)
20. Durieu, O.: Independence of four projective criteria for the weak invariance principle. Aléa **5**, 21–27 (2009)
21. Gaposhkin, V.F.: Spectral criteria for the existence of generalized ergodic transformations. Theory Probab. Appl. **41**, 247–264 (1996)
22. Gordin, M.I.: The central limit theorem for stationary processes. Dokl. Akad. Nauk SSSR **188**, 739–741 (1969)
23. Gordin, M.I., Lifshitz, B.: A remark about a Markov process with normal transition operator. Third Vilnius Conf. Proba. Stat. Akad. Nauk Litovsk (in Russian) Vilnius **1**, 147–148 (1981)
24. Gordin, M., Peligrad, M.: . On the functional CLT via martingale approximation. Bernoulli **17**, 424–440 (2011)
25. Hall, P., Heyde, C.C.: Martingale limit theory and its application. Academic, New York/London (1980)
26. Hannan, E.J.: Central limit theorems for time series regression. Z. Wahrsch. Verw. Gebiete **26**, 157–170 (1973)
27. Hannan, E.J.: The central limit theorem for time series regression. Stoch. Proc. Appl. **9**, 281–289 (1979)
28. Heyde, C.C.: On the central limit theorem for stationary processes. Z. Wahrsch. verw. Gebiete **30**, 315–320 (1974)
29. Kipnis, C., Varadhan, S.R.S.: Central limit theorem for additive functionals of reversible Markov processes. Comm. Math. Phys. **104**, 1–19 (1986)
30. Liverani, C.: Central limit theorem for deterministic systems. In: International Conference on Dynamical systems (Montevideo, 1995). Pitman Research Notes in Mathematics Series, vol. 362, pp. 56–75. Longman, Harlow (1996)
31. Maxwell, M., Woodroofe, M.: Central limit theorems for additive functionals of Markov chains. Ann. Probab. **28**, 713–724 (2000)
32. Merlevède, F., Peligrad, M., Utev, S.: Recent advances in invariance principles for stationary sequences. Probab. Surv. **3**, 1–36 (2006)
33. Merlevède, F., Peligrad, M.: On the weak invariance principle for stationary sequences under projective criteria. J. Theor. Probab. **19**, 647–689 (2006)
34. Merlevède, F., Peligrad, C., Peligrad, M.: Almost Sure Invariance Principles via Martingale Approximation. Stoch. Proc. Appl. **122**, 70–190 (2011)
35. Peligrad, M., Utev, S.: A new maximal inequality and invariance principle for stationary sequences. Ann. Probab. **33**, 798–815 (2005)
36. Peligrad, M., Utev, S., Wu, W.B.: A maximal L(p)-Inequality for stationary sequences and application. Proc. Am. Math. Soc. **135**, 541–550 (2007)
37. Peligrad, M., Wu, W.B.: Central limit theorem for Fourier transform of stationary processes. Ann. Probab. **38**, 2009–2022 (2010)
38. Peligrad, M.: Conditional central limit theorem via martingale approximation. In: Dependence in analysis, probability and number theory (The Phillipp memorial volume), pp. 295–311. Kendrick Press, Heber City (2010)
39. Philipp, W., Stout, W.: Almost sure invariance principles for partial sums of weakly dependent random variables. Mem. Am. Math. Soc. 2 **161** (1975)

40. Rassoul-Agha, F., Seppäläinen, T.: Quenched invariance principle for multidimensional ballistic random walk in a random environment with a forbidden direction. Ann. Probab. **35**, 1–31 (2007)
41. Rassoul-Agha, F., Seppäläinen, T.: An almost sure invariance principle for additive functionals of Markov chains. Stat. Probab. Lett. **78**, 854–860 (2008)
42. Rio, E.: Almost sure invariance principles for mixing sequences of random variables. Stoch. Proc. Appl. **48**, 319–334 (1993)
43. Rio, E.: Moment inequalities for sums of dependent random variables under projective conditions. J. Theor. Probab. **22**, 146–163 (2009)
44. Rootzén, H.: Fluctuations of sequences which converge in distribution. Ann. Probab. **4**, 456–463 (1976)
45. Shao, Q.M.: Maximal inequalities for partial sums of ρ-mixing sequences. Ann. Probab. **23**, 948–965 (1995)
46. Volný, D.: A nonadapted version of the invariance principle of Peligrad and Utev. C. R. Math. Acad. Sci. Paris **345**, 167–169 (2007)
47. Volný, D., Woodroofe, M.: An example of non-quenched convergence in the conditional central limit theorem for partial sums of a linear process. In: Dependence in Analysis, Probability and Number Theory (The Phillipp Memorial Volume), pp. 317–323. Kendrick Press, Heber City (2010)
48. Volný, D., Woodroofe, M.: Quenched central limit theorems for sums of stationary processes. Stat. and Prob. Lett. **85**, 161–167 (2014)
49. Zhao, O., Woodroofe, M.: . Law of the iterated logarithm for stationary processes. Ann. Probab. **36**, 127–142 (2008)
50. Zhao, O., Woodroofe, M.: On martingale approximations. Ann. Appl. Probab. **18**, 1831–1847 (2008)
51. Wu, W.B., Woodroofe, M.: Martingale approximations for sums of stationary processes. Ann. Probab. **32**, 1674–1690 (2004)

An Extended Martingale Limit Theorem with Application to Specification Test for Nonlinear Co-integrating Regression Model

Qiying Wang

Dedicated to Miklós Csörgő on the occasion of his 80th birthday.

1 Introduction

Let $\{M_{ni}, \mathscr{F}_{ni}, 1 \leq i \leq k_n\}$ be a zero-mean square integrable martingale array, having difference y_{ni} and nested σ-fields structure, that is, $\mathscr{F}_{n,i} \subseteq \mathscr{F}_{n+1,i}$ for $1 \leq i \leq k_n, n \geq 1$. Suppose that, as $n \to \infty, k_n \to \infty$,

$$\sum_{i=1}^{k_n} E[y_{ni}^2 I(|y_{ni}| \geq \epsilon) \mid \mathscr{F}_{n,i-1}] \to_P 0,$$

for all $\epsilon > 0$, and the conditional variance

$$\sum_{i=1}^{k_n} E[y_{ni}^2 \mid \mathscr{F}_{n,i-1}] \to_P M^2,$$

where M^2 is an a.s. finite random variable. The classical martingale limit theorem (MLT) shows that $M_{n,k_n} = \sum_{i=1}^{k_n} y_{ni} \to_D Z$, where the r.v. Z has characteristic function $Ee^{itZ} = Ee^{-M^2 t^2/2}$, $t \in R$. If M^2 is a constant, the nested structure of the σ-fields $\mathscr{F}_{n,i}$ is not necessary. See, e.g., Chapter 3 of Hall and Heyde (1980).

The classical MLT is a celebrated result and one of the main conventional tools in statistics, econometrics and other fields. In many applications, however, the convergence in probability for the conditional variance required in the classical MLT seems to be too restrictive. Illustrations can be found in Wang and Phillips [8–10], Wang and Wang [13].

Q. Wang (✉)
School of Mathematics and Statistics, The University of Sydney, Sydney, NSW, Australia
e-mail: qiying@maths.usyd.edu.au

© Springer Science+Business Media New York 2015
D. Dawson et al. (eds.), *Asymptotic Laws and Methods in Stochastics*,
Fields Institute Communications 76, DOI 10.1007/978-1-4939-3076-0_10

Motivated by econometrics applications, Wang [7] currently provided an extension of the classical MLT. In the paper, it is shown that, for a certain class of martingales, the convergence in probability for the conditional variance in the classical MLT can be reduced to less restrictive: the convergence in distribution. As noticed in Wang [7], this kind of extensions removes a main barrier in the applications of the classical MLT to non-parametric estimates with non-stationarity. Indeed, using this extended MLT as a main tool, Wang [7] improved the existing results on the asymptotics for the conventional kernel estimators in a non-linear cointegrating regression model.

The aim of this paper is to show that Wang's extended MLT can also be used to the inference with non-stationarity. In particular, when it is used to a specification test for a nonlinear co-integrating regression model, a neat proof can be provided for the main result in Wang and Phillips [11].

This paper is organized as follows. In next section, we state Wang's extended MLT, but with some improvements. Specification test for a nonlinear co-integrating regression model is considered in Sect. 3, where Wang's extended MLT is connected to the main results. Finally, in Sect. 4, we finish the proof of the main results by checking the conditions on Wang's extended MLT.

Throughout the paper, we denote by $C, C_1, \ldots$ the constants, which may change at each appearance. The notation $\to_D$ ($\to_P$ respectively) denotes the convergence in distribution (in probability respectively) for a sequence of random variables (or vectors). If $\alpha_n^{(1)}, \alpha_n^{(2)}, \ldots, \alpha_n^{(k)}$ ($1 \le n \le \infty$) are random elements on $D[0, 1]$ or $D[0, \infty)$ or R, we will understand the condition

$$(\alpha_n^{(1)}, \alpha_n^{(2)}, \ldots, \alpha_n^{(k)}) \Rightarrow (\alpha_\infty^{(1)}, \alpha_\infty^{(2)}, \ldots, \alpha_\infty^{(k)})$$

to mean that for all $\alpha_\infty^{(1)}, \alpha_\infty^{(2)}, \ldots, \alpha_\infty^{(k)}$-continuity sets $A_1, A_2, \ldots, A_k$

$$P\big(\alpha_n^{(1)} \in A_1, \alpha_n^{(2)} \in A_2, \ldots, \alpha_n^{(k)} \in A_k\big) \to P\big(\alpha_\infty^{(1)} \in A_1, \alpha_\infty^{(2)} \in A_2, \ldots, \alpha_\infty^{(k)} \in A_k\big).$$

[see Billingsley ([1], Theorem 3.1) or Hall [5]]. As usual, $D[0, 1]$ ($D[0, \infty)$ respectively) denotes the space of cadlag functions on $[0, 1]$ ($[0, \infty)$ respectively), which is equipped with Skorohod topology.

2   An Extended Martingale Limit Theorem

This section states a current work on the martingale limit theorem by Wang [7], but with some improvement. We only provide a simplified version of the paper, as it is sufficient for the purpose of this paper.

Let $\{(\epsilon_k, \eta_k), \mathscr{F}_k\}_{1 \le k \le n}$, where $\mathscr{F}_k = \sigma(\epsilon_j, \eta_j, \xi_j, j \le k)$ and ξ_j is a sequence of random variables, form a martingale difference, satisfying

$$E(\epsilon_{k+1}^2 \mid \mathscr{F}_k) \to_{a.s.} 1, \qquad E(\eta_{k+1}^2 \mid \mathscr{F}_k) \to_{a.s.} 1,$$

and as $K \to \infty$,

$$\sup_{k \geq 1} E\{[\epsilon_{k+1}^2 I(|\epsilon_{k+1}| \geq K) + \eta_{k+1}^2 I(|\eta_{k+1}| \geq K)] \mid \mathscr{F}_k\} \to 0.$$

Consider a class of martingale defined by

$$S_n = \sum_{k=1}^{n} x_{n,k} \epsilon_{k+1}, \tag{1}$$

where, for a real function $f_n(\ldots)$ of its components,

$$x_{n,k} = f_n(\epsilon_1, \epsilon_2, \ldots, \epsilon_k; \eta_1, \eta_2, \ldots, \eta_k; \xi_k, \xi_{k-1}, \ldots).$$

Let $W_n(t) = \frac{1}{\sqrt{n}} \sum_{j=1}^{[nt]} \eta_{j+1}$ and $G_n^2 = \sum_{k=1}^{n} x_{n,k}^2$. Recalling the definition of η_k, $W_n(t) \Rightarrow W(t)$ on $D[0,1]$, where $W(t)$ is a standard Winner process. The following theorem comes from Theorem 2.1 of Wang [7], but with some improvements. A outline on the proof of this theorem will be given in Sect. 4.

Theorem 1. *Suppose that (a)* $\max_{1 \leq k \leq n} |x_{n,k}| = o_P(1)$ *and for any* $|\beta_k| \leq C$

$$\frac{1}{\sqrt{n}} \sum_{k=1}^{n} \beta_k x_{n,k} E(\eta_{k+1} \epsilon_{k+1} \mid \mathscr{F}_k) = o_P(1); \tag{2}$$

(b) there exists an a.s. finite functional $g_1^2(W)$ *of* $W(s), 0 \leq s \leq 1$, *such that*

$$\{W_n(t), G_n^2\} \Rightarrow \{W(t), g_1^2(W)\}. \tag{3}$$

Then, as $n \to \infty$,

$$\{S_n, G_n^2\} \to_D \{g_1(W) N, g_1^2(W)\}, \tag{4}$$

where N is a standard normal variate independent of $g_1(W)$.

We mention that (2) is weaker than (2.3) of Wang [7], where a stronger version is used:

$$\frac{1}{\sqrt{n}} \sum_{k=1}^{n} |x_{n,k}| |E(\eta_{k+1} \epsilon_{k+1} \mid \mathscr{F}_k)| = o_P(1). \tag{5}$$

It is interesting to notice that, for certain classes of $x_{n,k}$, the condition (2) is satisfied, but it is hard to verify (5). Illustration can be found in (17), (18), and (19) of Sect. 4.

3 Specification Test for a Nonlinear Cointegrating Regression Model

Consider a nonlinear cointegrating regression model:

$$y_{t+1} = f(x_t) + \epsilon_{t+1}, \quad t = 1, 2, \ldots, n, \tag{6}$$

where ϵ_t is a stationary error process and x_t is a nonstationary regressor. We are interested in testing the null hypothesis:

$$H_0 : f(x) = f(x, \theta), \quad \theta \in \Omega_0,$$

for $x \in R$, where $f(x, \theta)$ is a given real function indexed by a vector θ of unknown parameters which lie in the parameter space Ω_0.

To test H_0, Wang and Phillips [11] [also see Gao et al [3, 4]] made use of the kernel-smoothed self-normalized test statistic S_n/V_n, where

$$S_n = \sum_{s,t=1, s \neq t}^{n} \hat{u}_{t+1} \hat{u}_{s+1} K\big[(x_t - x_s)/h\big], \quad V_n^2 = \sum_{s,t=1, s \neq t}^{n} \hat{u}_{t+1}^2 \hat{u}_{s+1}^2 K^2\big[(x_t - x_s)/h\big],$$

involving the parametric regression residuals $\hat{u}_{t+1} = y_{t+1} - f(x_t, \hat{\theta})$, where $K(x)$ is a non-negative real kernel function, h is a bandwidth satisfying $h \equiv h_n \to 0$ as the sample size $n \to \infty$ and $\hat{\theta}$ is a parametric estimator of θ under the null H_0, that is consistent whenever $\theta \in \Omega_0$.

Under certain conditions on the x_t and ϵ_t (see Assumptions 1, 2, 3, and 4 below), it was proved in Wang and Phillips [11] that

$$S_n = 2 \sum_{t=2}^{n} \epsilon_{t+1} Y_{nt} + o_P(n^{3/4} \sqrt{h}), \tag{7}$$

$$V_n^2 = \sigma^4 \sum_{\substack{t,s=1 \\ t \neq s}}^{n} K^2\big[(x_t - x_s)/h\big] + o_P(n^{3/2} h)$$

$$= 2\sigma^2 \sum_{t=2}^{n} Y_{nt}^2 + o_P(n^{3/2} h). \tag{8}$$

where $Y_{nt} = \sum_{i=1}^{t-1} \epsilon_{i+1} K\big[(x_t - x_i)/h\big]$ and σ^2 is given as in Assumption 2. It follows easily from (7) and (8) that the limit distribution of S_n/V_n is determined by the joint asymptotics for $\sum_{t=2}^{n} \epsilon_{t+1} Y_{nt}$ and $\sum_{t=2}^{n} Y_{nt}^2$, under a suitable standardization.

Using the extended MLT in Sect. 2, this section investigates the joint limit distribution of $\sum_{t=2}^{n} \epsilon_{t+1} Y_{nt}$ and $\sum_{t=2}^{n} Y_{nt}^2$ (under a suitable standardization), and hence the limit distribution of S_n/V_n. We use the following assumptions in our development.

Assumption 1. *(i)* $\{\eta_t\}_{t\in\mathbf{Z}}$ *is a sequence of independent and identically distributed (iid) continuous random variables with* $E\eta_0 = 0$, $E\eta_0^2 = 1$, *and with the characteristic function* $\varphi(t)$ *of* η_0 *satisfying* $|t||\varphi(t)| \to 0$, *as* $|t| \to \infty$. *(ii)*

$$x_t = \rho x_{t-1} + \xi_t, \quad x_0 = 0, \quad \rho = 1 + \kappa/n, \quad 1 \le t \le n, \tag{9}$$

where κ *is a constant and* $\xi_t = \sum_{k=0}^{\infty} \phi_k \eta_{t-k}$ *with* $\phi \equiv \sum_{k=0}^{\infty} \phi_k \neq 0$ *and* $\sum_{k=0}^{\infty} k^{1+\delta}|\phi_k| < \infty$ *for some* $\delta > 0$.

Assumption 2. $\{\epsilon_t, \mathscr{F}_t\}_{t\ge 1}$, *where* $\mathscr{F}_t$ *is a sequence of increasing* σ-*fields which is independent of* $\eta_k, k \ge t + 1$, *forms a martingale difference satisfying* $E(\epsilon_{t+1}^2|\mathscr{F}_t) \to_{a.s.} \sigma^2 > 0$, $E(\epsilon_{t+1}\eta_{t+1}|\mathscr{F}_t) \to_{a.s.} C_0$ *as* $t \to \infty$ *and* $\sup_{t\ge 1} E(|\epsilon_{t+1}|^4 | \mathscr{F}_t) < \infty$.

Assumption 3. $K(x)$ *is a nonnegative real function satisfying* $\sup_x K(x) < \infty$ *and* $\int K(x)dx < \infty$.

Define,

$$L_G(t, u) = \lim_{\varepsilon \to 0} \frac{1}{2\varepsilon} \int_0^t \int_0^t \mathbf{1}[|(G(x) - G(y)) - u| < \varepsilon]dxdy$$

$$= \int_0^t \int_0^t \delta_u[G(x) - G(y)]dxdy, \tag{10}$$

where δ_u is the dirac function. $L_G(t, u)$ characterizes the amount of time over the interval $[0, t]$ that the process $G(t)$ spends at a distance u from itself, and is well defined as shown in Section 5 of Wang and Phillips [11, 12]. We have the following theorem.

Theorem 2. *Under Assumptions 1, 2, and 3,* $nh^2 \to \infty$ *and* $h \log^2 n \to 0$, *we have*

$$\left(\frac{1}{\sigma d_n} \sum_{t=2}^n \epsilon_{t+1} Y_{nt}, \frac{1}{d_n^2} \sum_{t=2}^n Y_{nt}^2\right) \to_D (\eta N, \eta^2), \tag{11}$$

where $d_n^2 = (2\phi)^{-1}\sigma^2 n^{3/2}h \int_{-\infty}^{\infty} K^2(x)dx$, $\eta^2 = L_G(1, 0)$ *is the self intersection local time generated by the process* $G = \int_0^t e^{\kappa(t-s)}dW(s)$ *and* N *is a standard normal variate which is independent of* η^2.

Comparing to Theorem 3.3 of Wang and Phillips [11], Theorem 2 reduces the condition (2.4) of the paper, i.e., the requirement on the joint convergence of $\sum_{k=1}^n \epsilon_k/\sqrt{n}$ and $\sum_{k=1}^n \eta_k/\sqrt{n}$. More interestingly, proof of Theorem 2 is quite neat, since calculations involving in verification of Theorem 1 are also used in the proofs of (7) and (8). In comparison, the proof of Theorem 3.3 in Wang and Phillips [11] requires some different techniques, introducing some new calculations like their Propositions 6–8 which are quite complexed.

For the sake of completeness, we present the following theorem, which is the same as Theorem 3.1 of Wang and Phillips [11], providing the limit distribution of S_n/V_n under some additional assumptions on the regression function $f(x)$, the kernel function $K(x)$ and the η_k.

Assumption 4. (i) *There is a sequence of positive real numbers δ_n satisfying $\delta_n \to 0$ as $n \to \infty$ such that $\sup_{\theta \in \Omega_0} \|\hat{\theta} - \theta\| = o_P(\delta_n)$, where $\|\cdot\|$ denotes the Euclidean norm.*

(ii) *There exists some $\varepsilon_0 > 0$ such that $\frac{\partial^2 f(x,t)}{\partial t^2}$ is continuous in both $x \in R$ and $t \in \Theta_0$, where $\Theta_0 = \{t : \|t - \theta\| \le \varepsilon_0, \theta \in \Omega_0\}$.*

(iii) *Uniformly for $\theta \in \Omega_0$,*

$$\left|\frac{\partial f(x,t)}{\partial t}\Big|_{t=\theta}\right| + \left|\frac{\partial^2 f(x,t)}{\partial t^2}\Big|_{t=\theta}\right| \le C(1 + |x|^\beta),$$

for some constants $\beta \ge 0$ and $C > 0$.

(iv) *Uniformly for $\theta \in \Omega_0$, there exist $0 < \gamma' \le 1$ and $\max\{0, 3/4 - 2\beta\} < \gamma \le 1$ such that*

$$|g(x+y, \theta) - g(x, \theta)| \le C|y|^\gamma \begin{cases} 1 + |x|^{\beta-1} + |y|^\beta, & \text{if } \beta > 0, \\ 1 + |x|^{\gamma'-1}, & \text{if } \beta = 0, \end{cases} \tag{12}$$

for any $x, y \in R$, where $g(x, t) = \frac{\partial f(x,t)}{\partial t}$.

Assumption 5. $nh^2 \to \infty$, $\delta_n^2 n^{1+\beta}\sqrt{h} \to 0$ and $nh^4 \log^2 n \to 0$, where β and δ_n^2 are defined as in Assumption 4. Also, $\int (1 + |x|^{2\beta+1})K(x)dx < \infty$ and $E|\epsilon_0|^{4\beta+2} < \infty$.

As noticed in Wang and Phillips [11], the sequence δ_n in Assumption 4(i) may be chosen as $\delta_n^2 = n^{-(1+\beta)/2}h^{-1/8}$, due to the fact that δ_n also satisfies Assumption 5. Hence, by Park and Phillips [10], Assumption 4(i) is achievable under Assumption 4(ii)–(iv). Assumptions 4(ii)–(iv) are quite weak and include a wide class of functions. Typical examples include polynomial forms like $f(x, \theta) = \theta_1 + \theta_2 x + \ldots + \theta_k x^{k-1}$, where $\theta = (\theta_1, \ldots, \theta_k)$, power functions like $f(x, a, b, c) = a + b x^c$, shift functions like $f(x, \theta) = x(1 + \theta x)I(x \ge 0)$, and weighted exponentials such as $f(x, a, b) = (a + b e^x)/(1 + e^x)$.

Theorem 3. *Under Assumptions 1–5 and the null hypothesis, we have*

$$\frac{S_n}{\sqrt{2}\, V_n} \to_D N, \tag{13}$$

where N is a standard normal variate.

The limit in Theorem 3 is normal and does not depend on any nuisance parameters. As a test statistic, $Z_n = S_n/\sqrt{2}\, V_n$ has a big advantage in applications. As for the asymptotic power of the test, we refer to Wang and Phillips [11].

4 Proofs of Main Results

Proof of Theorem 1. Let $x^*_{n,k} = x_{n,k} I(\sum_{j=1}^k x^2_{j,n} \leq \lambda)$. Note that, on the set $G^2_n \leq n\lambda$,

$$\frac{1}{\sqrt{n}} \sum_{k=1}^n \beta_k x_{n,k} E(\eta_{k+1}\epsilon_{k+1} \mid \mathcal{F}_k) = \frac{1}{\sqrt{n}} \sum_{k=1}^n \beta_k x^*_{n,k} E(\eta_{k+1}\epsilon_{k+1} \mid \mathcal{F}_k)$$

and $P(G^2_n \geq n\lambda) \to 0$ as $n \to \infty$. The condition (2) implies that

$$\frac{1}{\sqrt{n}} \sum_{k=1}^n \beta_k x^*_{n,k} E(\eta_{k+1}\epsilon_{k+1} \mid \mathcal{F}_k) = o_P(1),$$

for any $\lambda > 0$. Hence (3.24) of Wang [7] still holds true if we replace (5), which is required in the proof of Wang [7], by (2). The remaining part in the proof of Theorem 1 is the same as that of Theorem 2.1 in Wang [7]. We omit the details. □

Proof of Theorem 2. Without loss of generality, assume $\sigma^2 = 1$. Furthermore, we may let $|\epsilon_k| \leq C$. This restriction can be removed by using the similar arguments as in Wang and Phillips ([11], pages 754–756). Write $x_{nt} = Y_{nt}/d_n$, where $Y_{nt} = \sum_{i=1}^{t-1} \epsilon_{i+1} K[(x_t - x_i)/h]$. Due to Assumptions 1 and 2, it is readily seen that

$$\left\{ \sum_{k=1}^n x_{n,k}\epsilon_{k+1}, \ \mathcal{F}_{n+1} \right\}_{n \geq 1},$$

where $\mathcal{F}_k = \sigma(\epsilon_1, \ldots, \epsilon_k; \eta_1, \ldots, \eta_k; \eta_0, \eta_{-1}, \ldots)$, forms a class of martingale defined as in Sect. 2. Hence, to prove Theorem 2, it suffices to verify conditions (a) and (b) of Theorem 1, which is given as follows. We first introduce the following proposition.

Proposition 1. *Under Assumptions 1, 2, and 3 and $h \log^2 n \to 0$, we have*

$$EY^2_{nt} \leq C(1 + h\sqrt{t}), \tag{14}$$

$$\sum_{k=1}^n \beta_k Y_{nk} = O_P(n^{5/4}h^{3/4}), \tag{15}$$

for any $|\beta_k| \leq C$, and if in addition $|\epsilon_k| \leq C$, then

$$EY^4_{nt} \leq C(1 + h^3 t^{3/2}). \tag{16}$$

The proof of Proposition 1 is given in Section 6 of Wang and Phillips [11]. Explicitly, (15) follows from (6.3) of the paper with $g(x) = 1$ and a minor improvement, as $|\beta_k| \leq C$. Equations (14) and (16) come from (6.6) and (6.8) of the paper, respectively.

Due to Proposition 1, it is routine to verify the condition (a) of Theorem 1. Indeed, by (16), it follows that

$$\max_{1\le t\le n} |x_{nt}| \le \frac{1}{d_n} \left(\sum_{k=1}^n Y_{nk}^4 \right)^{1/4} = O_P\big[(n^{5/2}h^3)^{1/4}/(n^{3/2}h)^{1/2}\big] = o_P(1),$$

which yields the first part of the condition (a). To prove the remaining part, we split the left hand of the (2) as

$$\frac{1}{\sqrt{n}} \sum_{k=1}^n \beta_k\, x_{n,k}\, E(\eta_{k+1}\epsilon_{k+1} \mid \mathscr{F}_k)$$

$$= \frac{1}{\sqrt{n}} \sum_{k=1}^n \beta_k\, x_{n,k}\, (V_k - C_0) + \frac{C_0}{\sqrt{n}} \sum_{k=1}^n \beta_k\, x_{n,k}$$

$$:= \Lambda_{1n} + \Lambda_{2n}, \tag{17}$$

where $V_k = E(\eta_{k+1}\epsilon_{k+1} \mid \mathscr{F}_k)$. It follows from (15) that, for any $|\beta_k| \le C$,

$$\Lambda_{2n} = \frac{1}{\sqrt{n}d_n} \sum_{k=1}^n \beta_k\, Y_{nk} = O_P(h^{1/4}) = o_P(1), \tag{18}$$

due to $h \to 0$. On the other hand, by recalling $V_k \to_{a.s} C_0$ and noting that

$$\frac{1}{\sqrt{n}} \sum_{k=1}^n |\beta_k|\, E|x_{n,k}| \le \frac{C}{\sqrt{n}d_n} \sum_{k=1}^n (EY_{nt}^2)^{1/2} \le \frac{C}{\sqrt{n}d_n} \sum_{k=1}^n (1 + h\sqrt{t})^{1/2} \le C, \tag{19}$$

simple calculations show that $\Lambda_{1n} = o_P(1)$. Taking these estimates into (17), we get the required (2), and hence the condition (a) of Theorem 1 is identified.

To verify the condition (b) of Theorem 1, by recalling (8) and $\sigma^2 = 1$, it suffices to show that

$$\left\{ \frac{1}{\sqrt{n}} \sum_{j=1}^{[nt]} \eta_{j+1},\; \frac{1}{2d_n^2} \sum_{t,s=1}^n K^2\big[(x_t - x_s)/h\big] \right\} \;\Rightarrow\; \{W(t),\, L_G(1,0)\}. \tag{20}$$

The individual convergence of two components in (20) has been established in Section 5 of Wang and Phillips [11]. The joint convergence can be established in a similar way, which is outlined as follows.

Write $g(x) = K^2(x)$, $x_{k,n} = x_k/(\sqrt{n}\phi)$ and $c_n = \sqrt{n}\phi/h$. It follows from these notation that

$$\frac{\int_{-\infty}^{\infty} K^2(x)dx}{2d_n^2} \sum_{k,j=1}^n K^2\big[(x_k - x_j)/h\big] = \frac{c_n}{n^2} \sum_{k,j=1}^n g\big[c_n\,(x_{k,n} - x_{j,n})\big]$$

We first prove the result (20) under an additional condition:

Con : $g(x)$ is continuous and $\hat{g}(t)$ has a compact support,

$$\text{where } \hat{g}(x) = \int_{-\infty}^{\infty} e^{ixt} g(t) dt. \tag{21}$$

To start, noting that $g(x) = \frac{1}{2\pi} \int_{-\infty}^{\infty} e^{itx} \hat{g}(-t) dt$, we have

$$\frac{c_n}{n^2} \sum_{k,j=1}^{n} g\big[c_n (x_{k,n} - x_{j,n})\big] = \frac{1}{2\pi n^2} \sum_{k,j=1}^{n} \int_{-\infty}^{\infty} \hat{g}(-s/c_n) e^{is (x_{k,n} - x_{j,n})} ds$$

$$= R_{1n} + R_{2n}, \tag{22}$$

where, for some $A > 0$,

$$R_{1n} = \frac{1}{2\pi n^2} \sum_{k,j=1}^{n} \int_{|s| \le A} \hat{g}(-s/c_n) e^{is (x_{k,n} - x_{j,n})} ds,$$

$$R_{2n} = \frac{1}{2\pi n^2} \sum_{k,j=1}^{n} \int_{|s| > A} \hat{g}(-s/c_n) e^{is (x_{k,n} - x_{j,n})} ds.$$

Furthermore, R_{1n} can be written as

$$R_{1n} = \frac{1}{2\pi} \int_{|s| \le A} \hat{g}(-s/c_n) \int_0^1 \int_0^1 e^{is (x_{[nu],n} - x_{[nv],n})} du \, dv \, ds + o_P(1).$$

Recall $c_n \to \infty$. It is readily seen that $\sup_{|s| \le A} |\hat{g}(-s/c_n) - \hat{g}(0)| \to 0$ for any fixed $A > 0$. Hence, by noting that the classical invariance principle gives

$$\left\{ \frac{1}{\sqrt{n}} \sum_{j=1}^{[nt]} \eta_{j+1}, \ x_{[nt],n} \right\} \Rightarrow \{W(t), \ G(t)\},$$

where $G(t) = \int_0^t e^{\kappa(t-s)} dW(s)$ (see, e.g., Phillips [6] and/or Buchmann and Chan [2]), it follows from the continuous mapping theorem that, for any $A > 0$,

$$\left\{ \frac{1}{\sqrt{n}} \sum_{j=1}^{[nt]} \eta_{j+1}, \ R_{1n} \right\} \to_D \{W(t), \ \hat{g}(0) g_A(G)\}, \tag{23}$$

where $g_A(G) = \frac{1}{2\pi} \int_{|s| \le A} \int_0^1 \int_0^1 e^{is[G(u) - G(v)]} du \, dv \, ds$, as $n \to \infty$. Note that $\hat{g}(0) = \int_{-\infty}^{\infty} K^2(x) dx$ and $g_A(G)$ converges to $L_G(1, 0)$ in L_2, as $A \to \infty$. See (7.2) of Wang and Phillips [12]. The results (20) will follow under the additional condition (21), if we prove

$$E|R_{2n}|^2 \to 0, \tag{24}$$

as $n \to \infty$ first and then $A \to \infty$. This follows from (7.8) of Wang and Phillips [12], and hence the details are omitted.

To remove the additional condition (21), it suffices to construct a $g_{\delta_0}(x)$ so that it satisfies (21), $\int_{-\infty}^{\infty} |g_{\delta_0}(x)| dx < \infty$, $\int_{-\infty}^{\infty} |g(x) - g_{\delta_0}(x)| dx < \epsilon$, and we may prove

$$\frac{c_n}{n^2} \sum_{k,j=1}^{n} \left| g\left[c_n \left(x_{k,n} - x_{j,n}\right)\right] - g_{\delta_0}\left[c_n \left(x_{k,n} - x_{j,n}\right)\right] \right| = O_P(\epsilon). \tag{25}$$

This is exactly the same as in the proof of Theorem 5.1 in Wang and Phillips [11, 12], and hence the details are omitted.   □

References

1. Billingsley, P.: Convergence of Probability Measures. Wiley, New York (1968)
2. Buchmann, B., Chan, N.H.: Asymptotic theory of least squares estimators for nearly unstable processes under strong dependence. Ann. Statist. **35**, 2001–2017 (2007)
3. Gao, J., Maxwell, K., Lu, Z., Tjøstheim, D.: Nonparametric specification testing for nonlinear time series with nonstationarity. Econ. Theory **25**, 1869–1892 (2009)
4. Gao, J., Maxwell, K., Lu, Z., Tjøstheim, D.: Specification testing in nonlinear and nonstationary time series autoregression. Ann. Statist. **37**, 3893–3928 (2009)
5. Hall, P.: Martingale invariance principles. Ann. Probab. **5**, 875–887 (1977)
6. Phillips, P.C.B.: Towards a unified asymptotic theory for autoregression. Biometrika **74**, 535–547 (1987)
7. Wang, Q.: Martingale limit theorem revisit and non-linear cointegrating regression. Econ. Theory **30**, 509–535 (2014)
8. Wang, Q., Phillips, P.C.B.: Asymptotic theory for local time density estimation and nonparametric cointegrating regression. Econ. Theory **25**, 710–738 (2009)
9. Wang, Q., Phillips, P.C.B.: Structural nonparametric cointegrating regression. Econometrica **77**, 1901–1948 (2009)
10. Wang, Q., Phillips, P.C.B.: Asymptotic theory for zero energy functionals with nonparametric regression applications. Econ. Theory **2**, 235–359 (2011)
11. Wang, Q., Phillips, P.C.B.: A specification test for nonlinear nonstationary models. Ann. Stat. **40**, 727–758 (2012)
12. Wang, Q., Phillips, P.C.B.: Supplement to – a specification test for nonlinear nonstationary models (2012). doi:10.1214/12-AOS975SUPP
13. Wang, Q., Wang, R.: Non-parametric cointegrating regression with NNH errors. Econ. Theory **19**, 1–27 (2013)

Part IV
Change-Point Problems

Change Point Detection with Stable AR(1) Errors

Alina Bazarova, István Berkes, and Lajos Horváth

1 Introduction and Results

In this paper we are interested to detect possible changes in the location model

$$X_j = c_j + e_j, \quad 1 \leq j \leq n. \tag{1}$$

We wish to test the null hypothesis of stability of the location parameter, i.e.,

$$H_0 : \quad c_1 = c_2 = \ldots = c_n$$

against the one change alternative

$$H_A : \quad \text{there is } k^* \text{ such that } c_1 = \ldots = c_{k^*} \neq c_{k^*+1} = \ldots = c_n.$$

We say that k^* is the time of change under the alternative. The time of change as well as the location parameters before and after the change are unknown. The most popular methods to test H_0 against H_A are based on the CUSUM process

A. Bazarova
Warwick Systems Biology Centre, Senate House,
University of Warwick, Coventry, United Kingdom CV4 7AL
e-mail: bazarova@tugraz.at; a.bazarova@warwick.ac.uk

I. Berkes (✉)
Institute of Statistics, Graz University of Technology,
Kopernikusgasse 24, 8010 Graz, Austria
e-mail: berkes@tugraz.at

L. Horváth
Department of Mathematics, University of Utah, Salt Lake City, UT 84112–0090, USA
e-mail: horvath@math.utah.edu

© Springer Science+Business Media New York 2015
D. Dawson et al. (eds.), *Asymptotic Laws and Methods in Stochastics*,
Fields Institute Communications 76, DOI 10.1007/978-1-4939-3076-0_11

$$U_n(x) = \sum_{i=1}^{\lfloor nx \rfloor} X_i - \frac{\lfloor nx \rfloor}{n} \sum_{i=1}^{n} X_i.$$

Clearly, if H_0 is true, then $U_n(t)$ does not depend on the common but unknown location parameter. It is well known that if $X_1, \ldots, X_n$ are independent and identically distributed random variables with a finite second moment, then

$$\frac{1}{(n\mathrm{var}(X_1))^{1/2}} U_n(x) \xrightarrow{\mathscr{D}[0,1]} B(x),$$

where $B(x)$ is a Brownian bridge. Throughout this paper $\mathscr{D}[0, 1]$ denotes the space of right continuous functions on $[0, 1]$ with left limits; $\xrightarrow{\mathscr{D}[0,1]}$ means weak convergence in $\mathscr{D}[0, 1]$ with respect to the Skorohod J_1 topology (cf. Billingsley [10]). Of course, $\mathrm{var}(X_1)$ can be consistently estimated by the sample variance in this case, resulting in

$$\frac{1}{\sigma_n^* n^{1/2}} U_n(x) \xrightarrow{\mathscr{D}[0,1]} B(x) \tag{2}$$

with

$$\sigma_n^* = \left\{ \frac{1}{n} \sum_{i=1}^{n} (X_i - \bar{X}_n)^2 \right\}^{1/2} \quad \text{with} \quad \bar{X}_n = \frac{1}{n} \sum_{i=1}^{n} X_i.$$

Assuming that $X_1, X_2, \ldots, X_n$ are independent and identically distributed random variables in the domain of attraction of a stable law of index $\alpha \in (0, 2)$, Aue et al. [3] showed that

$$\frac{1}{n^{1/\alpha} \hat{L}(n)} U_n(x) \xrightarrow{\mathscr{D}[0,1]} B_\alpha(x),$$

where $\hat{L}$ is a slowly varying function at ∞ and $B_\alpha(x)$ is an α–stable bridge. (The α–stable bridge is defined as $B_\alpha(x) = W_\alpha(x) - x W_\alpha(1)$, where W_α is a Lévy α–stable motion.) Since nothing is known on the distributions of the functionals of α–stable bridges, Berkes et al. [9] suggested the trimmed CUSUM process

$$T_n(x) = \sum_{i=1}^{\lfloor nx \rfloor} X_i I\{|X_i| \le \eta_{n,d}\} - \frac{\lfloor nx \rfloor}{n} \sum_{i=1}^{n} X_i I\{|X_i| \le \eta_{n,d}\},$$

where $\eta_{n,d}$ is the dth largest among $|X_1|, |X_2|, \ldots, |X_n|$. Assuming that the X_i's are independent and identically distributed and are in the domain of attraction of a stable law, they proved

$$\frac{1}{\hat{\sigma}_n n^{1/2}} T_n(x) \xrightarrow{\mathscr{D}[0,1]} B(x),$$

where

$$\hat{\sigma}_n^2 = \frac{1}{n} \sum_{i=1}^{n} \left(X_i I\{|X_i| \leq \eta_{n,d}\} - \frac{1}{n} \sum_{j=1}^{n} X_j I\{|X_j| \leq \eta_{n,d}\} \right)^2,$$

and $B(t)$ is a Brownian bridge. Roughly speaking, the classical CUSUM procedure in (2) can be used on the trimmed variables $X_j I\{|X_j| \leq \eta_{n,d}\}, 1 \leq j \leq n$. The CUSUM process has also been widely used in case of dependent variables, but it is nearly always assumed that the observations have high moments and the dependence in the sequence is weak, i.e. the limit distributions of the proposed statistics are derived from normal approximations. For a review we refer to Aue and Horváth [5]. However, very few papers consider the instability of time series models with heavy tails.

Fama [16] and Mandelbrot [23, 24] pointed out that the distributions of commodity and stock returns are often heavy tailed with possibly infinite variance and they started the investigation of time series models where the marginal distributions have regularly varying tails. Davis and Resnick [14, 15] investigated the properties of moving averages with regularly varying tails and obtained non–Gaussian limits for the sample covariances and correlations. Their results were extended to heavy tailed ARCH by Davis and Mikosch [13]. The empirical periodogram was studied by Mikosch et al. [25]. Andrews et al. [1] estimated the parameters of autoregressive processes with stable innovations.

In this paper we study testing H_0 against H_A when the error terms form an autoregressive process of order 1, i.e., e_i is a $\sigma(\varepsilon_j, j \leq i)$ measurable solution of

$$e_i = \rho e_{i-1} + \varepsilon_i \quad -\infty < i < \infty. \tag{3}$$

We assume throughout this paper that

$$\varepsilon_j, -\infty < j < \infty \text{ are independent and identically distributed,} \tag{4}$$

$$\varepsilon_0 \text{ belongs to the domain of attraction of a stable} \tag{5}$$

random variable $\xi^{(\alpha)}$ with parameter $0 < \alpha < 2$,

and

$$\varepsilon_0 \text{ is symmetric when } \alpha = 1. \tag{6}$$

Assumption (5) means that

$$\left(\sum_{j=1}^{n} \varepsilon_j - a_n \right) \bigg/ b_n \xrightarrow{\mathcal{D}} \xi^{(\alpha)} \tag{7}$$

for some numerical sequences a_n and b_n. The necessary and sufficient condition for this is

$$\lim_{t \to \infty} \frac{P\{\varepsilon_0 > t\}}{L_*(t)t^{-\alpha}} = p \quad \text{and} \quad \lim_{t \to \infty} \frac{P\{\varepsilon_0 \leq -t\}}{L_*(t)t^{-\alpha}} = q \tag{8}$$

for some numbers $p \geq 0, q \geq 0, p + q = 1$, where L_* is a slowly varying function at ∞. It is known that (3) has a unique stationary non–anticipative solution if and only if

$$-1 < \rho < 1. \tag{9}$$

Under assumptions (4), (5), (6), (7), (8), and (9), $\{e_j\}$ is a stationary sequence and $E|e_0|^\kappa < \infty$ for all $0 < \kappa < \alpha$ but $E|e_0|^\kappa = \infty$ for all $\kappa > \alpha$. The AR(1) process with stable innovations was considered by Chan and Tran [12], Chan [11], Aue and Horváth [4] and Zhang and Chan [28] who investigated the case when ρ is close to 1 and provided estimates for ρ and the other parameters when the observations do not have finite variances.

The convergence of the finite dimensional distributions of $U_n(x)$ is an immediate consequence of Phillips and Solo [26]. Let $\overset{fdd}{\longrightarrow}$ denote the convergence of the finite dimensional distributions.

Theorem 1. *If H_0, (3), (4), (5), (6) and (9) hold, then we have that*

$$\frac{1 - \rho}{n^{1/\alpha}L_*(n)} U_n(x) \overset{fdd}{\longrightarrow} B_\alpha(x),$$

where $B_\alpha(x), 0 \leq t \leq 1$ is an α–stable bridge.

It has been pointed out by Avram and Taqqu [6, 7] that the convergence of the finite dimensional distributions in Theorem 1 cannot be replaced with weak convergence in $\mathscr{D}[0, 1]$. Avram and Taqqu [6, 7] also proved that under further regularity conditions, the convergence of the finite dimensional distributions can be replaced with convergence in $\mathscr{D}[0, 1]$ with respect to the M_1 topology. However, the distributions of $\sup_{0 \leq x \leq 1} |B_\alpha(x)|dx$ and $\int_0^1 B_\alpha^2(x)dx$ depend on the unknown α and they are unknown for any $0 < \alpha < 2$.

The statistics used in this paper are based on $T_n(x)$ with a truncation parameter $d = d(n)$ satisfying

$$\lim_{n \to \infty} d(n)/n = 0 \tag{10}$$

and

$$d(n) \geq n^\delta \quad \text{with some} \quad 0 < \delta < 1. \tag{11}$$

Let $F(x) = P\{X_0 \leq x\}$, $H(x) = P\{|X_0| > x\}$ and let $H^{-1}(t)$ be the (generalized) inverse of H. We also assume that ε_0 has a density function $p(t)$ which satisfies

$$\int_{-\infty}^{\infty} |p(t+s) - p(t)| dt \leq C|s| \quad \text{with some} \quad C. \tag{12}$$

Let

$$A_n = d^{1/2} H^{-1}(d/n). \tag{13}$$

The following result was obtained by Bazarova et al. [8]:

Theorem 2. *If H_0, (3), (4), (5), (6) and (9), (10), (11), (12) hold, then we have that*

$$\left(\frac{2-\alpha}{\alpha}\right)^{1/2} \left(\frac{1-\rho}{1+\rho}\right)^{1/2} \frac{T_n(x)}{A_n} \xrightarrow{\mathscr{D}[0,1]} B(x),$$

where $B(x)$ is a Brownian bridge.

The weak convergence in Theorem 2 can be used to construct tests to detect possible changes in the location parameter in model (1). However, the normalizing sequence depends heavily on unknown parameters and they should be replaced with consistent estimators. We discuss this approach in Sect. 2. We show in Sect. 3 that ratio statistics can also be used so we can avoid the estimation of the long run variances.

2 Estimation of the Long Run Variance

The limit result in Theorem 2 is the same as one gets for the CUSUM process in case of weakly dependent stationary variables (cf. Aue and Horváth [5]). Hence we interpret the normalizing sequence as the long run variance of the sum of the trimmed variables. Based on this interpretation we suggest Bartlett type estimators as the normalization.

The Bartlett estimator computed from the trimmed variables $X_i^* = X_i I\{|X_i| \leq \eta_{n,d}\}$ is given by

$$\hat{s}_n^2 = \hat{\gamma}_0 + 2 \sum_{j=1}^{n-1} \omega \left(\frac{j}{h(n)}\right) \hat{\gamma}_j,$$

where

$$\hat{\gamma}_j = \frac{1}{n} \sum_{i=1}^{n-j} (X_i^* - \bar{X}_n^*)(X_{i+j}^* - \bar{X}_n^*), \quad \bar{X}_n^* = \frac{1}{n} \sum_{i=1}^{n} X_i^*,$$

$\omega(\cdot)$ is the kernel and $h(\cdot)$ is the length of the window. We assume that $\omega(\cdot)$ and $h(\cdot)$ satisfy the following standard assumptions:

$$\omega(0) = 1, \tag{14}$$

$$\omega(t) = 0 \quad \text{if} \quad t > a \quad \text{with some} \quad a > 0, \tag{15}$$

$$\omega(\cdot) \quad \text{is a Lipschitz function,} \tag{16}$$

$$\hat{\omega}(\cdot), \text{ the Fourier transform of } \omega(\cdot), \text{ is also Lipschitz and integrable} \tag{17}$$

and

$$h(n) \to \infty \quad \text{and} \quad h(n)/n \to \infty \quad \text{as} \quad n \to \infty. \tag{18}$$

For functions satisfying (14), (15), (16), and (17) we refer to Taniguchi and Kakizawa [27]. Following the methods in Liu and Wu [22] and Horváth and Reeder [18], the following weak law of large numbers can be established under H_0:

$$\frac{n\hat{s}_n^2}{A_n^2(1+\rho)\alpha/((1-\rho)(2-\alpha))} \xrightarrow{P} 1, \quad \text{as} \quad n \to \infty. \tag{19}$$

The next result is an immediate consequence of Theorem 2 and (19).

Corollary 1. *If H_0, (3), (4), (5), (6), (9), (10), (11), (12) and (19) hold, then we have that*

$$\frac{T_n(x)}{n^{1/2}\hat{s}_n} \xrightarrow{\mathscr{D}[0,1]} B(x),$$

where $B(x)$ is a Brownian bridge.

It follows immediately that under the no change null hypothesis

$$\hat{\mathscr{Q}}_n = \sup_{0 \le x \le 1} \frac{|T_n(x)|}{n^{1/2}\hat{s}_n} \xrightarrow{\mathscr{D}} \sup_{0 \le x \le 1} |B(x)|.$$

Simulations show that $\hat{s}_n$ performs well under H_0 but it overestimates the norming sequence under the alternative. Hence $\hat{\mathscr{Q}}_n$ has little power. The estimation of the long–run variance when a change occurs has been addressed in the literature. We follow the approach of Antoch et al. [2], who provided estimators for the long run variance which are asymptotically consistent under H_0 as well as under the one change alternative. Let x_0 denote the smallest value in $[0, 1]$ where $|T_n(x)|$ reaches its maximum and let $\tilde{k} = \lfloor x_0 n \rfloor$. The modified Bartlett estimator is defined as

$$\tilde{s}_n^2 = \hat{\gamma}_0' + 2 \sum_{j=1}^{n-1} \omega\left(\frac{j}{h(n)}\right) \tilde{\gamma}_j,$$

where

$$\tilde{\gamma}_j = \frac{1}{n-j} \sum_{\ell=1}^{n-j} \iota_\ell \iota_{\ell+j}, \qquad \iota_\ell = X_\ell^* - \frac{1}{\hat{k}} \sum_{\ell=1}^{\hat{k}} X_\ell^*, \quad \ell = 1, \ldots, \hat{k},$$

$$\iota_\ell = X_\ell^* - \frac{1}{n-\hat{k}} \sum_{\ell=\hat{k}+1}^{n} X_\ell^*, \quad \ell = \hat{k}+1, \ldots, n.$$

Combining the proofs in Antoch et al. [2] with Liu and Wu [22] and Horváth and Reeder [18] one can verify that

$$\frac{n\tilde{s}_n^2}{A_n^2(1+\rho)\alpha/((1-\rho)(2-\alpha))} \xrightarrow{P} 1, \quad \text{as} \quad n \to \infty \tag{20}$$

under H_0 as well as under the one change alternative H_A. Due to (20) we immediately have the following result:

Corollary 2. *If H_0, (3), (4), (5), (6), (9), (10), (11), (12) and (20) hold, then we have that*

$$\frac{T_n(x)}{n^{1/2}\tilde{s}_n} \xrightarrow{\mathscr{D}[0,1]} B(x),$$

where $B(x)$ is a Brownian bridge.

We suggest testing procedures based on

$$\tilde{\mathscr{Q}}_n = \frac{1}{n^{1/2}\tilde{s}_n} \sup_{0 \le x \le 1} |T_n(x)|.$$

It follows immediately from Corollary 2 that under H_0

$$\tilde{\mathscr{Q}}_n \xrightarrow{\mathscr{D}} \sup_{0 \le x \le 1} |B(x)|. \tag{21}$$

First we study experimentally the rate of convergence in Theorem 2. In this section we assume that the innovations ε_i in (3), (4), (5), (6), and (7) have the common distribution function

$$F(t) = \begin{cases} q(1-t)^{-3/2}, & \text{if} \quad -\infty < t \le 0, \\ 1 - p(1-t)^{-3/2}, & \text{if} \quad 0 < t < \infty, \end{cases}$$

where $p \ge 0$, $q \ge 0$ and $p + q = 1$. We present the results for the case of $\rho = p = q = 1/2$ based on 10^5 repetitions. We simulated the elements of an autoregressive sample $(e_1, \ldots, e_n)$ from the recursion (3) starting with some initial value and with a burn in period of 500, i.e. the first 500 generated variables were discarded and

Table 1 Simulated 95 % percentiles of the distribution of $\mathcal{Q}_n$ under H_0

n	400	600	800	1,000	∞
	1.29	1.32	1.33	1.34	1.36

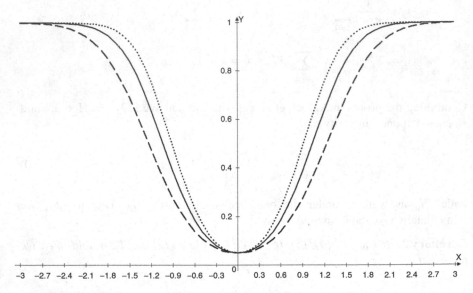

Fig. 1 Empirical power for $\mathcal{Q}_n$ with significance level 0.05, $n = 400$ (*dashed*), $n = 600$ (*solid*) and $n = 800$ (*dotted*) with $k_1 = n/2$

the next n give the sample $(e_1, \ldots, e_n)$. Thus $(e_1, \ldots, e_n)$ are from the stationary solution of (3). We trimmed the sample using $d(n) = \lfloor n^{0.45} \rfloor$ and computed

$$\mathcal{Q}_n = \left(\frac{2 - \alpha}{\alpha} \right)^{1/2} \left(\frac{1 - \rho}{1 + \rho} \right)^{1/2} \frac{1}{A_n} \sup_{0 \leq x \leq 1} |T_n(x)|.$$

Under H_0 we have

$$\mathcal{Q}_n \xrightarrow{\mathscr{D}} \sup_{0 \leq x \leq 1} |B(x)|.$$

The critical values in Table 1 provide information on the rate of convergence in Theorem 2.

Figures 1 and 2 show the empirical power of the test for H_0 against H_A based on the statistic $\mathcal{Q}_n$ for a change at time $k^* = n/4$ and $n/2$ and when the location changes from 0 to $c \in \{-3, -2.9, \ldots, 2.9, 3\}$ and the level of significance is 0.05. We used the asymptotic critical value 1.36. Comparing Figs. 1 and 2 we see that we have higher power when the change occurs in the middle of the data at $k^* = n/2$. We provided these results to illustrate the behaviour of functionals of T_n without introducing further noise due to the estimation of the norming sequence.

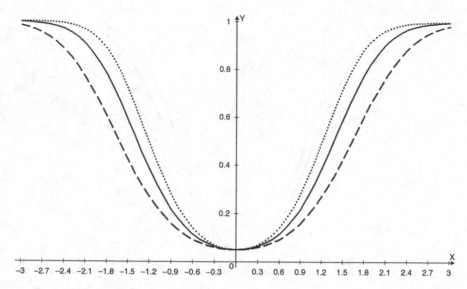

Fig. 2 Empirical power for $\mathcal{Q}_n$ with significance level 0.05, $n = 400$ (*dashed*), $n = 600$ (*solid*) and $n = 800$ (*dotted*) with $k_1 = n/4$

Table 2 Simulated 95 % percentiles of the distribution of $\tilde{\mathcal{Q}}_n$ under H_0

n	400	600	800	1,000	∞
	1.57	1.52	1.50	1.49	1.36

Next we study the applicability of (21) in case of small and moderate sample sizes. We used $h(n) = n^{1/2}$ as the window and the flat top kernel

$$\omega(t) = \begin{cases} 1 & 0 \le t \le .1 \\ 1.1 - |t| & .1 \le t \le 1.1 \\ 0 & t \ge 1.1 \end{cases}$$

Figures 3 and 4 show the empirical power of the test for H_0 against H_A based on the statistic $\tilde{\mathcal{Q}}_n$ for a change at time $k^* = n/4$ and $n/2$ and when the location changes from 0 to $c \in \{-3, -2.9, \ldots, 2.9, 3\}$ and the level of significance is 0.05. We used the asymptotic critical value 1.36 (Table 2). Comparing Figs. 3 and 4 we see that we have again higher power when the change occurs in the middle of the data at $k_1 = n/2$.

Figure 5 shows how the power of the test behaves depending on the value of $d = n^\epsilon$, $\epsilon \in \{0.3, 0.35, 0.42, 0.45, 0.5\}$ for $n = 400$. The bigger the d is, the better is the power curve.

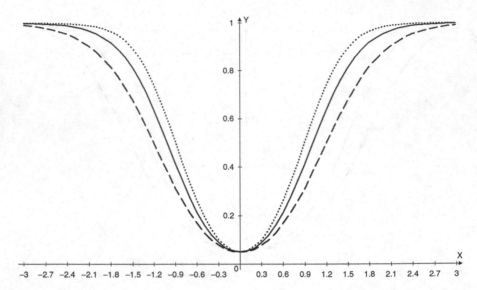

Fig. 3 Empirical power for $\tilde{\mathcal{D}}_n$ with significance level 0.05, $n = 400$ (*dashed*), $n = 600$ (*solid*) and $n = 800$ (*dotted*) with $k_1 = n/2$

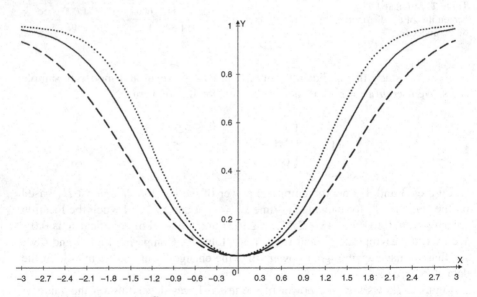

Fig. 4 Empirical power for $\tilde{\mathcal{D}}_n$ with significance level 0.05, $n = 400$ (*dashed*), $n = 600$ (*solid*) and $n = 800$ (*dotted*) with $k_1 = n/4$

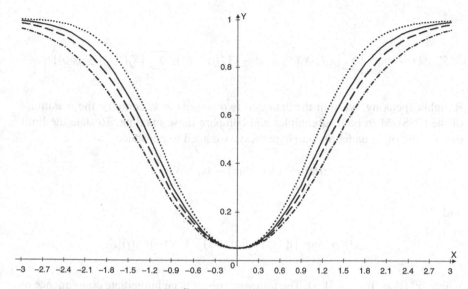

Fig. 5 Empirical power curves for $\tilde{\mathcal{Q}}_n$ with significance level 0.05 for $d = n^\epsilon$, $\epsilon = 0.35$ (*dash-dotted*), $\epsilon = 0.42$ (*dashed*), $\epsilon = 0.45$ (*solid*), $\epsilon = 0.5$ (*dotted*) with $n = 400$, $k_1 = n/2$

3 Ratio Statistics

The statistics $\hat{Q}_n$ as well as $\tilde{Q}_n$ are very sensitive to the behaviour of $\hat{s}_n$ and $\tilde{s}_n$. As we pointed out, $\hat{s}_n$ is the right norming only under H_0. The sequence $\tilde{Q}_n$ works under H_0 and under the one change alternative, but it could break down if multiple changes occur under the alternative. Even if the Bartlett type estimator is the asymptotically correct norming factor, the rate of convergence can be slow. Also, these estimators are very sensitive to the choice of the window $h = h(n)$. Following the work of Kim [19] (cf. also Kim et al. [20]) and Leybourne and Taylor [21], Horváth et al. [17] proposed ratio type statistics of functionals of CUSUM processes. We adapt their approach to the trimmed CUSUM process. Let $0 < \delta < 1$ and define

$$Z_n = \max_{n\delta \le k \le n - n\delta} \frac{Z_{n,1}(k)}{Z_{n,2}(k)},$$

where

$$Z_{n,1}(k) = \max_{1 \le i \le k} \left| \sum_{j=1}^{i}(X_j I\{|X_j| \le \eta_{n,d}\} - (1/k)\sum_{j=1}^{k}(X_j I\{|X_j| \le \eta_{n,d}\})) \right|$$

and

$$Z_{n,2}(k) = \max_{k<i\leq n} \left| \sum_{j=i}^{n}(X_j I\{|X_j| \leq \eta_{n,d}\} - (1/(n-k)) \sum_{j=k+1}^{n} (X_j I\{|X_j| \leq \eta_{n,d})\}) \right| .$$

Roughly speaking, we split the data into two subsets at k, compute the maximum of the CUSUM in both subsamples and compare these maxima. To state the limit distribution of Z_n under the null hypothesis, we need to introduce

$$z_1(t) = \sup_{0\leq s\leq t} |W(s) - (s/t)W(t)|$$

and

$$z_2(t) = \sup_{t\leq s\leq 1} |W^*(s) - ((1-s)/(1-t))W^*(t)|,$$

where $W^*(t) = W(1) - W(t)$. The following result is an immediate consequence of Theorem 2.

Theorem 3. *If H_0, (3), (4), (5), (6) and (9), (10), (11), (12) hold, then we have that*

$$Z_n \xrightarrow{\mathscr{D}} \sup_{\delta\leq t\leq 1-\delta} \frac{z_1(t)}{z_2(t)}. \tag{22}$$

We reject the no change null hypothesis if Z_n is large. Using Monte Carlo simulations, it is easy to obtain the distribution function of the limit in (22). Selected critical values can be found in Horváth et al. [17], where some probabilistic properties of the limit are also discussed.

Below we study the finite sample behaviour of Z_n. Table 3 contains simulated significance levels when $\delta = .2$, $n = 400, 600, 800, 1,000$ and $n = 5,000$. (Since the distribution function of the limit in (22) is unknown, we used $n = 5,000$ for the limit distribution.)

Figures 6 and 7 contain the empirical power curves of the test for H_0 against H_A based on the statistic Z_n for a change at time $k^* = n/4$ and $n/2$ and when the location changes from 0 to $c \in \{-5, -4.9, \ldots, 4.9, 5\}$ and the level of significance is 0.05. We used critical values from Table 3. Figure 8 shows how the power of the test behaves depending on the value of $d = n^\epsilon$, $\epsilon \in \{0.3, 0.35, 0.42, 0.45, 0.5\}$ for $n = 400$. The bigger the d is, the better is the power curve.

Table 3 Simulated 95 % percentiles of the distribution of Z_n under H_0

n	400	600	800	1,000	5,000
	5.90	5.67	5.49	5.43	5.03

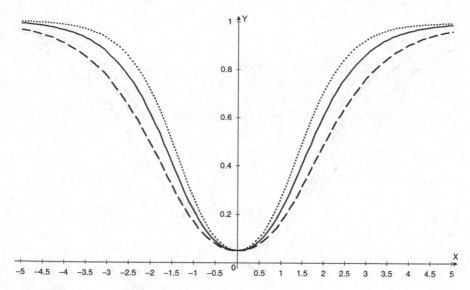

Fig. 6 Empirical power curves for Z_n with significance level 0.05, $n = 400$ (*dashed*), $n = 600$ (*solid*) and $n = 800$ (*dotted*) with $k_1 = n/2$

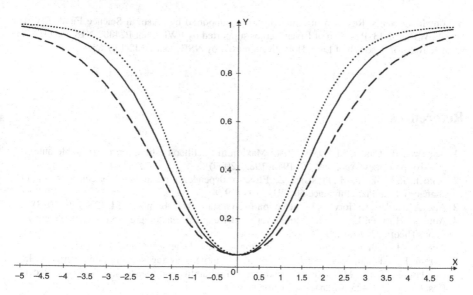

Fig. 7 Empirical power curves for Z_n with significance level 0.05, $n = 400$ (*dashed*), $n = 600$ (*solid*) and $n = 800$ (*dotted*) with $k_1 = n/4$

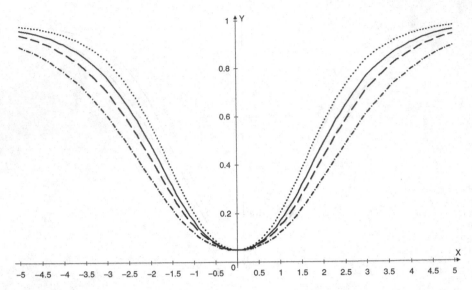

Fig. 8 Empirical power curves for Z_n with significance level 0.05 for $d = n^\epsilon$, $\epsilon = 0.35$ (*dash-dotted*), $\epsilon = 0.42$ (*dashed*), $\epsilon = 0.45$ (*solid*), $\epsilon = 0.5$ (*dotted*) with $n = 400$, $k_1 = n/2$

Acknowledgements Research of Alina Bazarova supported by Austrian Science Fund (FWF), Projekt W1230-N13. Research of István Berkes supported by FWF grant P24302-N18 and OTKA grant K 108615. Research of Lajor Horváth supported by NSF grant DMS 1305858.

References

1. Andrews, B., Calder, M., Davis, R.A.: Maximum likelihood estimation for α–stable autoregressive processes. Ann. Stat. **37**, 1946–1982 (2009)
2. Antoch, J., Hušková, M., Prašková, Z.: Effect of dependence on statistics for determination of change. J. Stat. Plan. Inference **60**, 291–310 (1997)
3. Aue, A., Berkes, I., Horváth, L.: Selection from a stable box. Bernoulli **14**, 125–139 (2008)
4. Aue, A., Horváth, L.: A limit theorem for mildly explosive autoregression with stable errors. Econ. Theory **23**, 201–220 (2007)
5. Aue, A., Horváth, L.: Stuctural breaks in time series. J. Time Ser. Anal. **34**, 1–16 (2013)
6. Avram, F.S., Taqqu, M.S.: Weak convergence of moving averages with infinite variance. In: Eberlain, E., Taqqu, M.S. (eds.) Dependence in Probability and Statistics: A Survey of Recent Results, pp. 399–425. Birkhäuser, Boston (1986)
7. Avram, F.S., Taqqu, M.S.: Weak convergence of moving averages in the α–stable domain of attraction. Ann. Probab. **20**, 483–503 (1992)
8. Bazarova, A., Berkes, I., Horváth, L.: Timmed stable AR(1) processes. Stoch. Process. Appl. **124**, 3441–3462 (2014)
9. Berkes, I., Horváth, L., Schauer, J.: Asymptotics trimmed CUSUM statistics. Bernoulli **17**, 1344–1367 (2011)
10. Billingsley, P.: Convergence of Probability Measures. Wiley, New York (1968)

11. Chan, N.H.: Inference for near–integrated time series with infinite variance. J. Am. Stat. Assoc. **85**, 1069–1074 (1990)
12. Chan, N.H., Tran, L.T.: On the first–order autoregressive processes with infinite variance. Econ. Theory **5**, 354–362 (1989)
13. Davis, R.A., Mikosch, T.: The sample autocorrelations of heavy–tailed processes with applications to ARCH. Ann. Stat. **26**, 2049–2080 (1998)
14. Davis, R., Resnick, S.: Limit theory for moving averages of random variables with regularly varying tail probabilities. Ann. Probab. **13**, 179–195 (1985)
15. Davis, R., Resnick, S.: Limit theory for the sample covariance and correlation functions of moving averages. Ann. Stat. **14**, 533–558 (1986)
16. Fama, E.: The behavior of stock market price. J. Bus. **38**, 34–105 (1965)
17. Horváth, L., Horváth, Zs., Hušková, M.: Ratio test for change point detection. IMS Collect. **1**, 293–304 (2008)
18. Horváth, L., Reeder, R.: Detecting changes in functional linear models. J. Multivar. Anal. **111**, 310–334 (2012)
19. Kim, J.-Y.: Detection of change in persistence of a linear time series. J. Econ. **95**, 97–116 (2000)
20. Kim, J.-Y., Belaire–Franch, J., Amador, R.: Corrigendum to "Detection of change in persistence of a linear time series". J. Econ. **109**, 389–392 (2002)
21. Leybourne, S., Taylor, A.: Persistence change tests and shifting stable autoregressions. Econ. Lett. **91**, 44–49 (2006)
22. Liu, W., Wu, W.B.: Asymptotics of spectral density estimates. Econ. Theory **26**, 1218–1245 (2010)
23. Mandelbrot, B.B.: The variation of certain speculative price. J. Bus. **36**, 394–419 (1963)
24. Mandelbrot, B.B.: The variation of other speculative prices. J. Bus. **40**, 393–413 (1967)
25. Mikosch, T., Resnick, S., Samorodnitsky, G.: The maximum of the periodogram for a heavy–tailed sequence. Ann. Probab. **28**, 885–908 (2000)
26. Phillips, P.C.B., Solo, V.: Asymptotics for linear processes. Ann. Statist. **20**, 971–1001 (1992)
27. Taniguchi, M., Kakizawa, Y.: Asymptotic Theory of Statistical Inference for Time Series. Springer, New York (2000)
28. Zhang, R.-M., Chan, N.H.: Maximum likelihood estimation for nearly non–stationary stable autoregressive processes. J. Time Ser. Anal. **33**, 542–553 (2012)

Change-Point Detection Under Dependence Based on Two-Sample U-Statistics

Herold Dehling, Roland Fried, Isabel Garcia, and Martin Wendler

1 Introduction

Change-point tests address the question whether a stochastic process is stationary during the entire observation period or not. In the case of independent data, there is a well-developed theory; see the book by Csörgő and Horváth [6] for an excellent survey. When the data are dependent, much less is known. The CUSUM statistic has been intensely studied, even for dependent data; see again Csörgő and Horváth [6]. The CUSUM test, however, is not robust against outliers in the data. In the present paper, we study a robust test which is based on the two-sample Wilcoxon test statistic. Simulations show that this test outperforms the CUSUM test in the case of heavy-tailed data.

In order to derive the asymptotic distribution of the test, we study the stochastic process

$$\sum_{i=1}^{[n\lambda]} \sum_{j=[n\lambda]+1}^{n} h(X_i, X_j), \ 0 \le \lambda \le 1, \tag{1}$$

H. Dehling (✉) • I. Garcia
Fakultät für Mathematik, Ruhr-Universität Bochum, 44780 Bochum, Germany
e-mail: herold.dehling@rub.de; isabel.garciaarboleda@rub.de

R. Fried
Fakultät für Statistik, Technische Universität Dortmund, 44221 Dortmund, Germany
e-mail: fried@statistik.tu-dortmund.de

M. Wendler
Institut für Mathematik und Informatik, Ernst-Moritz-Arndt-Universität,
17487 Greifswald, Germany
e-mail: martin.wendler@rub.de

© Springer Science+Business Media New York 2015
D. Dawson et al. (eds.), *Asymptotic Laws and Methods in Stochastics*,
Fields Institute Communications 76, DOI 10.1007/978-1-4939-3076-0_12

195

where $h : \mathbb{R}^2 \to \mathbb{R}$ is a kernel function. In the case of independent data, the asymptotic distribution of this process has been studied by Csörgő and Horváth [5]. In the present paper, we extend their result to short range dependent data $(X_i)_{i\geq 1}$. Similar results have been obtained for long range dependent data by Dehling, Rooch and Taqqu [10], albeit with completely different methods.

U-statistics have been introduced by Hoeffding [14], where the asymptotic normality was established both for the one-sample as well as the two-sample U-statistic in the case of independent data. The asymptotic distribution of one-sample U-statistics of dependent data was studied by Sen [18, 19], Yoshihara [22], Denker and Keller [12, 13] and by Borovkova, Burton and Dehling [3] in the so-called non-degenerate case, and by Babbel [1] and Leucht [16] in the degenerate case. For two-sample U-statistics, Dehling and Fried [8] established the asymptotic normality of $\sum_{i=1}^{n_1} \sum_{j=n_1+1}^{n_1+n_2} h(X_i, X_j)$ for dependent data, when $n_1, n_2 \to \infty$. The main theoretical result of the present paper is a functional version of this limit theorem.

In our paper, we focus on data that can be represented as functionals of a mixing process. In this way, we cover most examples from time series analysis, such as ARMA and ARCH processes, but also data from chaotic dynamical systems. For a survey of processes that have a representation as functional of a mixing process, see e.g. Borovkova, Burton and Dehling [3]. Earlier references can be found in Ibragimov and Linnik [15], Denker [11] and Billingsley [2].

2   Definitions and Main Results

Given the samples $X_1, \ldots, X_{n_1}$ and $Y_1, \ldots, Y_{n_2}$, and a kernel $h(x, y)$, we define the two-sample U-statistic

$$U_{n_1,n_2} := \frac{1}{n_1 n_2} \sum_{i=1}^{n_1} \sum_{j=1}^{n_2} h(X_i, Y_j). \tag{2}$$

More generally, one can define U-statistics with multivariate kernels $h : \mathbb{R}^k \times \mathbb{R}^l \to \mathbb{R}$. In the present paper, for the ease of exposition, we will restrict attention to bivariate kernels $h(x, y)$. The main results, however, can easily be extended to the multivariate case.

Assuming that $(X_i)_{i\geq 1}$ and $(Y_i)_{i\geq 1}$ are stationary processes with one-dimensional marginal distribution functions F and G, respectively, we can test the hypothesis $H : F = G$ using the two-sample U-statistic. E.g., the kernel $h(x, y) = y - x$ leads to the U-statistic

$$U_{n_1,n_2} = \frac{1}{n_1 n_2} \sum_{i=1}^{n_1} \sum_{j=1}^{n_2} (Y_j - X_i) = \frac{1}{n_2} \sum_{j=1}^{n_2} Y_j - \frac{1}{n_1} \sum_{i=1}^{n_1} X_i, \tag{3}$$

and thus to the familiar two-sample Gauß-test. Similarly, the kernel $h(x, y) = 1_{\{x \leq y\}}$ leads to the U-statistic

$$U_{n_1, n_2} = \frac{1}{n_1 \, n_2} \sum_{i=1}^{n_1} \sum_{j=1}^{n_2} 1_{\{X_i \leq X_j\}}, \tag{4}$$

and thus to the 2-sample Mann-Whitney-Wilcoxon test.

In the present paper, we investigate tests for a change-point in the mean of a stochastic process $(X_i)_{i \geq 1}$. We consider the model

$$X_i = \mu_i + \xi_i, \ i \geq 1, \tag{5}$$

where $(\mu_i)_{i \geq 1}$ are unknown constants and where $(\xi_i)_{i \geq 1}$ is a stochastic process. We want to test the hypothesis $H : \ \mu_1 = \ldots = \mu_n$ against the alternative that there exists $1 \leq k \leq n - 1$ such that $\mu_1 = \ldots = \mu_k \neq \mu_{k+1} = \ldots = \mu_n$.

Tests for the change-point problem are often derived from 2-sample tests applied to the samples $X_1, \ldots, X_k$ and $X_{k+1}, \ldots, X_n$, for all possible $1 \leq k \leq n - 1$. For two-sample tests based on U-statistics with kernel $h(x, y)$, this leads to the test statistic $\sum_{i=1}^{k} \sum_{j=k+1}^{n} h(X_i, X_j)$, $1 \leq k \leq n$, and thus to the process

$$U_n(\lambda) = \sum_{i=1}^{[n\lambda]} \sum_{j=[n\lambda]+1}^{n} h(X_i, X_j), \ 0 \leq \lambda \leq 1. \tag{6}$$

In this paper, we will derive a functional limit theorem for the process $(U_n(\lambda))_{0 \leq \lambda \leq 1}$, $n \geq 1$. Specifically, we will show that under certain technical assumptions on the kernel h and on the process $(X_i)_{i \geq 1}$, a properly centered and renormalized version of $(U_n(\lambda))_{0 \leq \lambda \leq 1}$ converges to a Gaussian process.

In our paper, we will assume that the process $(\xi_i)_{i \geq 0}$ is weakly dependent. More specifically, we will assume that $(\xi_i)_{i \geq 0}$ can be represented as a functional of an absolutely regular process.

Definition 1. (i) Given a stochastic process $(X_n)_{n \in \mathbb{Z}}$ we denote by $\mathscr{A}_l^k$ the σ−algebra generated by $(X_k, \ldots, X_l)$. The process is called absolutely regular if

$$\beta(k) = \sup_n \left\{ \sup \sum_{j=1}^{J} \sum_{i=1}^{I} |P(A_i \cap B_j) - P(A_i)P(B_j)| \right\} \to 0, \tag{7}$$

as $k \to \infty$, where the last supremum is over all finite $\mathscr{A}_1^n$−measurable partitions $(A_1, \ldots, A_I)$ and all finite $\mathscr{A}_{n+k}^{\infty}$−measurable partitions $(B_1, \ldots, B_J)$.

(ii) The process $(X_n)_{n \geq 1}$ is called a two-sided functional of an absolutely regular sequence if there exists an absolutely regular process $(Z_n)_{n \in \mathbb{Z}}$ and a measurable function $f : \mathbb{R}^{\mathbb{Z}} \to \mathbb{R}$ such that

$$X_i = f((Z_{i+n})_{n \in \mathbb{Z}}).$$

Analogously, $(X_n)_{n\geq 1}$ is called a one-sided functional if $X_i = f((Z_{i+n})_{n\geq 0})$.

(iii) The process $(X_n)_{n\geq 1}$ is called 1-approximating functional with coefficients $(a_k)_{k\geq 1}$ if

$$E\,|X_i - E(X_i|Z_{i-k},\ldots,Z_{i+k})| \leq a_k. \tag{8}$$

In addition to weak dependence conditions on the process $(X_i)_{i\geq 1}$, the asymptotic analysis of the process (6) requires some continuity assumptions on the kernel functions $h(x,y)$. We use the notion of 1-continuity, which was introduced by Borovkova, Burton and Dehling [3]. Alternative continuity conditions have been used by Denker and Keller [13].

Definition 2. The kernel $h(x,y)$ is called 1-continuous, if there exists a function $\phi : (0,\infty) \to (0,\infty)$ with $\phi(\epsilon) = o(1)$ as $\epsilon \to 0$ such that for all $\epsilon > 0$

$$E(|h(X',Y) - h(X,Y)|1_{\{|X-X'|\leq\epsilon\}}) \leq \phi(\epsilon) \tag{9}$$

$$E(|h(X,Y') - h(X,Y)|1_{\{|Y-Y'|\leq\epsilon\}}) \leq \phi(\epsilon) \tag{10}$$

for all random variables X, X', Y and Y' having the same marginal distribution as X_1, and such that X, Y are either independent or have joint distribution $P_{(X_1,X_k)}$, for some integer k.

The most important technical tool in the study of U-statistics is Hoeffding's decomposition, originally introduced by Hoeffding [14]. If $E|h(X,Y)| < \infty$ for two independent random variables X and Y with the same distribution as X_1, we can write

$$h(x,y) = \theta + h_1(x) + h_2(y) + g(x,y), \tag{11}$$

where the terms on the right-hand side are defined as follows:

$$\theta = \iint h(x,y)dF(x)dF(y)$$

$$h_1(x) = \int h(x,y)dF(y) - \theta$$

$$h_2(y) = \int h(x,y)dF(x) - \theta$$

$$g(x,y) = h(x,y) - h_1(x) - h_2(y) - \theta.$$

Here, F denotes the distribution function of the random variables X_i. Observe that, by Fubini's theorem,

$$E(h_1(X)) = E(h_2(X)) = 0.$$

In addition, the kernel $g(x, y)$ is degenerate in the sense of the following definition.

Definition 3. Let $(X_i)_{i \geq 1}$ be a stationary process, and let $g(x, y)$ be a measurable function. We say that $g(x, y)$ is degenerate if

$$E(g(x, X_1)) = E(g(X_1, y)) = 0, \tag{12}$$

for all $x, y \in \mathbb{R}$.

The following theorem, a functional central limit theorem for two-sample U-statistics of dependent data, is the main theoretical result of the present paper.

Theorem 1. *Let $(X_n)_{n \geq 1}$ be a 1-approximating functional with constants $(a_k)_{k \geq 1}$ of an absolutely regular process with mixing coefficients $(\beta(k))_{k \geq 1}$, and let $h(x, y)$ be a 1-continuous bounded kernel, satisfying*

$$\sum_{k=1}^{\infty} k^2 (\beta(k) + \sqrt{a_k} + \phi(a_k)) < \infty, \tag{13}$$

Then, as $n \to \infty$, the $D[0, 1]$-valued process

$$T_n(\lambda) := \frac{1}{n^{3/2}} \sum_{i=1}^{[\lambda n]} \sum_{j=[\lambda n]+1}^{n} (h(X_i, X_j) - \theta), \ 0 \leq \lambda \leq 1, \tag{14}$$

converges in distribution towards a mean-zero Gaussian process with representation

$$Z(\lambda) = (1 - \lambda) W_1(\lambda) + \lambda (W_2(1) - W_2(\lambda)), \ 0 \leq \lambda \leq 1, \tag{15}$$

where $(W_1(\lambda), W_2(\lambda))_{0 \leq \lambda \leq 1}$ is a two-dimensional Brownian motion with mean zero and covariance function $\mathrm{Cov}(W_k(s), W_l(t)) = \min(s, t)\sigma_{kl}$, where

$$\sigma_{kl} = E(h_k(X_0) h_l(X_0)) + 2 \sum_{j=1}^{\infty} \mathrm{Cov}(h_k(X_0), h_l(X_j)), \ k, l = 1, 2. \tag{16}$$

Remark 1. (i) In the case of i.i.d. data, Theorem 1 was established by Csörgő and Horváth [5]. In the case of long-range dependent data, weak convergence of the process $(T_n(\lambda))_{0 \leq \lambda \leq 1}$ has been studied by Dehling, Rooch and Taqqu [10] and by Rooch [17], albeit with a normalization different from $n^{3/2}$.

(ii) Using the representation (15), one can calculate the autocovariance function of the process $(Z(\lambda))_{0 \leq \lambda \leq 1}$. We obtain

$$\mathrm{Cov}(Z(\lambda), Z(\mu)) = \sigma_{11}[(1 - \lambda)(1 - \mu) \min\{\lambda, \mu\}]$$
$$+ \sigma_{22}[\lambda \mu (1 - \mu - \lambda + \min\{\lambda, \mu\})]$$
$$+ \sigma_{12}[\mu(1 - \lambda)(\lambda - \min\{\lambda, \mu\}) + \lambda(1 - \mu)(\mu - \min\{\lambda, \mu\})].$$

(iii) We conjecture that a similar theorem also holds for unbounded kernels under some moments conditions and faster mixing rates (similar to Theorem 2.7 of Sharipov, Wendler [20]). As our main application is the Wilcoxon test, where the kernel is bounded, we restrict the theorem to the case of bounded kernels.

(iv) For the kernel $h(x, y) = y - x$, we can analyze the asymptotic behavior of the process $T_n(\lambda)$ using the functional central limit theorem (FCLT). Note that, since $X_j - X_i = (X_j - E(X_j)) - (X_i - E(X_i))$, we may assume without loss of generality that X_i has mean zero. Then we get the representation

$$T_n(\lambda) = \frac{1}{n^{3/2}} \sum_{i=1}^{[n\lambda]} \sum_{j=[n\lambda]+1}^{n} (X_j - X_i)$$

$$= \frac{[n\lambda]}{n} \frac{1}{\sqrt{n}} \sum_{i=1}^{n} X_i - \frac{1}{\sqrt{n}} \sum_{i=1}^{[n\lambda]} X_i. \tag{17}$$

Thus, weak convergence of $(T_n(\lambda))_{0 \leq \lambda \leq 1}$ can be derived from the FCLT for the partial sum process $\frac{1}{\sqrt{n}} \sum_{i=1}^{[n\lambda]} X_i$. Such FCLTs have been proved under a wide range of conditions, e.g. for functionals of uniformly mixing data in Billingsley [2].

We finally want to state an important special case of Theorem 1, namely when the kernel is anti-symmetric, i.e. when $h(x, y) = -h(y, x)$. Kernels that occur in connection with change-point tests usually have this property. For anti-symmetric kernels, the limit process has a much simpler structure; moreover one can give a simpler direct proof in this case. Note that for independent random variables X, Y we have by anti-symmetry that $Eh(X, Y) = -Eh(Y, X) = -Eh(X, Y)$ and so $\theta = Eh(X, Y) = 0$.

Theorem 2. Let $(X_n)_{n \geq 1}$ be a 1-approximating functional with constants $(a_k)_{k \geq 1}$ of an absolutely regular process with mixing coefficients $(\beta(k))_{k \geq 1}$, and let $h(x, y)$ be a 1-continuous bounded anti-symmetric kernel, such that (13) holds. Then, as $n \to \infty$, the $D[0, 1]$-valued process

$$T_n(\lambda) := \frac{1}{n^{3/2}} \sum_{i=1}^{[\lambda n]} \sum_{j=[\lambda n]+1}^{n} h(X_i, X_j), \ 0 \leq \lambda \leq 1, \tag{18}$$

converges in distribution towards the mean-zero Gaussian process $\sigma W^{(0)}(\lambda)$, $0 \leq \lambda \leq 1$, where $(W^0(\lambda))_{0 \leq \lambda \leq 1}$ is a standard Brownian bridge and

$$\sigma^2 = \text{Var}(h_1(X_1)) + 2 \sum_{i=2}^{\infty} \text{Cov}(h_1(X_1), h_1(X_k)). \tag{19}$$

3 Application to Change Point Problems

In this section, we will apply Theorem 1 in order to derive the asymptotic distribution of two change-point test statistics. Specifically, we wish to test the hypothesis

$$H_0 : \mu_1 = \ldots = \mu_n \tag{20}$$

against the alternative of a level shift at an unknown point in time, i.e.

$$H_A : \mu_1 = \ldots = \mu_k \neq \mu_{k+1} = \ldots = \mu_n, \text{ for some } k \in \{1, \ldots, n-1\}. \tag{21}$$

We consider the following two test statistics,

$$T_{1,n} = \max_{1 \leq k < n} \left| \frac{1}{n^{3/2}} \sum_{i=1}^{k} \sum_{j=k+1}^{n} \left(1_{\{X_i < X_j\}} - 1/2 \right) \right| \tag{22}$$

$$T_{2,n} = \max_{1 \leq k < n} \left| \frac{1}{n^{3/2}} \sum_{i=1}^{k} \sum_{j=k+1}^{n} \left(X_j - X_i \right) \right|. \tag{23}$$

Theorem 3. *Let $(X_n)_{n \geq 1}$ be a 1-approximating functional with constants $(a_k)_{k \geq 1}$ of an absolutely regular process with mixing coefficients $(\beta(k))_{k \geq 1}$, satisfying (13), and assume that X_1 has a distribution function $F(x)$ with bounded density. Then, under the null hypothesis H_0,*

$$T_{1,n} \to \sigma_1 \sup_{0 \leq \lambda \leq 1} |W^{(0)}(\lambda)|, \tag{24}$$

where $(W^{(0)}(\lambda))_{0 \leq \lambda \leq 1}$ denotes the standard Brownian bridge process, and where

$$\sigma_1^2 = \mathrm{Var}(F(X_1)) + 2 \sum_{k=2}^{\infty} \mathrm{Cov}(F(X_1), F(X_k)). \tag{25}$$

Assuming that $E|X_i|^{2+\delta} < \infty$, $\beta(k) = O(k^{-(2+\delta)/\delta})$ and $a_k = O(k^{-(1+\delta)/2\delta})$, and under the null hypothesis H_0,

$$T_{2,n} \to \sigma_2 \sup_{0 \leq \lambda \leq 1} |W^{(0)}(\lambda)|, \tag{26}$$

where

$$\sigma_2^2 = \mathrm{Var}(X_1) + 2 \sum_{k=2}^{\infty} \mathrm{Cov}(X_1, X_k). \tag{27}$$

Proof. We will establish weak convergence of $T_{1,n}$. In order to do so, we will apply Theorem 1 to the kernel $h(x, y) = 1_{\{x<y\}}$. Borovkova, Burton and Dehling [3] showed that this kernel is 1-continuous. By continuity of the distribution function of X_1, we get that $\theta = \iint 1_{\{x<y\}} dF(x) dF(y) = 1/2$. Moreover, we get

$$h_1(x) = P(x < X_1) - \frac{1}{2} = \frac{1}{2} - F(x)$$

$$h_2(x) = P(X_1 < x) - \frac{1}{2} = F(x) - \frac{1}{2}.$$

Note that $h_2(x) = -h_1(x)$. Hence $W_2(\lambda) = -W_1(\lambda)$, and thus the limit process in Theorem 1 has the representation

$$Z(\lambda) = (1 - \lambda)W_1(\lambda) + \lambda(W_2(1) - W_2(\lambda)) = W_1(\lambda) - \lambda W_1(1).$$

Here $W_1(\lambda)$ is a Brownian motion with variance σ_1^2. Weak convergence of $T_{2,n}$ can be shown directly from the functional central limit theorem for the partial sum process; see Corollary 3.2 of Wooldridge and White [21]. We have to check the L_2-near epoch dependence. Note that by our assumptions

$$E\,|X_0 - E[X_0|Z_{-l}, \ldots, Z_l]|^2$$

$$= E\left[|X_0 - E[X_0|Z_{-l}, \ldots, Z_l]|^2 \, 1_{\{|X_0 - E[X_0|Z_{-l}, \ldots, Z_l]| \le a_l^{-\frac{1}{1+\delta}}\}}\right]$$

$$+ E\left[|X_0 - E[X_0|Z_{-l}, \ldots, Z_l]|^2 \, 1_{\{|X_0 - E[X_0|Z_{-l}, \ldots, Z_l]| > a_l^{-\frac{1}{1+\delta}}\}}\right]$$

$$\le a_l^{-\frac{1}{1+\delta}} E\,|X_0 - E[X_0|Z_{-l}, \ldots, Z_l]| + a_l^{\frac{\delta}{1+\delta}} E\,|X_0 - E[X_0|Z_{-l}, \ldots, Z_l]|^{2+\delta}$$

$$\le C a_l^{\frac{\delta}{1+\delta}} = O(l^{-1/2}), \qquad (28)$$

so the condition of Corollary 3.2 of Wooldridge and White [21] holds. Hence, the partial sum process $(\frac{1}{\sqrt{n}} \sum_{i=1}^{[nt]} X_i)_{0 \le t \le 1}$ converges in distribution to $(\sigma_2 W(t))_{0 \le t \le 1}$, where W is standard Brownian motion. Convergence in distribution of $T_{2,n}$ follows by an application of the continuous mapping theorem.

Remark 2. (i) The distribution of $\sup_{0 \le \lambda \le 1} |W(\lambda)|$ is the well-known Kolmogorov-Smirnov distribution. Quantiles of the Kolmogorov-Smirnov distribution can be found in most statistical tables.

(ii) In order to apply Theorem 3, we need to estimate the variances σ_1^2 and σ_2^2. Regarding σ_2^2 given in expression (27), we apply the non-overlapping subsampling estimator

$$\hat{\sigma}_2^2 = \frac{1}{[n/l_n]} \sum_{i=1}^{[n/l_n]} \frac{1}{l_n} \left(\sum_{j=(i-1)l_n+1}^{il_n} X_j - \frac{l_n}{n} \sum_{j=1}^{n} X_j \right)^2 \tag{29}$$

investigated by Carlstein [4] for α-mixing data. In case of AR(1)-processes, Carlstein derives

$$l_n = \max(\lceil n^{1/3} (2\rho/(1-\rho^2))^{2/3} \rceil, 1) \tag{30}$$

as the choice of the block length which minimizes the MSE asymptotically, with ρ being the autocorrelation coefficient at lag 1.

Regarding σ_1^2 given in (25), one faces the additional challenge that the distribution function F is unknown. This problem has been addressed, e.g. in Dehling, Fried, Sharipov, Vogel and Wornowizki [9], for the case of functionals of absolutely regular processes and F being estimated by the empirical distribution function F_n. The authors find the subsampling estimator for σ_1^2

$$\hat{\sigma}_1 = \frac{1}{[n/l_n]} \sqrt{\frac{\pi}{2}} \sum_{i=1}^{[n/l_n]} \frac{1}{\sqrt{l_n}} \left| \sum_{j=(i-1)l_n+1}^{il_n} F_n(X_j) - \frac{l_n}{n} \sum_{j=1}^{n} F_n(X_j) \right|, \tag{31}$$

employing non-overlapping subsampling, to give smaller biases, but somewhat larger MSEs than the corresponding overlapping subsampling estimator. The adaptive choice of the block length l_n proposed by Carlstein worked well in their simulations if the data were generated from a stationary ARMA(1,1) model and an estimate of ρ was plugged in. In the next section, we will explore this and other proposals in situations with level shifts and normally or heavy-tailed innovations.

4 Simulation Results

The assumptions regarding the underlying process (X_i) in Theorem 1 are satisfied by a wide range of time series, such as AR and ARMA processes. To illustrate the results and to investigate the finite sample behavior and the power of the tests based on $T_{1,n}$ and $T_{2,n}$, we will give some simulation results. We study the underlying change-point model

$$X_i = \begin{cases} \xi_i & \text{if } i = 1, \ldots, [n\lambda] \\ \mu + \xi_i & \text{if } i = [n\lambda] + 1, \ldots, n. \end{cases} \tag{32}$$

Within this model, the hypothesis of no change is equivalent to $\mu = 0$. We assume that the noise follows an AR(1) process, i.e. that

$$\xi_i = \rho \xi_{i-1} + \epsilon_i, \tag{33}$$

Table 1 Empirical level of the tests based on $T_{1,n}$ and $T_{2,n}$, for $n = 200$, with fixed or adaptive subsampling block length l_n and overlapping (ol) or non-overlapping (nol) subsampling. The results are for AR(1) observations with different lag-one autocorrelations ρ and different t_3-distributed innovations, and based on 4000 simulation runs each

| | | $T_{1,n}$ | | | | | $T_{2,n}$ | | | | |
| | | | l_n fixed | | Adaptive | | | l_n fixed | | Adaptive | |
ν	ρ	Unadj.	ol	nol	ol	nol	Unadj.	ol	nol	ol	nol
∞	0.0	2.8	2.0	2.9	2.0	2.2	4.5	2.9	3.9	3.7	3.8
∞	0.4	24.5	2.5	3.1	3.5	3.9	34.2	3.9	4.9	5.5	6.0
∞	0.8	81.6	6.2	6.5	1.9	2.5	91.5	10.5	10.6	3.4	4.0
3	0.0	3.1	2.2	2.9	2.2	2.9	3.8	2.5	3.5	3.1	3.1
3	0.4	26.9	2.4	3.0	3.2	3.0	32.0	3.3	3.8	4.3	4.9
3	0.8	82.7	6.9	7.0	2.0	2.8	90.6	10.2	10.5	3.2	3.9

where $-1 < \rho < 1$, and where the innovations ϵ_i are i.i.d. random variables with mean zero, bounded density and finite second moments. The innovations ϵ_i are generated from a standard normal or a t_ν-distribution with $\nu = 3$ degrees of freedom, scaled to have the same 84.13 % percentile as the standard normal, which is 1. The autoregression coefficient is varied in $\rho = \{0.0, 0.4, 0.8\}$, corresponding to zero, moderate or strong positive autocorrelation, and the sample size is $n = 200$. For the choice of the block length we used Carlstein's adaptive rule outlined above, or a fixed block length of $l_n = 9$, which is in good agreement with the empirical findings of Dehling et al. [10] for larger sample sizes, and their theoretical result that l_n should be chosen as $o(\sqrt{n})$ to achieve consistency. For comparison, we also include tests employing overlapping subsampling for estimation of the asymptotical variance, applying the same block lengths as the non-overlapping versions.

Table 1 contains the empirical levels (i.e. the fraction of rejections) of the tests with an asymptotic level of 5 %, obtained from 4000 simulation runs for each situation. Note that the tests developed under the assumption of independence, not adjusting for autocorrelation, become strongly oversized with an increasingly positive autocorrelation, i.e. they reject a true null hypothesis far too often, and are practically useless already for $\rho = 0.4$. The performance of the adjusted tests is much better in this respect and in a good agreement with the asymptotic results. Only if the autocorrelation is strong ($\rho = 0.8$), the tests with a fixed block length become somewhat anti-conservative (oversized), and even more so for the CUSUM-test. Longer block lengths are needed for stronger positive autocorrelations, and Carlstein's adaptive block length (30) adjusts for this. There is little difference between the tests employing overlapping and non-overlapping subsampling here.

In order to investigate the powers of the tests under the alternative, we consider shifts of increasing height μ, generating 400 data sets for each situation. The sample size is again $n = 200$, and the change point is at observation number $\tau = [\lambda n] = 100$.

Figure 1 illustrates the powers of the different versions of the tests in case of Gaussian or t_3-distributed innovations and several autocorrelation coefficients ρ. Under normality, the CUSUM test $T_{2,n}$ is somewhat more powerful than the test $T_{1,n}$ based on the Wilcoxon statistic, while under the t_3-distribution it is the other way round. The CUSUM test with the fixed block length considered here becomes strongly oversized if ρ is large, while this effect is less severe for the test based on the Wilcoxon statistic. Carlstein's adaptive choice of the block length increases the power if ρ is small and improves the size of the test substantially if ρ is large. The tests employing overlapping subsampling (not shown here) perform even slightly more powerful in case of zero or moderate autocorrelations, but much less powerful in case of strong autocorrelations. We have also considered the case of negative autocorrelation ($\rho = -0.4$, not shown here). We obtained similar results for the power of the test based on the Wilcoxon statistic relatively to that of the CUSUM test, and little difference between using a fixed or the adaptive block length.

The tests with Carlstein's adaptive choice of the block length could be improved further by using a more sophisticated estimate of ρ than the ordinary sample autocorrelation used here. The latter is positively biased in the presence of a shift, which leads to too large choices of the block length. This negative effect becomes more severe for larger values of ρ, since the plug-in-estimate of the asymptotically MSE-optimal choice of l_n increases more rapidly if $\hat{\rho}$ is close to 1, while it is rather stable for moderate and small values of $\hat{\rho}$. In our study, for $\rho = 0$ the average value chosen for l_n increases from about 2 to about 3, only, as the height of the shift increases, while it increases from about 6 to about 9 if $\rho = 0.4$, and even from about 16 to about 24 if $\rho = 0.8$. An estimate of the autocorrelation coefficient which resists shifts could be used, e.g. by applying a stepwise procedure which estimates the possible time of occurrence of a shift before calculating $\hat{\rho}$ from the corrected data, but this will not be pursued here.

5 Data Example

For illustration we apply the tests to time series data representing the monthly average daily minimum temperatures in Potsdam, Germany, measured between January 1893 and December 1992. The 1200 data points for these 100 years have been deseasonalized by subtracting the median value from each calendar month, see Fig. 2. Our interest is in whether the level of this time series is constant or whether there is a monotonic change. Such a systematic change is likely to show a trend-like behavior and not a sharp shift, but nevertheless we would like a change-point test to detect such a change if its null hypothesis is a constant level.

The empirical autocorrelation and partial autocorrelation functions suggest a first order autoregressive model with lag-one autocorrelation about 0.25 for the deseasonalized data. The test statistics take their maximum values after time point 595, i.e. rather in the middle of time series. The resulting p-values are 0.23 and 0.16 for the CUSUM test with the fixed and the adaptive block length, respectively.

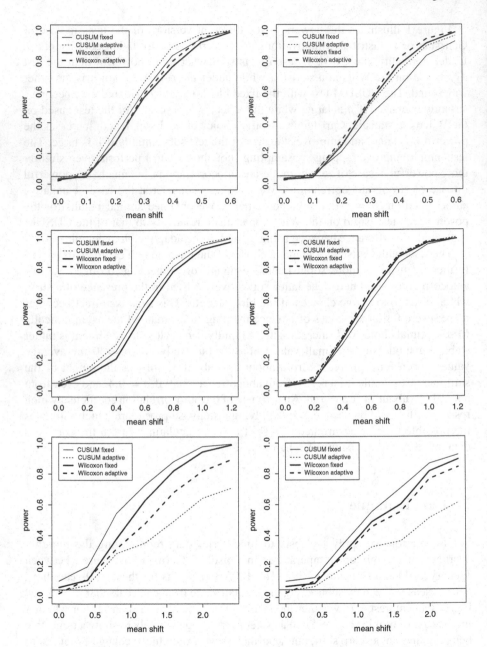

Fig. 1 Power of the tests in case of a shift in the middle of an AR(1) process with Gaussian (*left*) or t_3-innovations (*right*) and different lag one correlations $\rho = 0.0$ (*top*), $\rho = 0.4$ (*middle*) or $\rho = 0.8$ (*bottom*), $n = 200$. Wilcoxon test $T_{n,1}$ (*bold lines*) and CUSUM test $T_{n,2}$ (*thin lines*). Adjustment by non-overlapping subsampling with fixed (*black*) or adaptive block length (*dashed*)

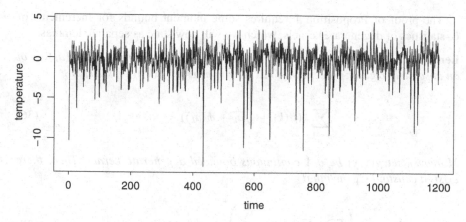

Fig. 2 Deseasonalized time series representing the monthly average daily minimum temperatures in Potsdam, Germany

As opposed to this, both versions of the Wilcoxon based test become significant as the corresponding p-values are 0.04 and 0.015, respectively. The differences between the results agree with the better power behavior of the Wilcoxon based test relatively to the CUSUM test in case of the (left-)skewed distributions of minimum temperatures, and the better power of the versions employing the adaptive block length over those with the fixed block length considered here in case of small positive autocorrelations. The sample median of the second time period is about 0.4 degrees larger than that of the first period.

6 Auxiliary Results

In this section, we will prove some auxiliary results which will play a crucial role in the proof of Theorem 1. The main result of this section is the following proposition, which essentially shows that the degenerate part in the Hoeffding decomposition of the U-statistic $T_n(\lambda)$ is uniformly negligible.

Proposition 1. *Let $(X_n)_{n\geq 1}$ be a 1-approximating functional with constants $(a_k)_{k\geq 1}$ of an absolutely regular process with mixing coefficients $(\beta(k))_{k\geq 1}$, satisfying*

$$\sum_{k=1}^{\infty} k(\beta(k) + \sqrt{a_k} + \phi(a_k)) < \infty. \tag{34}$$

Moreover, let $g(x, y)$ be a 1-continuous bounded degenerate kernel. Then, as $n \to \infty$,

$$\frac{1}{n^{3/2}} \sup_{0 \leq \lambda \leq 1} \left| \sum_{i=1}^{[n\lambda]} \sum_{j=[n\lambda]+1}^{n} g(X_i, X_j) \right| \to 0 \tag{35}$$

in probability.

The proof of Proposition 1 requires some moment bounds for increments of U-statistics of degenerate kernels, which we will now state as separate lemmas.

Lemma 1. *Let $(X_n)_{n\geq1}$ be a 1-approximating functional with constants $(a_k)_{k\geq1}$ of an absolutely regular process with mixing coefficients $(\beta(k))_{k\geq1}$, satisfying*

$$\sum_{k=1}^{\infty} k(\beta(k) + \sqrt{a_k} + \phi(a_k)) < \infty. \tag{36}$$

Moreover, let $g(x,y)$ be a 1-continuous bounded degenerate kernel. Then, there exists a constant C_1 such that

$$E\left(\sum_{i=1}^{[n\lambda]} \sum_{j=[n\lambda]+1}^{n} g(X_i, X_j)\right)^2 \leq C_1[n\lambda](n - [n\lambda]). \tag{37}$$

Proof. We can write

$$E\left(\sum_{i=1}^{[n\lambda]} \sum_{j=[n\lambda]+1}^{n} g(X_i, X_j)\right)^2 = \sum_{i=1}^{[n\lambda]} \sum_{j=[n\lambda]+1}^{n} E(g(X_i, X_j))^2$$

$$+ 2 \sum_{1 \leq i_1 \neq i_2 \leq [n\lambda]} \sum_{[n\lambda]+1 \leq j_1 \neq j_2 \leq n} E\left(g(X_{i_1}, X_{j_1}) g(X_{i_2}, X_{j_2})\right) \tag{38}$$

The elements of the first sum all are bounded, hence

$$\sum_{i=1}^{[n\lambda]} \sum_{j=[n\lambda]+1}^{n} E(g(X_i, X_j))^2 \leq C[n\lambda](n - [n\lambda]). \tag{39}$$

Concerning the second sum, by Lemma 5, we get

$$\sum_{1 \leq i_1 < i_2 \leq [n\lambda]} \sum_{[n\lambda]+1 \leq j_1 < j_2 \leq n} E\left(g(X_{i_1}, X_{j_1}) g(X_{i_2}, X_{j_2})\right)$$

$$\leq 4S \sum_{1 \leq i_1 < i_2 \leq [n\lambda]} \sum_{[n\lambda]+1 \leq j_1 \leq j_2 \leq n} \phi(a_{[k/3]})$$

$$+ 8S^2 \sum_{1 \leq i_1 < i_2 \leq [n\lambda]} \sum_{[n\lambda]+1 \leq j_1 \leq j_2 \leq n} (\sqrt{a_{[k/3]}} + \beta([k/3])) \tag{40}$$

with $k = \max\{|i_2 - i_1|, |j_2 - j_1|\}$. We will first treat the summands with $k = i_2 - i_1$. Suppose for one moment that k is fixed and we will bound the number of indices that appear in the sum. Observe that in this case we have $[n\lambda]$ ways to choose i_1, once i_1 is chosen we have one way to pick i_2 because $i_2 = i_1 + k$. For j_1 we have as before $n - [n\lambda]$ ways to pick this index and then for each j_1, j_2 need to be in the interval $[j_1, j_1 + k]$ and there are exactly k integers in such interval.

$$\sum_{1 \leq i_1 < i_2 \leq [n\lambda]} \sum_{[n\lambda]+1 \leq j_1 < j_2 \leq n} \left(4S\phi(a_{[k/3]}) + 8S^2 \sqrt{a_{[k/3]}} + 8S^2 \beta([k/3]) \right)$$

$$\leq C[n\lambda](n - [n\lambda]) \left(\sum_{k=1}^{n} k\phi(a_k) + \sum_{k=1}^{n} k\sqrt{a_k} + \sum_{k=1}^{n} k\beta(k) \right) \leq C[n\lambda](n - [n\lambda])$$

$$(41)$$

Analogously we can find the bounds for the terms with $k = i_1 - i_2$, $k = j_2 - j_1$ and $k = j_1 - j_2$ using the conditions of summability.

We now define the process $G(\lambda)$, $0 \leq \lambda \leq 1$, by

$$G_n(\lambda) := n^{-3/2} \sum_{i=1}^{[n\lambda]} \sum_{j=[n\lambda]+1}^{n} g(X_i, X_j), \quad 0 \leq \lambda \leq 1. \tag{42}$$

Lemma 2. *Under the conditions of Lemma 1, there exists a constant C such that*

$$E(|G_n(\eta) - G_n(\mu)|^2) \leq \frac{C}{n}(\eta - \mu), \tag{43}$$

for all $0 \leq \mu \leq \eta \leq 1$.

Proof. We can write

$$E(|G_n(\eta) - G_n(\mu)|^2) \tag{44}$$

$$\leq \frac{2}{n^3} E \left(\sum_{i=1}^{[n\mu]} \sum_{j=[n\mu]+1}^{[n\eta]} g(X_i, X_j) \right)^2 + \frac{2}{n^3} E \left(\sum_{i=[n\mu]+1}^{[n\eta]} \sum_{j=[n\eta]+1}^{n} g(X_i, X_j) \right)^2$$

$$= \frac{2}{n^3} E \left(\sum_{i=1}^{[n\mu]} \sum_{j=[n\mu]+1}^{[n\eta]} g(X_i, X_j) \right)^2 + \frac{2}{n^3} E \left(\sum_{i=1}^{[n\eta]-[n\mu]} \sum_{j=[n\eta]-[n\mu]+1}^{n-[n\mu]} g(X_i, X_j) \right)^2$$

$$\leq C \frac{1}{n^3} \left([n\mu]([n\eta] - [n\mu]) + ([n\eta] - [n\mu])(n - [n\eta]) \right) \leq \frac{C}{n}(\eta - \mu)$$

using the stationarity of the process $(X_n)_{n \in \mathbb{N}}$ and Lemma 1.

Proof of Proposition 1. From Lemma 2 we obtain, using Chebyshev's inequality,

$$P\left(|G_n(\eta) - G_n(\mu)| \geq \epsilon\right) \leq \frac{1}{\epsilon^2}\frac{C}{n}(\eta - \mu),\tag{45}$$

for all $\epsilon > 0$. Thus we get for $0 \leq k \leq m \leq n$ with $k, m, n \in \mathbb{N}$

$$P\left(\left|G_n\left(\frac{m}{n}\right) - G_n\left(\frac{k}{n}\right)\right| \geq \epsilon\right) \leq \frac{1}{\epsilon^2}E\left(G_n\left(\frac{m}{n}\right) - G_n\left(\frac{k}{n}\right)\right)^2$$

$$\leq \frac{1}{\epsilon^2}\frac{C}{n^2}(m - k) \leq \frac{1}{\epsilon^2}\frac{C}{n^{5/3}}(m - k)^{4/3}\tag{46}$$

as $m - k \leq n$. Now consider the variables

$$\zeta_i = \begin{cases} G_n\left(\frac{i}{n}\right) - G_n\left(\frac{i-1}{n}\right) & \text{if } i = 1, \ldots, n - 1 \\ 0 & \text{else} \end{cases}\tag{47}$$

and suppose that $S_i = \zeta_1 + \zeta_2 + \ldots + \zeta_i$ with $S_0 = 0$, then $S_i = G_n(\frac{i}{n})$. In consequence the inequality (46) is equivalent to

$$P(|S_m - S_k| \geq \epsilon) \leq \frac{1}{\epsilon^2}\left[\frac{C^{3/4}}{n^{5/4}}(m - k)\right]^{4/3} \quad \text{for} \quad 0 \leq k \leq m \leq n.\tag{48}$$

So the assumption of Theorem 7 are satisfied with the variables (47) in the role of the ξ_i, $\beta = 1/2$, $\alpha = 2/3$ and $u_l = C^{3/4}/n^{5/4}$, $u_o = 0$ and hence

$$P\left(\max_{1 \leq i \leq n-1}|S_i| \geq \epsilon\right) \leq \frac{K}{\epsilon^2}\left[\frac{C^{3/4}}{n^{5/4}}(n - 1)\right]^{4/3} \leq \frac{KC}{\epsilon^2 n^{1/3}}\tag{49}$$

where K depends only of α and β. Thus, (35) holds as $n \to \infty$. $\qquad\qquad\square$

7 Proof of Main Results

In this section, we will prove Theorems 1 and 2. Note that Theorem 2 is a direct consequence of Theorem 1, applied to anti-symmetric kernels. We will nevertheless present a direct proof of Theorem 2, since this proof is much simpler than the proof in the general case. Moreover, Theorem 2 covers those cases that are most relevant in applications.

The first part of the proof is identical for both Theorems 1 and 2. Note that, for each $\lambda \in [0, 1]$, the statistic $T_n(\lambda)$ is a two-sample U-statistic. Thus, using the Hoeffding decomposition (11), we can write $T_n(\lambda)$ as

$$T_n(\lambda) = \frac{1}{n^{3/2}} \left(\sum_{i=1}^{[\lambda n]} \sum_{j=[\lambda n]+1}^{n} (h_1(X_i) + h_2(X_j) + g(X_i, X_j)) \right)$$

$$= \frac{1}{n^{3/2}} \left((n - [n\lambda]) \sum_{i=1}^{[n\lambda]} h_1(X_i) + [n\lambda] \sum_{j=[n\lambda]+1}^{n} h_2(X_j) + \sum_{i=1}^{[\lambda n]} \sum_{j=[\lambda n]+1}^{n} g(X_i, X_j) \right)$$

$$(50)$$

By Proposition 1, we know that

$$\frac{1}{n^{3/2}} \sup_{0 \le \lambda \le 1} \left| \sum_{i=1}^{[\lambda n]} \sum_{j=[\lambda n]+1}^{n} g(X_i, X_j) \right| \to 0$$

in probability. Thus, by Slutsky's lemma, it suffices to show that the sum of the first two terms, i.e.

$$\left(\frac{n - [n\lambda]}{n^{3/2}} \sum_{i=1}^{[n\lambda]} h_1(X_i) + \frac{[n\lambda]}{n^{3/2}} \sum_{j=[n\lambda]+1}^{n} h_2(X_j) \right)_{0 \le \lambda \le 1} \qquad (51)$$

converges in distribution to the desired limit process.

Proof of Theorem 2. It remains to show that (51) converges in distribution to $\sigma W^{(0)}(\lambda), 0 \le \lambda \le 1$, where $(W^{(0)}(\lambda))_{0 \le \lambda \le 1}$ is standard Brownian bridge on $[0, 1]$, and where σ^2 is defined in (19). By antisymmetry of the kernel $h(x, y)$, we obtain that $h_2(x) = -h_1(x)$. Hence, in this case, (51) can be rewritten as

$$\frac{n - [n\lambda]}{n^{3/2}} \sum_{i=1}^{[n\lambda]} h_1(X_i) - \frac{[n\lambda]}{n^{3/2}} \sum_{i=[n\lambda]+1}^{n} h_1(X_i) = \frac{1}{n^{1/2}} \sum_{i=1}^{[n\lambda]} h_1(X_i) - \frac{[n\lambda]}{n^{3/2}} \sum_{i=1}^{n} h_1(X_i).$$

By Proposition 2.11 and Lemma 2.15 of Borovkova, Burton and Dehling [3], the sequence $(h_1(X_i))_{i \ge 1}$ is a 1-approximating functional with approximating constant $C\sqrt{a_k}$. Since $h_1(X_i)$ is bounded, the L_2-near epoch dependence in the sense of Wooldridge and White [21] also holds, with the same constants. Moreover, the underlying process $(Z_n)_{n \ge 1}$ is absolutely regular, and hence also strongly mixing. Thus we may apply the invariance principle in Corollary 3.2 of Wooldridge and White [21], and obtain that the partial sum process

$$\left(\frac{1}{n^{1/2}} \sum_{i=1}^{[n\lambda]} h_1(X_i) \right)_{0 \le \lambda \le 1} \qquad (52)$$

converges weakly to Brownian motion $(W(\lambda))_{0\leq\lambda\leq 1}$ with $\text{Var}(W(1)) = \sigma^2$. The statement of the Theorem follows with the continuous mapping theorem for the mapping $x(t) \mapsto x(t) - tx(1)$, $0 \leq t \leq 1$.

The proof of Theorem 1 requires an invariance principle for the partial sum process of $\mathbb{R}^2$-valued dependent random variables; see Proposition 2 below. For mixing processes, such invariance principles have been established even for partial sums of Hilbert space valued random vector, e.g. by Dehling [7]. In this paper, we provide an extension of these results to functionals of mixing processes.

Proposition 2. *Let $(X_n)_{n\in\mathbb{N}}$ be a 1-approximating functional of an absolutely regular process with mixing coefficients $(\beta(k))$ and let $h_1(\cdot)$, $h_2(\cdot)$ be bounded1– continuous functions with mean zero, such that*

$$\sum_k k^2(\beta(k) + a_k + \phi(a_k)) < \infty. \tag{53}$$

Then, as $n \to \infty$,

$$\left(\frac{1}{\sqrt{n}} \sum_{i=1}^{[nt]} \binom{h_1(X_i)}{h_2(X_i)}\right)_{0\leq t\leq 1} \longrightarrow \binom{W_1(t)}{W_2(t)}_{0\leq t\leq 1} \tag{54}$$

where $(W_1(t), W_2(t))_{0\leq t\leq 1}$ is a two-dimensional Brownian motion with mean zero and covariance $E(W_k(s)\, W_l(t)) = \min(s,t)\sigma_{kl}$, where $\sigma_{k,l}$ as defined in (16).

Proof. To prove (54), we need to establish finite dimensional convergence and tightness. Concerning finite-dimensional convergence, by the Cramér-Wold device it suffices to show the convergence in distribution of a linear combination of the coordinates of the vector

$$\left(\frac{1}{\sqrt{n}} \sum_{i=1}^{[nt_1]} h_1(X_i), \frac{1}{\sqrt{n}} \sum_{i=1}^{[nt_1]} h_2(X_i), \ldots, \frac{1}{\sqrt{n}} \sum_{i=1}^{[nt_j]} h_1(X_i), \frac{1}{\sqrt{n}} \sum_{i=1}^{[nt_j]} h_2(X_i),\right.$$

$$\left. \ldots, \frac{1}{\sqrt{n}} \sum_{i=1}^{n} h_1(X_i), \frac{1}{\sqrt{n}} \sum_{i=1}^{n} h_2(X_i)\right), \tag{55}$$

for $0 = t_0 < t_1 < \ldots < t_j < \ldots < t_k = 1$. Any such linear combination can be expressed as

$$\sum_{j=1}^{k} \frac{1}{\sqrt{n}} \sum_{i=[nt_{j-1}]+1}^{[nt_j]} (a_j h_1(X_i) + b_j h_2(X_i)), \tag{56}$$

for $(a_j, b_j)_{j=1}^k \in \mathbb{R}^{2k}$. By using the Cramér-Wold device again, the weak convergence of this sum is equivalent to the weak convergence of the vector

$$
\left(\frac{1}{\sqrt{n}} \sum_{i=1}^{[nt_1]} (a_1 h_1(X_i) + b_1 h_2(X_i)), \ldots, \frac{1}{\sqrt{n}} \sum_{i=[nt_{j-1}]+1}^{[nt_j]} (a_j h_1(X_i) + b_j h_2(X_i)), \right.
$$

$$
\left. \ldots, \frac{1}{\sqrt{n}} \sum_{i=[nt_{k-1}]+1}^{n} (a_k h_1(X_i) + b_k h_2(X_i)) \right) \tag{57}
$$

to

$$
\left(a_1(W_1(t_1) - W_1(t_0)) + b_1(W_2(t_1) - W_2(t_0)), \ldots, \right.
$$

$$
a_k(W_1(t_k) - W_1(t_{k-1})) + b_k(W_2(t_k) - W_2(t_{k-1}))). \tag{58}
$$

Since $(X_n)_{n\geq 1}$ is a 1-approximating functional, it can be coupled with a process consisting of independent blocks. Given integers $L := L_n = [n^{3/4}]$ and $l_n = [n^{1/2}]$, we introduce the (l, L) blocking $(B_m)_{m\geq 0}$ of the variables $(a_j h_1(X_i) + b_j h_2(X_i))$ with $i = [nt_{j-1}] + 1, \ldots, [nt_j], j = 0, \ldots, k$ and

$$
B_m := \sum_{i=(m-1)(L_n+l_n)+1}^{m(L_n+(m-1)l_n)} (a_j h_1(X_i) + b_j h_2(X_i)) \tag{59}
$$

and separating blocks

$$
\tilde{B}_m := \sum_{i=mL_n+(m-1)l_n+1}^{m(L_n+l_n)} (a_j h_1(X_i) + b_j h_2(X_i)). \tag{60}
$$

By Theorem 5 there exists a sequence of independent blocks (B_m') with the same blockwise marginal distribution as (B_m) and such that

$$
P\left(|B_m - B_m'| \leq 2\alpha_l \right) \geq 1 - \beta(l) - 2\alpha_l,
$$

where $\alpha_l := (2 \sum_{k=[l_n/3]}^{\infty} a_k)^{1/2}$. We can express the components of our vector (57) as a sum of blocks

$$
\sum_{i=[nt_j]+1}^{[nt_{j+1}]} (a_j h_1(X_i) + b_j h_2(X_i))
$$

$$
= \sum_{m=\left[\frac{nt_j}{L+l}\right]+1}^{\left[\frac{nt_{j+1}}{L+l}\right]} B_m + \sum_{m=\left[\frac{nt_j}{L+l}\right]+1}^{\left[\frac{nt_{j+1}}{L+l}\right]} \tilde{B}_m + \sum_{R_j} (a_j h_1(X_i) + b_j h_2(X_i)), \tag{61}
$$

where R_j denotes the set of indices not contained in the blocks. Observe that by the Lemma 3 for any set $A \subset \{1, \ldots, n\}$

$$E \left(\sum_{i \in A} (a_j h_1(X_i) + b_j h_2(X_i)) \right)^2 \leq C \#A \tag{62}$$

and hence

$$E \left(\sum_{m=\left[\frac{nt_j}{L+l}\right]+1}^{\left[\frac{nt_{j+1}}{L+l}\right]} \tilde{B}_m \right)^2 \leq C \frac{n}{L_n + l_n} l_n \leq C n^{3/4}, \tag{63}$$

so it follows with the Chebyshev inequality that this term is negligible. For the last summand, we have that

$$E \left(\sum_{R_j} (a_j h_1(X_i) + b_j h_2(X_i)) \right)^2 \leq C 2 (L_n + l_n) \leq C n^{3/4}. \tag{64}$$

Furthermore, we need to show that we can replace the blocks B_m by the independent coupled blocks B'_m:

$$P \left(\left| \frac{1}{\sqrt{n}} \sum_{m=\left[\frac{nt_j}{L+l}\right]+1}^{\left[\frac{nt_{j+1}}{L+l}\right]} (B_m - B'_m) \right| > \epsilon \right) \leq \sum_{m=\left[\frac{nt_j}{L+l}\right]+1}^{\left[\frac{nt_{j+1}}{L+l}\right]} P \left(|B_m - B'_m| > \frac{\epsilon \sqrt{n}}{n^{1/4}} \right)$$

$$\leq n^{\frac{1}{4}} \left(\beta([\frac{l_n}{3}]) + \alpha_{[\frac{l_n}{3}]} \right) \to 0$$

as $n \to \infty$ by our conditions on the mixing coefficients and approximation constants. Here we used that fact that $\alpha_n \to 0$ and thus, for almost all $n \in \mathbb{N}$,

$$P \left(|B_m - B'_m| > \epsilon n^{1/4} \right) \leq P \left(|B_m - B'_m| > 2\alpha_{l_n} \right). \tag{65}$$

With the above arguments the result holds if we show the convergence of

$$\frac{1}{\sqrt{n}} \left(\sum_{m=\left[\frac{nt_1}{L+l}\right]+1}^{\left[\frac{nt_1}{L+l}\right]} B'_m, \ldots, \sum_{m=\left[\frac{nt_k}{L+l}\right]+1}^{\left[\frac{nt_{k+1}}{L+l}\right]} B'_m \right). \tag{66}$$

Since this vector has independent components, we only need to show the one-dimensional convergence, which is a consequence of Theorem 4, using the summability condition (53).

We now turn to the question of tightness and show that, for each ϵ and η, there exist a δ, $0 < \delta < 1$, and an integer n_0 such that, for $0 \leq t \leq 1$,

$$\frac{1}{\delta} P \left(\sup_{t \leq s \leq t+\delta} |Y_n(s) - Y_n(t)| \geq \epsilon \right) \leq \eta, \quad n \geq n_0 \qquad (67)$$

with

$$Y_n(t) = \frac{1}{\sigma \sqrt{n}} \sum_{i=1}^{[nt]} h_1(X_i) + (nt - [nt]) \frac{1}{\sigma \sqrt{n}} h(X_{[nt]+1}) \qquad (68)$$

(h_2 can be treated in the same way) and by Theorem 8, this condition reduces to: For each positive ϵ there exist a $\alpha > 1$ and an integer n_0, s. t.

$$P \left(\max_{i \leq n} \left| \sum_{j=1}^{i} h_1(X_j) \right| \geq \lambda \sqrt{n} \right) \leq \frac{\epsilon}{\lambda^2}, \quad n \geq n_0. \qquad (69)$$

Let $t \geq s, s, t \in [0, 1]$. By Lemma 4 we get

$$E \left(\left| \frac{1}{\sqrt{n}} \sum_{i=1}^{[nt]} h_1(X_i) - \frac{1}{\sqrt{n}} \sum_{i=1}^{[ns]} h_1(X_i) \right|^4 \right) = \frac{1}{n^2} E \left(\sum_{[ns]+1}^{[nt]} h_1(X_i) \right)^4$$

$$\leq \frac{1}{n^2} (([nt] - [ns])^2 C) \qquad (70)$$

and this implies

$$P \left(\left| \frac{1}{\sqrt{n}} \sum_{i=1}^{m} h_1(X_i) - \frac{1}{\sqrt{n}} \sum_{i=1}^{k} h_1(X_i) \right| \geq \epsilon \right) \leq \frac{1}{\epsilon^4} \left(\frac{C^{1/2}}{n} (m - k) \right)^2. \qquad (71)$$

By Theorem 7

$$P \left(\max_{i \leq n} \left| \sum_{j=1}^{i} h_1(X_j) \right| \geq \epsilon \sqrt{n} \right) \leq \frac{K}{\epsilon^4} \left(\frac{C^{1/2}}{n} (n - 1) \right)^2 \qquad (72)$$

and we get the assertion. Thus we have established tightness of each of the two coordinates of the partial sum process, which implies tightness of the vector-valued process.

Proof of Theorem 1. From Proposition 2 we obtain that

$$\left(\frac{1}{\sqrt{n}} \sum_{i=1}^{[n\lambda]} \begin{pmatrix} h_1(X_i) \\ h_2(X_i) \end{pmatrix} \right)_{0 \le \lambda \le 1} \longrightarrow \begin{pmatrix} W_1(\lambda) \\ W_2(\lambda) \end{pmatrix}_{0 \le \lambda \le 1}, \tag{73}$$

in distribution on the space $(D([0, 1]))^2$. We consider the functional given by

$$\begin{pmatrix} x_1(t) \\ x_2(t) \end{pmatrix} \mapsto (1 - t)x_1(t) + t(x_2(1) - x_2(t)), \quad 0 \le t \le 1. \tag{74}$$

This is a continuous mapping from $(D[0, 1])^2$ to $D[0, 1]$, so we may apply the continuous mapping theorem to (73), and obtain

$$\left(\frac{n - [n\lambda]}{n^{3/2}} \sum_{i=1}^{[n\lambda]} h_1(X_i) + \frac{[n\lambda]}{n^{3/2}} \sum_{j=[n\lambda]+1}^{n} h_2(X_j) \right)_{0 \le \lambda \le 1}$$

$$\longrightarrow ((1 - \lambda)W_1(\lambda) + \lambda(W_2(1) - W_2(\lambda)))_{0 \le \lambda \le 1}.$$

Together with the remarks at the beginning of this section, this proves Theorem 1.

Appendix: Some Auxiliary Results from the Literature

In this section, we collect some known lemmas and theorems for weakly dependent data. We start with some results on the behaviour of partials sums:

Lemma 3 (Borovkova, Burton, Dehling [3], Lemma 2.23). *Let $(X_k)_{k \in \mathbb{Z}}$ be a 1-approximating functional with constants $(a_k)_{k \ge 0}$ of an absolutely regular process with mixing coefficients $(\beta(k))_{k \ge 0}$. Suppose moreover that $EX_i = 0$ and that one of the following two conditions holds:*

1. *X_0 is bounded a.s. and $\sum_{k=0}^{\infty}(a_k + \beta(k)) < \infty$.*
2. *$E|X_0|^{2+\delta} < \infty$ and $\sum_{k=0}^{\infty}(a_k^{\frac{\delta}{1+\delta}} + \beta^{\frac{\delta}{1+\delta}}(k)) < \infty$.*

Then, as $N \to \infty$,

$$\frac{1}{N}ES_N^2 \to EX_0^2 + 2\sum_{j=1}^{\infty} E(X_0 X_j) \tag{75}$$

and the sum on the r.h.s. converges absolutely.

Lemma 4 (Borovkova, Burton, Dehling [3], Lemma 2.24). *Let $(X_k)_{k \in \mathbb{Z}}$ be a 1-approximating functional with constants (a_k) of an absolutely regular process with mixing coefficients $(\beta(k))_{k \geq 0}$. Suppose moreover that $EX_i = 0$ and that one of the following two conditions holds:*

1. X_0 is bounded a.s. and $\sum_{k=0}^{\infty} k^2 (a_k + \beta(k)) < \infty$.

2. $E|X_0|^{4+\delta} < \infty$ and $\sum_{k=0}^{\infty} k^2 (a_k^{\frac{\delta}{3+\delta}} + \beta^{\frac{\delta}{4+\delta}}(k)) < \infty$.

Then there exits a constant C such that

$$ES_N^4 \leq CN^2. \tag{76}$$

Theorem 4 (Borovkova, Burton, Dehling [3], Theorem 4). *Let $(X_k)_{k \in \mathbb{Z}}$ be a 1-approximating functional with constants $(a_k)_{k \geq 0}$ of an absolutely regular process with mixing coefficients $(\beta(k))_{k \geq 0}$. Suppose moreover that $EX_i = 0$, $E|X_0|^{4+\delta} < \infty$ and that*

$$\sum_{k=0}^{\infty} k^2 (a_k^{\frac{\delta}{3+\delta}} + \beta^{\frac{\delta}{4+\delta}}(k)) < \infty, \tag{77}$$

for some $\delta > 0$. Then, as $n \to \infty$,

$$\frac{1}{\sqrt{n}} \sum_{i=1}^{n} X_i \to \mathcal{N}(0, \sigma^2), \tag{78}$$

where $\sigma^2 = EX_0^2 + 2 \sum_{j=1}^{\infty} E(X_0 X_j)$. In case $\sigma^2 = 0$, $\mathcal{N}(0,0)$ denotes the point mass at the origin. If X_0 is bounded, the CLT continues to hold if (77) is replaced by the condition that $\sum_{k=0}^{\infty} k^2 (a_k + \beta(k)) < \infty$.

An important tool to derive asymptotic results for weakly dependent data are coupling methods. We will apply this method in the proof of Proposition 2.

Theorem 5 (Borovkova, Burton, Dehling [3], Theorem 3). *Let $(X_n)_{n \in \mathbb{N}}$ be a 1-approximating functional with summable constants $(a_k)_{k \geq 0}$ of an absolutely regular process with mixing rate $(\beta(k))_{k \geq 0}$. Then given integers K, L and N, we can approximate the sequence of $(K + 2L, N)$-blocks $(B_s)_{s \geq 1}$ by a sequence of independent blocks $(B_s')_{s \geq 1}$ with the same marginal distribution in such a way that*

$$P(\|B_s - B_s'\| \leq 2\alpha_L) \geq 1 - \beta(K) - 2\alpha_L, \tag{79}$$

where $\alpha_L := \left(2 \sum_{l=L}^{\infty} a_l\right)^{1/2}$.

In statistical application, the question of how to estimate σ^2 is important. In the situation when the observations are a functional of α-mixing process, Dehling et al. [10] propose the estimation of the variance of partial sums of dependent processes by the subsampling estimator

$$\hat{D}_n = \frac{1}{[n/l_n]} \sqrt{\frac{\pi}{2}} \sum_{i=1}^{[n/l_n]} \frac{|\hat{T}_i(l_n) - l_n \tilde{U}_n|}{\sqrt{l_n}} \tag{80}$$

with $\hat{T}_i(l) = \sum_{j=(i-1)l+1}^{il} F_n(X_j)$ and $\tilde{U}_n = \frac{1}{n} \sum_{j=1}^{n} F_n(X_j)$, where $F_n(\cdot)$ is the empirical distribution function.

Theorem 6 (Dehling, Fried, Sharipov, Vogel, Wornowizki [9], Theorem 1.2).
Let $(X_k)_{k\geq 1}$ be a stationary, 1-approximating functional of an α-mixing processes. Suppose that for some $\delta > 0$, $E|X_1|^{2+\delta} < \infty$, and that the mixing coefficients $(\alpha_k)_{k\geq 1}$ and the approximation constants $(a_k)_{k\geq 1}$ satisfy

$$\sum_{k=1}^{\infty} (\alpha_k)^{\frac{2}{2+\delta}} < \infty, \quad \sum_{k=1}^{\infty} (a_k)^{\frac{1+\delta}{2+\delta}} < \infty. \tag{81}$$

In addition, we assume that F is Lipschitz-continuous, that $\alpha_k = O(n^{-8})$ and that $a_m = O(m^{-12})$. Then, as $n \to \infty$, $l_n \to \infty$ and $l_n = o(\sqrt{n})$, we have $\hat{D}_n \longrightarrow \sigma$ in L_2.

To deal with the degenerate kernel g, we need to find upper bounds for the expectations $E\left(g(X_{i_1}, X_{j_1})g(X_{i_2}, X_{j_2})\right)$, in terms of the maximal distance among the indices. Since $1 \leq i_1 < i_2 \leq [n\lambda]$ and $[n\lambda] + 1 \leq j_1 < j_2 \leq n$, we get $i_1 < i_2 < j_1 < j_2$.

Lemma 5 (Dehling, Fried [8], Proposition 6.1). *Let $(X_n)_{n\geq 1}$ be a 1-approximating functional with constants $(a_k)_{k\geq 1}$ of an absolutely regular process with mixing coefficients $(\beta(k))_{k\geq 1}$ and let $g(x, y)$ be a 1-continuous bounded degenerate kernel. Then we have*

$$|E(g(X_{i_1}, X_{j_1})g(X_{i_2}, X_{j_2}))| \leq 4S\phi(a_{[k/3]}) + 8S^2(\sqrt{a_{[k/3]}} + \beta([k/3])) \tag{82}$$

where $S = |\sup_{x,y} g(x, y)|$ and $k = \max\{i_2 - i_1, j_1 - i_2, j_2 - j_1\}$.

The following two results are useful for proving tightness of a stochastic process. The first one is used to control the fluctuation of maximum. Let $\xi_1, \ldots, \xi_n$ be random variables, and define $S_k = \xi_1 + \ldots + \xi_k$ ($S_0 = 0$), and $M_n = \max_{0\leq k\leq n} |S_k|$.

Theorem 7 (Billingsley [2], Theorem 10.2). *Suppose that $\beta \geq 0$ and $\alpha > 1/2$ and that there exist nonnegative numbers $u_1, \ldots, u_n$ such that for all positive λ*

$$P\left(|S_j - S_i| \geq \lambda\right) \leq \frac{1}{\lambda^{4\beta}} \left(\sum_{i<l\leq j} u_l\right)^{2\alpha}, \quad 0 \leq i \leq j \leq n \quad, \tag{83}$$

then for all positive λ

$$P(M_n \geq \lambda) \leq \frac{K_{\beta,\alpha}}{\lambda^{4\beta}} \left(\sum_{0<l\leq n} u_l \right)^{2\alpha}, \tag{84}$$

where $K_{\beta,\alpha}$ is a constant depending only on β and α.

Theorem 8 (Billingsley [2], Theorem 8.4). *The sequence $\{Y_n\}$, defined by*

$$Y_n(t) = \frac{1}{\sigma\sqrt{n}}S_{[nt]} + (nt - [nt])\frac{1}{\sigma\sqrt{n}}\xi_{[nt]+1} \tag{85}$$

is tight if for each $\epsilon > 0$ there exist a $\lambda > 1$ and a $n_0 \in \mathbb{N}$ such that for $n \geq n_0$

$$P\left(\max_{i\leq n} |S_{k+i} - S_k| \geq \lambda\sigma\sqrt{n} \right) \leq \frac{\epsilon}{\lambda^2}. \tag{86}$$

Acknowledgements The authors wish to thank the referees for their very careful reading of an earlier version of this manuscript, and for their many thoughtful comments that helped to improve the presentation of the paper. This research was supported by the Collaborative Research Center 823, Project C3 *Analysis of Structural Change in Dynamic Processes*, of the German Research Foundation DFG.

References

1. Babbel, B.: Invariance principles for U-statistics and von Mises functionals. J. Stat. Plan. Inference **22**, 337–354 (1989)
2. Billingsley, P.: Convergence of Probability Measures. 2nd edn. Wiley, New York (1999)
3. Borovkova, S.A., Burton, R.M., Dehling, H.G.: Limit theorems for functionals of mixing processes with applications to U-statistics and dimension estimation. Trans. Am. Math. Soc. **353**, 4261–4318 (2001)
4. Carlstein, E.: The use of subseries values for estimating the variance of a general statistic from a stationary sequence. Ann. Stat. **14**, 1171–1179 (1986)
5. Csörgő, M., Horvath, L.: Invariance principles for changepoint problems. J. Multivar. Anal. **27**, 151–168 (1988)
6. Csörgő, M., Horvath, L.: Limit Theorems in Change Point Analysis. Wiley, New York (1997)
7. Dehling, H.: Limit theorems for sums of weakly dependent Banach space valued random variables. Z. für Wahrscheinlichkeitstheorie und verwandte Gebiete **63**, 393–432 (1983)
8. Dehling, H., Fried, R.: Asymptotic distribution of two-sample empirical U-quantiles with applications to robust tests for shifts in location. J. Multivar. Anal. **105**, 124–140 (2012)
9. Dehling, H., Fried, R., Sharipov, O.Sh., Vogel, D., Wornowizki, M.: Estimation of the variance of partial sums of dependent processes. Stat. Probab. Lett. **83**, 141–147 (2013)
10. Dehling, H., Rooch, A., Taqqu, M.S.: Nonparametric change-point tests for long-range dependent data. Scand. J. Stat. **40**, 153–173 (2013)
11. Denker, M.: Asymptotic Distribution Theory in Nonparametrics Statistics. Vieweg Verlag, Braunschweig/Wiesbaden (1985)

12. Denker, M., Keller, G.: On U-statistics and v. Mises' statistics for weakly dependent processes. Z. für Wahrscheinlichkeitstheorie und verwandte Gebiete **64**, 505–522 (1983)
13. Denker, M., Keller, G.: Rigorous statistical procedures for data from dynamical systems. J. Stat. Phys. **44**, 67–93 (1986)
14. Hoeffding, W.: A class of statistics with asymptotically normal distribution. Ann. Math. Stat. **19**, 293–325 (1948)
15. Ibragimov, I.A., Linnik, Yu.V.: Independent and Stationary Sequences of Random Variables. Wolters-Noordhoff, Groningen (1971)
16. Leucht, A.: Degenerate U- and V-statistics under weak dependence: asymptotic theory and bootstrap consistency. Bernoulli **18**, 552–585 (2012)
17. Rooch, A.: Change-point tests for long-range dependent data. Dissertation, Ruhr-Universität Bochum (2012)
18. Sen, P.K.: On the properties of U-statistics when the observations are not independent. I. Estimation of non-serial parameters in some stationary stochastic processes. Calcutta Stat. Assoc. Bull. **12**, 69–92 (1963)
19. Sen, P.K.: Limiting behavior of regular functionals of empirical distributions for stationary *-mixing processes. Z. für Wahrscheinlichkeitstheorie und Verwandte Gebiete **25**, 71–82 (1972)
20. Sharipov, O.Sh., Wendler, M.: Bootstrap for the sample mean and for U-statistics of mixing and near-epoch dependent processes. J. Nonparametric Stat. **24**, 317–342 (2012)
21. Wooldridge, J.M., White, H.: Some invariance principles and central limit theorems for dependent heterogeneous processes. Econ. Theory **4**, 210–230 (1988)
22. Yoshihara, K.-I.: Limiting behavior of U-statistics for stationary, absolutely regular processes. Z. für Wahrscheinlichkeitstheorie und verwandte Gebiete **35**, 237–252 (1976)

Binary Time Series Models in Change Point Detection Tests

Edit Gombay

1 Introduction

Consider a binary time series, $\{Y_t\}$, with probability of success $\pi_t(\beta)$ and an accompanying vector of covariates $\{Z_t\}$ defined below in (1) and (2). This model has a lot of applications in various fields. Kedem and Fokianos [6] contains many examples that include rainfall data in environmental studies; mortality data in biostatistics; stock prices in financial studies, to name a few. Fokianos et al. [5] studies change detection algorithms for such models restricting attention to logistic regression. It is not practical to consider a general class of link functions as then the conditions would be too cumbersome hence an obstacle in applications. So it is customary to consider the frequently used link functions separately. In this note the work of Fokianos et al. [5] will be supplemented by considering the link function leading to the probit model, the loglog and complementary link functions.

As in Kedem and Fokianos [6] we denote denote the history of the binary process and its past covariate vector values by $\{\mathscr{F}_{t-1}\}$ that is a filtration generated by $\{Y_{t-1}, Y_{t-2}, \ldots, Z_{t-1}, Z_{t-2}, \ldots\}$. It is convenient to think that the vector of covariates Z_t may contain lagged values of the binary response itself, thus permitting an $AR(p)$-type serial dependence over time, and in this case $\{\mathscr{F}_{t-1}\}$ is the sigma field generated by $\{Z_{t-1}, Z_{t-2}, \ldots\}$. The conditional density of the series $\{Y_t\}$ is the Bernoulli probability function

$$f(y_t; \beta | \mathscr{F}_{t-1}) = \exp\left\{y_t \log\left(\frac{\pi_t(\beta)}{1 - \pi_t(\beta)}\right) + \log(1 - \pi_t(\beta))\right\}, \tag{1}$$

E. Gombay (✉)
Department of Mathematical and Statistical Sciences,
University of Alberta, Edmonton, AB, Canada
e-mail: egombay@ualberta.ca

© Springer Science+Business Media New York 2015
D. Dawson et al. (eds.), *Asymptotic Laws and Methods in Stochastics*,
Fields Institute Communications 76, DOI 10.1007/978-1-4939-3076-0_13

221

where $\beta \in \mathcal{R}^p$ ($p \geq 1$) is the parameter vector, and the dependence on the covariate vector $Z_{t-1} \in \mathcal{R}^p$ is expressed with the help of a general inverse link Φ as

$$\pi_t(\beta) \equiv P(Y_t = 1 \mid \mathcal{F}_{t-1}) = \Phi(\beta' Z_{t-1}) = \Phi(\eta_t), \tag{2}$$

where Z_t is assumed to be $\mathcal{F}_t$-measurable. The logit link function $\eta_t = \beta' Z_{t-1} = \log[\pi_t(\beta)/(1 - \pi_t(\beta))]$ is using the standard logistic distribution

$$\Phi(x) = \frac{e^x}{1 + e^x} = \frac{1}{1 + e^{-x}},$$

leading to logistic regression.

The standard normal distribution function Φ gives the probit model with link function $\beta' Z_{t-1} = \Phi^{-1}(\pi_t(\beta))$.

Other link functions in the literature are the following: the double exponential distribution $\Phi(x) = exp(-exp(-x))$ leading to the log-log link function $\beta' Z_{t-1} = -\log[-\log(\pi(\beta))]$; the double exponential distribution $\Phi(x) = 1 - exp(-exp(x))$ leading to the complementary log-log link function $\beta' Z_{t-1} = \log - \log(1 - \pi(\beta))$; and the identity link uses the uniform distribution. See Kedem and Fokianos [6] for a discussion and comparisons.

We study the important problem of stability of the parameter vector β over time, hence, to formulate the problem we will index it as β_t when necessary. However, for simplicity we will omit the subscript t when we work under the hypothesis of no change in its value.

Retrospective change-point detection assumes that a series of observations $y_1, \ldots, y_n$ generated by this model is available and tests hypotheses

$$H_0 : \beta_t = \beta_0, \text{ for } t = 1, 2, \ldots, n, \quad \beta_0 \text{ unknown}, \tag{3}$$

$$H_a : \beta_t = \beta_0, \text{ for } t = 1, 2, \ldots, \tau - 1, \text{ and } \beta_t \neq \beta_0 \text{ for } t \geq \tau,$$

where τ, $1 < \tau < n$, is the unknown time when a change occurs in a component of vector β.

2   Conditions and Results

Our test statistic is based on the standardized score obtained via a partial likelihood function. In general, inferences concerning the binary time series model introduced in Sect. 1 are based on the so-called partial likelihood function defined as

$$\prod_{t=1}^{n} f(y_t; \beta \mid \mathcal{F}_{t-1}) = \prod_{t=1}^{n} (\pi_t(\beta))^{y_t} (1 - \pi_t(\beta))^{(1-y_t)},$$

or equivalently on the log-partial likelihood function,

$$L(\beta) = \sum_t l_t(\beta) = \sum_{t=1}^n [y_t \log \frac{\pi_t(\beta)}{1 - \pi_t(\beta)} + \log(1 - \pi_t(\beta))]. \qquad (4)$$

The score vector of this log-partial likelihood is

$$S_n(\beta) = \sum_t \nabla_\beta l_t(\beta) = \sum_{t=1}^n Z_{t-1} \left(Y_t - \Phi(\beta' Z_{t-1}) \right) \frac{\phi(\beta' Z_{t-1})}{\Phi(\beta' Z_{t-1})(1 - \Phi(\beta' Z_{t-1}))}, \qquad (5)$$

where $\phi(u) = \frac{\partial}{\partial u} \Phi(u)$.

In case of the logit link this has the simple form

$$S_n(\beta) = \sum_{t=1}^n Z_{t-1} (Y_t - \pi_t(\beta)) = \sum_{t=1}^n Z_{t-1} \left(Y_t - \frac{\exp(\beta' Z_{t-1})}{1 + \exp(\beta' Z_{t-1})} \right),$$

which is the reason why it is so often used.

The (cumulative) conditional information matrix $T_n(\beta)$ is defined on p. 12 of Kedem and Fokianos [6] by the formula

$$T_n(\beta) = \sum_{t=1}^n Cov(\nabla_\beta l_t(\beta) | \mathscr{F}_{t-1}).$$

As $E((Y_t - \pi(\beta))^2 | \mathscr{F}_{t-1}) = \pi_t(\beta)(1 - \pi_t(\beta))$ in our model, we obtain its alternative expression

$$T_n(\beta) = \sum_{t=1}^n Z_{t-1} Z_{t-1}' \frac{\phi^2(\beta' Z_{t-1})}{\Phi(\beta' Z_{t-1})(1 - \Phi(\beta' Z_{t-1}))}. \qquad (6)$$

2.1 Null Hypothesis of No Change

Under the null hypothesis of no change we need the following conditions on the covariate process.

(C1) It is ergodic and stationary in the sense that for all $k \geq 0$ $(Z_{k+1}, Z_{k+2}, \dots)$ has the same distribution as $(Z_0, Z_1, \dots)$.

(C2) $E|Z_k^i|^6 < \infty$, $i = 1, \dots, p$, where Z_k^i, $1 \leq i \leq p$, are the components of vector Z_k.

(C3) The true value of β is in an open subset of the parameter space Ω, $\Omega \subset \mathfrak{R}^p$.

We note here that condition (C1) is strong stationarity, which is needed in the proofs below.

From (C2) we have

$$\frac{1}{n}\sum_{t=1}^{n} Z_t^i Z_t^j \xrightarrow{a.s} E(Z_t^i Z_t^j), \quad n \to \infty,$$

$$\frac{1}{n}\sum_{t=1}^{n} Z_t^i Z_t^j Z_t^l \xrightarrow{a.s} E(Z_t^i Z_t^j Z_t^l), \quad n \to \infty,$$

for all $i, j, l \in \{1, 2, \ldots, p\}$ by the ergodic theorem for strongly stationary stochastic processes (cf. Doob [3]).

The L_1 norm of a random variable X is defined as $\|X\|_1 = E(|X|)$, and for a random vector X it is the sum of the L_1 norms of its components. With this notation, if

$$E\|Z_{t-1}Z_{t-1}' \frac{\phi^2(\beta' Z_{t-1})}{\Phi(\beta' Z_{t-1})(1 - \Phi(\beta' Z_{t-1}))}\|_1 < \infty, \tag{7}$$

then by the ergodic theorem, as $n \to \infty$

$$\frac{1}{n}T_n(\beta) = \frac{1}{n}\sum_{t=1}^{n} Z_{t-1}^i Z_{t-1}^j \frac{\phi^2(\beta' Z_{t-1})}{\Phi(\beta' Z_{t-1})(1 - \Phi(\beta' Z_{t-1}))} \xrightarrow{a.s} T,$$

where

$$T = E\left(Z_{t-1}^i Z_{t-1}^j \frac{\phi^2(\beta' Z_{t-1})}{\Phi(\beta' Z_{t-1})(1 - \Phi(\beta' Z_{t-1}))}\right)_{i,j=1,\ldots,p}.$$

We will show that

$$\gamma(\beta' Z_{t-1}) = \frac{\phi^2(\beta' Z_{t-1})}{\Phi(\beta' Z_{t-1})(1 - \Phi(\beta' Z_{t-1}))}$$

is a bounded function, so (7) holds, and by assumption (C2) covariance matrix T exists.

We verify (7) separately for the various link functions.

CASE 1: Probit model.

We can use the tail approximation for the normal distribution function. On the right tail $1 - \Phi(x) \sim \phi(x)/x$ as $x \to \infty$. ($\sim$ means that the ratio converges to a constant.) Hence,

$$\frac{\phi^2(x)}{\Phi(x)(1 - \Phi(x))} \sim \frac{\phi^2(x)x}{\Phi(x)\phi(x)} = \frac{\phi(x)x}{\Phi(x)}, \quad x \to \infty. \tag{8}$$

On the left tail as $x \to -\infty$, by symmetry, $\Phi(x) \sim -\phi(x)/x$ giving

$$\frac{\phi^2(x)}{\Phi(x)(1 - \Phi(x))} \sim \frac{\phi^2(x)(-x)}{(1 - \Phi(x))\phi(x)} = \frac{-\phi(x)x}{1 - \Phi(x)}, \quad x \to \infty. \tag{9}$$

As $x\phi(x) \to 0, x \to \pm\infty$, by (8), (9), and with $x = \beta' Z_{t-1}$ an upper bound for (7) can be obtained by using

$$E|Z_{t-1}^i Z_{t-1}^j g(\beta' Z_{t-1})| < \infty,$$

where $g(x)$ is a bounded function.

CASE 2: Log-log Link.

Now $\Phi(x) = \exp(-\exp(-x))$ and $\phi(x) = \exp(-\exp(-x))\exp(-x)$, and calculations show that

$$\frac{(\exp(-\exp(-x))\exp(-x))^2}{\exp(-\exp(-x))(1 - \exp(-\exp(-x)))} \to 0 \quad x \to \pm\infty.$$

So by condition (C2) on the moments of Z_t we get (7).

CASE 3: Complementary Log-log Link.

The verification of (7) is done by calculations similar to CASE 2 with the different different Φ and ϕ functions.

Note. The identity link uses the uniform distribution function $\Phi(u)$ on a finite interval (a, b), so it will truncate the vector Z_t through $\eta_t = \beta' Z_t$ at finite values, and the score vector (5) is also using truncated Z_t values through the uniform density. Hence, under condition (C1) condition (C2) will follow for random variables with finite support. Under (C1) and (C3) it is easy to see that all results derived in the rest of this paper will be valid, so we shall not write out the details of the case of the uniform link.

For our procedures to work we need the strong approximation of the score process by a Brownian motion. Several applicable theorems are available in the literature. We shall use Theorem 1 of Eberlein [4] which requires the verification of five properties of the partial sum process.

Let

$$X_{t-1} = Z_{t-1}(Y_t - \pi_t(\beta))\left[\frac{\phi(\beta' Z_{t-1})}{\Phi(\beta' Z_{t-1})(1 - \Phi(\beta' Z_{t-1}))}\right] =$$
$$= Z_{t-1}(Y_t - \pi_t(\beta))\psi(\beta' Z_{t-1}),$$

and $S_n(m) = \sum_{t=m+1}^{m+n} X_t$.

We list the five conditions as (E1)–(E5).

(E1) $E(X_t) = 0, \ t \geq 0$.

It is clear that $E(X_t) = E(E(X_t|\mathscr{F}_{t-1})) = 0$, hence condition (E1) is satisfied.
Next, we need that uniformly in m
(E2) $\|E(S_n(m)|\mathscr{F}_m)\|_1 = O(n^{1/2-\theta})$ for some $0 < \theta < 1/2$.
As X_t are martingale differences $E(S_n(m)|\mathscr{F}_m) = 0$, so (E2) is clearly satisfied.
The third condition (E3) ((1.5) of Theorem 1 in Eberlein [4]) is that uniformly in m
(E3) $\|E[S_n(m)_i S_n(m)_j|\mathscr{F}_m] - E[S_n(m)_i S_n(m)_j]\|_1 = O(n^{1-\theta})$ for some $\theta > 0$ and
for all $1 \le i, j \le p$.
For condition (E4) we need that for some $M > 0$ and $\delta > 0$
(E4) $E(\|X_t\|^{2+\delta}) < M$.
Finally, let

$$(T_n(m))_{i,j} = \frac{1}{n} E[S_n(m)_i S_n(m)_j].$$

We need that there is a covariance matrix Γ such that uniformly in m
(E5) $(T_n(m))_{i,j} - \Gamma_{i,j} = O(n^{-\rho})$, for some $\rho > 0$ and all $1 \le i, j \le d$.

2.2 Proof of Condition (E3)

Note that X_k and X_l are uncorrelated if $k \ne l$. By stationarity it is sufficient to
consider $m = 0$. With notation

$$\gamma(\beta' Z_{k-1}) = \frac{\phi^2(\beta' Z_{k-1})}{\Phi(\beta' Z_{k-1})(1 - \Phi(\beta' Z_{k-1}))}$$

we have for $1 \le i, j, \le p$, that

$$E(S_n(0)_i S_n(0)_j) = E\left(\sum_{k=1}^n Z_k^i Z_k^j \gamma(\beta' Z_{k-1})\right).$$

Similarly, we calculate that

$$E(S_n(0)_i S_n(0)_j|\mathscr{F}_0) = E\left(\sum_{k=1}^n Z_k^i Z_k^j \gamma(\beta' Z_{k-1})|\mathscr{F}_0\right).$$

Hence, for the validity of (E3) we have to analyze

$$\sum_{k=1}^n \left[E\left(Z_k^i Z_k^j \gamma(\beta' Z_{k-1})|\mathscr{F}_0\right) - E\left(Z_k^i Z_k^j \gamma(\beta' Z_{k-1})\right)\right].$$

If the Central Limit Theorem holds for its non-conditional version, that is, if

$$n^{-1/2} \sum_{k=1}^{n} \left[Z_k^i Z_k^j \gamma(\beta' Z_{k-1}) - E\left(Z_k^i Z_k^j \gamma(\beta' Z_{k-1}) \right) \right] \to^D N(0, \sigma^2), \ n \to \infty, \quad (10)$$

with $\sigma^2 < \infty$, then, as the limit is almost surely finite, by multiplying with $n^{-\theta}$, $\theta > 0$, we get

$$n^{-(1/2+\theta)} \left| \sum_{k=1}^{n} \left[Z_k^i Z_k^j \gamma(\beta' Z_{k-1}) - E\left(Z_k^i Z_k^j \gamma(\beta' Z_{k-1}) \right) \right] \right| \to^{a.s.} 0. \quad (11)$$

The sequence on the left hand side of (11) is a non-negative sequence of random variables, that is, nonnegative measurable functions almost surely converging to zero, hence by the bounded convergence theorem their integral is converging to zero, which gives

$$n^{-(1/2+\theta)} \left| \sum_{k=1}^{n} \left[E\left(Z_k^i Z_k^j \gamma(\beta' Z_{k-1}) | \mathcal{F}_0 \right) - E\left(Z_k^i Z_k^j \gamma(\beta' Z_{k-1}) \right) \right] \right| \to^{a.s.} 0, \quad (12)$$

and from (12) condition (E3) will follow. To see that the the Central Limit Theorem in (10) holds we have to verify the conditions for its validity. There are several versions in the literature, see, for example, Serfling [8] for a detailed study. A basic set of conditions is that the zero mean terms in the partial sums have finite $2+\delta$-order absolute moments for some $\delta > 0$, and the existence of the asymptotic variance $0 < \sigma^2 < \infty$. This is clearly satisfied by condition (C2) as the function $\gamma(\beta' Z_{t-1})$ is bounded. Some additional regularity conditions are required, such as, for example, the Ibragimov condition. "These conditions are not severe additional restrictions, but are not, ..., very amenable to verification, although they have some intuitive appeal", remarks Serfling [8]. We assume they are valid, but will not formulate them, as it would not contribute to our main purpose.

2.3 Proof of Condition (E4)

We will do the calculations separately for the various Φ functions.

CASE 1: Probit model.

Again, we use the well-known tail approximation for the normal distribution function: $1 - \Phi(x) \sim \phi(x)/x$ as $x \to \infty$. For the one-dimensional case we have

$$E(\|X_t\|^{2+\delta}) = E\left(\left| \frac{Z_{t-1}\phi(\eta_t)}{\Phi(\eta_t)(1 - \Phi(\eta_t))} \right|^{2+\delta} E\left(|Y_t - \pi_t|^{2+\delta} | \mathcal{F}_{t-1} \right) \right).$$

As

$$E\left(|Y_t - \pi_t|^{2+\delta}|\mathscr{F}_{t-1}\right) = |1 - \Phi(\eta_t)|^{2+\delta}\Phi(\eta_t) + (\Phi(\eta_t))^{2+\delta}(1 - \Phi(\eta_t)),$$

$(\eta_t = \beta'Z_{t-1})$, we have

$$E(\|X_t\|^{2+\delta}) = E\left(|Z_{t-1}|^{2+\delta}\left[\frac{\phi^{2+\delta}(\eta_t)}{\Phi^{1+\delta}(\eta_t)} + \frac{\phi^{2+\delta}}{(1 - \Phi(\eta_t))^{1+\delta}}\right]\right). \tag{13}$$

At the tails, in the first term as $x \to -\infty$ for the normal distribution we have

$$\frac{\phi^{2+\delta}(x)}{(\Phi(x))^{1+\delta}} \sim \frac{\phi^{2+\delta}|x|^{1+\delta}}{\phi^{1+\delta}} = \phi(x)|x|^{1+\delta}.$$

Similar approximation can be used for the second term in (13) when $x \to \infty$. For the normal density $\phi(x)x^{1+\delta} \to 0, x \to \pm\infty$, so we obtain an upper bound for (13) that is the expected value of $|Z_{t-1}|^{2+\delta}$ multiplied by a bounded function, hence finite by condition (C2). For higher dimensions note that each component of the Z_{t-1} vector is multiplied by the same function, and we use Minkowski's Inequality to arrive to (E4).

CASE 2: Log-log Link.
Now $\Phi(x) = \exp(-\exp(-x))$ and $\phi(x) = \exp(-\exp(-x))\exp(-x)$. To show that

$$E\|X_t\|^{2+\delta} < \infty$$

we use in our calculations that

$$E\left(\|Z_{t-1}\frac{\phi(\eta_t)}{\Phi(\eta_t)(1 - \Phi(\eta_t))}(Y_t - \pi_t(\beta))\|^{2+\delta}|\mathscr{F}_{t-1}\right)$$

$$= E\left(\|Z_{t-1}\|^{2+\delta}\left[\frac{\phi(\eta_t)}{\Phi(\eta_t)(1 - \Phi(\eta_t))}\right]^{2+\delta}\right.$$

$$\times[|1 - \Phi(\eta_t)|^{2+\delta}\Phi(\eta_t) + (\Phi(\eta_t))^{2+\delta}(1 - \Phi(\eta_t))]\right).$$

Putting into this expression the current distribution function Φ and density ϕ, we can directly calculate the tail behavior of the multiplier function of $[Z_t^i]^{2+\delta}$, and obtain that it is a bounded function. So we can see by condition $E(|Z_t^i|^{2+\delta}|) < \infty$, $i = 1, \ldots, p$, that (E4) is satisfied.

CASE 3: Complementary Log-Log Link.
For $\Phi(x) = 1 - \exp(-\exp(x))$ the density is $\phi(x) = \exp(x)\exp(-\exp(x))$. Again, we can use in the calculations the the explicit form of $E(\|(Y_t - \pi_t(\beta))\|^{2+\delta}|\mathscr{F}_{t-1})$. It can be expressed as a function of $\Phi(\eta_t)$. The the integrand is $\|Z_{t-1}\|^{2+\delta}$ multiplied by a bounded function, so it will follow that (E4) is true. This concludes the proof of (E4).

2.4 Proof of Condition (E5)

We need that uniformly in m

$$(T_n(m))_{i,j} - \Gamma_{i,j} = O(n^{-\rho})$$

for some $\rho > 0$ and all $1 \leq i,j \leq d$. By stationarity $T_n(m) = T$ all m. As (7) holds, we can take $\Gamma = T$.

Hence we have the following result for our three models.

Theorem 1. *Under our assumptions (C1)–(C3), there exists a vector of Brownian motions $(W_t)_{t \geq 0}$ with covariance matrix $E(T_n(\beta))$, with $T_n(\beta)$ defined in (6), such that, if β is the true vector of coefficients in the regression model (1), then the score vector in (5) admits the following approximation*

$$S_n(\beta) - W(n) = O(n^{1/2 - \delta}) \quad a.s.$$

for some $\delta > 0$.

Next we consider the problem of consistently estimating the parameter vector β. It will be shown that the maximizer of the partial log-likelihood function

$$L(\beta) = \sum_{t=1}^{n} [y_t \log \frac{\pi_t}{1 - \pi_t} + \log(1 - \pi_t)]$$

can serve this purpose. Let

$$\psi(\eta_t) = \frac{\phi(\beta' Z_{t-1})}{\Phi(\beta' Z_{t-1})(1 - \Phi(\beta' Z_{t-1}))}, \quad \eta_t = \beta' Z_{t-1}.$$

Taylor expansion about the true value β_0 gives

$$\frac{1}{n} L(\beta) - \frac{1}{n} L(\beta_0) = S_1 + S_2 + S_3 = \left[\sum_{j=1}^{p} (\beta^j - \beta_0^j) \frac{1}{n} \sum_{t=1}^{n} Z_{t-1}^j (Y_t - \Phi(\eta_t)) \psi(\eta_t) \right]$$

$$+ \left[\frac{1}{2} \sum_{j=1}^{p} \sum_{k=1}^{p} (\beta^j - \beta_0^j)(\beta^k - \beta_0^k) \frac{1}{n} \sum_{t=1}^{n} Z_{t-1}^j Z_{t-1}^k [(Y_t - \Phi(\eta_t)) \psi'(\eta_t) - \phi(\eta_t)\psi(\eta_t)] \right]$$

$$+ \left[\frac{1}{6} \sum_{j=1}^{p} \sum_{k=1}^{p} \sum_{l=1}^{p} (\beta^j - \beta_0^j)(\beta^k - \beta_0^k)(\beta^l - \beta_0^l) \frac{1}{n} \sum_{t=1}^{n} Z_{t-1}^j Z_{t-1}^k Z_{t-1}^l c_t^{jkl} \right],$$

where

$$c_t^{jkl} = (Y_t - \Phi(\eta_t)) \psi''(\eta_t) - \phi(\eta_t)\psi'(\eta_t) - \phi'(\eta_t)\psi(\eta_t) - \phi(\eta)\psi'(\eta).$$

Terms in S_1 have mean zero. To calculate their variance we consider

$$E\left(Z_{t-1}^j\,(Y_t - \Phi(\eta_t))\,\psi(\eta_t)\right)^2$$

$$= E\left((Z_{t-1}^j)^2\Phi(\eta_t)(1 - \Psi(\eta_t))\frac{\phi^2(\eta_t)}{(\Phi(\eta_t)(1 - \Psi(\eta_t)))^2}\right)$$

$$= E\left((Z_{t-1}^j)^2\gamma(\eta_t)\right).$$

Function $\gamma(x)$ was shown to be bounded in the proof of (7), so by stationarity and the moment conditions on Z_t we get that

$$\frac{1}{n}\sum_{t=1}^n Z_{t-1}^j\,(Y_t - \Phi(\eta_t))\,\psi(\eta_t) \to^{a.s.} 0,\ n \to \infty. \tag{14}$$

If $\Phi(x)(1 - \Phi(x))\psi'(x)$ is a bounded function, then by arguments and calculations as above we get that

$$\frac{1}{n}\sum_{t=1}^n Z_{t-1}^j Z_{t-1}^k (Y_t - \Phi(\eta_t))\psi'(\eta_t) \to^{a.s.} 0,\ n \to \infty.$$

Furthermore,

$$\frac{1}{n}\sum_{t=1}^n Z_{t-1}^j Z_{t-1}^k (-\phi(\eta_t)\psi(\eta_t)) = (-1)\frac{1}{n}\sum_{t=1}^n Z_{t-1}^j Z_{t-1}^k \frac{\phi^2(\eta_t)}{\Phi(x)(1 - \Phi(x))} \to^{a.s.} -T^{j,k},$$

as was shown in the proof of (7), giving

$$S_2 \to^{a.s.} -\frac{1}{2}(\beta - \beta_0)'T(\beta - \beta_0),\ n \to \infty. \tag{15}$$

Calculations to prove that $\Phi(x)(1 - \Phi(x))\psi'(x)$ is a bounded function are done in the Appendix.

Finally, we can show that for all $i, j, k = 1,\dots,p$ there exists a function $M_{ijk}(x)$ such that

$$\left|\frac{1}{n}\sum_{t=1}^n Z_{t-1}^j Z_{t-1}^k Z_{t-1}^l c_t^{jkl}\right| \le M_{ijk}(x), \tag{16}$$

for which we have

$$E(M_{ijk}(x)) < \infty.$$

Again, these calculations are done in the Appendix.

We now prove that the maximizer of the partial log-likelihood function in a neighborhood of the true value β_0 is consistent as $n \to \infty$. For this we follow the ideas of Lehmann [7] for the multidimensional case. (cf. Cramér [2] for the one-dimensional case.)

Consider a small a-radius neighborhood Q_a of β_0. It is sufficient to show that with probability converging to one $L(\beta) < L(\beta_0)$ in Q_a. By (14) in Q_a

$$|S_1| < pa^3,$$

because if n is large enough, then

$$\left| \frac{1}{n} \sum_{t=1}^{n} Z_{t-1}^j \left(Y_t - \Phi(\eta_t) \right) \psi(\eta_t) \right| < a^2,$$

and $|\beta^i - \beta_0^j| < a$.

By (16) in Q_a for large n and probability close to one $|S_3| < a^3 c_1$, c_1 a constant. The quadratic form in (15) can be expressed with the help of an orthogonal transformation in a form $\sum_{i=1}^{p} \lambda_i \xi_i^2$ with $\sum \xi_i^2 = a^2$ on Q_a, where $\{\lambda_i\}$ are positive, hence we get that with a constant c_2

$$P(S_2 < -c_2 a^2) \to 1, \ n \to \infty.$$

Putting these together we have with probability arbitrarily close to one, that

$$S_1 + S_2 + S_3 < -c_2 a^2 + c_3 a^3,$$

where c_3 is a constant. This means that for a small a, if $\beta \in Q_a$, then $L(\beta) < L(\beta_0)$. Thus we have proven

Theorem 2. *Let $\hat{\beta}_n$ be the maximizer of the partial log-likelihood function. If assumptions (C1)–(C3) hold for our models, then*

$$P(\lim_{n \to \infty} \hat{\beta}_n = \beta_0) \to 1, \ n \to \infty.$$

The next result proves the asymptotic normality of estimator $\hat{\beta}_n$.

Theorem 3. *Let $\hat{\beta}_n$ be the maximizer of the partial log-likelihood function. Under our assumptions (C1)–(C3) for our models*

$$n^{1/2} T^{1/2} (\hat{\beta}_n - \beta) \to^d N(0, I),$$

where $I_{p \times p}$ is the identity matrix.

Finally, we need the following theorem for the large sample behavior of the test statistic $S_k(\hat{\beta}_n)$, which is the score vector function evaluated at $\beta = \hat{\beta}_n$.

Theorem 4. *Under assumptions (C1)–(C3) for our models we have that the statistics process*

$$n^{-1/2}\hat{T}_n^{-1/2}S_k(\hat{\beta}_n), \; k = 1, \ldots, n,$$

converges in distribution to $B(t)$, a p-dimensional vector of independent Brownian bridges.

The proofs of the last two theorems are very similar to the proofs of the corresponding statements in Fokianos et al. [5] with the appropriate replacement in the formulas. As all the necessary differences in the details have been considered above, those proofs will be omitted.

These theoretical results allow us to extend the use of the tests of H_0 for our current models. Note, that the test statistic is the score vector $S_k(\beta)$ of (5), evaluated at $\beta = \hat{\beta}_n$, hence only one parameter estimation using all available data is required.

Test 1: *(one-sided) The null hypothesis of no change is rejected if for some i, i = 1, 2, \ldots, p, the maximum of the standardized score component corresponding to the ith coefficient crosses a boundary $C_1(\alpha^*)$. That is, as soon as for some i, i = 1, 2, \ldots, p,*

$$\left(\hat{T}^{-1/2} \max_{1 < k \leq n} n^{-1/2} S_k(\hat{\beta}_n) \right)^i \geq C_1(\alpha^*).$$

In this testing procedure, $\alpha^* = 1 - (1 - \alpha)^{1/p}$ is the probability of false alarm in monitoring the ith regression coefficient, while α is the overall probability of false alarm for a change in any coefficient. The threshold $C_1(.)$ is obtained from the distribution of the supremum of the one-dimensional Brownian bridge $B(u)$ using

$$P\left(\sup_{0 \leq u \leq 1} B(u) \geq C_1(\alpha^*) \right) = \alpha^*.$$

Similarly, a two-sided test can be defined as follows:

Test 2: *(two-sided) The null hypothesis is rejected if for some i, i = 1, 2, \ldots, p, the maximum of the absolute value of the standardized score component corresponding to the ith coefficient crosses a boundary $C_2(\alpha^*)$. That is, as soon as for some i, i = 1, 2, \ldots, p,*

$$\left| \left(\hat{T}^{-1/2} \max_{1 < k \leq n} n^{-1/2} S_k(\hat{\beta}_n) \right)^i \right| \geq C_2(\alpha^*).$$

The threshold $C_2(.)$ is computed from the equation

$$P\left(\sup_{0 \leq u \leq 1} |B(u)| \geq C_1(\alpha^*)\right) = \alpha^*,$$

where $B(u)$ denotes a one-dimensional Brownian bridge. Values of $C_1(\alpha)$ and $C_2(\alpha)$ are readily available in the literature.

Paper Fokianos et al. [5] has simulation studies and data applications that show the excellent performance of the above tests in case of the logit link function. We expect that the change in the link function will not make the practical applications any different.

2.5 Alternative Hypothesis of One Change

Recall that the test statistic process is the standardized score vector $S_k(\hat{\beta}_n) = \sum_{t=1}^{k} Z_{t-1}(Y_t - \Phi(\hat{\beta}_n' Z_{t-1}))\psi(\hat{\eta}_t)$, $\hat{\eta}_t = \hat{\beta}_n' Z_{t-1}$, $k = 1, \ldots, n$. To examine the behavior of this process under the alternative of one change (AMOC – At Most One Change), we need to separate the case when the source of change is the covariate vector from the case when the parameter vector changes. Tests for change in the covariate vector were defined in Berkes et al. [1], so we need to consider only the situation when the β vector changes but the $\{Z_t\}$ process is stationary. This rules out the auto-regressive type components in the Z_t vector.

As function $\psi(x)$ is bounded in out models, it is easily seen that the mean of the terms in $S_k(\hat{\beta}_n)$ is

$$E((Y_t - \Phi(\hat{\eta}_n))g(\hat{\eta}_t)),$$

where $g(x)$ is a bounded function. We assume that $E(\Phi(\beta_0' Z_t)) \neq E(\Phi(\beta_1' Z_t))$, where β_1 denotes the value of the parameter vector after change. We cannot have $E(Y_t - \Phi(\hat{\beta}_n')) = 0$ for all $t = 1, \ldots, n$ as $E(Y_t) = E(\Phi(\beta_0' Z_t))$, $t \leq \tau$, and $E(Y_t) = E(\Phi(\beta_1' Z_t))$, $t > \tau$. Let $\tau = n\delta$ for some $0 < \delta < 1$, and separate the various cases as follows.

1. $\hat{\beta}_n \to^p \beta_a$, $\beta_a \neq \beta_0$, $\beta_a \neq \beta_1$
2. $\hat{\beta}_n \to^p \beta_i$, $i = 0$ or $i = 1$
3. $\hat{\beta}_n$ is not convergent

In the first case if $1 \leq t \leq \tau$ then $E(Y_t - \Phi(\hat{\beta}_n' Z_{t-1})) = E(Y_t - \Phi(\beta_0' Z_{t-1})) + E(\Phi(\beta_0' Z_{t-1}) - \Phi(\hat{\beta}_n' Z_{t-1}))$, and if $\tau < t \leq n$ then $E(Y_t - \Phi(\hat{\beta}_n' Z_{t-1})) = E(Y_t - \Phi(\beta_1' Z_{t-1})) + E(\Phi(\beta_1' Z_{t-1}) - \Phi(\hat{\beta}_n' Z_{t-1}))$. These converge to different values, and from this we get that the size of drift at $t = \tau$ is $O(\sqrt{n})$.

In the second case, as the process is tied down $S_n(\hat{\beta}_n) \equiv 0$, so $E(S_n(\hat{\beta}_n)) = 0$. From this we get that $E(\Phi(\beta_0' Z_{t-1}) - \Phi(\hat{\beta}_n' Z_{t-1}))$, $t \leq \tau$ and $E(\Phi(\beta_1' Z_{t-1}) - \Phi(\hat{\beta}_n' Z_{t-1}))$, $t > \tau$ have different signs, and this leads to the same conclusion as in the first case.

Finally, if $\hat{\beta}_n$ is not convergent, then as $\beta_0 \neq \beta_1$, $\|\beta_0 - \beta_1\| \neq 0$, so there exists a $\delta > 0$ such that for any n and any $\hat{\beta}_n$ value, we have that $\|\hat{\beta}_n - \beta_i\| > \delta$ for $i = 0$ or $i = 1$. Considering, again $E(\Phi(\beta_0' Z_{t-1}) - \Phi(\hat{\beta}_n' Z_{t-1}))$ and $E(\Phi(\beta_1' Z_{t-1}) - \Phi(\hat{\beta}_n' Z_{t-1}))$ we can argue as in the first two cases to conclude that the drift is of size $O(\sqrt{n})$.

Appendix

To see that $\Phi(x)(1 - \Phi(x))\psi'(x)$ is a bounded function the three cases have to be treated separately.

CASE 1: Probit model.

We have

$$\Phi(x)(1 - \Phi(x))\psi'(x) = \frac{\phi'(x)\Phi(x) - \phi'(x)\Phi^2(x) - \phi^2(x)}{\Phi(x)(1 - \Phi(x))} = \phi'(x) - \gamma(x). \quad (17)$$

We use the normal density and distribution function and the tail approximation for the distribution function to show that $\phi'(x) \to 0$ as $x \to \pm\infty$. The two tails are looked at separately, and then by the earlier results for the $\gamma(x)$ function we can show that the limit of the function in (17) as $x \to \pm\infty$ is zero, so it is bounded.

CASE 2: Log-log link.

In (17) we use $\Phi(x) = \exp(-e^{-x})$ and $\phi(x) = \exp(-e^{-x})e^{-x}$ now, and straightforward calculation gives again separately for $x \to +\infty$ and $x \to -\infty$ that the function has limit zero, hence boundedness will follow.

CASE 3: Complementary log-log link.

We do the calculations as above with the different distribution and density function.

Next we show that (16) is true. Note, that

$$c_t^{jkl} = (Y_t - \Phi(\eta_t))\psi''(\eta_t) - 2\phi(\eta_t)\psi'(\eta_t) - \phi'(\eta_t)\psi(\eta_t).$$

By the previous results we need to consider the first term only. We use Cauchy's Inequality, and then it is sufficient to show that

$$E\left(E(Y_t - \Phi(\eta_t)|\mathscr{F}_{t-1})^2 (\psi''(\eta_t))^2\right)$$
$$= E\left(\Phi(\eta_t)(1 - \Phi(\eta_t))(\psi''(\eta_t)^2\right).$$

is finite. We can apply tail approximation formula for the normal distribution, and only straightforward calculations are needed for the other cases. These are very tiresome, but no new ideas are necessary, hence they are omitted. Note, that in condition (C2) $E|Z_k^i|^\kappa < \infty$, $i = 1, \ldots, p$, $\kappa = 6$ is needed for this part of the proofs only. In all other parts $\kappa = 4 + \delta$, $\delta > 0$, would suffice.

References

1. Berkes, I., Gombay, E., Horváth, L.: Testing for changes in the covariance stucture of linear processes. J. Stat. Plann. Inference **139**(6), 2044–2063 (2009)
2. Cramér, H.: Mathematical Methods of Statistics. Princeton University Press, Princeton (1946)
3. Doob, J.L.: Stochastic Processes. Wiley, New York (1953)
4. Eberlein, E.: On strong invariance principles under dependence assumptions. Ann. Probab. **14**(1), 260–270 (1986)
5. Fokianos, K., Gombay, E., Hussein, A.: Retrospective change detection for binary time series models. J. Stat. Plann. Infernce **145**, 102–112 (2014)
6. Kedem, B., Fokianos, K.: Regression Models for Time Series Analysis. Wiley, Hoboken (2002)
7. Lehmann, E.L.: Theory of Point Estimation. Wiley, New York (1983)
8. Serfling, R.J.: Contributions to central limit theory for dependent variables. Ann. Math. Stat. **39**, 1158–1175 (1968)

Part V
Short and Long Range Dependent
Time Series

Diagnostic Tests for Innovations of ARMA Models Using Empirical Processes of Residuals

Kilani Ghoudi and Bruno Rémillard

This paper is dedicated to Professor Miklós Csörgő on his 80th birthday.

1 Introduction

Measures and formal tests of lack of fit for time series models have attracted a lot of attention during the last sixty years. The first ad hoc procedure was based on correlograms, a term which, according to Kendall [22], was coined by Wold in his 1938 Ph.D. thesis. See, e.g., Wold [35]. Motivated by the pioneering work of Kendall [21, 22], the first rigorous results on the asymptotic covariance between sample autocorrelations were done by Bartlett [3] for autoregressive models. Then, Quenouille [32] proved the asymptotic normality of autocorrelations and proposed a test of goodness-of-fit for autoregressive models using linear combinations of autocorrelation coefficients. It was extended to moving average models by Wold [34]. The development of tests of goodness-of-fit using residuals for ARMA time series models with Gaussian innovations started in the 1970s, following the publication of Box and Jenkins [6] and the famous work of Box and Pierce [7], where the authors proposed a test of lack of fit using the sum of the squares of autocorrelation coefficients of the residuals, viz. $Q_n = n \sum_{k=1}^{m} r_e^2(k)$, where $r_e(k) = \sum_{t=k+1}^{n} e_{t,n} e_{t-k,n} / \sum_{t=1}^{n} e_{t,n}^2$ is the lag k autocorrelation coefficient and the $e_{t,n}$'s are the residuals of the ARMA model. Even if the authors warned the reader that the suggested chi-square approximation with $m - p - q$ degrees of freedom was "valid" for m large (and the sample size much larger), many researchers applied the test with m small. In fact, taking a simple AR(1) model

K. Ghoudi
Department of Statistics, United Arab Emirates University,
PO Box 17555, Al Ain, United Arab Emirates
e-mail: kghoudi@uaeu.ac.ae

B. Rémillard (✉)
CRM, GERAD and Department of Decision Sciences, HEC Montréal, 3000 chemin de la Côte Sainte-Catherine, Montréal, QC H3T 2A7, Canada
e-mail: bruno.rcmillard@hec.ca

© Springer Science+Business Media New York 2015
D. Dawson et al. (eds.), *Asymptotic Laws and Methods in Stochastics*,
Fields Institute Communications 76, DOI 10.1007/978-1-4939-3076-0_14

$X_t - \mu = \phi(X_{t-1} - \mu) + \epsilon_i$, it is easily seen that for m fixed, the limiting distribution Q of the sequence Q_n is not a chi-square. It is a quadratic form of Gaussian variables with $E(Q) = m - 1 + \phi^{2m}$. Due to that incorrect limiting distribution, modifications were suggested. See, e.g., Davies et al. [10] and Ljung and Box [28]. Nevertheless, as mentioned in Davies et al. [10] these corrections are far from optimal and their behavior, in most practical situations, still differs from the prescribed asymptotic. The right quadratic form finally appeared in Li and McLeod [26] where they considered general innovations. During that fruitful period, McLeod and Li [30] proposed a test based on autocorrelation on the squared residuals, viz $Q_n^* = n(n + 2) \sum_{k=1}^{m} r_{ee}^2(k)/(n - k)$, where $r_{ee}(k) = \sum_{t=k+1}^{n} (e_{t,n}^2 - \hat{\sigma}^2)(e_{t-k,n}^2 - \hat{\sigma}^2) / \sum_{t=1}^{n} (e_{t,n}^2 - \hat{\sigma}^2)^2$ and $\hat{\sigma}^2 = \sum_{t=1}^{n} e_{t,n}^2/n$. These authors proved that the joint limiting distribution of $r_{ee}(1), \ldots, r_{ee}(m)$ is Gaussian with zero mean and identity covariance matrix, so that the limiting distribution of the sequence Q_n^* is chi-square with m degrees of freedom. Until recently, the squared residuals were not used.

Because the tests based on autocorrelations of residuals or squared residuals are not consistent, i.e. when the null hypothesis is false, the power does not always tend to one as the sample size n goes to infinity, Bai [2] investigated the sequential empirical processes based on the (unnormalized) residuals of *mean zero* ARMA processes. He showed that these processes have the same asymptotic behavior whether the model parameters are estimated or not. His result was then cited in many subsequent works dealing with empirical process based on residuals. Unfortunately many of these authors forget to specify that Bai's result is only valid for mean zero ARMA processes. Even if the results of Bai [2] are theoretically interesting, they are of limited practical use. First, in applications, the mean is rarely known. When the mean must be estimated, the limiting distribution of the empirical process is much more complicated and there is a significant effect on the test statistics. Second, he did not consider the important case of standardized residuals. Building the empirical process with standardized residuals yield a much different limiting process. These two problems were solved by Lee [25] in the case of AR(p) models; however he did not consider the sequential empirical process. Surprisingly his results are still ignored. Last but not least, for testing independence, one needs to study the behavior of the empirical process based on successive residuals. Even if the assumption of Bai is kept, it will be shown that the limiting distribution is no longer parameter free.

The rest of this paper is organized as follows. The main results appear in Sect. 2, where one considers multivariate serial empirical processes of residuals, including results for the empirical copula process and associated Möbius transforms. Motivated by the findings of Genest et al. [15], one studies in Sect. 3 the asymptotic behavior of empirical processes based on squared residuals, including the associated copula processes. These results shed some light on the surprising findings of McLeod and Li [30]. Under the additional assumption of symmetry about 0 of the innovations, it is shown that the limiting processes are parameter free. Section 4 contains tests statistics for diagnostic of ARMA models, using empirical processes constructed from the underlying residuals. In particular, one proposes nonparametric tests of change-point in the distribution of the innovations, tests of goodness-of-fit for the law of innovations, and tests of serial independence for m consecutive innovations. Simulations are also carried out to assess the finite sample properties of the proposed tests and give tables of critical values. Section 5 contains an example of application of the proposed methodologies. The proofs are given in Sect. 6.

2 Empirical Processes of Residuals

Consider an ARMA(p, q) model given by

$$X_i - \mu - \sum_{k=1}^{p} \phi_k(X_{i-k} - \mu) = \varepsilon_i - \sum_{k=1}^{q} \theta_k \varepsilon_{i-k}$$

where the innovations (ε_i) are independent and identically distributed with continuous distribution F with mean zero and variance σ^2. Suppose that μ, ϕ, θ are estimated respectively by $\hat{\mu}_n, \hat{\phi}_n, \hat{\theta}_n$. The residuals $e_{i,n}$ are defined by $e_{i,n} = 0$ for $i = 1, \ldots, \max(p, q)$, while for $i = \max(p, q) + 1, \ldots, n$,

$$e_{i,n} = X_i - \hat{\mu}_n - \sum_{k=1}^{p} \hat{\phi}_{k,n}(X_{i-k} - \hat{\mu}_n) + \sum_{k=1}^{q} \hat{\theta}_{k,n} e_{i-k,n}.$$

2.1 Asymptotic Behavior of the Multivariate Sequential Empirical Process

In this section, one studies the asymptotic behavior of empirical processes needed to define the tests statistics proposed in Sect. 4.

Let m be a fixed integer and for all $(s, \mathbf{x}) \in [0, 1] \times [-\infty, +\infty]^m$, define the multivariate sequential empirical process by

$$\mathbb{H}_n(s, \mathbf{x}) = \frac{1}{\sqrt{n}} \sum_{i=1}^{\lfloor ns \rfloor} \left\{ \prod_{j=1}^{m} \mathbf{1}(e_{i+j-1,n} \leq x_j) - \prod_{j=1}^{m} F(x_j) \right\},$$

where $e_{n+i,n} = e_{i,n}$ for $i \geq 1$. Since only a finite number of residuals is affected by this circular definition, the asymptotic behavior of $\mathbb{H}_n$ is not altered.

Similarly, the univariate sequential empirical process is given by

$$\mathbb{F}_n(s, y) = \frac{1}{\sqrt{n}} \sum_{i=1}^{\lfloor ns \rfloor} \{\mathbf{1}(e_{i,n} \leq y) - F(y)\} = \mathbb{H}_n(s, y, \infty, \ldots, \infty),$$

for all $(s, y) \in [0, 1] \times [-\infty, +\infty]$. It was studied by Bai [2] under the assumption that μ is known. One will see that it makes a difference for the asymptotic distribution of $\mathbb{H}_n$ and $\mathbb{F}_n$.

To simplify the proof, one can assume without loss of generality that $m \leq q$. For if $m > q$, set $\theta_k = 0$ for all $k \in \{q + 1, \ldots, m\}$. This assumption is not needed in practice.

Next, define $B_\theta = \theta_1$ if $m = 1$ and for any $m > 1$,

$$
B_\theta = \begin{pmatrix}
0 & 1 & 0 & \cdots & 0 \\
0 & 0 & 1 & \cdots & 0 \\
\vdots & \vdots & \vdots & \ddots & 0 \\
0 & \cdots & \cdots & 0 & 1 \\
\theta_m & \theta_{m-1} & \cdots & \theta_2 & \theta_1
\end{pmatrix} \in \mathbb{R}^{m \times m}.
$$

Further set $\mathscr{O} = \{\boldsymbol{\theta} \in \mathbb{R}^m; \rho(B_\theta) < 1\}$, where $\rho(B)$ is the spectral radius of B. Assume that stationarity and invertibility conditions are met, i.e. the roots of the polynomials $1 - \sum_{k=1}^p \phi_k z^k$ and $1 - \sum_{k=1}^q \theta_k z^k$ are all outside the complex unit circle. The latter condition is equivalent to $\theta \in \mathscr{O}$.

For all $(s, \mathbf{x}) \in [0, 1] \times [-\infty, +\infty]^m$, set

$$
\overset{\circ}{\mathbb{H}}_n (s, \mathbf{x}) = \frac{1}{\sqrt{n}} \sum_{i=1}^{\lfloor ns \rfloor} \left\{ \prod_{j=1}^m \mathbf{1}(\varepsilon_{i+j-1} \leq x_j) - \prod_{j=1}^m F(x_j) \right\}
$$

$$
= \mathbb{E}_n \{s, F(x_1), \ldots, F(x_m)\},
$$

where

$$
\mathbb{E}_n(s, u_1, \ldots, u_m) = \frac{1}{\sqrt{n}} \sum_{i=1}^{\lfloor ns \rfloor} \left[\prod_{j=1}^m \mathbf{1}\{F(\varepsilon_{i+j-1}) \leq u_j\} - \prod_{j=1}^m u_j \right],
$$

for $(s, \mathbf{u}) \in [0, 1]^{1+m}$. Note that each $F(\varepsilon_{i+j-1})$ is uniformly distributed over $(0, 1)$.

The price to pay for having to estimate the parameters (μ, ϕ, θ) is to make the following assumptions, described in terms of the estimation errors $\Phi_n = \sqrt{n} \, (\hat{\phi}_n - \phi)$, $\Theta_n = \sqrt{n} \, (\hat{\theta}_n - \theta)$ and $M_n = \sqrt{n} \, (\hat{\mu}_n - \mu) \left(1 - \sum_{k=1}^p \hat{\phi}_{k,n}\right)$.

(A1) Let S_F denotes the interior of the support of F, i.e., $S_F = \{x \in \mathbb{R}; 0 < F(x) < 1\}$, and assume that F admits a uniformly continuous bounded density f on its support $\bar{S}_F$, and such that f is positive on S_F.

(A2) $\theta \in \mathscr{O}$ and, as $n \to \infty$, $(\overset{\circ}{\mathbb{H}}_n, M_n, \Phi_n, \Theta_n) \rightsquigarrow (\overset{\circ}{\mathbb{H}}, \mathscr{M}, \Phi, \Theta)$ in $\mathscr{D}_m \times \mathbb{R}^{1+p+q}$, where $(\overset{\circ}{\mathbb{H}}, \mathscr{M}, \Phi, \Theta)$ is a centered and continuous Gaussian process. Here $\mathscr{D}(A)$ is the Skorokod space of càdlàg processes on A, and $\mathscr{D}_m = \mathscr{D}([0, 1] \times [-\infty, \infty]^m)$.

Remark 1. Because our limiting process is a function of $\overset{\circ}{\mathbb{H}}, \mathscr{M}, \varPhi, \varTheta$, the joint convergence of $(\overset{\circ}{\mathbb{H}}_n, M_n, \varPhi_n, \varTheta_n)$ is needed for its representation. First, it is easy to check that $\mathbb{E}_n \rightsquigarrow \mathbb{E}$ in $\mathscr{D}([0,1]^{1+m})$, and $\overset{\circ}{\mathbb{H}}_n \rightsquigarrow \overset{\circ}{\mathbb{H}}$ in $\mathscr{D}_m$, where $\overset{\circ}{\mathbb{H}}(\mathbf{x}) = \mathbb{E}\{F(x_1), \dots, F(x_m)\}$, using the results of Bickel and Wichura [4]; see also Ghoudi et al. [18]. The joint convergence of the parameters $(M_n, \varPhi_n, \varTheta_n) \rightsquigarrow (\mathscr{M}, \varPhi, \varTheta)$ in $\mathbb{R}^{1+p+q}$ is also a formality in general, so (A2) will hold true if one can show that any linear combination of a finite number of random variables $\overset{\circ}{\mathbb{H}}_n(s, \mathbf{x})$, $(s, \mathbf{x}) \in [0,1] \times [-\infty, +\infty]^m$ and $(M_n, \varPhi_n, \varTheta_n)$ converges in law to the appropriate limit. That would be the case for example if one could write

$$(M_n, \varPhi_n, \varTheta_n) = \frac{1}{\sqrt{n}} \sum_{i=1}^{n} \xi_i + o_P(1), \tag{1}$$

where $\xi_i = \xi_i(\varepsilon_i, \varepsilon_{i-1}, \dots)$ is a stationary ergodic sequence of square integrable martingale differences, i.e., $E(\xi_i | \varepsilon_{i-1}, \varepsilon_{i-2}, \dots) = 0$. For if the latter is true, then the joint convergence to a centered Gaussian variable follows from the CLT for martingales [13]. Note that typically, the weak convergence of the estimators is proven using a representation like (1). In particular, this is true for OLS estimators and for many robust estimators as well.

Before stating the convergence results, one needs to define the following elements:

For any $\mathbf{x} \in [-\infty, +\infty]^m$ and any $j, k \in \mathscr{J}_m = \{1, \dots, m\}, j \neq k$, set

$$\mathfrak{F}_{j,k}(\mathbf{x}) = H(x_k) \prod_{l \in \mathscr{J}_m \setminus \{j,k\}} F(x_l),$$

where $H(y) = E\{\varepsilon_1 \mathbf{1}(\varepsilon_1 \leq y)\}$. Note that $H(\infty) = H(-\infty) = 0$, and one can verify that $\int_{-\infty}^{+\infty} H(y) dy = -\sigma^2$. Next, for all $\mathbf{x} \in [-\infty, +\infty]^m$ and $\mathbf{P} = (M, \varPhi, \varTheta, \theta) \in \mathbb{R}^{1+p+q} \times \mathscr{O}$ and for $j \in \{1, \dots, m\}$, define

$$\begin{aligned}
\upsilon_j(\mathbf{x}, \mathbf{P}) = {} & \frac{M}{1 - \sum_{l=1}^{q} \theta_l} \prod_{k=1, k \neq j}^{m} F(x_k) \\
& - \sum_{l=1}^{\min(q, j-1)} \varTheta_l \sum_{k=m-j}^{m-1-l} \left(B_\theta^k\right)_{jm} \mathfrak{F}_{j, m-k-l}(x) \\
& + \sum_{l=1}^{\min(p, j-1)} \varPhi_l \sum_{k=m-j}^{m-1-l} \left(B_\theta^k\right)_{jm} \sum_{t=0}^{m-1-k-l} \psi_t \mathfrak{F}_{j, m-k-l-t}(x),
\end{aligned}$$

where the coefficients $\psi_0, \psi_1, \ldots$ are uniquely determined by the equation

$$\sum_{k=0}^{\infty} \psi_k z^k = \frac{1 - \sum_{k=1}^{q} \theta_k z^k}{1 - \sum_{k=1}^{p} \phi_k z^k}, \qquad |z| \le 1.$$

Remark 2. Note that for any $j \in \mathscr{J}_m$, $v_j(\mathbf{x}, \mathbf{P})$ does not depend on x_j. Also, for an AR(p) model, $q = 0$, so $(B_\theta^k)_{jm} = 1$ if $j = m - k$, and $(B_\theta^k)_{jm} = 0$ otherwise. It follows that

$$v_j(\mathbf{x}, \mathbf{P}) = M \prod_{k=1, k \neq j}^{m} F(x_k) + \sum_{l=1}^{\min(p, j-1)} \Phi_l \sum_{t=0}^{j-1-l} \psi_t \mathfrak{F}_{j, j-l-t}(x).$$

In particular, when $p = 1$, then $\psi_t = \phi^t$, so $v_1(\mathbf{x}, \mathbf{P}) = M \prod_{k=2}^{m} F(x_k)$ and for all $j = 2, \ldots, m$,

$$v_j(\mathbf{x}, \mathbf{P}) = M \prod_{k=1, k \neq j}^{m} F(x_k) + \Phi \sum_{t=0}^{j-2} H(x_{j-1-t}) \phi^t \prod_{k \neq j, j-1-t} F(x_k). \tag{2}$$

Recall that from Remark 1, $\mathbb{E}_n \rightsquigarrow \mathbb{E}$ in $\mathscr{D}([0, 1]^{1+m})$, and $\mathring{\mathbb{H}}_n \rightsquigarrow \mathring{\mathbb{H}}$ in $\mathscr{D}_m$, where $\mathring{\mathbb{H}}(\mathbf{x}) = \mathbb{E}\{F(x_1), \ldots, F(x_m)\}$. Also, $\mathbb{K}(s, u) = \mathbb{E}(s, u, 1, \ldots, 1)$, $(s, u) \in [0, 1]^2$, is the well-known Kiefer process, i.e. a continuous centered Gaussian process with covariance function $\mathrm{Cov}\{\mathbb{K}(s, u), \mathbb{K}(t, v)\} = \min(s, t)\{\min(u, v) - uv\}$, $s, u, t, y \in [0, 1]$. As a result, $\mathring{\mathbb{F}}(s, x) = \mathbb{K}\{s, F(x)\}$, for all $(s, x) \in [0, 1] \times [-\infty, \infty]$. One can now state the main result of the paper about the convergence of $\mathbb{H}_n$ in $\mathscr{D}_{m, F} = \mathscr{D}\left([0, 1] \times \overline{S_F}^m\right)$.

Theorem 1. *Under assumptions (A1–A2), $\mathbb{H}_n \rightsquigarrow \mathbb{H}$ in $\mathscr{D}_{m, F}$, where*

$$\mathbb{H}(s, \mathbf{x}) = \mathring{\mathbb{H}}(s, \mathbf{x}) + s \sum_{j=1}^{m} f(x_j) v_j(\mathbf{x}, \mathscr{P}), \qquad (s, \mathbf{x}) \in [0, 1] \times \overline{S_F}^m,$$

with $\mathscr{P} = (\mathscr{M}, \Phi, \Theta, \boldsymbol{\theta})$. In particular, $\mathbb{F}_n \rightsquigarrow \mathbb{F}$ in $\mathscr{D}_{1, F}$, where

$$\mathbb{F}(s, y) = \mathring{\mathbb{F}}(s, y) + s f(y) \mathscr{E}, \qquad (s, y) \in [0, 1] \times \overline{S_F},$$

with $\mathscr{E} = \mathscr{M} / \left(1 - \sum_{k=1}^{q} \theta_k\right)$. If in addition $\hat{\mu}_n = \bar{X}_n + o_P(1/\sqrt{n})$, then $\sqrt{n} \, \bar{\varepsilon}_n \rightsquigarrow \mathscr{E} \sim N(0, \sigma^2)$, and $\hat{\mathbb{F}}_n \rightsquigarrow \mathbb{F}$ in $\mathscr{D}_{1, F}$, where

$$\hat{\mathbb{F}}_n(s, y) = \frac{1}{\sqrt{n}} \sum_{i=1}^{\lfloor ns \rfloor} \{\mathbf{1}(\varepsilon_i - \bar{\varepsilon}_n \le y) - F(y)\}$$

and $\mathrm{cov}\left\{\mathring{\mathbb{F}}(s, y), \mathscr{E}\right\} = s H(y)$, $(s, y) \in [0, 1] \times \overline{S_F}$.

Remark 3. To recover the result of Bai [2], note that if μ is known, then $\mathscr{E} = 0$, so $\mathbb{F}_n \rightsquigarrow \overset{\circ}{\mathbb{F}}$ by Theorem 1. However, if $m > 1$, $\mathbb{H}_n \not\rightsquigarrow \overset{\circ}{\mathbb{H}}$, even in the simple case of AR(1) models, as seen from (2). The result of Lee [25] for AR(p) models is obtained by setting $m = 1$ and $s = 1$.

2.2 Empirical Process of Standardized Residuals

When testing for goodness-of-fit, it is often necessary to consider standardized residuals. To this end, let $\hat{\sigma}_n^2 = \frac{1}{n-p} \sum_{i=p+1}^n e_{i,n}^2$. It follows from the proof of Theorem 1 that $\hat{\sigma}_n^2 = \sigma^2 s_{\epsilon,n}^2 + o_P(n^{-1/2})$, where $s_{\epsilon,n}^2 = \sum_{i=1}^n \epsilon_i^2/n$, with $\varepsilon_i = \sigma\epsilon_i$. As a result, under the assumption that the kurtosis β_2 of ε_i exists, i.e., $\beta_2 = \frac{E(\varepsilon_i^4)}{\sigma^4} = E(\epsilon_i^4) < \infty$, one has $\mathscr{S}_n^* = \sqrt{n} \left(\frac{\hat{\sigma}_n^2}{\sigma^2} - 1 \right) = \frac{1}{\sqrt{n}} \sum_{i=1}^n (\epsilon_i^2 - 1) + o_P(1) \rightsquigarrow \mathscr{S}^*$ where $\mathscr{S}^* \sim N(0, \beta_2 - 1)$. For $y \in [-\infty, +\infty]$, set

$$\mathbb{F}_n^*(y) = \frac{1}{\sqrt{n}} \sum_{i=1}^n \{\mathbf{1}(e_{i,n}/\hat{\sigma}_n \leq y) - F^*(y)\},$$

where $F^*(y) = F(\sigma y)$, and $f^*(y) = \sigma f(\sigma y)$ are respectively the distribution function and the density of $\epsilon_i = \varepsilon_i/\sigma$. Further set $\mathscr{E}^* = \mathscr{E}/\sigma$.

Corollary 1. *If* $E(\varepsilon_i^4) < \infty$ *and (A1–A2) hold, then* $\mathbb{F}_n^* \rightsquigarrow \mathbb{F}^*$ *in* $\mathscr{D}_{1,F}$, *where*

$$\mathbb{F}^*(y) = \mathbb{K}\{1, F^*(y)\} + f^*(y)\mathscr{E}^* + \frac{y}{2} f^*(y)\mathscr{S}^*, \quad y \in \overline{S_F}.$$

Furthermore, if $\hat{\mu}_n = \bar{X}_n + o_P(1/\sqrt{n})$, *set*

$$\hat{\mathbb{F}}_n^*(y) = \frac{1}{\sqrt{n}} \sum_{i=1}^n \{\mathbf{1}(Z_{i,n} \leq y) - F^*(y)\}, \quad y \in [-\infty, +\infty],$$

where for any $i = 1, \ldots, n$, $Z_{i,n} = \frac{\epsilon_i - \bar{\epsilon}_n}{s_{\epsilon,n}} = \frac{\varepsilon_i - \bar{\varepsilon}_n}{s_{\varepsilon,n}}$. *Then* $\hat{\mathbb{F}}_n^* \rightsquigarrow \mathbb{F}^*$ *in* $\mathscr{D}_{1,F}$.

As noted by Durbin [12] and detailed in Sect. 4, the last result can be used to test the null hypothesis that $\epsilon_i = \varepsilon_i/\sigma \sim F^*$. The possibility of constructing goodness-of-fit tests using $\mathbb{F}_n^*$ was also mentioned in passing by Lee [25].

Remark 4. To illustrate the inadequacy of using Bai [2] results when the mean is estimated, consider doing a goodness-of-fit test of normality for the following simple model: $X_i = 1 + \varepsilon_i$, where $\varepsilon_i \sim N(0, 1)$ are independent, $i = 1, \ldots, n$. For testing $H_0 : \varepsilon_i \sim N(0, 1)$, one applies the Kolmogorov-Smirnov test based on

Table 1 Percentages of rejection of the standard Gaussian hypothesis for $N = 10,000$ replications of the Kolmogorov-Smirnov and Lilliefors tests, using samples of size $n = 100$

Kolmogorov-Smirnov test		Lilliefors test	
ε_i	e_i	ε_i	e_i
4.89	0.03	4.96	4.96

the statistic $\sup_{x \in \mathbb{R}} |\mathbb{F}_n(1, x)|$, and the Lilliefors tests based on $\sup_{x \in \mathbb{R}} |\mathbb{F}_n^*(x)|$. Both tests are evaluated with $\varepsilon_i = X_i - 1$ and $e_i = X_i - \bar{X}_n$, respectively, $i = 1, \ldots, n$.

Because Lilliefors test is a corrected version of the Kolmogorov-Smirnov test in case of estimated parameters, one could get around 5 % of rejection whether one uses ε_i or e_i. If estimation of the mean would not matter, the same should be true for the Kolmogorov-Smirnov test. The results of $N = 10,000$ replications of that experiment are displayed in Table 1 where samples of size $n = 100$ were used. As predicted, both tests are correct when one uses ε_i corresponding to a known mean, while the results differ a lot when using residuals e_i corresponding to an estimated mean. In fact, for the ε_i, Kolmogorov-Smirnov statistic $KS_n = \sup_{x \in \mathbb{R}} |\mathbb{F}_n(1, x)|$ converges in law to $\sup_{x \in \mathbb{R}} | \overset{\circ}{\mathbb{F}} (1, x)| = \sup_{u \in [0,1]} |\mathbb{K}(1, u)|$, as predicted by Bai [2] and Theorem 1, while for the residuals e_i, KS_n converges in law to $\sup_{x \in \mathbb{R}} |\mathbb{F}(1, x)|$.

2.3   Empirical Processes for Testing Randomness

When testing for randomness, defined here as the independence of m consecutive innovations, the marginal distribution F is unknown, so one cannot use directly the empirical process $\mathbb{H}_n$. It is then suggested to estimate F by its empirical analog F_n defined by $F_n(y) = \frac{1}{n} \sum_{i=1}^n \mathbf{1}(e_{i,n} \leq y)$, $y \in [-\infty, +\infty]$. One can base the inference on the empirical process

$$\mathbb{A}_n(\mathbf{x}) = \frac{1}{\sqrt{n}} \sum_{i=1}^n \left\{ \prod_{j=1}^m \mathbf{1}(e_{i+j-1,n} \leq x_j) - \prod_{j=1}^m F_n(x_j) \right\}$$

$$= \mathbb{H}_n(1, \mathbf{x}) - \sqrt{n} \left\{ \prod_{j=1}^m F_n(x_j) - \prod_{j=1}^m F(x_j) \right\}.$$

The following result is a direct consequence of Theorem 1 and the multinomial formula [18]. Before stating it, recall that v_j is defined by (2), and set

$$v_j^{ser}(\mathbf{x}, \mathbf{P}) = v_j(\mathbf{x}, \mathbf{P}) - \frac{M}{1 - \sum_{l=1}^{q} \theta_l} \prod_{k=1, k \neq j}^{m} F(x_k), \quad x_j \in \overline{S_F}, \, j = 1, \dots, m.$$

Corollary 2. *Under Assumptions (A1–A2),* $\mathbb{A}_n \rightsquigarrow \mathbb{A}$ *in* $\mathscr{D}_{m,F}$, *where*

$$\mathbb{A}(\mathbf{x}) = \mathbb{H}(1, \mathbf{x}) - \sum_{j=1}^{m} \mathbb{F}(1, x_j) \prod_{k=1, k \neq j}^{m} F(x_k) = \overset{\circ}{\mathbb{A}}(\mathbf{x}) + \sum_{j=1}^{m} f(x_j) v_j^{ser}(\mathbf{x}, \mathscr{P}),$$

and $\overset{\circ}{\mathbb{A}}(\mathbf{x}) = \overset{\circ}{\mathbb{H}}(\mathbf{x}) - \sum_{j=1}^{m} \mathbb{F}(1, x_j) \prod_{k=1, k \neq j}^{m} F(x_k),$ $\mathbf{x} \in \overline{S_F}^m$.

As suggested by many authors, e.g., Genest and Rémillard [14], one can also base tests of randomness on the residuals empirical copula process defined by

$$\mathbb{C}_n(\mathbf{u}) = \frac{1}{\sqrt{n}} \sum_{i=1}^{n} \left[\prod_{j=1}^{m} \mathbf{1} \left\{ F_n(e_{i+j-1,n}) \le u_j \right\} - \prod_{j=1}^{m} u_j \right], \quad \mathbf{u} \in [0, 1]^m.$$

To obtain more powerful tests, it may be appropriate to use the Möbius decomposition of the copula [14], defined for any subset A in $\mathscr{I}_m = \{ B \subset \{1, \dots, m\}, B \ni 1$ and $|B| > 1 \}$, by

$$\mathbb{C}_{A,n}(\mathbf{u}) = \frac{1}{\sqrt{n}} \sum_{i=1}^{n} \prod_{j \in A} \left[\mathbf{1} \left\{ F_n(e_{i+j-1,n}) \le u_j \right\} - u_j \right],$$

The asymptotic behavior of these processes is given next. It is a consequence of Theorem 1, and the fact that the Möbius transformed process $\mathbb{C}_{A,n}$ is a continuous function of $\mathbb{C}_n$.

Before stating the result, define $\overset{\circ}{\mathbb{C}}(\mathbf{u}) = \mathbb{E}(1, \mathbf{u}) - \sum_{j=1}^{m} \mathbb{K}(1, u_j) \prod_{k=1, k \neq j}^{m} u_k$, $\mathbf{u} \in [0, 1]^m$, and note that $\overset{\circ}{\mathbb{A}}(\mathbf{x}) = \overset{\circ}{\mathbb{C}} \{F(x_1), \dots, F(x_m)\}$, $\mathbf{x} \in [-\infty, +\infty]^m$. $\overset{\circ}{\mathbb{C}}$ is the limiting distribution of the serial empirical process defined in Genest and Rémillard [14].

Corollary 3. *Under Assumptions (A1–A2),* $\mathbb{C}_n \rightsquigarrow \mathbb{C}$ *in* $\mathscr{D}([0, 1]^m)$, *where*

$$\mathbb{C}(\mathbf{u}) = \mathbb{A}\{F^{-1}(u_1), \dots, F^{-1}(u_m)\}, \quad \mathbf{u} \in [0, 1]^m.$$

Moreover, $\{\mathbb{C}_{A,n}\}_{A \in \mathscr{I}_m} \rightsquigarrow \{\mathbb{C}_A\}_{A \in \mathscr{I}_m}$, *where*

$$\mathbb{C}_A(\mathbf{u}) = \overset{\circ}{\mathbb{C}}_A(\mathbf{u}) + f \circ F^{-1}(u_\ell) \, H \circ F^{-1}(u_1) \mathscr{A}_\ell \, \mathbf{1}(A = \{1, \ell\}), \quad \mathbf{u} \in [0, 1]^m,$$

with

$$\mathscr{A}_\ell = \sum_{t=1}^{\min(p,\ell-1)} \Phi_t \sum_{k=m-\ell}^{m-1-t} \left(B_\theta^k\right)_{\ell m} \Psi_{m-k-t-1}$$

$$- \sum_{t=1}^{\min(q,\ell-1)} \Theta_t \left(B_\theta^{m-t-1}\right)_{\ell m}. \qquad (3)$$

The processes $\{\overset{\circ}{\mathbb{C}}_A\}_{A \in \mathscr{I}_m}$ *are independent Wiener sheets, i.e., independent centered Gaussian processes with covariance function*

$$\Gamma_A(\mathbf{u}, \mathbf{v}) = \mathrm{Cov}\left\{\overset{\circ}{\mathbb{C}}_A(\mathbf{u}), \overset{\circ}{\mathbb{C}}_A(\mathbf{v})\right\} = \prod_{j \in A} \left\{\min(u_j, v_j) - u_j v_j\right\},$$

$\mathbf{u}, \mathbf{v} \in [0, 1]^m$.

Remark 5. In Genest and Rémillard [14], where there was no estimation of parameters, the distribution free processes $\{\overset{\circ}{\mathbb{C}}_A\}_{A \in \mathscr{I}_m}$ were used to construct powerful tests of serial independence. Applications of these processes in the present context would require resampling techniques, such as weighted bootstrap, to obtain independent copies of $(\overset{\circ}{\mathbb{C}}_{1,\ell}, \mathscr{A}_\ell)$, for $2 \le \ell \le m$. Such techniques are being investigated.

3 Empirical Processes of Squared Residuals

Let G be the distribution function of ε_i^2, i.e., for any $y \ge 0$, $G(y) = F(\sqrt{y}) - F(-\sqrt{y})$. Assume here that the open support S_F is symmetric about 0 and define accordingly the open support S_G in $\mathbb{R}_+ = [0, \infty)$ of G. As before, let m be a fixed integer. In this section, one omits the parameter s which is only is used for change-point tests. As one will see later, basing test statistics of change-point on $\mathbb{F}_n$ produces parameter-free asymptotic limits, so there is no need of considering change-point tests based on squared residuals. For all $\mathbf{x} \in [0, \infty]^m$, set

$$\mathbb{L}_n(\mathbf{x}) = \frac{1}{\sqrt{n}} \sum_{i=1}^{n} \left\{\prod_{j=1}^{m} \mathbf{1}(e_{i+j-1,n}^2 \le x_j) - \prod_{j=1}^{m} G(x_j)\right\}.$$

Next, for any $\mathbf{x} \in \mathbb{R}_+^m$ and any $A \subset \mathscr{I}_m$, set $(\mathbf{x}_A)_j = \begin{cases} \sqrt{x_j} & \text{if } j \notin A \\ -\sqrt{x_j} & \text{if } j \in A \end{cases}$, and define the (continuous) linear operator Ψ_m from $\mathscr{D}_{m,F}$ to $\mathscr{D}_{m,G}$ by

$$\Psi_m(g)(\mathbf{x}) = \sum_{A \subset \mathscr{I}_m} (-1)^{|A|} g(1, \mathbf{x}_A), \quad \mathbf{x} \in \overline{S_G}^m,$$

where $|A|$ is the cardinal A. In particular, when $m = 1$, one has $\Psi_1(g)(x) = g(1, \sqrt{x}) - g(1, -\sqrt{x})$. The usefulness of this operator can be easily seen from the relation $\mathbb{L}_n = \Psi_m(\mathbb{H}_n)$, that holds almost everywhere.

Before stating the main result, define

$$\overset{\circ}{\mathbb{L}}_n(\mathbf{x}) = \frac{1}{\sqrt{n}} \sum_{i=1}^{n} \left\{ \prod_{j=1}^{m} \mathbf{1}(\varepsilon_{i+j-1}^2 \leq x_j) - \prod_{j=1}^{m} G(x_j) \right\}, \quad \mathbf{x} \in [0, \infty]^m,$$

$$\mathbb{G}_n(y) = \frac{1}{\sqrt{n}} \sum_{i=1}^{n} \left\{ \mathbf{1}(e_{i,n}^2 \leq y) - G(y) \right\}, \quad y \in [0, \infty],$$

and

$$\overset{\circ}{\mathbb{G}}_n(y) = \frac{1}{\sqrt{n}} \sum_{i=1}^{n} \left\{ \mathbf{1}(\varepsilon_i^2 \leq y) - G(y) \right\}, \quad y \in [0, \infty].$$

The next result is a consequence of Theorem 1 and the continuity of Ψ_m.

Theorem 2. *Under assumptions (A1–A2), $\mathbb{L}_n \rightsquigarrow \mathbb{L}$ in $\mathcal{D}_{m,G}$, and $\overset{\circ}{\mathbb{L}}_n \rightsquigarrow \overset{\circ}{\mathbb{L}}$ in $\mathcal{D}_{m,[0,\infty]}$, where $\mathbb{L} = \Psi_m(\mathbb{H})$, $\overset{\circ}{\mathbb{L}} = \Psi_m(\overset{\circ}{\mathbb{H}})$, and*

$$\mathbb{L}(\mathbf{x}) = \overset{\circ}{\mathbb{L}}(\mathbf{x}) + \sum_{j=1}^{m} \left\{ f(\sqrt{x_j}) - f(-\sqrt{x_j}) \right\} \sum_{A \subset \mathcal{J}_m \setminus \{j\}} (-1)^{|A|} v_j(\mathbf{x}_A, \mathscr{P}),$$

for $\mathbf{x} \in \overline{S_G}^m$, where v_j is defined by (2). In particular, $\mathbb{G}_n \rightsquigarrow \mathbb{G} \; \mathcal{D}_{1,G}$, where

$$\mathbb{G}(y) = \overset{\circ}{\mathbb{G}}(y) + \left\{ f(\sqrt{y}) - f(-\sqrt{y}) \right\} \mathscr{E}, \quad y \in [0, \infty],$$

and $\overset{\circ}{\mathbb{G}} = \Psi_1(\overset{\circ}{\mathbb{F}})$ is a G-Brownian bridge.

In addition, if the law of ε is symmetric about 0, then $\mathbb{L} = \overset{\circ}{\mathbb{L}}$ is parameter free. In particular, $\mathbb{G}_n$ and $\overset{\circ}{\mathbb{G}}_n$ both converge in $\mathcal{D}_{1,G}$ to $\overset{\circ}{\mathbb{G}}_0$.

3.1 Empirical Processes of Pairs of Lagged Squared Residuals

For $\ell \in \{2, \ldots, m\}$, set

$$\mathbb{L}_{1,\ell,n}(x, y) = \frac{1}{\sqrt{n}} \sum_{i=1}^{n} \left\{ \mathbf{1}(e_{i,n}^2 \leq x, e_{i+\ell-1,n}^2 \leq y) - G(x)G(y) \right\}, \quad x, y \in [0, \infty].$$

Further set $\overset{\circ}{\mathbb{L}}_{1,\ell}(x,y) = \overset{\circ}{\mathbb{L}}(x, \infty, \ldots, y, \infty, \ldots)$. Note that $\overset{\circ}{\mathbb{L}}_{1,\ell}$ is the limiting process of

$$\frac{1}{\sqrt{n}} \sum_{i=1}^{n} \left\{ \mathbf{1}(\varepsilon_i^2 \leq x, \varepsilon_{i+\ell-1}^2 \leq y) - G(x)G(y) \right\}.$$

Corollary 4. *Under assumptions (A1–A2),*

$$(\mathbb{L}_{1,2,n}, \ldots, \mathbb{L}_{1,m,n}) \rightsquigarrow (\mathbb{L}_{1,2}, \ldots, \mathbb{L}_{1,m}) \quad in \; \mathscr{D}_{2,G}^{\otimes (m-1)},$$

where, for all $x, y \in \overline{S_G}$,

$$\begin{aligned}
\mathbb{L}_{1,\ell}(x,y) = \; &\overset{\circ}{\mathbb{L}}_{1,\ell}(x,y) + \left\{ f(\sqrt{x}) - f(-\sqrt{x}) \right\} G(y)\mathscr{E} \\
&+ \left\{ f(\sqrt{y}) - f(-\sqrt{y}) \right\} G(x)\mathscr{E} \\
&+ \left\{ f(\sqrt{y}) - f(-\sqrt{y}) \right\} \left\{ H(\sqrt{x}) - H(-\sqrt{x}) \right\} \mathscr{A}_\ell,
\end{aligned}$$

where $\mathscr{A}_\ell$ is defined by (3). Moreover, if the law of ϵ_i is symmetric about 0, then $\mathbb{L}_{1,\ell} = \overset{\circ}{\mathbb{L}}_{1,\ell}$ is parameter free.

Now for all $(x,y) \in [0, \infty]^2$, set

$$\mathscr{R}_{1,\ell,n}(x,y) = \frac{1}{\sqrt{n}} \sum_{i=1}^{n} \left\{ \mathbf{1}(e_{i,n}^2 \leq x, e_{i+\ell-1,n}^2 \leq y) - G_n(x)G_n(y) \right\},$$

where $G_n(x) = \sum_{i=1}^{n} \mathbf{1}(e_{i,n}^2 \leq x)/n$. It is easy to check that the processes

$$\overset{\circ}{\mathscr{R}}_{1,\ell,n}(x,y) = \frac{1}{\sqrt{n}} \sum_{i=1}^{\lfloor ns \rfloor} \left\{ \mathbf{1}(\epsilon_i^2 \leq x) - G(x) \right\} \left\{ \mathbf{1}(\varepsilon_{i+\ell-1}^2 \leq y) - G(y) \right\},$$

with $\ell \in \{2, \ldots, m\}$, converge jointly to $\overset{\circ}{\mathscr{R}}_{1,2}, \ldots, \overset{\circ}{\mathscr{R}}_{1,m}$, where

$$\overset{\circ}{\mathscr{R}}_{1,\ell}(x,y) = \overset{\circ}{\mathbb{L}}_{1,\ell}(x,y) - G(x) \overset{\circ}{\mathbb{G}}(y) - G(y) \overset{\circ}{\mathbb{G}}(x), \quad x, y \in [0, \infty].$$

Moreover the processes $\overset{\circ}{\mathscr{R}}_{1,2}, \ldots, \overset{\circ}{\mathscr{R}}_{1,m}$ are independent copies of each other. Combining Theorem 2 and Corollary 4, one obtains the following result.

Corollary 5. *Under assumptions (A1–A2), $\mathscr{R}_{1,\ell,n} \rightsquigarrow \mathscr{R}_{1,\ell}$ in $\mathscr{D}_{2,G}$, where*

$$\mathscr{R}_{1,\ell}(x,y) = \overset{\circ}{\mathscr{R}}_{1,\ell}(x,y) + \left\{ f(\sqrt{y}) - f(-\sqrt{y}) \right\} \left\{ H(\sqrt{x}) - H(-\sqrt{x}) \right\} \mathscr{A}_\ell,$$

for $x, y \in \overline{S_G}$.

Remark 6. If $\mu_4 = E(\varepsilon_i^4) < \infty$, it follows from Höffding's equality that

$$\int_{\mathbb{R}^2} \mathscr{R}_{1,\ell,n}(1,x,y)dxdy = \frac{1}{\sqrt{n}} \sum_{i=1}^n (e_{i,n}^2 - \hat{\sigma}_n^2)(e_{i+\ell,n}^2 - \hat{\sigma}_n^2). \tag{4}$$

Following McLeod and Li [30], let $r_{ee,\ell,n}$ be the correlation between pairs of lagged squared residuals $(e_{i,n}^2, e_{i+\ell,n}^2)$. It then follows from (4), Corollary 5 and the calculations in Ghoudi et al. [18] and Genest and Rémillard [14], that, as $n \to \infty$, the variables $\sqrt{n}\, r_{ee,\ell,n}$ converge jointly to independent variables $r_{ee,\ell}$, where $(\mu_4 - \sigma^4)r_{ee,\ell} = \int_{\mathbb{R}^2} \mathscr{R}_{1,\ell+1}(x,y)dxdy = \int_{\mathbb{R}^2} \overset{\circ}{\mathscr{R}}_{1,\ell+1}(x,y)dxdy$, $\ell = 1,\ldots,m-1$. The equality follows from the facts that $I_1(x) = f(\sqrt{x}) - f(-\sqrt{x})$ and $I_2(x) = H(\sqrt{x}) - H(-\sqrt{x})$ are integrable, $\int_0^\infty I_1(x)dx = 2\int_{\mathbb{R}} xf(x)dx = 2E(\epsilon_i) = 0$ and $\int_0^\infty I_2(x)dx = -E\{\epsilon_i^3\}$. This sheds new light on the results of McLeod and Li [30].

3.2 Empirical Process of Standardized Squared Residuals

Assume that $E(\varepsilon_i^4) < \infty$, and set $\mathbb{G}_n^*(y) = \frac{1}{\sqrt{n}} \sum_{i=1}^n \left\{ \mathbf{1}\left(\frac{e_{i,n}^2}{\hat{\sigma}_n^2} \leq y \right) - G^*(y) \right\}$, where $G^*(y) = G(\sigma^2 y)$, $y \in [0,\infty]$, is the distribution function of $\epsilon_i^2 = \varepsilon_i^2/\sigma^2$.

Corollary 6. *Suppose that (A1–A2) hold and that the law of ε_i is symmetric about 0. Then $\mathbb{G}_n^* \rightsquigarrow \mathbb{G}^*$ in $\mathscr{D}_{1,G}$, where $\mathbb{G}^*(y) = \tilde{\mathbb{K}}\{G^*(y)\} + yg^*(y)\mathscr{S}^*$, $y \in \overline{S_G}$. Furthermore, set $\hat{\mathbb{G}}_n^*(y) = \frac{1}{\sqrt{n}} \sum_{i=1}^n \{\mathbf{1}(Z_{i,n}^2 \leq y) - G^*(y)\}$, $y \in \mathbb{R}$, where $Z_{i,n} = \varepsilon_i/s_{\varepsilon,n}$. Then $\hat{\mathbb{G}}_n^* \rightsquigarrow \hat{\mathbb{G}}^*$ in $\mathscr{D}_{1,G}$.*

3.3 Empirical Copula for Squared Residuals

The empirical copula process for squared residuals is defined by

$$\mathbb{D}_n(\mathbf{u}) = \frac{1}{\sqrt{n}} \sum_{i=1}^n \left[\prod_{j=1}^m \mathbf{1}\left\{ G_n(e_{i+j-1,n}^2) \leq u_j \right\} - \prod_{j=1}^m u_j \right], \quad \mathbf{u} \in [0,1]^m.$$

Next, for any $A \in \mathscr{I}_m$, set

$$\mathbb{D}_{A,n}(\mathbf{u}) = \frac{1}{\sqrt{n}} \sum_{i=1}^n \prod_{j \in A} \left[\mathbf{1}\left\{ G_n(e_{i+j-1,n}^2) \leq u_j \right\} - u_j \right], \mathbf{u} \in [0,1]^m.$$

Further set

$$
\overset{\circ}{\mathbb{D}}_n(\mathbf{u}) = \frac{1}{\sqrt{n}} \sum_{i=1}^{n} \left[\prod_{j=1}^{m} \mathbf{1}\{G(\varepsilon_{i+j-1,n}^2) \le u_j\} - \prod_{j=1}^{m} u_j \right], \quad \mathbf{u} \in [0,1]^m,
$$

and for any $A \in \mathscr{I}_m$, define

$$
\overset{\circ}{\mathbb{D}}_{A,n}(\mathbf{u}) = \frac{1}{\sqrt{n}} \sum_{i=1}^{n} \prod_{j\in A} \left[\mathbf{1}\{G(\varepsilon_{i+j-1,n}^2) \le u_j\} - u_j \right], \quad \mathbf{u} \in [0,1]^m.
$$

Note that $\overset{\circ}{\mathbb{D}}_n \rightsquigarrow \overset{\circ}{\mathbb{D}}$ in $\mathscr{D}([0,1]^m)$, while the processes $\overset{\circ}{\mathbb{D}}_{A,n}$ converge jointly to processes $\overset{\circ}{\mathbb{D}}_A$ that are independent Wiener sheets for all $A \in \mathscr{I}_m$ [14].

Corollary 7. *Under assumptions (A1–A2), $\mathbb{D}_n \rightsquigarrow \mathbb{D}$ in $\mathscr{D}([0,1]^m)$, where*

$$
\mathbb{D}(\mathbf{u}) = \mathbb{L}\{G^{-1}(u_1),\dots,G^{-1}(u_m)\} - \sum_{j=1}^{m} \mathbb{G}\{G^{-1}(u_j)\} \prod_{\substack{\ell=1 \\ \ell\ne j}}^{m} u_\ell.
$$

Moreover $\{\mathbb{D}_{A,n}\}_{A\in\mathscr{I}_m}$ converge jointly to $\{\mathbb{D}_A\}_{A\in\mathscr{I}_m}$ having representation

$$
\mathbb{D}_A(\mathbf{u}) = \overset{\circ}{\mathbb{D}}_A(\mathbf{u}) + \left[f\left\{ \sqrt{G^{-1}(u_\ell)} \right\} - f\left\{ -\sqrt{G^{-1}(u_\ell)} \right\} \right]
$$
$$
\times \left[H\left\{ \sqrt{G^{-1}(u_1)} \right\} - H\left\{ -\sqrt{G^{-1}(u_1)} \right\} \right] \mathscr{A}_\ell \, \mathbf{1}(A = \{1,\ell\}),
$$

where $\{\overset{\circ}{\mathbb{D}}_A\}_{A\in\mathscr{I}_m}$ are independent Wiener sheets, and $\mathscr{A}_\ell$ is defined by (3). Furthermore, if the law of ϵ_i is symmetric about 0, then $\mathbb{D} = \overset{\circ}{\mathbb{D}}$, and $\mathbb{D}_A = \overset{\circ}{\mathbb{D}}_A$, with $A \in \mathscr{I}_m$, are parameter and distribution free.

Remark 7. Since the autocorrelations $r_{ee}(\ell)$ are distribution free, even if the law of ε_i is not symmetric about 0, one can ask if the same is true for standard nonparametric measures of dependence like Spearman's rho. The answer is no in general. To see that, suppose that $\int_{\mathbb{R}} f^2(x)dx$ is finite. Then

$$
I_1(u) = f\left\{ \sqrt{G^{-1}(u)} \right\} - f\left\{ -\sqrt{G^{-1}(u)} \right\}
$$

and

$$
I_2(u) = V\left\{ \sqrt{G^{-1}(u)} \right\} - V\left\{ -\sqrt{G^{-1}(u)} \right\}
$$

are integrable, $J_1 = \int_0^1 I_1(u)du = \int_0^\infty \{f^2(x) - f^2(-x)\}\,dx$ and $J_2 = \int_0^1 I_2(u)du = -E\left[|\epsilon_i|\{F(\epsilon_i) - F(-\epsilon_i)\}\right]$. Since $\sqrt{n}\,\hat{\rho}_{1,\ell,n} = 12\int_{[0,1]^2} \mathbb{D}_{\{1,\ell\},n}(u,v)dudv$, it follows from Corollary 7 that

$$\sqrt{n}\,\hat{\rho}_{1,\ell,n} \rightsquigarrow 12\int_{[0,1]^2} \overset{\circ}{\mathbb{D}}_{\{1,\ell\}}(u,v)dudv + 12J_1J_2\mathscr{A}_\ell$$

$$\neq 12\int_{[0,1]^2} \overset{\circ}{\mathbb{D}}_{\{1,\ell\}}(u,v)dudv,$$

unless $J_1J_2 = 0$. Note that $12\int_{[0,1]^2} \overset{\circ}{\mathbb{D}}_{\{1,\ell\}}(u,v)dudv$ are i.i.d. $N(0,1)$, and $J_1 = 0$ if the distribution of ϵ_i is symmetric about 0. One can check that the same results will hold if $e_{i,n}^2$ is replaced by $|e_{i,n}|$ or any even function $\mathscr{H}$ which is increasing on $[0, \infty)$. However, the results of McLeod and Li [30] do no extend unless $E\{\mathscr{H}'(\varepsilon_i)\} = 0$.

4 Diagnostic Tests for ARMA Models

To carry on diagnostic of ARMA models, one may consider to test several hypotheses, such as change-point analysis, tests of goodness-of-fit, and tests of randomness. Tests statistics for these hypotheses are defined next, based on the empirical processes defined previously.

4.1 Change-point Problems for Innovations

To test for change-point in the distribution of the innovations, that is whether there exists $\tau \in \{1, \ldots, n-1\}$ such that $\varepsilon_1 \ldots, \varepsilon_\tau$ follow a distribution F_1 and $\varepsilon_{\tau+1} \ldots, \varepsilon_n$ follow a distribution $F_2 \neq F_1$, Bai [2] proposed statistics based on the sequential empirical process

$$\mathscr{B}_n(s,y) = \frac{1}{\sqrt{n}} \sum_{i=1}^{\lfloor ns \rfloor} \{\mathbf{1}(e_{i,n} \leq y) - F_n(y)\} = \mathbb{B}_n\{s, F_n(y)\},$$

with $F_n(y) = \frac{1}{n}\sum_{i=1}^n \mathbf{1}(e_{i,n} \leq y)$, $y \in [-\infty, +\infty]$,

$$\mathbb{B}_n(s,u) = \frac{1}{\sqrt{n}} \sum_{i=1}^{\lfloor ns \rfloor} \left\{\mathbf{1}\left(\frac{R_{i,n}}{n} \leq u\right) - \frac{\lfloor nu \rfloor}{n}\right\}, \quad u \in [0,1],$$

and $R_{i,n}$ is the rank of $e_{i,n}$, $i =, \ldots, n$. Note that $\mathscr{B}_n(s, y) = \mathbb{F}_n(s, y) - \frac{\lfloor ns \rfloor}{n} \mathbb{F}_n(1, y)$ and $\mathbb{B}_n(s, u) = \mathbb{K}_n(s, u) - \frac{\lfloor ns \rfloor}{n} \mathbb{K}_n(1, u)$, so it follows from Theorem 1 that $\mathscr{B}_n$ and $\mathbb{B}_n$ converge respectively in $\mathscr{D}_{1,F_1}$ and $\mathscr{D}\left([0, 1]^2\right)$ to $\mathscr{B}$ and $\mathbb{B}$, where $\mathbb{B}(s, u) = \mathbb{K}(s, u) - s\mathbb{K}(1, u)$, $s, u \in [0, 1]$ and $\mathscr{B}(s, y) = \mathbb{B}\{s, F_1(y)\}$, $s \in [0, 1]$, $y \in [-\infty, \infty]$. The process $\mathbb{B}$ does not depend on the estimated parameters nor the marginal distribution F_1. It is a continuous centered Gaussian process with covariance given by $\text{Cov}\{\mathbb{B}(s, u), \mathbb{B}(t, v)\} = \{\min(s, t) - st\}\{\min(u, v) - uv\}$. In fact, $\mathbb{B}$ appears as the limit of many other processes used in tests of change-point [9, 31] and tests of independence [5, 18]. A natural statistic for testing for change-point is the Kolmogorov-Smirnov statistic

$$T_{1n} = \sup_{s \in [0,1], y \in \mathbb{R}} |\mathscr{B}_n(s, y)| = \sup_{s, u \in [0,1]} |\mathbb{B}_n(s, u)|.$$

Carlstein [9] suggested to consider statistics of the form $\sup_{s \in [0,1]} \varphi\{\mathscr{B}_n(s, \cdot)\}$. For instance, the Cramér-von Mises statistic leads to

$$T_{2n} = \max_{1 \leq k \leq n} \int_0^1 \{\mathbb{B}_n(k/n, u)\}^2 du$$

$$= \max_{1 \leq k \leq n} \left[\frac{k^2}{n^2} \frac{(n+1)(2n+1)}{6n} + \frac{k}{n} \sum_{i=1}^k \frac{R_{i,n}(R_{i,n} - 1)}{n^2} \right.$$

$$\left. - \sum_{i=1}^k \sum_{j=1}^k \frac{\max(R_{i,n}, R_{j,n})}{n^2} \right].$$

Remark 8. Carlstein [9] suggests to estimate the first time τ of a change-point by $\tau_{1n} = \inf\{j; \sup_{0 \leq u \leq 1} |\mathbb{B}_n(j/n, u)| = T_{1n}\}$, related to the Kolmogorov-Smirnov statistic, or by $\tau_{2n} = \inf\{j; \sup_{0 \leq u \leq 1} \int_0^1 \{\mathbb{B}_n(j/n, u)\}^2 du = T_{2n}\}$, related to the Cramér-von Mises statistic.

Quantiles of T_{1n} and T_{2n}, appearing in Table 2, were computed using $N = 100,000$ replications of the statistics applied to

$$\hat{\mathbb{B}}_n(s, u) = \frac{1}{\sqrt{n}} \sum_{i=1}^{\lfloor ns \rfloor} \left\{ \mathbf{1}(U_i \leq u) - \hat{F}_n(u) \right\}, \quad s, u \in [0, 1],$$

where $\hat{F}_n(u) = \frac{1}{n} \sum_{i=1}^n \mathbf{1}(U_i \leq u)$, and where $U_1, \ldots, U_n$ are i.i.d. uniformly distributed over $[0, 1]$.

Table 2 Quantiles of order 95 % and 99 % for the statistic $\hat{T}_{1n}$ and $\hat{T}_{2n}$

	$\hat{T}_{1n}$				$\hat{T}_{2n}$			
	Sample size				Sample size			
Level (%)	50	100	250	500	50	100	250	500
95	0.775	0.795	0.814	0.822	0.181	0.188	0.194	0.196
99	0.888	0.911	0.934	0.943	0.253	0.263	0.272	0.278

Table 3 Percentage of detection of change-point for the first experiment, with $N = 10{,}000$ replications

	$n = 100$						$n = 250$					
	$\sigma = 1.5$		$\sigma = 2$		$\sigma = 5$		$\sigma = 1.5$		$\sigma = 2$		$\sigma = 5$	
p	T_{1n}	T_{2n}	T_{1n}	T_{2n}	T_{1n}	T_{2n}	T_{1n}	T_{2n}	T_{1n}	T_{2n}	T_{1n}	T_{2n}
0	5.05	4.96	5.51	5.32	5.2	5.21	6.16	5.53	5.71	5.15	5.93	5.31
0.1	5.51	5.42	6	5.78	7.64	7.48	7.25	6.27	9.05	8.2	20.47	18.73
0.3	8.39	7.21	17.43	14.03	84.17	84.83	18.74	13.74	58.64	56.46	100	100
0.5	11.19	8.75	29.48	22.36	99.12	99.37	29.59	22.04	84.87	85.25	100	100

4.1.1 Simulation Results

One considers two experiments. In the first experiment, $X_1, \ldots, X_n$ are independent, with $X_i \sim N(0, 1)$, for $i = 1, \ldots, n(1 - p)$, while $X_i \sim N(0, \sigma^2)$, for $i = n(1 - p) + 1, \ldots, n$. Here $p \in \{0, 0.1, 0.3, 0.5\}$, $\sigma \in \{1.5, 2, 5\}$, and $n \in \{100, 250\}$. The residuals are defined as if the X_i's were independent. As seen in Table 3, the Kolmogorov-Smirnov test (based on T_{1n}) seems more powerful especially for detecting small changes in the structure. As expected, the maximum power is attained when $p = 0.5$. The values for $p > 0.5$ are omitted since the power is symmetric about $p = 0.5$.

In the second experiment, $\varepsilon_1, \ldots, \varepsilon_n$ are independent, with $\varepsilon_i \sim N(0, 1)$, and $X_i = 0.2 + 0.5X_{i-1} + \varepsilon_i$, $i = 1, \ldots, np$, while $X_i = 0.2 + 0.5X_{i-1} + \varepsilon_i - \theta\varepsilon_{i-1}$, for $i = np + 1, \ldots, n$, $p \in \{0, 0.1, 0.3, 0.5, 0.7, 0.9\}$, $\theta \in \{0.1, 0.25, 0.5\}$, and $n \in \{100, 250\}$. One fits an AR(1) model to the data. Contrary to the first experiment, the power of the tests should not be symmetric about $p = 0.5$. That is reflected in Table 4. Surprisingly, the Cramér-von Mises test (based on T_{2n}) seems more powerful for detecting the type of changes modeled here.

4.2 Goodness-of-Fit Tests for Innovations

Two familiar scenarios could be considered: F is equal to a specific distribution F_0 or F belongs to a scale family of distributions. Only the second scenario is discussed next. Applications to the first scenario are straightforward.

Table 4 Percentage of detection of change-point for the second experiment, with $N = 10,000$ replications

	$n = 100$						$n = 250$					
	$\theta = 0.10$		$\theta = 0.25$		$\theta = 0.5$		$\theta = 0.10$		$\theta = 0.25$		$\theta = 0.5$	
p	T_{1n}	T_{2n}	T_{1n}	T_{2n}	T_{1n}	T_{2n}	T_{1n}	T_{2n}	T_{1n}	T_{2n}	T_{1n}	T_{2n}
0.0	4.46	4.31	4.53	4.02	4.59	4.25	6	5.51	6.16	5.17	5.94	5.03
0.1	4.93	4.5	4.93	4.84	5.53	5.4	7.39	6.54	7.23	6.81	7.76	7.72
0.3	7.98	8.68	8.56	9.69	11.38	12.58	18.63	20.57	20.54	22.52	26.05	28.77
0.5	11.56	12.39	14.26	15.99	18.94	22	27.45	30.21	33.5	37.69	43.78	49.27
0.7	8.99	9.88	12.64	15.2	17.85	22.59	20	23.42	28.68	33.01	38.35	44.8
0.9	5.75	6.07	7.68	8.23	7.5	8.53	9.1	8.95	11.03	12	12.35	14.41

4.2.1   Testing $H_0 : F = F_0(\cdot/\sigma)$ for Some $\sigma > 0$

Next assume that one wants to test the hypothesis that the error distribution belongs to a scale family, that is, the ε_i's have distribution $F(\cdot) = F_0(\cdot/\sigma)$ for some $\sigma > 0$ and some standardized distribution F_0. To this end, define $\mathbb{F}_n^*(y) = \frac{1}{\sqrt{n}} \sum_{i=1}^n \{\mathbf{1}(e_{i,n}/\hat{\sigma}_n \leq y) - F_0(y)\}$, $y \in [-\infty, +\infty]$, and let $u_{(1:n)}^*, \ldots, u_{(n:n)}^*$ be the order statistics of the pseudo-observations $F_0(e_{1,n}/\hat{\sigma}_n), \ldots, F_0(e_{n,n}/\hat{\sigma}_n)$.

One can then use the statistics $T_{3n}^* = \|\mathbb{F}_n^*\| = \mathrm{KS}_n\left(u_{(1:n)}^*, \ldots, u_{(n:n)}^*\right)$ and $T_{4n}^* = \mathscr{T}_{F_0}(\mathbb{F}_n^*) = \mathrm{CVM}_n\left(u_{(1:n)}^*, \ldots, u_{(n:n)}^*\right)$, where

$$\mathrm{KS}_n\left\{u_{(1:n)}, \ldots, u_{(n:n)}\right\} = \sqrt{n} \max_{1 \leq k \leq n} \left\{\left|u_{(k:n)} - \frac{(k-1)}{n}\right|, \left|u_{(k:n)} - \frac{k}{n}\right|\right\}$$

and

$$\mathrm{CVM}_n\left\{u_{(1:n)}, \ldots, u_{(n:n)}\right\} = \sum_{i=1}^n \left(u_{(i:n)} - \frac{2i-1}{2n}\right)^2 + \frac{1}{12n}.$$

- If $\hat{\mu}_n = \bar{X}_n + o_P(1)$, critical values or P-values for T_{3n}^* or T_{4n}^* can be obtained via Monte-Carlo simulation. In fact, recalling the construction of $\hat{\mathbb{F}}_n^*$ in Corollary 1, one obtains that both T_{3n}^* and $\hat{T}_{3n}^* = \|\hat{\mathbb{F}}_n^*\|$ converge in law to the same limit, while T_{4n}^* and $\hat{T}_{4n}^* = \mathscr{T}_{F_0}(\hat{\mathbb{F}}_n^*)$ converge in law to the same limit.

  In particular, when F_0 is standard Gaussian distribution, the limit law of $\hat{T}_{3n}^*$ is the same as that obtained by Lilliefors [27]. For instance, the Lilliefors test, based on the Kolmogogov-Smirnov statistic, is available in many statistical packages and can be applied to residuals of ARMA models without any change, whenever $\hat{\mu}_n = \bar{X}_n + o_P(1)$. Table 5 provides critical values of the Kolmogorov-Smirnon statistic T_{3n}^* and Cramér-von Mises statistic $\hat{T}_{4n}^*$ for different levels. These quantiles are computed for a sample size $n = 250$ for each statistics, $N = 100,000$ replications. Table 6 shows that these quantiles are quite precise for almost any sample size.

Table 5 Quantiles of order 90 %, 95 % and 99 % for the statistics $\hat{T}_{3n}^*$, $\hat{T}_{4n}^*$, $\hat{T}_{5n}^*$ and T_{6n}^* for sample sizes larger than 40

Order (%)	Statistic			
	$\hat{T}_{3n}^*$	$\hat{T}_{4n}^*$	$\hat{T}_{5n}^*$	$\hat{T}_{6n}^*$
90	0.8200	0.1035	1.0300	0.2058
95	0.8900	0.1258	1.1400	0.2656
99	1.0500	0.1770	1.3600	0.4252

Table 6 Percentage of rejection for statistics $\hat{T}_{3n}^*$, $\hat{T}_{4n}^*$, $\hat{T}_{5n}^*$, $\hat{T}_{6n}^*$, for different sample sizes, using 10,000 replications based on the quantiles in Table 5

Statistic	Level (%)	Length of series				
		50	100	250	500	1000
T_{3n}^*	10	8.93	9.59	9.89	10.88	10.88
	5	4.64	4.53	5.36	5.69	5.82
	1	0.77	0.78	0.92	1.17	1.13
T_{4n}^*	10	9.59	9.36	9.87	10.31	9.97
	5	4.93	4.50	4.80	5.40	5.11
	1	1.07	1.00	1.03	1.11	1.05
T_{5n}^*	10	9.91	10.55	10.89	10.37	10.69
	5	4.85	5.51	5.33	5.43	5.37
	1	1.03	1.04	1.03	1.21	1.22
T_{6n}^*	10	10.05	9.93	10.20	9.53	10.12
	5	5.07	4.88	5.13	4.76	5.08
	1	0.85	0.95	0.78	0.91	0.98

- If the law F_0 is symmetric about zero (whether $\hat{\mu}_n = \bar{X}_n + o_P(1)$ or not), one can use test statistics based on the empirical process of squared residuals

$$\mathbb{G}_n^*(y) = \frac{1}{\sqrt{n}} \sum_{i=1}^n \left\{ \mathbf{1} \left(\frac{e_{i,n}^2}{\hat{\sigma}_n^2} \leq y \right) - G_0(y) \right\}, \quad y \in [0, \infty].$$

Further let $v_{(1:n)}^*, \ldots, v_{(n:n)}^*$ be the order statistics of the pseudo-observations $G_0(e_{i,n}^2/\hat{\sigma}_n^2)$, $i = 1, \ldots, n$, and set $T_{5n}^* = \|\mathbb{G}_n^*\| = \mathrm{KS}_n(v_{(1:n)}^*, \ldots, v_{(n:n)}^*)$ and $T_{6n}^* = \mathscr{T}_{G_0}(\mathbb{G}_n^*) = \mathrm{CVM}_n(v_{(1:n)}^*, \ldots, v_{(n:n)}^*)$. According to Corollary 6, the limiting behavior of the statistics is not distribution free. However, they have the same limiting distributions as $\hat{T}_{5n}^* = \|\hat{\mathbb{G}}_n^*\|$ and $\hat{T}_{6n}^* = \mathscr{T}_{G_0}(\hat{\mathbb{G}}_n^*)$, where

$$\hat{\mathbb{G}}_n^*(y) = \frac{1}{\sqrt{n}} \sum_{i=1}^n \left\{ \mathbf{1} \left(\varepsilon_i^2 / s_{\varepsilon,n}^2 \leq y \right) - G_0(y) \right\}, \quad y \in [0, \infty].$$

As a result, the statistics $\hat{T}_{5n}^*$ and $\hat{T}_{6n}^*$ can be easily simulated, and then used to estimate P-values for T_{5n}^* and T_{6n}^* respectively. Quantiles for $\hat{T}_{5n}^*$ and $\hat{T}_{6n}^*$ are computed in Table 5 for a sample size $n = 250$ using $N = 100,000$ replications. Table 6 shows that these quantiles are quite precise for almost any sample size.

Table 7 Percentage of rejection of the null hypothesis of Gaussianity for an AR(1) model with $n \in \{100, 250\}$ and $N = 1000$ replications, when the innovations are Student with $v \in \{\infty, 20, 15, 10, 5\}$

Statistic	$n = 100$					$n = 250$				
	v					v				
	∞	20	15	10	5	∞	20	15	10	5
T_{3n}^*	4.2	7.3	8.3	10.2	31.4	5.8	8.3	10.1	16.1	62.4
T_{4n}^*	5.1	7.7	9.7	13.4	41.3	5.7	9.3	13.8	21.2	75.9
T_{5n}^*	5.3	7.2	9.3	13.5	40.9	6.1	11.3	14.0	26.2	79.0
T_{6n}^*	5.2	8.9	10.7	16.8	49.1	5.7	13.0	16.8	31.0	84.2

4.2.2 Simulation Results

Consider the following experiment for measuring the power of a test of Gaussianity: Assume that $\varepsilon_1, \ldots, \varepsilon_n$ are independent with Student distribution with parameter $v \in \{5, 10, 15, 20, \infty\}$, and $X_i = 0.2 + 0.5X_{i-1} + \varepsilon_i$, $i = 1, \ldots, n$. The null hypothesis is that the distribution is AR(1) with Gaussian innovations. The results of 1000 replications of the experiment for samples sizes $n \in \{100, 250\}$ are displayed in Table 7. As seen from the case $v = \infty$ corresponding the null hypothesis, the levels of the tests are respected. As expected, the power of the tests increases as the degree of freedom v decreases. Also, for each test statistic, the power increases with the sample size. The best test statistic seems to be T_{6n}^*, for all alternatives considered, although T_{4n}^* and T_{5n}^* are close contenders. From a practical point of view, statistics T_{4n}^* and T_{6n}^* are easier to compute.

4.3 Tests of Serial Independence

One could define the empirical copula process $\mathbb{C}_n$ of the residuals. However, as shown in Proposition 3, its limiting behavior is not distribution free, even if Möbius transforms were used. Fortunately, it was shown in Corollary 7 that when the law of the innovations is symmetric about 0, the limiting distribution of the empirical copula process $\mathbb{D}_n$ of the squared residuals defined in Sect. 3.3, does not depend on the estimated parameters, nor the underlying distribution function F. However, as suggested in Genest and Rémillard [14], it is recommended to use the slightly modified process $\tilde{\mathbb{D}}_n$, defined for $u = (u_1, \ldots, u_m) \in [0, 1]^m$ by

$$\tilde{\mathbb{D}}_n(\mathbf{u}) = \frac{1}{\sqrt{n}} \sum_{i=1}^n \left\{ \prod_{j=1}^m \mathbf{1}\left(\frac{R_{i+j-1,n}}{n} \leq u_j \right) - \prod_{j=1}^m \frac{\lfloor nu_j \rfloor}{n} \right\}.$$

Here $R_{i,n}$ is the rank of $e_{i,n}^2$ amongst $e_{1,n}^2, \ldots, e_{n,n}^2$, and $R_{n+i,n} = R_{i,n}$ for $i \geq 1$. In addition, to produce critical values or P-values for statistics based on $\tilde{\mathbb{D}}_n$, it is worth noticing that under the assumption of symmetry, $\tilde{\mathbb{D}}_n$ has the same limiting distribution as

$$\hat{\mathbb{D}}_n(\mathbf{u}) = \frac{1}{\sqrt{n}} \sum_{i=1}^{n} \left\{ \prod_{j=1}^{m} \mathbf{1} \left(\frac{R_{i+j-1}}{n} \leq u_j \right) - \prod_{j=1}^{m} \frac{\lfloor n u_j \rfloor}{n} \right\},$$

$\mathbf{u} = (u_1, \ldots, u_m) \in [0,1]^m$, where R_i is the rank of U_i amongst the i.i.d. uniform variates $U_1, \ldots, U_n$. Consequently the methodology developed in Genest and Rémillard [14] and Genest et al. [16] could be applied here, including optimal tests combining Möbius transforms. In the sequel, consider the test statistics $W_{m,n} = \sum_{|A|>1, A \subset \mathscr{I}_m} \pi^{2|A|} B_{A,n}$ and

$$B_{m,n} = \int_{[0,1]^m} \tilde{\mathbb{D}}_n^2(u) du$$

$$= \frac{1}{n} \sum_{i=1}^{n} \sum_{j=1}^{n} \prod_{k=1}^{m} \left\{ 1 - \frac{\max(R_{i+k-1,n}, R_{j+k-1,n})}{n} \right\}$$

$$+ n \left\{ \frac{(n-1)(2n-1)}{6n^2} \right\}^m$$

$$- 2 \sum_{i=1}^{n} \prod_{k=1}^{m} \left\{ \frac{n(n-1) - R_{j+k-1,n}(R_{j+k-1,n} - 1)}{2n^2} \right\},$$

where $B_{A,n} = \int_{[0,1]^m} \tilde{\mathbb{D}}_{A,n}^2(u) du = \frac{1}{n} \sum_{i=1}^{n} \sum_{j=1}^{n} \prod_{k \in A} D_n(R_{i+k-1,n}, R_{j+k-1,n})$, for any $A \subset$ $\mathscr{I}_m$, and $D_n(s,t) = \frac{(n+1)(2n+1)}{6n^2} + \frac{s(s-1)}{2n^2} + \frac{t(t-1)}{2n^2} - \frac{\max(s,t)}{n}$. Approximate quantiles for statistics $B_{m,n}$ and $W_{m,n}$ for $n = 100$ and $m \in \{2, \ldots, 6\}$ can be found in Table 8.

Table 8 Quantiles of order 90 %, 95 % and 99 % for statistics $B_{m,n}$ and $W_{m,n}$ for $m \in \{2,3,4,5,6\}$, $n = 100$, using $N = 100,000$ replications

Statistic	Order (%)	m				
		2	3	4	5	6
$B_{m,n}$	90	0.046897	0.060624	0.049211	0.032736	0.019022
	95	0.058246	0.075242	0.061684	0.040947	0.024196
	99	0.085475	0.111449	0.093319	0.066445	0.042824
$W_{m,n}$	90	4.568240	13.237556	33.925330	87.203099	229.239008
	95	5.673710	14.741794	36.038715	90.624271	237.633039
	99	8.326064	18.169826	40.856984	98.200949	255.363139

Table 9 List of models with Gaussian noise u_i

Model	Name	Equation
A1	I.I.D.	$\varepsilon_i = u_i$
A2	AR(1)	$\varepsilon_i = 0.3\,\varepsilon_{i-1} + u_i$
A3	ARCH(1)	$\varepsilon_i = h_i^{1/2} u_i,\ h_i = 1 + 0.8\,\varepsilon_{i-1}^2$
A4	Threshold GARCH(1, 1)	$\varepsilon_i = h_i^{1/2} u_i,$ with $h_i^2 = 0.25 + 0.6\,h_{i-1}^2$ $+ 0.5\,\varepsilon_{i-1}^2 \mathbf{1}(u_{i-1} < 0) + 0.2\,\varepsilon_{i-1}^2 \mathbf{1}(u_{i-1} \geq 0)$
A5	Bilinear AR(1)	$\varepsilon_i = 0.8\,\varepsilon_{i-1} u_{i-1} + u_i$
A6	Nonlinear MA(1)	$\varepsilon_i = 0.8\,u_{i-1}^2 + u_i$
A7	Threshold AR(1)	$\varepsilon_i = 0.4\,\varepsilon_{i-1}\mathbf{1}(\varepsilon_{i-1} > 1) - 0.5\,\varepsilon_{i-1}\mathbf{1}(\varepsilon_{i-1} \leq 1) + u_i$
A8	Fractional AR(1)	$\varepsilon_i = 0.8\,|\varepsilon_{i-1}|^{1/2} + u_i$
A9	Sign AR(1)	$\varepsilon_i = \text{sign}(\varepsilon_{i-1}) + 0.43\,u_i$

Table 10 Percentage of rejection of the null hypothesis of serial independence for the alternatives described in Table 9 using samples of size $n = 100$ and $N = 10,000$ replications

Model	$m = 2$ $B_{m,n}$	$m = 2$ $W_{m,n}$	$m = 4$ $B_{m,n}$	$m = 4$ $W_{m,n}$	$m = 6$ $B_{m,n}$	$m = 6$ $W_{m,n}$
I.I.D.	4.92	4.92	4.68	4.93	4.79	4.37
AR(1)	71.79	71.79	64.41	43.65	55.13	20.31
ARCH(1)	11.94	11.94	9.33	36.85	7.77	80.45
Threshold GARCH(1,1)	9.36	9.36	9.10	32.03	7.54	78.91
Bilinear AR(1)	72.67	72.67	41.08	63.96	30.65	83.55
Nonlinear MA(1)	40.83	40.83	11.12	22.81	9.50	24.51
Threshold AR(1)	45.63	45.63	7.62	20.01	6.20	12.32
Fractional AR(1)	59.16	59.16	47.32	30.27	39.73	13.18
Sign AR(1)	58.13	58.13	59.33	60.70	59.28	61.09

To assess the finite sample power of these two statistics, one uses the same models as in Hong and White [19] and Genest et al. [15]. Those models, listed in Table 9 are all of the form $\varepsilon_i = \varphi(\varepsilon_{i-1}, u_i, u_{i-1})$. As in Hong and White [19], the white noise u_i was taken to be Gaussian. The percentage of rejection of the null hypothesis of serial independence are given in Table 10 for samples of size $n = 100$, using the tests statistics $B_{m,n}$ and $W_{m,n}$ with $m \in \{2, 4, 6\}$. As seen from that Table, the test based on $B_{m,n}$ is quite good for all alternatives but stochastic volatility models, when $m = 2$. Another characteristic is that its power seems to decrease sometimes dramatically as m increases. For the test statistic $W_{m,n}$, the power seems to increase with m for stochastic volatility models, while it seems to decrease for constant volatility models. It outperforms the test based on $B_{m,n}$ when $m > 2$ for all models but the AR(1) and the fractional AR(1).

5 Example of Application

As an example of application of the proposed tests, consider the Indian sugarcane annual production data studied in Mandal [29] who suggested an ARIMA(2,1,0) model for these data. Note also that other studies of sugarcane production showed that ARIMA models were quite appropriate, see, e.g., Suresh and Krishna Priya [33] and references therein. The data consisted of 53 values representing the annual sugar production (million tonnes) from 1951 to 2003. As suggested in Mandal [29], an ARIMA(2,1,0) model was fitted to the data and the diagnostic tests described in Sect. 4 were applied to the series of residuals. No change-point was detected in the series of residuals. In fact, for the change-point test statistics T_{1n} and T_{2n}, their respective values are 0.7894 and 0.1233, yielding P-values of 4.40 % and 18.35 % respectively, using $N = 10,000$ replications. Thus the null hypothesis is barely rejected at the 5 % level and is accepted at the 1 % level.

Next, for testing that the innovations have a Gaussian distribution, the tests based on T_{3n}^*, T_{4n}^*, T_{5n}^* and T_{6n}^* clearly reject the null hypothesis since the largest P-value is 0.4 %. Finally, for tests of serial dependence based on $B_{6,n}$ and $W_{6,n}$, both tests accept the null hypothesis with P-values of 46 % and 24 % respectively.

Rejecting the Gaussian distribution hypothesis for the innovations might indicate that the OLS estimation of the parameters is not be the optimal choice. To double check, a robust estimation of the parameters using the LAD method leads to P-values of 5.9 % and 23.4 % for the the change-point tests while both the null hypothesis of a Gaussian distribution and Laplace distribution are rejected at the 5 % level. As for the test of serial dependence with these residuals, the null hypothesis is accepted with P-values of 61 % and 29.4 % for $B_{6,n}$ and $W_{6,n}$ respectively. So basically, the two methods of estimation provide similar conclusions.

6 Proofs

The proofs extend the techniques used by Bai [2] and Ghoudi and Rémillard [17] and are given after introducing some useful notations and auxiliary results.

Let $Y_i = X_i - \mu$, and recall that $M_n = \sqrt{n} \ (\hat{\mu}_n - \mu)\left(1 - \sum_{k=1}^p \hat{\phi}_{k,n}\right)$, $\Phi_{k,n} = \sqrt{n} \ \left(\hat{\phi}_{k,n} - \phi_k\right)$, $1 \le k \le p$, and $\Theta_{k,n} = \sqrt{n} \ \left(\hat{\theta}_{k,n} - \theta_k\right)$, $1 \le k \le q$. To simplify the notations let $\mathbf{P}_n = (M_n, \Phi_n, \Theta_n, \theta_n)$ and for $i \ge 1$, define $\omega_i = (\varepsilon_i, \dots, \varepsilon_{i+m-1})^\top$, $\mathbf{w}_{i,n} = (e_{i,n}, \dots, e_{i+m-1,n})^\top$ and $D_{i,n} = D_{i,n}(\mathbf{P}_n) = \omega_i - \mathbf{w}_{i,n} = (d_{i,n}, \dots, d_{i+m-1,n})^\top$. By setting $V_{i,n} = (0, \dots, 0, v_{i+m-1,n})^\top \in \mathbb{R}^m$, with $v_{i,n} = M_n + \sum_{k=1}^p \Phi_{k,n} Y_{i-k} - \sum_{k=1}^q \Theta_{k,n}\varepsilon_{i-k}$, one writes $D_{i,n} = V_{i,n}/\sqrt{n} + B_{\theta_n}D_{i-1,n}$, for $i > 1$. By iteration, one obtains $D_{i,n} = B_{\theta_n}^{i-1}D_{1,n} + \sum_{k=0}^{i-2} B_{\theta_n}^k V_{i-k,n}/\sqrt{n}$, for $i > 1$.

Now for $\mathbf{P} = (M, \Phi, \Theta, \theta) \in \mathbb{R}^{1+p+q} \times \mathcal{O}$ define $L_i(\mathbf{P}) = \sum_{k=0}^{i-2} \left(B_\theta^k\right) V_{i-k}(\mathbf{P})$ and

$$D_{i,n}(\mathbf{P}) = B_\theta^{i-1} D_1 + \frac{1}{\sqrt{n}} L_i(\mathbf{P}) = B_\theta^{i-1} D_1 + \frac{1}{\sqrt{n}} \sum_{k=0}^{i-2} B_\theta^k V_{i-k}(\mathbf{P}), \quad i > 1,$$

with $D_1 = (\varepsilon_1, \ldots, \varepsilon_m)^\top$ and

$$V_i(\mathbf{P}) = (0, \ldots, 0, M + \sum_{k=1}^{p} \Phi_k Y_{i+m-1-k} - \sum_{k=1}^{q} \Theta_k \varepsilon_{i+m-1-k})^\top.$$

Observe that $D_{i,n} = D_{i,n}(\mathbf{P}_n)$ and $d_{i+j-1,n} = d_{i+j-1,n}(\mathbf{P}_n)$ where $d_{i+j-1,n}(\mathbf{P})$ denotes the jth component of the vector $D_{i,n}(\mathbf{P})$.

The next subsection provides some auxiliary results needed for the main proof.

6.1 Auxiliary Results

For any $\mathbf{t} = (t_1, \ldots, t_m)^\top \in [0, 1]^m$ and any $j, k \in \mathscr{J}_m = \{1, \ldots, m\}, j \neq k$, define $\tilde{\mathfrak{F}}_{j,k}(\mathbf{t}) = \tilde{H}(t_k) \prod_{l \in \mathscr{J}_m \setminus \{j,k\}} t_l$, where $\tilde{H}(y) = H\{F^{-1}(y)\}$ and

$$\tilde{v}_j(\mathbf{t}, \mathbf{P}) = \frac{M}{1 - \sum_{l=1}^{m} \theta_l} \prod_{k=1, k\neq j}^{m} t_k - \sum_{l=1}^{\min(q,j-1)} \Theta_l \sum_{k=m-j}^{m-1-l} \left(B_\theta^k\right)_{jm} \tilde{\mathfrak{F}}_{j,m-k-l}(\mathbf{t})$$

$$+ \sum_{l=1}^{\min(p,j-1)} \Phi_l \sum_{k=m-j}^{m-1-l} \left(B_\theta^k\right)_{jm} \sum_{t=0}^{m-1-k-l} \psi_t \tilde{\mathfrak{F}}_{j,m-k-l-t}(\mathbf{t}).$$

Note that $\tilde{v}_j(\mathbf{t}, \mathbf{P}) = v_j(\mathbf{x}, \mathbf{P})$, with $\mathbf{x} = (F^{-1}(t_1), \ldots, F^{-1}(t_m))^\top$.

Since $\theta \in \mathcal{O}$ satisfies $\rho(\mathbf{B}_\theta) < 1$ then $\|\mathbf{B}_\theta\|_\rho < \tau < 1$ for some natural matrix norm $\|.\|_\rho$ [20, p. 14]. Next, for any $\gamma, \lambda \in \mathbb{R}$, let $\Gamma_n(i, j, \gamma, \lambda, \mathbf{P}) = \gamma d_{i+j-1,n}(\mathbf{P}) + \lambda \Lambda_n(i, j)$ where $\Lambda_n(i, j) = i\tau^i \|D_1\|_\infty + R_{i,j}/\sqrt{n}$ with

$$R_{i,j} = \sum_{k=1}^{i+j-m-1} k\tau^k (1 + \sum_{l=1}^{p} |Y_{i+j-k-l}| + \sum_{l=1}^{q} |\varepsilon_{i+j-k-l}|).$$

Observe that $\Gamma_n(i,j,0,\lambda,\mathbf{P}) = \lambda\Lambda_n(i,j)$ does not depend on $\mathbf{P}$. Define also

$$U^j_{i,n}(\mathbf{t},\gamma,\lambda,\mathbf{P}) = \{\mathbf{1}(\varepsilon_{i+j-1} \le F^{-1}(t_j) + \Gamma_n(i,j,\gamma,\lambda,\mathbf{P}))$$
$$- \mathbf{1}(\varepsilon_{i+j-1} \le F^{-1}(t_j))\} \prod_{k=1,k\neq j}^{m} \mathbf{1}(\upsilon_{i+k-1} \le t_k),$$

where $\upsilon_i = F(\varepsilon_i)$ has a uniform distribution. We also let

$$\bar{U}^j_{i,n}(\mathbf{t},\gamma,\lambda,\mathbf{P}) = \left\{\prod_{k>j} t_k\right\} \left[F\{F^{-1}(t_j) + \Gamma_n(i,j,\gamma,\lambda,\mathbf{P})\} - t_j\right]$$
$$\times \prod_{k<j} \mathbf{1}(\upsilon_{i+k-1} \le t_k),$$

$$\mathbb{U}^j_n(s,\mathbf{t},\gamma,\lambda,\mathbf{P}) = \frac{1}{\sqrt{n}} \sum_{i=1}^{\lfloor ns \rfloor} \left\{U^j_{i,n}(\mathbf{t},\gamma,\lambda,\mathbf{P}) - \bar{U}^j_{i,n}(\mathbf{t},\gamma,\lambda,\mathbf{P})\right\}$$

and $\bar{\mathbb{U}}^j_n(s,\mathbf{t},\gamma,\lambda,\mathbf{P}) = \frac{1}{\sqrt{n}} \sum_{i=1}^{\lfloor ns \rfloor} \bar{U}^j_{i,n}(\mathbf{t},\gamma,\lambda,\mathbf{P})$.

Now for $b > 0$, let

$$\mathfrak{D}_b = \{\mathbf{P} = (M,\Phi,\Theta,\theta) : \|\mathbf{B}_\theta\|_\rho \le \tau, \ \max\{\|M\|_\infty, \|\Phi\|_\infty, \|\Theta\|_\infty\} \le b\}.$$

The next lemmas are used to prove Theorem 1.

Lemma 1. *For any $\delta > 0$, $i \ge 1$ and $1 \le j \le m$ if $\mathbf{P}, \mathbf{P}' \in \mathfrak{D}_b$ are such that $\|\mathbf{P} - \mathbf{P}'\|_\infty \le \delta$ then there exists a constant C_b depending on m, τ and b such that*

(i) $|\Gamma_n(i,j,\gamma,\lambda,\mathbf{P}) - \Gamma_n(i,j,\gamma,\lambda,\mathbf{P}')| \le |\gamma|\delta C_b \Lambda_n(i,j).$
(ii) $|d_{i+j-1,n}(\mathbf{P}) - d_{i+j-1,n}(\mathbf{P}')| \le \delta C_b \Lambda_n(i,j).$
(iii) $|d_{i+j-1,n}(\mathbf{P})| \le C_b \Lambda_n(i,j).$
(iv) $|\Gamma_n(i,j,\gamma,\lambda,\mathbf{P})| \le (|\gamma| + |\lambda|)C_b \Lambda_n(i,j).$

The righthand-sides of the four inequalities given above do not depend on $\mathbf{P}$.

Lemma 2. *If the ε_i's are independent and identically distributed with zero mean and finite variance then*

(i) $\sum_{i=1}^{n} E(\Lambda_n(i,j)^2) \le C.$
(ii) $\max_{1 \le i \le n} \max_{1 \le j \le m} |R_{i,j}|/\sqrt{n} \xrightarrow{Pr} 0$ *as n goes to infinity.*
(iii) $\frac{1}{n}\sum_{i=1}^{n} R_{i,j} = O_p(1)$ *and* $\frac{1}{n}\sum_{i=1}^{n} R_{i,j}^2 = O_p(1).$

Lemma 3. *Under the conditions of Theorem 1 for any $\gamma, \lambda \in \mathbb{R}$*

$$\sup_{\mathbf{P}\in\mathfrak{D}_b} \sup_{\mathbf{t}\in[0,1]^m} \sup_{s\in[0,1]} \left|\mathbb{U}^j_n(s,\mathbf{t},\gamma,\lambda,\mathbf{P})\right| \xrightarrow{Pr} 0 \qquad (5)$$

and for any $\epsilon, \eta > 0$ there exists $\delta > 0$ such that

$$P\left\{\sup_{\mathbf{P}\in\mathfrak{D}_b}\sup_{(\mathbf{t},\mathbf{t}')\in\mathscr{N}_j^\delta}\sup_{s\in[0,1]}|\bar{\mathbb{U}}_n^j((s,\mathbf{t},\gamma,\lambda,\mathbf{P})) - \bar{\mathbb{U}}_n^j((s,\mathbf{t}',\gamma,\lambda,\mathbf{P}))| > \epsilon\right\} < \eta, \qquad (6)$$

where $\mathscr{N}_j^\delta = \{(\mathbf{t},\mathbf{t}') \in [0,1]^m \times [0,1]^m : t_j = t_j' \text{ and } \|\mathbf{t}-\mathbf{t}'\|_\infty \leq \delta\}$.

Next, set $\mathbb{S}_n^j(s,\mathbf{t},\mathbf{P}) = \frac{1}{n}\sum_{i=1}^{\lfloor ns\rfloor}(L_i(\mathbf{P}))_j\prod_{k=j+1}^m t_k\prod_{\ell=1}^{j-1}\mathbf{1}\{u_{i+\ell-1} \leq t_\ell\} - s\tilde{v}_j(\mathbf{t},\mathbf{P})$, for $t \in [0,1]^m$. The next Lemmas establishes the asymptotics of $\mathbb{S}_n^j$ and $\bar{\mathbb{U}}_n^j$.

Lemma 4. *Under the conditions of Theorem 1*

$$\sup_{\mathbf{t}\in[0,1]^m}\sup_{s\in[0,1]}\sup_{\mathbf{P}\in\mathfrak{D}_b}|E\{\mathbb{S}_n^j(s,\mathbf{t},\mathbf{P})\}| \longrightarrow 0. \qquad (7)$$

and

$$\sup_{s\in[0,1]}\sup_{\mathbf{t}\in[0,1]^m}\sup_{\mathbf{P}\in\mathfrak{D}_b}|\mathbb{S}_n^j(s,\mathbf{t},\mathbf{P})| \xrightarrow{Pr} 0. \qquad (8)$$

Lemma 5. *If the conditions of Theorem 1 are satisfied then for any $\gamma \in \mathbb{R}$*

$$\sup_{\mathbf{P}\in\mathfrak{D}_b}\sup_{\mathbf{t}\in[0,1]^m}\sup_{s\in[0,1]}\left|\bar{\mathbb{U}}_n^j(s,\mathbf{t},\gamma,0,\mathbf{P}) - sf(F^{-1}(t_j))\gamma\tilde{v}_j(\mathbf{t},\mathbf{P})\right| \xrightarrow{Pr} 0. \qquad (9)$$

6.1.1 Proof of Lemma 1

One can easily check that the matrix B_θ satisfies $(B_\theta^k)_{jm} = 0$ if $k < m-j$. Recall that, from the equivalence of norms, there exits a constant $C > 0$ such that $\|V\|_\rho/C \leq \|V\|_\infty \leq C\|V\|_\rho$ holds for any vector $V \in \mathbb{R}^m$. Using these facts and the definition of Γ_n one sees that

$$|\Gamma_n(i,j,\gamma,\lambda,\mathbf{P}) - \Gamma_n(i,j,\gamma,\lambda,\mathbf{P}')|$$

$$\leq |\gamma|\left\|(B_\theta^{i-1} - B_{\theta'}^{i-1})D_1 + \frac{1}{\sqrt{n}}\sum_{k=m-j}^{i-2} B_\theta^k V_{i-k}(\mathbf{P}) - B_{\theta'}^k V_{i-k}(\mathbf{P}')\right\|_\infty$$

$$\leq |\gamma|C\left\|(B_\theta^{i-1} - B_{\theta'}^{i-1})D_1 + \frac{C}{\sqrt{n}}\sum_{k=m-j}^{i-2} B_\theta^k V_{i-k}(\mathbf{P}) - B_{\theta'}^k V_{i-k}(\mathbf{P}')\right\|_\rho$$

$$\leq |\gamma|C^2\|B_\theta^{i-1} - B_{\theta'}^{i-1}\|_\rho\|D_1\|_\infty$$

$$+\frac{|\gamma|C^2}{\sqrt{n}}\sum_{k=m-j}^{i-2}\|B_\theta^k - B_{\theta'}^k\|_\rho\|V_{i-k}(\mathbf{P'})\|_\infty$$

$$+\|B_\theta^k\|_\rho\|V_{i-k}(\mathbf{P}) - V_{i-k}(\mathbf{P'})\|_\infty.$$

Now recall that since $\mathbf{P}, \mathbf{P'} \in \mathfrak{D}_b$ one has $\|B_\theta\|_\rho \leq \tau$ and $\|B_{\theta'}\|_\rho \leq \tau$. Moreover, from $\|\mathbf{P} - \mathbf{P'}\|_\infty \leq \delta$ one can verify that $\|B_\theta - B_{\theta'}\|_\infty \leq \delta$. Combining these facts one sees that

$$\|B_\theta^k - B_{\theta'}^k\|_\rho = \left\|\sum_{r=0}^{k-1} B_{\theta'}^r(B_\theta - B_{\theta'})B_\theta^{k-r-1}\right\|_\rho$$

$$\leq \sum_{r=0}^{k-1}\|B_{\theta'}\|_\rho^r\|B_\theta - B_{\theta'}\|_\rho\|B_\theta\|_\rho^{k-r-1}$$

$$\leq C\sum_{r=0}^{k-1}\|B_{\theta'}\|_\rho^r\|B_\theta - B_{\theta'}\|_\infty\|B_\theta\|_\rho^{k-r-1} \leq C\delta k\tau^{k-1},$$

$\|V_i(\mathbf{P}) - V_i(\mathbf{P'})\|_\infty \leq \|\mathbf{P} - \mathbf{P'}\|_\infty\left(1 + \sum_{l=1}^p|Y_{i+m-1-l}| + \sum_{l=1}^q|\varepsilon_{i+m-1-l}|\right)$ and that $\|V_i(\mathbf{P})\|_\infty \leq b\left(1 + \sum_{l=1}^p|Y_{i+m-1-l}| + \sum_{l=1}^q|\varepsilon_{i+m-1-l}|\right)$. Collecting these terms yields $|\Gamma_n(i,j,\gamma,\lambda,\mathbf{P}) - \Gamma_n(i,j,\gamma,\lambda,\mathbf{P'})| \leq C_b\delta|\gamma|\Lambda_n(i,j)$, where $C_b = C^2\max\{C/\tau^2, mbC/\tau + 1\}$. This proves ($i$). Inequality ($ii$) follows immediately from the fact that $d_{i+j-1,n}(\mathbf{P}) = \Gamma_n(i,j,1,0,\mathbf{P})$. The proofs of ($iii$) and ($iv$) are omitted since they easily follow from the proof of (i).

6.1.2 Proof of Lemma 2

First recall that the random variables Y_i's and ε_i's are stationary with finite variances. To prove (i) one easily verifies that

$$E(R_{i,j}^2) \leq 3\{1 + p^2E(Y_1^2) + q^2E(\varepsilon_1^2)\}\sum_{k=1}^{i+j-m-1}\sum_{h=1}^{i+j-m-1}kh\tau^{k+h} \leq C_1$$

for some constant C_1. Therefore $\sum_{i=1}^n E(\Lambda_n(i,j)^2) \leq 2\sum_{i=1}^n i^2\tau^{2i}E(\varepsilon_1^2) + 2C_1 \leq C$ for some constant C. To prove (ii), notice that $|R_{i,j}| \leq (1 + p\max_{1\leq i\leq n}|Y_i| + q\max_{1\leq i\leq n}|\varepsilon_i|)\sum_{k=1}^n k\tau^k$. The above sum converges since $\tau < 1$. The random variables Y_i and ε_i are stationary with finite variances, so an application of Bonferroni's inequality shows that $\max_{1\leq i\leq n}|Y_i|/\sqrt{n}$ and $\max_{1\leq i\leq n}|\varepsilon_i|/\sqrt{n}$ converge to zero in probability. To complete the proof, note that (iii) follows from the fact that $(E|R_{i,j}|)^2 \leq E(R_{i,j}^2) \leq C$.

6.1.3   Proof of Lemma 3

First, define

$$\mathbb{U}_n^{j+}(s, \mathbf{t}, \gamma, \lambda, \mathbf{P}) = \sum_{i=1}^{\lfloor ns \rfloor} \{U_{i,n}^{j+}(\mathbf{t}, \gamma, \lambda, \mathbf{P}) - \bar{U}_{i,n}^{j+}(\mathbf{t}, \gamma, \lambda, \mathbf{P})\}/\sqrt{n}$$

and $\bar{\mathbb{U}}_n^{j+}(s, \mathbf{t}, \gamma, \lambda, \mathbf{P}) = \sum_{i=1}^{\lfloor ns \rfloor} \bar{U}_{i,n}^{j+}(\mathbf{t}, \gamma, \lambda, \mathbf{P})/\sqrt{n}$, where

$$U_{i,n}^{j+}(\mathbf{t}, \gamma, \lambda, \mathbf{P}) = \prod_{k=1, k \neq j}^{m} \mathbf{1}\{u_{i+k-1} \leq t_k\}\mathbf{1}\{F^{-1}(t_j) < \varepsilon_{i+j-1} \leq t_{ij,n}^{+}\},$$

$$\bar{U}_{i,n}^{j+}(\mathbf{t}, \gamma, \lambda, \mathbf{P}) = [F(t_{ij,n}^{+}) - t_j] \prod_{k>j} t_k \prod_{k<j} \mathbf{1}\{\varepsilon_{i+k-1} \leq F^{-1}(t_k)\}$$

and $t_{ij,n}^{+} = F^{-1}(t_j) + \max[0, \Gamma_n(i, j, \gamma, \lambda, \mathbf{P})]$. One also defines

$$\mathbb{U}_n^{j-}(s, \mathbf{t}, \gamma, \lambda, \mathbf{P}) = \sum_{i=1}^{\lfloor ns \rfloor} \{U_{i,n}^{j-}(\mathbf{t}, \gamma, \lambda, \mathbf{P}) - \bar{U}_{i,n}^{j-}(\mathbf{t}, \gamma, \lambda, \mathbf{P})\}/\sqrt{n}$$

and $\bar{\mathbb{U}}_n^{j-}(s, \mathbf{t}, \gamma, \lambda, \mathbf{P}) = \sum_{i=1}^{\lfloor ns \rfloor} \bar{U}_{i,n}^{j-}(\mathbf{t}, \gamma, \lambda, \mathbf{P})/\sqrt{n}$, where

$$U_{i,n}^{j-}(\mathbf{t}, \gamma, \lambda, \mathbf{P}) = -\mathbf{1}\{t_{ij,n}^{-} < \varepsilon_{i+j-1} \leq F^{-1}(t_j)\} \times \prod_{k=1, k \neq j}^{m} \mathbf{1}\{u_{i+k-1} \leq t_k\},$$

$$\bar{U}_{i,n}^{j-}(\mathbf{t}, \gamma, \lambda, \mathbf{P}) = -\prod_{k<j} \mathbf{1}\{\varepsilon_{i+k-1} \leq F^{-1}(t_k)\} \times \prod_{k>j} t_k \times [t_j - F(t_{ij,n}^{-})]$$

and $t_{ij,n}^{-} = F^{-1}(t_j) + \min[0, \Gamma_n(i, j, \gamma, \lambda, \mathbf{P})]$.

It is easy to see that $\mathbb{U}_n^{j}(s, \mathbf{t}, \gamma, \lambda, \mathbf{P}) = \mathbb{U}_n^{j+}(s, \mathbf{t}, \gamma, \lambda, \mathbf{P}) + \mathbb{U}_n^{j-}(s, \mathbf{t}, \gamma, \lambda, \mathbf{P})$, and $\bar{\mathbb{U}}_n^{j}(s, \mathbf{t}, \gamma, \lambda, \mathbf{P}) = \bar{\mathbb{U}}_n^{j+}(s, \mathbf{t}, \gamma, \lambda, \mathbf{P}) + \bar{\mathbb{U}}_n^{j-}(s, \mathbf{t}, \gamma, \lambda, \mathbf{P})$, therefore the Lemma will follows if one shows that statement (11) holds with $\mathbb{U}_n^{j+}$ and $\mathbb{U}_n^{j-}$ in place of $\mathbb{U}_n^{j}$ and that statement (6) holds with $\bar{\mathbb{U}}_n^{j+}$ and $\bar{\mathbb{U}}_n^{j-}$ in place of $\bar{\mathbb{U}}_n^{j}$. The proofs for $j+$ exponent and for $j-$ exponent are very similar, therefore only the proof with $j+$ exponent is presented next. The limit in (6) will be established first. It will be shown that the convergence is also uniform for all bounded λ and γ. Since $t_j = t_j'$ one easily verifies that

$$\bar{\mathbb{U}}_n^{j+}(s,\mathbf{t},\gamma,\lambda,\mathbf{P}) - \bar{\mathbb{U}}_n^{j+}(s,\mathbf{t}',\gamma,\lambda,\mathbf{P})| = \frac{1}{\sqrt{n}} \left| \sum_{i=1}^{\lfloor ns \rfloor} \left[\prod_{k>j} t_k \prod_{k<j} \mathbf{1}\{v_{i+k-1} \le t_k\} \right. \right.$$

$$\left. \left. - \prod_{k>j} t'_k \prod_{k<j} \mathbf{1}\{v_{i+k-1} \le t'_k\} \right] \left[F(t^+_{ij,n}) - t_j \right] \right|,$$

which is bounded by

$$\frac{1}{\sqrt{n}} \sum_{i=1}^{n} \left| F(t^+_{ij,n}) - t_j \right| \left| \prod_{k>j} t_k \prod_{k<j} \mathbf{1}\{v_{i+k-1} \le t_k\} \right.$$

$$\left. - \prod_{k>j} t'_k \prod_{k<j} \mathbf{1}\{v_{i+k-1} \le t'_k\} \right|.$$

Using Lemma 1 and the fact that F is uniformly continuous and admits a bounded density, one finds that

$$|\bar{\mathbb{U}}_n^{j+}(s,\mathbf{t},\gamma,\lambda,\mathbf{P}) - \bar{\mathbb{U}}_n^{j+}(s,\mathbf{t}',\gamma,\lambda,\mathbf{P})|$$

$$\le \frac{C}{\sqrt{n}} \sum_{i=1}^{n} \Lambda_n(i,j) \left| \left(\prod_{k>j} t_k - \prod_{k>j} t'_k \right) \prod_{k<j} \mathbf{1}\{v_{i+k-1} \le t_k\} \right.$$

$$\left. + \prod_{k>j} t'_k \left(\prod_{k<j} \mathbf{1}\{v_{i+k-1} \le t_k\} - \prod_{k<j} \mathbf{1}\{v_{i+k-1} \le t'_k\} \right) \right|.$$

With few straightforward algebraic manipulations, one sees that

$$|\bar{\mathbb{U}}_n^{j+}(s,\mathbf{t},\gamma,\lambda,\mathbf{P}) - \bar{\mathbb{U}}_n^{j+}(s,\mathbf{t}',\gamma,\lambda,\mathbf{P})|$$

$$\le \frac{C}{\sqrt{n}} \sum_{i=1}^{n} \Lambda_n(i,j) \left\{ \left| \prod_{k>j} t_k - \prod_{k>j} t'_k \right| \right.$$

$$\left. + \left[\prod_{k<j} \mathbf{1}\{v_{i+k-1} \le \max(t_k,t'_k)\} - \prod_{k<j} \mathbf{1}\{v_{i+k-1} \le \min(t_k,t'_k)\} \right] \right\}.$$

One then easily verifies that

$$|\bar{\mathbb{U}}_n^{j+}(s, \mathbf{t}, \gamma, \lambda, \mathbf{P}) - \bar{\mathbb{U}}_n^{j+}(s, \mathbf{t}', \gamma, \lambda, \mathbf{P})| \leq C \left(\sum_{i=1}^{n} \Lambda_n(i,j)^2 \right)^{\frac{1}{2}}$$

$$\times \left\{ (m-j)\delta + \left(\frac{1}{n} \sum_{i=1}^{n} \left[\prod_{k<j} \mathbf{1}\{v_{i+k-1} \leq \max(t_k, t_k')\} \right. \right. \right.$$

$$\left. \left. \left. - \prod_{k<j} \mathbf{1}\{v_{i+k-1} \leq \min(t_k, t_k')\} \right] \right)^{\frac{1}{2}} \right\},$$

which in turn is bounded by

$$C \left(\sum_{i=1}^{n} \Lambda_n(i,j)^2 \right)^{\frac{1}{2}} \left\{ (m-j)\delta + \left(\prod_{k<j} \max(t_k, t_k') - \prod_{k<j} \min(t_k, t_k') \right)^{1/2} \right.$$

$$\left. + n^{-1/4} \sup_{\|\mathbf{t}-\mathbf{t}'\| \leq \delta} |\beta_n(1, \mathbf{t}) - \beta_n(1, \mathbf{t}')|^{\frac{1}{2}} \right\}$$

$$\leq C \left(\sum_{i=1}^{n} \Lambda_n(i,j)^2 \right)^{\frac{1}{2}} \left\{ m\delta + \sqrt{m\delta} + n^{-1/4} \sup_{\|\mathbf{t}-\mathbf{t}'\| \leq \delta} |\beta_n(1, \mathbf{t}) - \beta_n(1, \mathbf{t}')|^{\frac{1}{2}} \right\},$$

where β_n is the serial Kiefer process defined for all $(s, \mathbf{t}) \in [0, 1] \times [0, 1]^m$ by

$$\beta_n(s, \mathbf{t}) = \frac{1}{\sqrt{n}} \sum_{i=1}^{\lfloor ns \rfloor} \left\{ \prod_{j=1}^{m} \mathbf{1}(v_{i+j-1} \leq t_j) - \prod_{j=1}^{m} t_j \right\}. \tag{10}$$

By Lemma 2, $\sum_{i=1}^{n} \Lambda_n(i,j)^2 = O_p(1)$ and since β_n is tight [11], it follows that for any $\epsilon, \eta > 0$ the probability that the above right-hand-side is greater than ϵ can be made less than η by choosing the appropriate δ.

To prove (5) note that for any $b > 0$, the set $\mathfrak{D}_b$ is compact. Therefore for any $\delta > 0$ the set $\mathfrak{D}_b$ can be covered by a finite number of balls $(\mathscr{B}_1, \ldots, \mathscr{B}_K)$ with diameters less or equal to δ. Denote $\mathbf{P}_1, \ldots, \mathbf{P}_K$ the centers of these balls. Now if $\mathbf{P} \in \mathfrak{D}_b$, then $\mathbf{P} \in \mathscr{B}_r$ for some $1 \leq r \leq K$. It follows from the definitions of Γ, U and $\bar{U}$ and Lemma 1 that

$$\Gamma_n(i, j, \gamma, \lambda - C_1\delta, \mathbf{P}_r) \leq \Gamma_n(i, j, \gamma, \lambda, \mathbf{P}) \leq \Gamma_n(i, j, \gamma, \lambda + C_1\delta, \mathbf{P}_r),$$

$$U_{i,n}^{j+}(\mathbf{t}, \gamma, \lambda - C_1\delta, \mathbf{P}_r) \leq U_{i,n}^{j+}(\mathbf{t}, \gamma, \lambda, \mathbf{P}) \leq U_{i,n}^{j+}(\mathbf{t}, \gamma, \lambda + C_1\delta, \mathbf{P}_r),$$

$$\bar{U}_{i,n}^{j+}(\mathbf{t}, \gamma, \lambda - C_1\delta, \mathbf{P}_r) \leq \bar{U}_{i,n}^{j+}(\mathbf{t}, \gamma, \lambda, \mathbf{P}) \leq \bar{U}_{i,n}^{j+}(\mathbf{t}, \gamma, \lambda + C_1\delta, \mathbf{P}_r),$$

for some constant $C_1 > 0$. One also gets

$$|\mathbb{U}_n^{j+}(s, \mathbf{t}, \gamma, \lambda, \mathbf{P})|$$

$$\leq |\mathbb{U}_n^{j+}(s, \mathbf{t}, \gamma, \lambda + C_1\delta, \mathbf{P}_r)| + |\mathbb{U}_n^{j+}(s, \mathbf{t}, \gamma, \lambda - C_1\delta, \mathbf{P}_r)|$$

$$+ |\bar{\mathbb{U}}_n^{j+}(s, \mathbf{t}, \gamma, \lambda + C_1\delta, \mathbf{P}_r) - \bar{\mathbb{U}}_n^{j+}(s, \mathbf{t}, \gamma, \lambda - C_1\delta, \mathbf{P}_r)|.$$

Upon calling on condition (A1) one sees that

$$|\bar{\mathbb{U}}_n^{j+}(s, \mathbf{t}, \gamma, \lambda + C_1\delta, \mathbf{P}_r) - \bar{\mathbb{U}}_n^{j+}(s, \mathbf{t}, \gamma, \lambda - C_1\delta, \mathbf{P}_r)|$$

$$\leq \frac{C\delta}{\sqrt{n}} \sum_{i=1}^{n} \Lambda_n(i,j) \leq C\delta \left(\sum_{i=1}^{n} \Lambda_n(i,j)^2 \right)^{1/2}.$$

Note that the right hand side does not depend on s, $\mathbf{t}$ or $\mathbf{P}$. Taking the supremum and then the expectation, it follows upon calling on Lemma 2 that

$$E \left\{ \sup_{\mathbf{P} \in \mathfrak{D}_b} \sup_{\mathbf{t} \in [0,1]^m} \sup_{s \in [0,1]} |\bar{\mathbb{U}}_n^{j+}(s, \mathbf{t}, \gamma, \lambda + C_1\delta, \mathbf{P}_r) - \bar{\mathbb{U}}_n^{j+}(s, \mathbf{t}, \gamma, \lambda - C_1\delta, \mathbf{P}_r)| \right\}$$

is bounded by $C'\delta$, which can be made arbitrarily small by choosing δ. So it remains to show that

$$\sup_{\mathbf{t} \in [0,1]^m} \sup_{s \in [0,1]} |\mathbb{U}_n^{j+}(s, \mathbf{t}, \gamma, \lambda, \mathbf{P})| \xrightarrow{Pr} 0 \tag{11}$$

for any fixed $\mathbf{P} \in \mathfrak{D}_b$. For let $\delta > 0$, and $\Delta_n > 0$ be such that $\lim_{n \to \infty} \Delta_n \sqrt{n} + (n\Delta_n)^{-1} = 0$ and let $K = \lfloor 1/\delta \rfloor + 1$, and let $0 = a_0 < a_1 < \cdots < a_K = 1$ be a partition of $[0, 1]$ with mesh less or equal to δ and set $K_n = \lfloor 1/\Delta_n \rfloor + 1$ and let $0 = b_0 < b_1 < \cdots < b_{K_n} = 1$ be a partition of $[0, 1]$ with mesh less or equal to Δ_n. For any given $1 \leq j \leq m$, note that for any $\mathbf{t} \in [0, 1]^m$ one has $b_{r_j} < t_j \leq b_{r_j+1}$ for some $1 \leq r_j \leq K_n - 1$ and for any $k : 1 \leq k \leq m; k \neq j$ $a_{r_k} < t_k \leq a_{r_k+1}$ for some $1 \leq r_k \leq K - 1$. It follows that

$$U_{i,n}^{j+}(\mathbf{t}, \gamma, \lambda, \mathbf{P}) \leq U_{i,n}^{j+}(\mathbf{t}^+, \gamma, \lambda, \mathbf{P})$$

$$+ \mathbf{1}(b_{r_j} < v_{i+j-1} \leq b_{r_j+1}) \prod_{k \neq j} \mathbf{1}(v_{i+k-1} \leq a_{r_k+1}),$$

and

$$U_{i,n}^{j+}(\mathbf{t}, \gamma, \lambda, \mathbf{P}) \geq U_{i,n}^{j+}(\mathbf{t}^-, \gamma, \lambda, \mathbf{P})$$

$$- \mathbf{1}(b_{r_j} < v_{i+j-1} \leq b_{r_j+1}) \prod_{k \neq j} \mathbf{1}(v_{i+k-1} \leq a_{r_k}),$$

where $\mathbf{t}^+ = (a_{r_1+1}, \ldots, a_{r_{j-1}+1}, b_{r_j}+1, a_{r_{j+1}+1}, \ldots, a_{r_m+1})^\top$ and $\mathbf{t}^- = (a_{r_1}, \ldots, a_{r_{j-1}}, b_{r_j}, a_{r_{j+1}}, \ldots, a_{r_m})^\top$. One can also verify that

$$\bar{U}_{i,n}^{j+}(\mathbf{t}, \gamma, \lambda, \mathbf{P}) \leq \bar{U}_{i,n}^{j+}(\mathbf{t}^+, \gamma, \lambda, \mathbf{P})$$

$$+ \{b_{r_j}+1 - b_{r_j}\} \prod_{k<j} \mathbf{1}(v_{i+k-1} \leq a_{r_k+1}) \prod_{k>j} a_{r_k+1},$$

and

$$\bar{U}_{i,n}^{j+}(\mathbf{t}, \gamma, \lambda, \mathbf{P}) \geq \bar{U}_{i,n}^{j+}(\mathbf{t}^-, \gamma, \lambda, \mathbf{P}) - \{b_{r_j}+1 - b_{r_j}\} \prod_{k<j} \mathbf{1}(v_{i+k-1} \leq a_{r_k}) \prod_{k>j} a_{r_k}.$$

Straightforward computations show that

$$\sup_{\mathbf{t} \in [0,1]^m} \sup_{s \in [0,1]} \left| \mathbb{U}_n^{j+}(s, \mathbf{t}, \gamma, \lambda, \mathbf{P}) \right|$$

$$\leq \max_{1 \leq r_j \leq K_n} \max_{1 \leq r_k \leq K, k \neq j} \sup_{0 \leq s \leq 1} \left\{ |\mathbb{U}_n^{j+}(s, \mathbf{t}^+, \gamma, \lambda, \mathbf{P})| + |\mathbb{U}_n^{j+}(s, \mathbf{t}^-, \gamma, \lambda, \mathbf{P})| \right.$$

$$\left. + |\bar{\mathbb{U}}_n^{j+}(s, \mathbf{t}^+, \gamma, \lambda, \mathbf{P}) - \bar{\mathbb{U}}_n^{j+}(s, \mathbf{t}^-, \gamma, \lambda, \mathbf{P})| \right\}$$

$$+ 2 \sup_{\|\mathbf{t}-\mathbf{t}'\| \leq \Delta_n} \sup_{0 \leq s \leq 1} |\beta_n(s, \mathbf{t}) - \beta_n(s, \mathbf{t}')| + 2\Delta_n \sqrt{n}.$$

The last two terms go to zero in probability from the definition of Δ_n and because of the tightness of β_n [11].

Next, $|\bar{\mathbb{U}}_n^{j+}(s, \mathbf{t}^+, \gamma, \lambda, \mathbf{P}) - \bar{\mathbb{U}}_n^{j+}(s, \mathbf{t}^-, \gamma, \lambda, \mathbf{P})|$ is bounded by

$$|\bar{\mathbb{U}}_n^{j+}(s, \mathbf{t}^+, \gamma, \lambda, \mathbf{P}) - \bar{\mathbb{U}}_n^{j+}(s, \mathbf{t}^\star, \gamma, \lambda, \mathbf{P})|$$

$$+ |\bar{\mathbb{U}}_n^{j+}(s, \mathbf{t}^\star, \gamma, \lambda, \mathbf{P}) - \bar{\mathbb{U}}_n^{j+}(s, \mathbf{t}^-, \gamma, \lambda, \mathbf{P})|,$$

where $\mathbf{t}^\star$ is such that $t_k^\star = t_k^-$ for all $1 \leq k \neq j \leq m$ and $t_j^\star = t_j^+$. Now by (6), the sup of $|\bar{\mathbb{U}}_n^{j+}(s, \mathbf{t}^+, \gamma, \lambda, \mathbf{P}) - \bar{\mathbb{U}}_n^{j+}(s, \mathbf{t}^\star, \gamma, \lambda, \mathbf{P})|$ converges in probability to zero. In addition

$$|\bar{\mathbb{U}}_n^{j+}(s, \mathbf{t}^\star, \gamma, \lambda, \mathbf{P}) - \bar{\mathbb{U}}_n^{j+}(s, \mathbf{t}^-, \gamma, \lambda, \mathbf{P})|$$

$$= \frac{1}{\sqrt{n}} \left| \sum_{i=1}^{\lfloor ns \rfloor} \left[F(F^{-1}(b_{r_j}+1) + \Gamma_n(i, j, \gamma, \lambda, \mathbf{P})) - b_{r_j}+1 \right. \right.$$

$$\left. \left. - F(F^{-1}(b_{r_j}) + \Gamma_n(i, j, \gamma, \lambda, \mathbf{P})) + b_{r_j} \right] \prod_{k>j} a_{r_k} \prod_{k<j} \mathbf{1}(v_{i+k-1} \leq a_{r_k}) \right|$$

$$\leq \frac{1}{\sqrt{n}} \sum_{i=1}^{\lfloor ns \rfloor} \left| F(F^{-1}(b_{r_j+1}) + \Gamma_n(i,j,\gamma,\lambda,\mathbf{P})) - b_{r_j+1} \right.$$

$$\left. - F(F^{-1}(b_{r_j}) + \Gamma_n(i,j,\gamma,\lambda,\mathbf{P})) + b_{r_j} \right|,$$

converges in probability to zero upon using $(A1)$ and mimicking the proof of Lemma 2.1 of Koul [23]. Finally, to complete the proof, note that the behavior of $|\mathbb{U}_n^{j+}(s,\mathbf{t}^+,\gamma,\lambda,\mathbf{P})|$ and that of $|\mathbb{U}_n^{j+}(s,\mathbf{t}^-,\gamma,\lambda,\mathbf{P})|$ are identical, so only the proof for $|\mathbb{U}_n^{j+}(s,\mathbf{t}^-,\gamma,\lambda,\mathbf{P})|$ will be presented. Note that $\sup_{0\leq s\leq 1} |\mathbb{U}_n^{j+}(s,\mathbf{t}^-,\gamma,\lambda,\mathbf{P})| = \sup_{1\leq l\leq n} |\mathbb{U}_n^{j+}(l/n,\mathbf{t}^-,\gamma,\lambda,\mathbf{P})|$. To study the behavior of the above as n goes to infinity, let $1 \leq h \leq m$ and define

$$\tilde{\mathbb{U}}_n^{j+}(h,s,\mathbf{t},\gamma,\lambda,\mathbf{P})$$

$$= \frac{1}{\sqrt{n}} \sum_{i=1}^{\lfloor ns/m \rfloor} \left\{ U_{(i-1)*m+h,n}^{j+}(\mathbf{t},\gamma,\lambda,\mathbf{P}) - \bar{U}_{(i-1)*m+h,n}^{j+}(\mathbf{t},\gamma,\lambda,\mathbf{P}) \right\}.$$

Observe that $|\mathbb{U}_n^{j+}(s,\mathbf{t},\gamma,\lambda,\mathbf{P}) - \sum_{h=1}^{m} \tilde{\mathbb{U}}_n^{j}(h,s,\mathbf{t},\gamma,\lambda,\mathbf{P})| \leq m/\sqrt{n}$. Since m is fixed and finite, the proof will be complete if one shows that

$$\max_{1\leq r_j\leq K_n} \max_{1\leq r_k\leq K, k\neq j} \sup_{1\leq \ell\leq \lfloor n/m \rfloor} |\tilde{\mathbb{U}}_n^{j+}(h,m\ell/n,\mathbf{t}^-,\gamma,\lambda,\mathbf{P})| \xrightarrow{Pr} 0$$

for any $1 \leq h \leq m$. Set $\mathscr{F}_i = \sigma(\varepsilon_0,\ldots,\varepsilon_{i*m+h+j-2})$. We can drop the first m terms without affecting the limit.

For the rest one can easily verifies that $\{\tilde{\mathbb{U}}_n^{j+}(h,m\ell/n,\mathbf{t}^-,\gamma,\lambda,\mathbf{P}),\mathscr{F}_\ell\}$ is a martingale and $\kappa_{i,n}(\mathbf{t}) = U_{(i-1)*m+h,n}^{j+}(\mathbf{t},\gamma,\lambda,\mathbf{P}) - \bar{U}_{(i-1)*m+h,n}^{j+}(\mathbf{t},\gamma,\lambda,\mathbf{P})$ are martingale differences. Applying Doob's inequality followed by Rosenthal's inequality, one shows that

$$P\{\sup_{1\leq \ell\leq \lfloor n/m \rfloor} |\tilde{\mathbb{U}}_n^{j}(h,m\ell/n,\mathbf{t},\gamma,\lambda,\mathbf{P})| > \epsilon\}$$

$$\leq C\epsilon^{-4}n^{-2}E\left\{ \sum_{i=1}^{\lfloor n/m \rfloor} E(\kappa_{i,n}(\mathbf{t})^2|\mathscr{F}_{i-1}) \right\}^2 + C\epsilon^{-4}n^{-2} \sum_{i=1}^{\lfloor n/m \rfloor} E(\kappa_{i,n}(\mathbf{t})^4).$$

As a result, $E(\kappa_{i,n}(\mathbf{t})^2|\mathscr{F}_{i-1}) \leq |\bar{U}_{(i-1)*m+h,n}^{j}(\mathbf{t},\gamma,\lambda,\mathbf{P})| \leq |F(F^{-1}(t_j) + \Gamma_n((i-1)*m+h,j,\gamma,\lambda,\mathbf{P})) - t_j| \leq \|f\| |\Gamma_n((i-1)*m+h,j,\gamma,\lambda,\mathbf{P})|$, which implies that

$$E\left\{ \sum_{i=1}^{\lfloor n/m \rfloor} E(\kappa_{i,n}(\mathbf{t})^2|\mathscr{F}_{i-1}) \right\}^2 \leq n\|f\| \sum_{i=1}^{n} E(|\Gamma_n(i,j,\gamma,\lambda,\mathbf{P})|^2) \leq Cn.$$

The last inequality follows from Lemmas 1 and 2 and hypothesis (A1). Note also that $\sum_{i=1}^{\lfloor n/m \rfloor} E(\kappa_{i,n}(\mathbf{t})^4)$ is bounded by n since $|\kappa_{i,n}(\mathbf{t})| \leq 1$. Collecting the terms shows that $P\{\sup_{1 \leq \ell \leq \lfloor n/m \rfloor} |\tilde{\mathbb{U}}_n^j(h, m\ell/n, t, \gamma, \lambda, \mathbf{P})| > \epsilon\} \leq C_1 \epsilon^{-4} n^{-1}$ for some constant $C_1 > 0$ that does not depend $\mathbf{t}$. The proof is then complete upon noting that

$$P\left\{ \max_{1 \leq r_j \leq K_n} \max_{1 \leq r_k \leq K, k \neq j} \sup_{1 \leq \ell \leq \lfloor n/m \rfloor} |\tilde{\mathbb{U}}_n^j(h, m\ell/n, \mathbf{t}^-, \gamma, \lambda, \mathbf{P})| > \epsilon \right\}$$

$$\leq K_n K^{m-1}$$

$$\times \max_{1 \leq r_j \leq K_n} \max_{1 \leq r_k \leq K, k \neq j} P\left\{ \sup_{1 \leq \ell \leq \lfloor n/m \rfloor} |\tilde{\mathbb{U}}_n^j(h, m\ell/n, \mathbf{t}^-, \gamma, \lambda, \mathbf{P})| > \epsilon \right\}$$

$$\leq C_1 \epsilon^{-4} (n \Delta_n)^{-1} K^{m-1} \longrightarrow 0.$$

$\square$

6.1.4 Proof of Lemma 4

Observe that

$$E\{\mathbb{S}_n^j(s, \mathbf{t}, \mathbf{P})\} = \frac{1}{n} \sum_{i=1}^{\lfloor ns \rfloor} E\left[\{(L_i(\mathbf{P}))_j\} \prod_{\ell > j} t_\ell \prod_{\ell < j} \mathbf{1}(v_{i+\ell-1} \leq t_\ell) \right] - s\tilde{v}_j(\mathbf{t}, \mathbf{P})$$

$$= \frac{1}{n} \sum_{i=1}^{\lfloor ns \rfloor} \left[\sum_{k=0}^{i-2} \left(B_\theta^k E\left\{ V_{i-k}(\mathbf{P}) \prod_{\ell < j} \mathbf{1}(v_{i+\ell-1} \leq t_\ell) \prod_{\ell > j} t_\ell \right\} \right)_j \right]$$

$$- s\tilde{v}_j(\mathbf{t}, \mathbf{P}).$$

First note that there exists $C > 0$ such that for any $\mathbf{P} \in \mathfrak{D}_b$, $E|V_{i-k}(\mathbf{P})| \leq Cb$. For $i \geq m$ one easily checks that $E\{V_{i-k}(\mathbf{P}) \prod_{\ell < j} \mathbf{1}\{u_{i+\ell-1} \leq t_\ell\}\} \prod_{\ell > j} t_\ell$ is equal to $(0, \ldots, 0, \mu_{j,k})^\top$ where

$$\mu_{j,k} = \prod_{\ell > j} t_\ell E\left[\{M + \sum_{l=1}^{p} \Phi_l Y_{i-k+m-l-1} \right.$$

$$\left. - \sum_{l=1}^{q} \Theta_l \varepsilon_{i-k+m-l-1}\} \prod_{\ell < j} \mathbf{1}(v_{i+\ell-1} \leq t_\ell) \right]$$

$$= \prod_{\ell > j} t_\ell \left(M \prod_{\ell < j} t_\ell + \sum_{l=1}^{p} \Phi_l E \left[Y_{m-k-l} \prod_{\ell < j} \mathbf{1}(\upsilon_\ell \leq t_\ell) \right] \right.$$

$$\left. - \sum_{l=1}^{q} \Theta_l E \left[\varepsilon_{m-k-l} \prod_{\ell < j} \mathbf{1}(\upsilon_\ell \leq t_\ell) \right] \right).$$

Now note that $E\left\{ \prod_{\ell=1}^{j-1} \mathbf{1}(\upsilon_\ell \leq t_\ell) \varepsilon_{m-k-l} \right\} = 0$ for $m - k - l < 1$ and
$E\left\{ \prod_{\ell=1}^{j-1} \mathbf{1}(\upsilon_\ell \leq t_\ell) \varepsilon_{m-k-l} \right\} = \tilde{H}(t_{m-k-l}) \prod_{\ell=1;\ell \neq m-k-l}^{j-1} t_\ell$ for $1 \leq m-k-l \leq j-1$
and $E\left\{ \prod_{\ell=1}^{j-1} \mathbf{1}(\upsilon_\ell \leq t_\ell) Y_{m-k-l} \right\} = \sum_{\alpha=0}^{m-1-k-l} \psi_\alpha \tilde{H}(t_{m-k-l-\alpha}) \prod_{\ell=1;\ell \neq m-k-l-\alpha}^{j-1} t_\ell$
for $m - k - l \geq 1$ and 0 otherwise. Using these facts, one sees that $\mu_{j,k}$ simplifies
to $\mu_{j,k} = M \prod_{l \neq j} t_\ell + \sum_{l=1}^{p} \Phi_l \sum_{\alpha=0}^{m-1-k-l} \psi_\alpha \tilde{\mathfrak{F}}_{j,m-k-l-\alpha}(\mathbf{t}) - \sum_{l=1}^{q} \Theta_l \tilde{\mathfrak{F}}_{j,m-k-l}(\mathbf{t})$
which implies that

$$\left| E\{\mathbb{S}_n^j(s,\mathbf{t},\mathbf{P})\} \right| \leq \left| -s\tilde{\upsilon}_j(\mathbf{t},\mathbf{P}) + \frac{1}{n} \sum_{i=m}^{\lfloor ns \rfloor} \left\{ \sum_{k=0}^{i-2} (B_\theta^k)_{jm} M \prod_{l \neq j} t_\ell \right. \right.$$

$$+ \sum_{k=0}^{m-l-1} \sum_{l=1}^{p} (B_\theta^k)_{jm} \Phi_l \sum_{\alpha=0}^{m-1-k-l} \psi_\alpha \tilde{\mathfrak{F}}_{j,m-k-l-\alpha}(\mathbf{t})$$

$$\left. \left. - \sum_{k=0}^{m-l-1} \sum_{l=1}^{q} (B_\theta^k)_{jm} \Theta_l \tilde{\mathfrak{F}}_{j,m-k-l}(\mathbf{t}) \right\} \right| + \frac{C_1 b}{n},$$

for some positive constant C_1. Since $(B_\theta^k)_{jm} = 0$ if $k < m-j$, the above simplifies to

$$\left| E\{\mathbb{S}_n^j(s,\mathbf{t},\mathbf{P})\} \right| \leq \left| -s\tilde{\upsilon}_j(\mathbf{t},\mathbf{P}) + \frac{\lfloor ns \rfloor - m + 1}{n} \left\{ (I - B_\theta)_{jm}^{-1} M \prod_{l \neq j} t_\ell \right. \right.$$

$$+ \sum_{k=m-j}^{m-l-1} \sum_{l=1}^{p} (B_\theta^k)_{jm} \Phi_l \sum_{\alpha=0}^{m-1-k-l} \psi_\alpha \tilde{\mathfrak{F}}_{j,m-k-l-\alpha}(\mathbf{t})$$

$$\left. \left. - \sum_{k=m-j}^{m-l-1} \sum_{l=1}^{q} (B_\theta^k)_{jm} \Theta_l \tilde{\mathfrak{F}}_{j,m-k-l}(\mathbf{t}) \right\} \right|$$

$$+ \frac{C_1 b}{n} + \frac{1}{n} \left| \sum_{i=m}^{\lfloor ns \rfloor} \sum_{k=i-1}^{\infty} (B_\theta^k)_{jm} M \prod_{l \neq j} t_\ell \right|.$$

Using the definition of $\tilde{v}$ and the fact that $((I - B_\theta)^{-1})_{jm} = 1 \Big/ \left(1 - \sum_{k=1}^{m} \theta_k\right)$, the above simplifies to

$$|E\{\mathbb{S}_n^j(s, \mathbf{t}, \mathbf{P})\}| \leq \left|\frac{\lfloor ns \rfloor - ns - m + 1}{n}\right| |\tilde{v}_j(\mathbf{t}, \mathbf{P})| + \frac{Cb}{n} \sum_{i=m}^{\lfloor ns \rfloor} \sum_{k=i-1}^{\infty} \tau^k.$$

Straightforward computations show that $|\tilde{v}_j(\mathbf{t}, \mathbf{P})| \leq Cb$ for any $\mathbf{P} \in \mathfrak{D}_b$ and any $\mathbf{t} \in [0, 1]^m$, and reduce the above to

$$\sup_{\mathbf{t} \in [0,1]^m} \sup_{s \in [0,1]} \sup_{\mathbf{P} \in \mathfrak{D}_b} |E\{\mathbb{S}_n^j(s, \mathbf{t}, \mathbf{P})\}| \leq \frac{Cb}{n},$$

for some constant $C > 0$. This goes to zero as n goes to infinity and completes the proof of the first part of the Lemma.

To complete the proof and establish (8), note that since F is continuous, an application of Schwartz's Inequality yields $|\tilde{H}(x) - \tilde{H}(y)| \leq \sigma \sqrt{|x - y|}$ for all $x, y \in [0, 1]$. That is, the function $\tilde{H}$ is uniformly continuous. Using this fact one also sees that for any $\eta > 0$ there exists a $\delta > 0$ such that $\sup_{\|\mathbf{t}-\mathbf{t}'\|_\infty \leq \delta} \sup_{\mathbf{P} \in \mathfrak{D}_b} |\tilde{v}_j(\mathbf{t}, \mathbf{P}) - \tilde{v}_j(\mathbf{t}', \mathbf{P})| \leq \eta$. Straightforward algebraic manipulations show that $|\mathbb{S}_n^j(s, \mathbf{t}, \mathbf{P}) - \mathbb{S}_n^j(s, \mathbf{t}', \mathbf{P})|$ is bounded by

$$|\tilde{v}_j(\mathbf{t}, \mathbf{P}) - \tilde{v}_j(\mathbf{t}', \mathbf{P})|$$

$$+ \frac{1}{n} \left| \sum_{i=1}^{\lfloor ns \rfloor} (L_i(\mathbf{P}))_j \left(\prod_{k=j+1}^{m} t_k - \prod_{k=j+1}^{m} t'_k \right) \prod_{\ell=1}^{j-1} \mathbf{1}(v_{i+\ell-1} \leq t_\ell) \right|$$

$$+ \frac{1}{n} \left| \sum_{i=1}^{\lfloor ns \rfloor} (L_i(\mathbf{P}))_j \prod_{k=j+1}^{m} t'_k \left\{ \prod_{\ell=1}^{j-1} \mathbf{1}(v_{i+\ell-1} \leq t_\ell) - \prod_{\ell=1}^{j-1} \mathbf{1}(v_{i+\ell-1} \leq t'_\ell) \right\} \right|,$$

which in turn is bounded by

$$\sup_{\|\mathbf{t}-\mathbf{t}'\|_\infty \leq \delta} \sup_{\mathbf{P} \in \mathfrak{D}_b} |\tilde{v}_j(\mathbf{t}, \mathbf{P}) - \tilde{v}_j(\mathbf{t}', \mathbf{P})| + \frac{Cb}{n} \left| \prod_{k=j+1}^{m} t_k - \prod_{k=j+1}^{m} t'_k \right| \sum_{i=1}^{n} R_{i,j}$$

$$+ \frac{Cb}{n} \sum_{i=1}^{n} R_{i,j} \left[\prod_{\ell=1}^{j-1} \mathbf{1}\{v_{i+\ell-1} \leq \max(t_\ell, t'_\ell)\} \right.$$

$$\left. - \prod_{\ell=1}^{j-1} \mathbf{1}\{v_{i+\ell-1} \leq \min(t_\ell, t'_\ell)\} \right]$$

$$\leq \sup_{\|t-t'\|_\infty \leq \delta} \sup_{\mathbf{P} \in \mathfrak{D}_b} |\tilde{v}_j(\mathbf{t}, \mathbf{P}) - \tilde{v}_j(\mathbf{t}', \mathbf{P})| + \frac{C_b(m-j)\delta}{n} \sum_{i=1}^{n} R_{i,j}$$

$$+ C_b \left(\frac{1}{n} \sum_{i=1}^{n} R_{i,j}^2 \right)^{\frac{1}{2}} \left[\frac{1}{n} \sum_{i=1}^{n} \left[\prod_{\ell=1}^{j-1} \mathbf{1}\{v_{i+\ell-1} \leq \max(t_\ell, t'_\ell)\} \right. \right.$$

$$\left. \left. - \prod_{\ell=1}^{j-1} \mathbf{1}\{v_{i+\ell-1} \leq \min(t_\ell, t'_\ell)\} \right] \right]^{\frac{1}{2}}.$$

Therefore $\sup_{s \in [0,1]} \sup_{\|t-t'\|_\infty \leq \delta} \sup_{\mathbf{P} \in \mathfrak{D}_b} |\mathbb{S}_n^j(s, \mathbf{t}, \mathbf{P}) - \mathbb{S}_n^j(s, \mathbf{t}', \mathbf{P})|$ is smaller than

$$\frac{C_b(m-j)\delta}{n} \sum_{i=1}^{n} R_{i,j} + \sup_{\|t-t'\|_\infty \leq \delta} \sup_{\mathbf{P} \in \mathfrak{D}_b} |\tilde{v}_j(\mathbf{t}, \mathbf{P}) - \tilde{v}_j(\mathbf{t}', \mathbf{P})|$$

$$+ C_b \left(n^{-1/4} \sup_{\|t-t'\|_\infty \leq \delta} |\beta_n(1, \mathbf{t}) - \beta_n(1, \mathbf{t}')|^{\frac{1}{2}} + (m\delta)^{\frac{1}{2}} \right) \left(\frac{1}{n} \sum_{i=1}^{n} R_{i,j}^2 \right)^{\frac{1}{2}},$$

which can be made arbitrarily small by choosing δ and calling on the tightness of β_n and Lemma 2. Moreover, using the same arguments as in the proof of Lemma 1, one can easily verify that

$$\sup_{s \in [0,1] \; \mathbf{P}, \mathbf{P}' \in \mathfrak{D}_b: \; \|\mathbf{P}-\mathbf{P}'\|_\infty < \delta} |\mathbb{S}_n^j(s, \mathbf{t}, \mathbf{P}) - \mathbb{S}_n^j(s, \mathbf{t}, \mathbf{P}')| \xrightarrow{Pr} 0.$$

Since $[0, 1]^m$ and $\mathfrak{D}_b$ are compact, the proof will be complete if one shows that for any $\mathbf{t} \in [0, 1]^m$ and any $\mathbf{P} \in \mathfrak{D}_b$,

$$\sup_{0 \leq s \leq 1} |\mathbb{S}_n^j(s, \mathbf{t}, \mathbf{P})| \xrightarrow{Pr} 0. \tag{12}$$

To prove (12), set

$$S_{1n}(s, \mathbf{t}, \mathbf{P}) = \frac{M}{n} \sum_{i=1}^{\lfloor ns \rfloor} \sum_{k=0}^{i+j-m-2} (B_\theta^{m-j+k})_{jm} \prod_{h<j} \mathbf{1}(v_{i+h-1} \leq t_h)$$

$$S_{2n}(s, \mathbf{t}, \mathbf{P}) = \sum_{l=1}^{p} \Phi_l \frac{1}{n} \sum_{i=1}^{\lfloor ns \rfloor} \sum_{k=0}^{i+j-m-2} (B_\theta^{m-j+k})_{jm} Y_{i+j-k-1-l} \prod_{h<j} \mathbf{1}(v_{i+h-1} \leq t_h)$$

$$S_{3n}(s, \mathbf{t}, \mathbf{P}) = \sum_{l=1}^{q} \Theta_l \frac{1}{n} \sum_{i=1}^{\lfloor ns \rfloor} \sum_{k=0}^{i+j-m-2} (B_\theta^{m-j+k})_{jm} \varepsilon_{i+j-k-1-l} \prod_{h<j} \mathbf{1}(v_{i+h-1} \leq t_h).$$

Because of (7), and since $\mathbb{S}_n^j(s, \mathbf{t}, \mathbf{P}) = \sum_{k=1}^{3} S_{kn}^j(s, \mathbf{t}, \mathbf{P}) - s\tilde{v}_j(\mathbf{t}, \mathbf{P})$, the limit in (12) follows if one shows that $\sup_{0 \le s \le 1} |S_{kn}^j(s, \mathbf{t}, \mathbf{P}) - E\{S_{kn}^j(s, \mathbf{t}, \mathbf{P})\}| \xrightarrow{Pr} 0$ for $k \in \{1, 2, 3\}$.

Here only the asymptotic of S_{2n}^j shall be established. Those of S_{1n}^j and S_{3n}^j use the same arguments and are much simpler. Since $\|\Phi\|_\infty \le b$ and p is finite, the convergence of S_{2n}^j will follows from that of $\kappa_{l,n}$, where

$$\kappa_{l,n} = \frac{1}{n} \sum_{i=1}^{\lfloor ns \rfloor} \sum_{k=0}^{i+j-m-2} (B_\theta^{m-j+k})_{jm} Y_{i+j-k-1-l} \prod_{h<j} \mathbf{1}(\upsilon_{i+h-1} \le t_h).$$

Note that $\kappa_{l,n} = \tilde{\kappa}_{l,n} - \bar{\kappa}_{l,n}$ where

$$\tilde{\kappa}_{l,n} = \frac{1}{n} \sum_{i=1}^{\lfloor ns \rfloor} \sum_{k=0}^{\infty} (B_\theta^{m-j+k})_{jm} Y_{i+j-k-1-l} \prod_{h<j} \mathbf{1}(\upsilon_{i+h-1} \le t_h)$$

and

$$\bar{\kappa}_{l,n} = \frac{1}{n} \sum_{i=1}^{\lfloor ns \rfloor} \sum_{k=i+j-m-2}^{\infty} (B_\theta^{m-j+k})_{jm} Y_{i+j-k-1-l} \prod_{h<j} \mathbf{1}(\upsilon_{i+h-1} \le t_h).$$

Using the stationarity of Y, one sees that

$$E(\sup_{0 \le s \le 1} |\bar{\kappa}_{l,n}|) \le \frac{C}{n} \sum_{i=1}^{n} \sum_{k=i-1}^{\infty} \tau^k E|Y_1| = \frac{CE|Y_1|}{n} \sum_{i=1}^{n} \frac{\tau^{i-1}}{1-\tau} \le \frac{CE|Y_1|}{n(1-\tau)^2},$$

which goes to zero as n goes to infinity. Therefore is just remains to establish $\sup_{s \in [0,1]} |\tilde{\kappa}_{l,n} - E(\tilde{\kappa}_{l,n})|$ converges to zero in probability. To do so, one uses the invertibility of Y and gets

$$\tilde{\kappa}_{l,n} = \frac{1}{n} \sum_{i=1}^{\lfloor ns \rfloor} \sum_{k=0}^{\infty} (B_\theta^{m-j+k})_{jm} Y_{i+j-k-1-l} \prod_{h<j} \mathbf{1}(\upsilon_{i+h-1} \le t_h) = \frac{1}{n} \sum_{i=1}^{\lfloor ns \rfloor} \zeta_i,$$

where

$$\zeta_i = h(\varepsilon_{i+j-1}, \varepsilon_{i+j-2}, \ldots) = \sum_{h=0}^{\infty} \chi_h \varepsilon_{i+j-1-h} \prod_{k=1}^{j-1} \mathbf{1}(\upsilon_{i+j-1-k} \le t_{j-k}),$$

with $\chi_h = \sum_{\ell=0}^{h} \psi_\ell (B_\theta^{m-j+h-\ell})_{jm}$. Observe that

$$E|\zeta_0| \le E|\varepsilon_0| \sum_{h=0}^{\infty} |\chi_h| \le CE|\varepsilon_0| \sum_{h=0}^{\infty} \sum_{\ell=0}^{h} |\psi_\ell| \tau^{m-j+h-\ell}$$

$$= \frac{C}{1-\tau} E|\varepsilon_0| \tau^{m-j} \sum_{\ell=0}^{\infty} |\psi_\ell| < \infty.$$

Since $\sum_{\ell=0}^{\infty} |\psi_\ell|$ converges see Bai Bai [1] or Brockwell and Davis Brockwell and Davis [8], Lemma 3.6 of Kulperger and Yu Kulperger and Yu [24] yields the invariance property, that is $\sup_{0 \le s \le 1} \left| \frac{1}{n} \sum_{i=1}^{\lfloor ns \rfloor} \zeta_i - sE(\zeta_0) \right| = o_p(1)$ and completes the proof. $\square$

6.1.5 Proof of Lemma 5

To prove Lemma 5, observe that $|\bar{\mathbb{U}}_n^j(s, \mathbf{t}, \gamma, 0, \mathbf{P}) - s\gamma f \circ F^{-1}(t_j)\tilde{v}_j(\mathbf{t}, \mathbf{P})|$ is equal to

$$\left| \frac{1}{\sqrt{n}} \sum_{i=1}^{\lfloor ns \rfloor} \left[\left[F\left\{ F^{-1}(t_j) + \Gamma_n(i,j,\gamma,0,\mathbf{P}) \right\} - F\left\{ F^{-1}(t_j) + \frac{\gamma}{\sqrt{n}} (L_i(\mathbf{P}))_j \right\} \right] \right. \right.$$

$$\left. + \left[F\left\{ F^{-1}(t_j) + \frac{\gamma}{\sqrt{n}} (L_i(\mathbf{P}))_j \right\} - t_j \right] \right] \prod_{k=j+1}^{m} t_k \prod_{\ell=1}^{j-1} \mathbf{1}(v_{i+\ell-1} \le t_\ell)$$

$$\left. - s\gamma f \circ F^{-1}(t_j)\tilde{v}_j(\mathbf{t}, \mathbf{P}) \right|.$$

Applications of the mean value theorem shows that the above is bounded by

$$\frac{C|\gamma| \|f\| \|D_1\|_\infty}{\sqrt{n}} \sum_{i=1}^{n} \tau^{i-1} + \frac{C_b|\gamma|}{n} \left| \sum_{i=1}^{n} |f(\xi_{i,j}) - f\{F^{-1}(t_j)\}|R_{i,j} \right.$$

$$+ \frac{|\gamma| \|f\|}{n} \left| \sum_{i=1}^{\lfloor ns \rfloor} (L_i(\mathbf{P}))_j \prod_{k=j+1}^{m} t_k \prod_{\ell=1}^{j-1} \mathbf{1}(v_{i+\ell-1} \le t_\ell) - \tilde{v}_j(\mathbf{t}, \mathbf{P}) \right|$$

$$+ \frac{|\lfloor ns \rfloor - ns|}{n} \|f\| |\tilde{v}_j(\mathbf{t}, \mathbf{P})|,$$

where $\xi_{i,j}$ is such that $\max |\xi_{i,j} - F^{-1}(t_j)| \le \max R_{i,j}/\sqrt{n} = o_p(1)$ by Lemma 2. By $(A1)$ and Lemma 2 the first two terms converge to zero in probability. The middle term goes in probability to zero by Lemma 4 and the last term converges to zero since f is bounded and $\sup_{t \in [0,1]^m} \sup_{\mathbf{P} \in \mathfrak{D}_b} |\tilde{v}_j(t, \mathbf{P})| \le Cb$ as mentioned in the proof of Lemma 4. $\square$

6.2   Proof of Theorem 1

To prove Theorem 1, note that $\mathbb{H}_n = \overset{\circ}{\mathbb{H}}_n + \tilde{\beta}_n$, where

$$\tilde{\beta}_n(s, \mathbf{x}, \mathbf{P}_n) = \frac{1}{\sqrt{n}} \sum_{i=1}^{\lfloor ns \rfloor} \{\mathbf{1}(\mathbf{w}_{i,n} \leq \mathbf{x}) - \mathbf{1}(\omega_i \leq \mathbf{x})\}, \quad (s, \mathbf{x}) \in [0, 1] \times \mathbb{R}^m.$$

Since the functions v_j are continuous, the proof of the theorem will be complete if one can show that, as $n \to \infty$,

$$\sup_{s \in [0,1]} \sup_{\mathbf{x} \in \mathbb{R}^m} \left| \tilde{\beta}_n(s, \mathbf{x}, \mathbf{P}_n) - s \sum_{j=1}^m f(x_j) v_j(\mathbf{x}, \mathbf{P}_n) \right| \overset{P}{\to} 0. \tag{13}$$

From the tightness of $\mathbf{P}_n$ one sees that the probability that $\mathbf{P}_n$ is outside $\mathfrak{D}_b$ can be made arbitrarily small by choosing b large enough. Therefore (13) will follows if one shows

$$\sup_{s \in [0,1]} \sup_{\mathbf{x} \in \mathbb{R}^m} \sup_{\mathbf{P} \in \mathfrak{D}_b} \left| \tilde{\beta}_n(s, \mathbf{x}, \mathbf{P}) - s \sum_{j=1}^m f(x_j) v_j(\mathbf{x}, \mathbf{P}) \right| \overset{P}{\to} 0. \tag{14}$$

The proof shall be given after adding few notations. For $1 \leq j \leq m$ let

$$\tilde{\beta}_{j,n}(s, \mathbf{x}, \mathbf{P}) = \frac{1}{\sqrt{n}} \sum_{i=1}^{[ns]} \left[\mathbf{1}\{\varepsilon_{i+j-1} \leq x_j + d_{i+j-1,n}(\mathbf{P})\} \right.$$

$$\left. - \mathbf{1}(\varepsilon_{i+j-1} \leq x_j) \right] \prod_{k=1, k \neq j}^m \mathbf{1}(\varepsilon_{i+k-1} \leq x_k).$$

Now Eq. (14) will be established by showing that

$$\sup_{s \in [0,1]} \sup_{\mathbf{x} \in \mathbb{R}^m} \sup_{\mathbf{P} \in \mathfrak{D}_b} \max_{1 \leq j \leq m} \left| \tilde{\beta}_{j,n}(s, \mathbf{x}, \mathbf{P}) - s f(x_j) v_j(\mathbf{x}, \mathbf{P}) \right| \overset{Pr}{\longrightarrow} 0 \tag{15}$$

and that

$$\sup_{s \in [0,1]} \sup_{\mathbf{x} \in \mathbb{R}^m} \sup_{\mathbf{P} \in \mathfrak{D}_b} \left| \tilde{\beta}_n(s, \mathbf{x}, \mathbf{P}) - \sum_{j=1}^m \tilde{\beta}_{j,n}(s, \mathbf{x}, \mathbf{P}) \right| \overset{Pr}{\longrightarrow} 0. \tag{16}$$

Upon setting $\mathbf{t} = (t_1, \ldots, t_m)^\top$ with $t_k = F(x_k)$ for $k = 1, \ldots, m$ and noticing that $\tilde{\beta}_{j,n}(s, x, \mathbf{P}) = \mathbb{U}_n^j(s, t, 1, 0, \mathbf{P}) + \bar{\mathbb{U}}_n^j(s, t, 1, 0, \mathbf{P})$, one concludes that (15) is just a consequence of Lemmas 3 and 5. To prove (16), observe that an application of the multinomial formula shows that $\tilde{\beta}_n(s, \mathbf{x}, , \mathbf{P}) = \sum_{A \subset \mathcal{J}_m; A \neq \emptyset} \tilde{\beta}_{A,n}(s, \mathbf{x}, \mathbf{P})$ where $\tilde{\beta}_{A,n}(s, \mathbf{x}, \mathbf{P}) = \sum_{i=1}^{\lfloor ns \rfloor} \tilde{\beta}_{i,A,n}(s, \mathbf{x}, \mathbf{P})/\sqrt{n}$ and

$$\tilde{\beta}_{i,A,n}(s, \mathbf{x}, \mathbf{P}) = \prod_{j \in A} \left[\mathbf{1}\{\varepsilon_{i+j-1} \le x_j + d_{i+j-1,n}(\mathbf{P})\} \right.$$

$$\left. -\mathbf{1}(\varepsilon_{i+j-1} \le x_j) \right] \prod_{k \in A^c} \mathbf{1}(\varepsilon_{i+k-1} \le x_k).$$

The proof will then be complete if one shows that

$$\sup_{s \in [0,1]} \sup_{\mathbf{x} \in \mathbb{R}^m} \sup_{\mathbf{P} \in \mathfrak{D}_b} |\tilde{\beta}_{A,n}(s, \mathbf{x}, \mathbf{P})| \xrightarrow{Pr} 0,$$

for all subsets A with $|A| \ge 2$. The rest of the proof mimics the arguments of Ghoudi and Rémillard [17] and is given next.

Let A be a subset of $\mathcal{J}_m$ with $|A| \ge 2$. Let $\delta > 0$ and

$$\tilde{\beta}_{A,\delta,n}(s, \mathbf{x}, \mathbf{P}) = \frac{1}{\sqrt{n}} \sum_{i=1}^{\lfloor ns \rfloor} \tilde{\beta}_{i,A,n}(s, \mathbf{x}, \mathbf{P}) \prod_{j=1}^m \mathbf{1}\{C_b \Lambda_n(i, j) \le \delta\},$$

where C_b and $\Lambda_n(i, j)$ are defined in Sect. 6.1. One easily verifies that

$$\sup_{s \in [0,1]} \sup_{\mathbf{x} \in \mathbb{R}^m} \sup_{\mathbf{P} \in \mathfrak{D}_b} \left| \tilde{\beta}_{A,n}(s, \mathbf{x}, \mathbf{P}) - \tilde{\beta}_{A,\delta,n}(s, \mathbf{x}, \mathbf{P}) \right|$$

$$\le \frac{1}{\sqrt{n}} \sum_{i=1}^n \mathbf{1} \left\{ \cup_{j=1}^m \{C_b \Lambda_n(i, j) > \delta\} \right\}.$$

The righthand-side of the above goes to zero in probability since

$$E \left| \frac{1}{\sqrt{n}} \sum_{i=1}^n \mathbf{1} \left\{ \cup_{j=1}^m \{C_b \Lambda_n(i, j) > \delta\} \right\} \right| = \frac{1}{\sqrt{n}} \sum_{j=1}^m \sum_{i=1}^n P\{C_b \Lambda_n(i, j) > \delta\}$$

$$\le \frac{C_b^2}{\delta^2 \sqrt{n}} \sum_{j=1}^m \sum_{i=1}^n E[\Lambda_n(i, j)^2].$$

The last bound, which is consequence of Markov inequality, goes to zero by Lemma 2. To complete the proof, it remains to show that for any $A \subset \mathcal{J}_m$ with $|A| \ge 2$,

$$\sup_{s\in[0,1]} \sup_{\mathbf{x}\in\mathbb{R}^m} \sup_{\mathbf{P}\in\mathfrak{D}_b} |\tilde{\beta}_{A,\delta,n}(s,\mathbf{x},\mathbf{P})| \xrightarrow{Pr} 0.$$

Since $|A| \geq 2$ one assumes that $j, j_0 \in A$ and uses Lemma 1 to verify that $|\tilde{\beta}_{A,\delta,n}(s,\mathbf{x},\mathbf{P})|$ is bounded the sum of the following two terms

$$\frac{1}{\sqrt{n}} \sum_{i=1}^{\lfloor ns \rfloor} \mathbf{1}\{x_j < \varepsilon_{i+j-1} \leq x_j + C_b \Lambda_n(i,j)\} \mathbf{1}\{x_{j_0} - \delta < \varepsilon_{i+j_0-1} \leq x_{j_0} + \delta\}$$

$$\times \prod_{l\in A\setminus\{j,j_0\}} \mathbf{1}(\varepsilon_{i+l-1} \leq x_l + \delta) \prod_{k\in A^c} \mathbf{1}(\varepsilon_{i+k-1} \leq x_k),$$

and

$$\frac{1}{\sqrt{n}} \sum_{i=1}^{\lfloor ns \rfloor} \mathbf{1}(x_j - C_b \Lambda_n(i,j) < \varepsilon_{i+j-1} \leq x_j) \mathbf{1}(x_{j_0} - \delta < \varepsilon_{i+j_0-1} \leq x_{j_0} + \delta)$$

$$\times \prod_{l\in A\setminus\{j,j_0\}} \mathbf{1}(\varepsilon_{i+l-1} \leq x_l + \delta) \prod_{k\in A^c} \mathbf{1}(\varepsilon_{i+k-1} \leq x_k).$$

Upon noting that $C_b \Lambda_n(i,j) = \Gamma_n(i,j,0,C_b,\mathscr{P}_0)$ for any arbitrary $\mathscr{P}_0 \in \mathfrak{D}_b$, one gets

$$\begin{aligned}
|\tilde{\beta}_{A,\delta,n}(s,\mathbf{x},\mathbf{P})| \leq\ & |\mathbb{U}_n^j(s,\mathbf{t}_\delta,0,C_b,\mathscr{P}_0)| + |\mathbb{U}_n^j(s,\mathbf{t}'_\delta,0,C_b,\mathscr{P}_0)| \\
& + |\mathbb{U}_n^j(s,\mathbf{t}_\delta,0,-C_b,\mathscr{P}_0)| + |\mathbb{U}_n^j(s,\mathbf{t}'_\delta,0,-C_b,\mathscr{P}_0)| \\
& + |\bar{\mathbb{U}}_n^j((s,\mathbf{t}_\delta,0,C_b,\mathscr{P}_0)) - \bar{\mathbb{U}}_n^j((s,\mathbf{t}'_\delta,0,C_b,\mathscr{P}_0))| \\
& + |\bar{\mathbb{U}}_n^j((s,\mathbf{t}_\delta,0,-C_b,\mathscr{P}_0)) - \bar{\mathbb{U}}_n^j((s,\mathbf{t}'_\delta,0,-C_b,\mathscr{P}_0))|,
\end{aligned}$$

where $(x_\delta)_l = x_l + \delta$ if $l \in A \setminus \{j\}$ and $(x_\delta)_l = x_l$ otherwise, $(x'_\delta)_l = x_l + \delta$ if $l \in A \setminus \{j,j_0\}$ while $(x'_\delta)_{j_0} = x_{j_0} - \delta$, $(x'_\delta)_j = x_j$ and $(x'_\delta)_l = x_l$ for $l \in A^c$. We also have $t_l = F(x_l)$ and $t'_l = F(x'_l)$ for all $1 \leq l \leq m$. The proof is concluded by using Lemma 3. $\square$

6.3 Proof of Corollary 3

First, note that

$$\left| \left[F\{F_n^{-1}(u)\} - u \right] - \left[F\{F_n^{-1}(u)\} - F_n\{F_n^{-1}(u)\} \right] \right| = |F_n\{F_n^{-1}(u)\} - u| \leq \frac{1}{n}.$$

Therefore $\sup_{0 < u < 1} |F\{F_n^{-1}(u)\} - u| \le \sup_{y \in \mathbb{R}} |\mathbb{F}_n(1, y)|/\sqrt{n} + 1/n$, which goes to zero in probability by Theorem 1. The above inequality also implies that the process $\sqrt{n}(F\{F_n^{-1}(u)\} - u)$ is tight and is asymptotically equivalent to $-\mathbb{F}_n(1, F_n^{-1}(u))$. To complete the proof observe that

$$
\begin{aligned}
\mathbb{C}_n(u_1, \ldots, u_m) &= \mathbb{H}_n(1, F_n^{-1}(u_1), \ldots, F_n^{-1}(u_m)) \\
&\quad + \sqrt{n}\left[\prod_{i=1}^{m} F\{F_n^{-1}(u_i)\} - \prod_{i=1}^{m} u_i\right] \\
&= \mathbb{H}_n(1, F_n^{-1}(u_1), \ldots, F_n^{-1}(u_m)) \\
&\quad + \sum_{j=1}^{m} \sqrt{n}(F\{F_n^{-1}(u_j) - u_j\}) \prod_{i \ne j} u_i + o_p(1).
\end{aligned}
$$

Now $\mathbb{H}_n(1, F_n^{-1}(u_1), \ldots, F_n^{-1}(u_m)) \rightsquigarrow \mathbb{H}(1, F^{-1}(u_1), \ldots, F^{-1}(u_m))$ and

$$
\sqrt{n}(F\{F_n^{-1}(u)\} - u) \rightsquigarrow -\mathbb{F}(1, F^{-1}(u)) = -\mathbb{H}(1, F^{-1}(u), \infty, \ldots, \infty).
$$

The convergence of $\mathbb{C}_{A,n}$ follows the convergence of $\mathbb{C}_n$ and the fact that $\mathbb{C}_{A,n} = M_A(\mathbb{C}_n)$ where M_A is the continuous Möbius transform discussed in Genest and Rémillard [14]. □

Acknowledgements Funding in partial support of this work was provided by the National Research Foundation of the United Arab Emirates, the Natural Sciences and Engineering Research Council of Canada, the Fonds québécois de la recherche sur la nature et les technologies, and Desjardins Global Asset Management.

References

1. Bai, J.: On the partial sums of residuals in autoregressive and moving average models. J. Time Ser. Anal. **14**(3), 247–260 (1993)
2. Bai, J.: Weak convergence of the sequential empirical processes of residuals in ARMA models. Ann. Stat. **22**, 2051–2061 (1994)
3. Bartlett, M.S.: On the theoretical specification and sampling properties of autocorrelated time-series. Suppl. J. R. Stat. Soc. **8**, 27–41 (1946)
4. Bickel, P.J., Wichura, M.J.: Convergence criteria for multiparameter stochastic processes and some applications. Ann. Math. Stat. **42**, 1656–1670 (1971)
5. Blum, J.R., Kiefer, J., Rosenblatt, M.: Distribution free test of independence based on the sample distribution function. Ann. Math. Stat. **32**, 485–498 (1961)
6. Box, G.E.P., Jenkins, G.M.: Times Series Analysis. Forecasting and Control. Holden-Day, San Francisco (1970)
7. Box, G.E.P., Pierce, D.A.: Distribution of residual autocorrelations in autoregressive-integrated moving average time series models. J. Am. Stat. Assoc. **65**, 1509–1526 (1970)
8. Brockwell, P.J., Davis, R.A.: Time Series: Theory and Methods, 2nd edn. Springer Series in Statistics. Springer, New York (1991)

9. Carlstein, E.: Nonparametric change-point estimation. Ann. Stat. **16**, 188–197 (1988)
10. Davies, N., Triggs, C.M., Newbold, P.: Significance levels of the Box-Pierce portmanteau statistic in finite samples. Biometrika **64**(3), 517–522 (1977)
11. Delgado, M.A.: Testing serial independence using the sample distribution function. J. Time Ser. Anal. **17**, 271–285 (1996)
12. Durbin, J.: Weak convergence of the sample distribution function when parameters are estimated. Ann. Stat. **1**(2), 279–290 (1973)
13. Durrett, R.: Probability: Theory and Examples, 2nd edn. Duxbury Press, Belmont (1996)
14. Genest, C., Rémillard, B.: Tests of independence or randomness based on the empirical copula process. Test **13**, 335–369 (2004)
15. Genest, C., Ghoudi, K., Rémillard, B.: Rank-based extensions of the Brock Dechert Scheinkman test for serial dependence. J. Am. Stat. Assoc. **102**, 1363–1376 (2007a)
16. Genest, C., Quessy, J.-F., Rémillard, B.: Asymptotic local efficiency of Cramér-von Mises tests for multivariate independence. Ann. Stat. **35**, 166–191 (2007b)
17. Ghoudi, K., Rémillard, B.: Empirical processes based on pseudo-observations. II. The multivariate case. In: Asymptotic Methods in Stochastics. Volume 44 of Fields Institute Communications, pp. 381–406. American Mathematical Society, Providence (2004)
18. Ghoudi, K., Kulperger, R.J., Rémillard, B.: A nonparametric test of serial independence for time series and residuals. J. Multivar. Anal. **79**, 191–218 (2001)
19. Hong, Y., White, H.: Asymptotic distribution theory for nonparametric entropy measures of serial dependence. Econometrica **73**, 837–901 (2005)
20. Isaacson, E., Keller, H.B.: Analysis of Numerical Methods. Wiley, New York (1966)
21. Kendall, M.G.: Oscillatory movements in English agriculture. J. R. Stat. Soc. (N. S.) **106**, 91–124 (1943)
22. Kendall, M.G.: On the analysis of oscillatory time-series. J. R. Stat. Soc. (N. S.) **108**, 93–141 (1945)
23. Koul, H.L.: A weak convergence result useful in robust autoregression. J. Stat. Plann. Inference **29**(3), 291–308 (1991)
24. Kulperger, R., Yu, H.: High moment partial sum processes of residuals in GARCH models and their applications. Ann. Stat. **33**(5), 2395–2422 (2005)
25. Lee, S.: A note on the residual empirical process in autoregressive models. Stat. Probab. Lett. **32**(4), 405–411 (1997)
26. Li, W.K., McLeod, A.I.: ARMA modelling with non-Gaussian innovations. J. Time Ser. Anal. **9**(2), 155–168 (1988)
27. Lilliefors, H.W.: On the Kolmogorov-Smirnov test for normality with mean and variance unknown. J. Am. Stat. Assoc. **62**, 399–402 (1967)
28. Ljung, G.M., Box, G.E.P.: On a measure of lack of fit in time series models. Biometrika **65**, 297–303 (1978)
29. Mandal, B.N.: Forecasting sugarcane production in India with ARIMA model. http://interstat.statjournals.net/YEAR/2005/articles/0510002.pdf (2005)
30. McLeod, A.I., Li, W.K.: Diagnostic checking ARMA time series models using squared-residual autocorrelations. J. Time Ser. Anal. **4**(4), 269–273 (1983)
31. Picard, D.: Testing and estimating change-points in time series. Adv. Appl. Probab. **17**(4), 841–867 (1985)
32. Quenouille, M.H.: A large-sample test for the goodness of fit of autoregressive schemes. J. R. Stat. Soc. (N.S.) **110**, 123–129 (1947)
33. Suresh, K., Krishna Priya, S.: Forecasting sugarcane yield of Tamilnadu using ARIMA models. Sugar Tech **13**, 23–26 (2011). doi:10.1007/s12355-011-0071-7
34. Wold, H.: A large-sample test for moving averages. J. R. Stat. Soc. Ser. B. **11**, 297–305 (1949)
35. Wold, H.: A Study in the Analysis of Stationary Time Series, 2nd edn. Almqvist and Wiksell, Stockholm (1954). With an appendix by Peter Whittle

Short Range and Long Range Dependence

Murray Rosenblatt

1 Introduction

In this section a discussion of the evolution of a notion of strong mixing as a measure of short range dependence and with additional restrictions a sufficient condition for a central limit theorem, is given. In the next section I will give a characterization of strong mixing for stationary Gaussian sequences. In Sect. 3 I will give a discussion of processes subordinated to Gaussian processes and in Sect. 4 results concerning the finite Fourier transform is noted. In Sect. 5 a number of open questions are considered.

In an effort to obtain a central limit theorem for a dependent sequence of random variables in [12], I made use of a blocking argument of S.N. Bernstein [1] and was led to what I called a strong mixing condition [2, 12]. In the blocking argument big blocks are separated by small blocks. Consider a sequence of random variables X_n, $n = \ldots, -1, 0, 1, \ldots$. Let $\mathscr{B}_n$ and $\mathscr{F}_m$ be the σ-fields generated by $X_j, j \leq n$ and X_j, $j \geq m$, respectively. If

$$\sup_{A \in \mathscr{B}_n, B \in \mathscr{F}_m} |P(A \cap B) - P(A)P(B)| \leq \alpha(m - n),$$

$m > n$ with $\alpha(k) \to 0$ as $k \to \infty$, the sequence $\{X_n\}$ is said to satisfy a *strong mixing condition*. Such a sequence needn't be stationary. A sequence with such a strong mixing condition can be thought of as one with short range dependence and its absence an indicator of long range dependence.

M. Rosenblatt (✉)
Department of Mathematics, University of California, San Diego,
La Jolla, CA 92093-0112, USA
e-mail: mrosenblatt@math.ucsd.edu

© Springer Science+Business Media New York 2015
D. Dawson et al. (eds.), *Asymptotic Laws and Methods in Stochastics*,
Fields Institute Communications 76, DOI 10.1007/978-1-4939-3076-0_15

The strong mixing condition together with the following assumptions are enough to obtain asymptotic normality for partial sums of the sequence. Assume that $EX_n = 0$ for all n. The critical additional assumptions are

1.

$$E\left|\sum_{j=a}^{b} X_j\right|^2 \sim h(b-a)$$

as $b - a \to \infty$ with $h(m) \uparrow \infty$ as $m \to \infty$, where $x(\theta) \sim y(\theta)$ means $x(\theta)/y(\theta) \to 1$ as $\theta \to \theta_0$ and

2.

$$E\left|\sum_{j=a}^{b} X_j\right|^{2+\delta} = O\left(h(b-a)\right)^{1+\delta/2}$$

as $b - a \to \infty$ for some $\delta > 0$.

The following theorem was obtained.

Theorem 1. *If* $\{X_n\}$, $E(X_n) = 0$, *is a sequence satisfying a strong mixing condition and assumptions 1. and 2., we can determine numbers* k_n, p_n, q_n *satisfying*

$$k_n(p_n + q_n) = n,$$

$$k_n, p_n, q_n \to \infty,$$

$$q_n/p_n \to 0$$

as $n \to \infty$ *such that*

$$\frac{S_n}{\sqrt{k_n \cdot h(p_n)}} \qquad (S_n = \sum_{j=1}^{n} X_j)$$

is asymptotically normally distributed with positive variance (see [12] and [2]).

In the argument the numbers $k_n\alpha(q_n)$ have to be made very small. An elegant statement of a result can be given in a stationary case (see Bradley [3] for a proof).

Theorem 2. *Let* $\{X_n\}$ *be a strictly stationary sequence with* $E(X_0) = 0$, $EX_0^2 < \infty$ *that is strongly mixing and let* $\sigma_n^2 = ES_n^2 \to \infty$ *as* $n \to \infty$. *The sequence* (S_n^2/σ_n^2) *is uniformly integrable if and only if* S_n/σ_n *is asymptotically normally distributed with mean zero and variance one.*

In the paper [9] Kolmogorov and Rozanov showed that a sufficient condition for a
stationary Gaussian sequence to be strongly mixing is that the spectral distribution
be absolutely continuous with positive continuous spectral density.

One should note that the concept of strong mixing here is more restrictive in the
stationary case than the ergodic theory concept of strong mixing. A very extensive
discussion of our notion of strong mixing as well as that of other related concepts is
given in the excellent three volume work of Richard Bradley [3].

In 1961 corresponding questions were taken up for what is sometimes referred
as narrow band-pass filtering in the engineering literature. These results are strong
enough to imply asymptotic normality for the real and imaginary parts of the
truncated Fourier transform of a continuous time parameter stationary process.
Let $X(t)$, $EX(t) = 0$, be a separable strongly mixing stationary process with
$EX^4(t) < \infty$ that is continuous in mean of fourth order. If the covariance and 4th
order cumulant function are integrable, it than follows that

$$\left(\frac{1}{2}T\right)^{-1/2} \int_0^T \cos(\lambda t) X(t) dt,$$

$$\left(\frac{1}{2}T\right)^{-1/2} \int_0^T \sin(\lambda t) X(t) dt, \qquad \lambda \neq 0,$$

are asymptotically normal with variance

$$\pi f(\lambda)$$

and independent as $T \to \infty$ ($f(\lambda)$ the spectral density of $X(t)$ at λ). This follows
directly from the results given in [14].

2 Gaussian Processes

In the 1961 paper [13] a Gaussian stationary sequence $\{Y_k\}$ with mean zero and
covariance

$$r_k = EY_0 Y_k = (1 + k^2)^{-D/2} \sim k^{-D} \qquad \text{as} \quad k \to \infty,$$

$0 < D < 1/2$, was considered. The normalized partial sums process

$$Z_n = n^{-1+D} \sum_{k=1}^n X_k$$

of the derived quadratic sequence

$$X_k = Y_k^2 - 1$$

was shown to have a limiting non-Gaussian distribution as $n \to \infty$. The characteristic function of the limiting distribution is

$$\phi(\theta) = \exp\left(\frac{1}{2}\sum_{k=2}^{\infty}(2i\theta)^k c_k/k\right)$$

with

$$c_k = \int_0^1 dx_1 \cdots \int_0^1 dx_k |x_1 - x_2|^{-D}|x_2 - x_3|^{-D} \cdots |x_{k-1} - x_k|^{-D}|x_k - x_1|^{-D}.$$

Since conditions 1. and 2. are satisfied by X_k, the fact that the limiting distribution is non-Gaussian implies that $\{X_k\}$ and $\{Y_k\}$ cannot be strongly mixing.

In their paper Helson and Sarason [6] obtained a necessary and sufficient condition for a Gaussian stationary sequence to be strongly mixing. This was that the spectral distribution of the sequence be absolutely continuous with spectral density w

$$w = |P|^2 \exp(u + \tilde{v})$$

with P a trigonometric polynomial and u and v real continuous functions on the unit circle and $\tilde{v}$ the conjugate function of v.

It is of some interest to note that the functions of the form

$$\exp(u + \tilde{v}) = w,$$

with u and w continuous are such that w^n is integrable for every positive or negative integer n. (The set of such functions w is W.) An example with a discontinuity at zero is noted in Ibragimov and Rozanov [7]

$$f(\lambda) = \exp\left\{\sum_{k=1}^{\infty}\frac{\cos(k\lambda)}{k(\ln k + 1)}\right\}.$$

Making use of results on trigonometric series with monotone coefficients, it is clear that

$$\sum_{k=1}^{\infty}\frac{\sin(k\lambda)}{k(\ln k + 1)}$$

is continuous and that

$$\ln f(\lambda) \sim \ln\ln\frac{1}{\lambda}$$

as $\lambda \to 0$. So, $f(\lambda)$ and $1/f(\lambda)$ are both spectral densities of strongly mixing Gaussian stationary sequences. $f(\lambda)$ has a discontinuity at $\lambda = 0$ while $1/f(\lambda)$ is continuous with a zero at $\lambda = 0$. Sarason has also shown in [15, 16] that the functions $\log w$, $w \in W$, have vanishing mean oscillation. Let f be a complex function on $(-\pi, \pi]$ and I an interval with measure $|I|$.

Let

$$f_I = |I|^{-1} \int_I f(x)dx$$

and

$$M_a(f) = \sup_{|I| \le a} |I|^{-1} \int_I |f(x) - f_I|dx.$$

f is said to be of bounded mean oscillation if $M_{2\pi}(f) < \infty$. Let

$$M_0(f) = \lim_{a \to 0} M_a(f).$$

f is said to be of vanishing mean oscillation if f is of bounded mean oscillation and $M_0(f) = 0$.

In the case of a vector valued (d-vector) stationary strong mixing Gaussian sequence there is $d_0 \le d$ such that the spectral density matrix $w(\lambda)$ has rank d_0 for almost all λ. If $d_0 = d$ the sequence is said to have full rank. The case of sequences of rank $d_0 < d$ can be reduced to that of sequences of full rank. A result of Treil and Volberg [20] in the full rank case is noted.

Theorem 3. *Assume that the spectral density w of a stationary Gaussian process is such that $w^{-1} \in L^1$. The process is strongly mixing if and only if*

$$\lim_{|\lambda| \to 1} \sup \left\{ \det(w(\lambda)) \exp\left(-[\log \det w](\lambda)\right) \right\} = 1,$$

where $\det(w(\lambda))$ and $[\log \det w](\lambda)$ are the harmonic extensions of w and $\log \det w$ on the unit circle at the point λ on the unit disc.

The harmonic extension u on the unit disc of a function f on the unit circle is given via the Poisson kernel

$$P_r(\theta) = \frac{1 - r^2}{1 - 2r\cos\theta + r^2} = \mathrm{Re}\left(\frac{1 + re^{i\theta}}{1 - re^{i\theta}}\right), \qquad 0 \le r \le 1,$$

$$u(re^{i\theta}) = \frac{1}{2\pi} \int_{-\pi}^{\pi} P_r(\theta - t)f(e^{it})dt.$$

3   Processes Subordinated to Gaussian Processes

In the paper [18] M. Taqqu considered the weak limit of the stochastic process

$$Z_n(t) = n^{-1+D} \sum_{k=1}^{[nt]} X_k$$

as $n \to \infty$ and noted various properties of the limit process. Here $[s]$ denotes the greatest integer less than or equal to s. M. Taqqu [19] and R. Dobrushin and P. Major [5] discovered about the same time that the simple example of M. Rosenblatt was a special case of an interesting broad class of nonlinear processes subordinated to the Gaussian stationary processes. Consider $\{X_n\}$, $EX_n = 0$, $EX_n^2 = 1$ a stationary Gaussian sequence with covariance

$$r(n) = n^{-\alpha}L(n), \qquad 0 < \alpha < 1,$$

where $L(t)$, $t \in (0, \infty)$ is slowly varying. Let $H(\cdot)$ be a function with

$$EH(X_n) = 0, \qquad EH^2(X_n) = 1.$$

$H_j(\cdot)$ is the jth Hermite polynomial with leading coefficient one. Then $H(\cdot)$ can be expanded in terms of the H_j's

$$H(X_n) = \sum_{j=1}^{\infty} c_j H_j(X_n)$$

with

$$\sum_{j=1}^{\infty} c_j^2 j! < \infty.$$

Assume that $\alpha < 1/k$ with k the smallest index such that $c_k \neq 0$ (H is then said to have rank k). Set

$$A_N = N^{1-k\alpha/2}(L(N))^{k/2}$$

and

$$Y_n^N = A_N^{-1} \sum_{j=N(n-1)}^{Nn-1} H(X_j),$$

$n = \ldots, -1, 0, 1, \ldots$ and $N = 1, 2, \ldots$. Then the finite dimensional distributions of $Y_n^N, n = \ldots, -1, 0, 1, \ldots$ as $N \to \infty$ tend to those of the sequence Y_n^*

$$Y_n^* = d^{-k/2} c_k \int e^{in(x_1 + \cdots + x_k)} \frac{e^{i(x_1 + \cdots + x_k)} - 1}{i(x_1 + \cdots + x_k)} |x_1|^{\frac{\alpha-1}{2}} \cdots |x_k|^{\frac{\alpha-1}{2}} dW(x_1) \cdots dW(x_k)$$

with $W(\cdot)$ the Wiener process on $(-\infty, \infty)$ where in the integration it is understood that the hyper-diagonals $x_i = x_j, i \neq j$ are excluded, and

$$d = \int \exp(ix)|x|^{\alpha-1} dx = 2\Gamma(x) \cos\left(\frac{\alpha\pi}{2}\right).$$

In [4] P. Breuer and P. Major obtained central limit theorems for nonlinear functions of Gaussian stationary fields. As in the discussion of results for noncentral limit theorem we shall consider the case of stationary sequences. Again, let

$$Y_n^N = A_N^{-1} \sum_{j=N(n-1)}^{Nn-1} H(X_j),$$

with X_n a stationary Gaussian sequence $EX_n = 0, EX_n^2 = 1$. $H(\cdot)$ is real-valued with

$$EH(X_n) = 0, \qquad EH^2(X_n) < \infty.$$

Assume that H has rank k and that

$$\sum_n |r(n)|^k < \infty$$

($r(\cdot)$ the covariance function of the X sequence). Let H_l be the lth Hermite polynomial. With $A_N = N^{1/2}$ the limits

$$\lim_{N \to \infty} E\left(Y_0^N(H_l)\right)^2 = \lim_{N \to \infty} A_N^{-2} l! \sum_{-N \leq i,j < 0} r^l(i-j) = \sigma_l^2 l!$$

exist for all $l \geq k$ and

$$\sigma^2 = \sum_{l=k}^{\infty} c_l^2 l! \sigma_l^2 < \infty.$$

The finite dimensional distributions of Y_n^N as $N \to \infty$ tend to the finite dimensional distributions of σZ_n with the Z_n i.i.d. standard normal random variables. T.C. Sun obtained the case of this result for $k = 2$ in [17].

4 Finite Fourier Transform

In 1961 paper [14] I showed that in the case of a separable continuous time parameter process a variety of filters amounting to narrow band-pass filtering, under the assumption of strong mixing, integrability of the covariance function and the 4th order cumulant function, stationarity and positivity of the spectral density imply asymptotic normality. This implies that

$$\int_0^T \cos(\lambda t) X(t) dt, \qquad \int_0^T \sin(\lambda t) X(t) dt$$

are asymptotically normal as $T \to \infty$ for all λ and independent for $\lambda \neq 0$ as $T \to \infty$.

A recent paper of Peligrad and Wu [11] is of considerable interest. They use a stationary ergodic Markov sequence ξ_n on the probability space $(\Omega, \mathscr{F}, P)$ with marginal distribution

$$\pi(A) = P(\xi_0 \in A).$$

Let

$$\mathscr{L}_0^2(\pi) = \left\{ h : \int h^2 d\pi < \infty, \int h d\pi = 0 \right\},$$

$$\mathscr{F}_k = \mathscr{B}\{\xi_j, j \leq k\}, \qquad X_j = h(\xi_j).$$

The condition

$$E(X_0 | \mathscr{F}_{-k}) = 0 \qquad P \text{ almost surely} \tag{1}$$

is of particular interest. They obtain the following theorem among others.

Theorem 4. *If (X_k) is stationary ergodic satisfying (1) then for almost all $\theta \in (0, 2\pi)$*

$$\lim_{n \to \infty} \frac{E|S_n(\theta)|^2}{n} = g(\theta), \qquad S_n(\theta) = \sum_{k=1}^{[n\theta]} X_k$$

with g integrable over $[0, 2\pi]$ and

$$\frac{1}{\sqrt{n}} [\text{Re}(S_n(\theta)), \text{Im}(S_n(\theta))] \Rightarrow [N_1(\theta), N_2(\theta)]$$

under P with $S_n(\theta)$ the Fast Fourier transform computed at θ and $N_1(\theta)$, $N_2(\theta)$ independent identically distributed normal random variables with mean zero and variance $g(\theta)/2$. One can always take X_k as a function of a Markov sequence $\xi_n = (X_k, k \leq n)$.

In a number of examples one considers derived sequences

$$Z_n^N = A_N^{-1} \sum_{j \in B_n^N} \xi_j \qquad N = 1, 2, \ldots,$$

with

$$B_n^N = \{j : nN \le j < (n+1)N\}$$

and A_N a norming constant (which needn't be $\sqrt{N}$). The interest is in convergence of the finite dimensional distributions of the sequence Z_n^N as $N \to \infty$ to finite dimensional distributions of a limit sequence Z_n^*. The object is to determine the appropriate norming constant A_N and the character of the nontrivial limit sequence Z_n^*. One is also led to the following question – for which sequences ξ_n does one have

$$(\xi_{n_1}, \ldots, \xi_{n_k}) \overset{d}{=} (Z_{n_1}^N, \ldots, Z_{n_k}^N)$$

(equality in distribution for all $N = 1, 2, \ldots$ and $n_1, \ldots, n_k$). If this is satisfied with $A_N = N^\alpha$, ξ_n is said to be a *self-similar* sequence with self-similarity parameter α.

In the case of the limit theorems of Taqqu [18, 19], Dobrushin and Major [5] the limit processes are self-similar with self-similarity parameter α.

It's of interest to note that if the covariances

$$r(n) = n^{-\alpha} L(n), \qquad \alpha \in (0, 1)$$

with $L(n)$ slowly varying are monotone

$$f_\alpha(x) = \sum_{n=1}^{\infty} r(n) \cos nx,$$

$$g_\alpha(x) = \sum_{n=1}^{\infty} r(n) \sin nx,$$

converge uniformly outside an arbitrarily small neighbourhood of $x = 0$ and

$$f_\alpha(x) \sim x^{\alpha-1} L(x^{-1}) \Gamma(1 - \alpha) \sin\left(\frac{1}{2}\pi\alpha\right),$$

$$g_\alpha(x) \sim x^{\alpha-1} L(x^{-1}) \Gamma(1 - \alpha) \cos\left(\frac{1}{2}\pi\alpha\right)$$

as $x \to 0+$. The real spectral density of the Gaussian stationary process with covariances $r(n)$ has a singularity at $x = 0$. Given the Hermite polynomial H_k consider the derived process $H_k(X_j)$ (X_j the Gaussian process). The covariance of the derived process

$$EH_k(X_0)H_k(X_j) = k!r(j)^k$$

so its spectral density will have a singularity at zero if and only if $k\alpha < 1$.

A limit theorem of Kesten and Spitzer [8] is of great interest.

$$S_n = X_1 + \cdots + X_n, \qquad n \geq 1,$$

is the simple random walk on the integers ($X_i = \pm 1$ with probability 1/2 and i.i.d.) with random sequence $\xi(x)$, x integer, i.i.d. with the same distribution as the X_i's but independent of them. The asymptotic behaviour of

$$U_n = \sum_{k=1}^{n} \xi(S_k)$$

is considered as $n \to \infty$. U_s is the linearly interpolated process. They show that

$$n^{-3/4}U_{nt}, \qquad t \geq 0, \, n = 1, 2, 3, \ldots$$

converges weakly to

$$\int_{-\infty}^{\infty} L_t(x)dZ(x), \qquad t \geq 0,$$

where $L_t(x)$ is the local time at x of Brownian motion B_t and $Z(x)$ is a Brownian motion with time $-\infty < x < \infty$.

$\xi(S_k)$, $k = 1, 2, \ldots$ can be extended to a two-sided stationary sequence as follows. Introduce $X_0, X_{-1}, X_{-2}, \ldots$ as i.i.d. random variables with the same distribution as the earlier random variables and independent of all the other variables. Let $\eta_0 = \xi(0)$,

$$\eta_i = \begin{cases} \xi\left(\sum_{j=0}^{i-1} X_j\right) & \text{if } i > 0 \\ \xi\left(-\sum_{j=-1}^{i} X_j\right) & \text{if } i < 0 \end{cases}.$$

The sequence η_i is stationary and we obtain an approximation to its spectral density

$$E(\eta_0\eta_i) = \begin{cases} 0 & \text{if } \sum_{j=0}^{i-1} X_j \neq 0, \quad i > 0 \\ E(\xi^2(0))\binom{2m}{m}\frac{1}{2^{2m}} & \text{if } \sum_{j=0}^{i-1} X_j = 0, \quad i = 2m \end{cases}$$

and

$$2^{-2m}\binom{2m}{m} \sim \frac{2^{1/2}}{\sqrt{2\pi}}\frac{1}{\sqrt{m}}$$

as $m \to \infty$. This suggests that the spectral density is of the form

$$\sum_m \frac{1}{\sqrt{m}} \cos mx$$

and this behaves like

$$(2\lambda)^{-1/2} \Gamma(1/2) \sim \frac{\pi}{4}$$

as $\lambda \to 0$.

5 Open Questions

The almost everywhere character (in θ) of the result of Peligrad and Wu indicates that the asymptotics of the finite Fourier transform at points where there is a singularity of the spectral density functions are not dealt with. This would, for example, be the case if we had a Gaussian stationary sequence (X_j) with covariance of the form

$$r(n) = \sum_j \beta_j |n|^{-\alpha_j} \cos n(\lambda - \lambda_j) L_j(n),$$

$\beta_j > 0, 0 < \alpha_j < 1, \lambda_j$ distinct, and wished to compute the finite Fourier transform of $H(X_k)$ at $\lambda = \lambda_j$ with the leading non-zero Fourier-Hermite coefficient k of $H(\cdot)$ such that $k\alpha_j < 1$. As before the $L_j(\cdot)$ are slowly varying. The variance of the finite Fourier transform and its limiting distributions when properly normalized as N tends to infinity are not determined. Of course this is just a particular example of interest under the assumptions made in the theorem of Peligrad and Wu.

The random sequences with covariances almost periodic functions contain a large class of interesting nonstationary processes. The harmonizable processes of this type have all their spectral mass concentrated on at most a countable number of

$$\lambda = \mu + b, \quad b = b_j, j = \ldots, -1, 0, 1, \ldots.$$

It would be of some interest to see whether one could characterize the Gaussian processes of this type which are strongly mixing. Assume that the spectra on the lines of support are absolutely continuous with spectral densities $f_b(u)$. Under rather strong conditions one can estimate the $f_b(\cdot)$ (see [10]). However, there are still many open questions.

Acknowledgements I thank Professor Rafal Kulik for his help in putting this paper into coherent form.

References

1. Bernstein, S.: Sur l'extension du théoreme limite du calcul des probabilités aux sommes de quantités dépendentes. Math. Ann. **97**, 1–59 (1926)
2. Blum, J., Rosenblatt, M.: A class of stationary processes and a central limit theorem. Proc. Natl. Acad. Sci. USA **42**, 412–413 (1956)
3. Bradley, R.: Introduction to Strong Mixing Conditions, vol. 1–3. Kendrick Press, Heber City (2007)
4. Breuer, P., Major, P.: Central limit theorems for non-linear functionals of Gaussian fields. J. Multivar. Anal. **13**, 425–441 (1983)
5. Dobrushin, R.L., Major, P.: Non-central limit theorems for nonlinear functionals of Gaussian fields. Z. Wahrsch. Verw. Gebiete **50**(1), 27–52 (1979)
6. Helson, H., Sarason, D.: Past and future. Math. Scand. **21**, 5–16 (1967)
7. Ibragimov, I., Rozanov, Y.: Gaussian Random Processes. Springer, New York (1978)
8. Kesten, H., Spitzer, F.: A limit theorem related to a new class of self similar processes. Z. Wahrsch. Verw. Gebiete **50**, 5–25 (1979)
9. Kolmogorov, A., Rozanov, Y.: Strong mixing conditions for stationary Gaussian processes. Theor. Probab. Appl. **5**, 204–208 (1960)
10. Lii, K.S., Rosenblatt, M.: Estimation for almost periodic processes. Ann. Stat. **34**, 1115–1130 (2006)
11. Peligrad, M., Wu, W.B.: Central limit theorem for Fourier transforms of stationary processes. Ann. Probab. **38**, 2007–2022 (2010)
12. Rosenblatt, M.: A central limit theorem and a strong mixing condition. Proc. Natl. Acad. Sci. USA **42**, 43–47 (1956)
13. Rosenblatt, M.: Independence and dependence. In: Proceedings of the 4th Berkeley Symposium on Mathematical Statistics and Probability, vol. II, pp. 431–443. University California Press, Berkeley (1961)
14. Rosenblatt, M.: Some comments on narrow band-pass filters. Q. Appl. Math. **18**, 387–393 (1961)
15. Sarason, D.: An addendum to "Past and future". Math. Scand. **30**, 62–64 (1972)
16. Sarason, D.: Functions of vanishing mean oscillations. Trans. Am. Math. Soc. **207**, 30 (1975)
17. Sun, T.C.: Some further results on central limit theorems for non-linear functionals of normal stationary processes. J. Math. Mech. **14**, 71–85 (1965)
18. Taqqu, M.S.: Weak convergence to fractional Brownian motion and to the Rosenblatt process. Z. Wahrsch. Verw. Gebiete **31**, 287–302 (1975)
19. Taqqu, M.S.: Convergence of integrated processes of arbitrary Hermite rank. Z. Wahrsch. Verw. Gebiete **50**, 53–83 (1979)
20. Treil, S., Volberg, A.: Completely regular multivariate stationary processes and Muchenhaupt condition. Pac. J. Math. **190**, 361–381 (1999)

Part VI
Applied Probability and Stochastic Processes

Kernel Method for Stationary Tails: From Discrete to Continuous

Hongshuai Dai, Donald A. Dawson, and Yiqiang Q. Zhao

1 Introduction

The kernel method proposed in this paper is an extension of the classical kernel method first introduced by Knuth [11], and later developed as the kernel method by Banderier et al. [1]. The key idea in the kernel method is very simple: consider a functional equation $K(x, y)F(x, y) = A(x, y)G(x) + B(x, y)$, where $F(x, y)$ and $G(x)$ are unknown functions. Through the kernel function K, we find a branch, say $y = y_0(x)$, such that $K(x, y_0(x)) = 0$. When substituting this branch to the right-hand side of the functional equation, we then obtain $G(x) = -B(x, y_0(x))/A(x, y_0(x))$, and therefore,

$$F(x, y) = \frac{-A(x, y)B(x, y_0(x))/A(x, y_0(x)) + B(x, y)}{K(x, y)},$$

through analytic continuation. Inspired by Fayolle, Iasnogorodski and Malyshev [7], Li and Zhao [13, 14] applied this method to study tail asymptotics of discrete reflected random walks in the quarter plane. The key challenge in the extension is that instead of one unknown function in the right-hand side of the functional

H. Dai
School of Mathematics and Statistics, Carleton University, Ottawa, ON K1S 5B6, Canada

School of Statistics, Shandong University of Finance and Economics, Jinan, 250014, P.R. China
e-mail: mathdsh@gmail.com

D.A. Dawson • Y.Q. Zhao (✉)
School of Mathematics and Statistics, Carleton University, Ottawa, ON K1S 5B6, Canada
e-mail: ddawson@math.carleton.ca; zhao@math.carleton.ca

© Springer Science+Business Media New York 2015
D. Dawson et al. (eds.), *Asymptotic Laws and Methods in Stochastics*,
Fields Institute Communications 76, DOI 10.1007/978-1-4939-3076-0_16

equation (referred to as the fundamental form in the case of random walks in the quarter plane), there are now two unknowns. Specifically, the fundamental form is of the form:

$$h(x, y)\pi(x, y) = h_1(x, y)\pi_1(x) + h_2(x, y)\pi_2(y) + h_0(x, y)\pi_{0,0},$$

where $\pi(x, y)$, $\pi_1(x)$ and $\pi_2(y)$ are unknown generating functions for joint and two boundary probabilities, respectively. Following the spirit in the kernel method, we find a branch $Y = Y_0(x)$ such that $h(x, Y_0(x)) = 0$, which only leads to a relationship between the two unknown boundary generating functions:

$$h_1(x, Y_0(x))\pi_1(x) + h_2(x, Y_0(x))\pi_2(Y_0(x)) + h_0(x, Y_0(x))\pi_{0,0} = 0,$$

instead of a determination of the unknown functions. Without such a determination, the analytic continuation of the branch and the unknown functions, and the interlace of the two unknown functions, allow us to carry out a singularity analysis for π_1 and π_2, which leads to not only a decay rate, but to also exact tail asymptotic properties of the boundary probabilities through a Tauberian-like theorem.

The purpose of this paper is to further extend the kernel method to study continuous random walks. It is well known that there is a close relationship between the discrete random walk and the continuous one. For example, some classical continuous models can be approached in law by discrete random walks, which is a natural motivation for the extension. The direct motivation is the recent work by Dai and Miyazawa [4, 5], in which the authors studied tail asymptotic properties for a semimartingale reflecting Brownian motion by extending the approach used in Miyazawa [16] for the discrete random walk.

Semimartingale reflecting Brownian motions (SRBM) are important models, often playing a fundamental role in both theoretical and applied issues (see for example, Dai and Harrison [3] and Williams [20, 21]). Their stationary behavior, such as properties of stationary distributions when they exist, is important, especially in applications. However, except for a very limited number of special cases, a simple closed expression for the stationary distribution is not available. Therefore, the asymptotic analysis, often used as a tool of approximation, becomes more important besides for its own interest. For example, Miyazawa and Rolski [17] considered asymptotics for a continuous tandem queueing system; Dai and Miyazawa [4] used an inverse-technique to study the tail behavior of the marginal distributions for the two-dimensional SRBM; and Dai and Miyazawa [5] combined an analytic method with geometric properties of the SRBM to study the tail asymptotic properties of a marginal measure, which is closely related to the kernel method surveyed here (also see our final note at the end of this paper).

The main focus of this paper is to provide a survey on how we can extend the kernel method, which is employed for two-dimensional discrete reflected random walks, to study asymptotic properties of stationary measures for continuous random walks. We take the SRBM as a concrete example to detail all key steps in the extension of the kernel method. One can find that the extension proved here is completely in parallel to the method for discrete random walks. In fact, the SRBM case is much simpler than a "typical discrete random walk case" (a non-singular genus one case). Specifically, the analytic continuation of a branch defined by the kernel equation and the meromorphic continuation of the unknown moment generating functions to the whole cut plane become straightforward for the SRBM case as shown later in this paper. Therefore, the interlace between the two unknown functions and the continuous version of the Tauberian-like theorem are among the key challenges, details of which will be provided.

The rest of this paper is organized as follows. In Sect. 2, we provide the model description of the semimartingale reflecting Brownian motion, and discuss some properties of this model. In Sect. 3, properties of the branch points and the two branches of the algebraic function defined by the kernel equation are studied. Section 4 is devoted to asymptotic analysis of the two unknown functions in the kernel method. In Sect. 5, we prove a continuous version of the Tauberian-like theorem. Section 6 is devoted to characterizing the exact tail asymptotic for a boundary measure of the model. A final note is provided to complete the paper.

2 SRBM

We first introduce the general SRBM models. SRBM models arise as an approximation for queueing networks of various kinds (see for example, Williams [20, 21]). The state space for a d-dimensional SRBM $Z = \{Z(t), t \geq 0\}$ is $\mathbb{R}^d_+$. The dynamics of the process consists of a drift vector μ, a non-singular covariance matrix Σ, and a $d \times d$ reflection matrix R that specifies the boundary behavior. In the interior of the orthant, Z is an ordinary Brownian motion with parameters μ and Σ, and Z is pushed in direction R^j, whenever the boundary surface $\{z \in \mathbb{R}^d : z_j = 0\}$ is hit, where R^j is the jth column of R, for $j = 1, \ldots, d$. The precise description of Z is given as follows:

$$Z(t) = X(t) + RY(t), \text{ for } t \geq 0, \tag{1}$$

where X is an unconstrained Brownian motion with drift vector μ, covariance matrix Σ and $Z(0) = X(0) \in \mathbb{R}^d_+$, and Y is a d-dimensional process with components $Y_1, \ldots, Y_d$ such that

(i) Y is continuous and non-decreasing with $Y(0) = 0$;
(ii) Y_j only increases at times t for which $Z_j(t) = 0, j = 1, \ldots, d$;
(iii) $Z(t) \in \mathbb{R}^d_+, t \geq 0$.

In order to study the existence of such a process, we introduce some definitions. We call a $d \times d$ matrix R an $\mathbb{S}$-matrix, if there exists a d-vector $\omega \geq 0$ such that $R\omega \geq 0$, or equivalently, if there exists $\omega > 0$ such that $R\omega > 0$. Furthermore, R is called completely $\mathbb{S}$ if each of its principal sub-matrices is an $\mathbb{S}$-matrix. Taylor and Williams [19] and Reiman and Williams [18] proved that for a given set of data (Σ, μ, R) with Σ being positive definite, there exists an SRBM for each initial distribution of $Z(0)$ if and only if R is completely $\mathbb{S}$. Furthermore, when R is completely $\mathbb{S}$, the SRBM is unique in distribution for each given initial distribution. A necessary condition of the existence of the stationary distribution for Z is

$$R \text{ is non-singular and } R^{-1}\mu < 0. \tag{2}$$

Recall that any matrix A of the form $A = sI - B$ with $s > 0$ and $B \geq 0$, for which $s \geq \rho(B)$, where $\rho(B)$ is the spectral radius of B, is called an $\mathbb{M}$-matrix. For more information, see [2]. Harrison and Williams [10] proved that if R is an $\mathbb{M}$-matrix, the existence and uniqueness of a stationary distribution of Z is equivalent to Eq. (2). They further explained how the $\mathbb{M}$-matrix structure arises naturally in queueing network applications. For a two-dimensional SRBM, Harrison and Hasenbein [9] showed that condition (2) and R being a $\mathbb{P}$-matrix are necessary and sufficient for the existence of a stationary distribution. Here, we call a square matrix M a $\mathbb{P}$-matrix if all of its principal minors are positive.

In this paper, we consider the same model as in Dai and Miyazawa [4]. It is a two-dimensional SRBM Z with data (Σ, μ, R), where $R = (r_{ij})_{2\times2}$ is a $\mathbb{P}-$ matrix, and (R, μ) satisfies the condition (2); namely

$$r_{11} > 0, \ r_{22} > 0, \ \text{and} \ r_{11}r_{22} - r_{12}r_{21} > 0; \tag{3}$$

and

$$r_{22}\mu_1 - r_{12}\mu_2 < 0, \ \text{and} \ r_{11}\mu_2 - r_{21}\mu_1 < 0. \tag{4}$$

Under conditions (3) and (4), the SRBM is well defined and has a unique stationary distribution π. Let $Z = (Z_1, Z_2)$ be a random vector that has the stationary distribution of the SRBM. We also introduce two boundary measures as they did in [4]. Let $\mathbb{E}_\pi(\cdot)$ denote the conditional expectation given that $Z(0)$ follows the stationary distribution π. By Proposition 3 of Dai and Harrison [3], we get that each component of $\mathbb{E}_\pi(Y(1))$ is finite. Therefore, define

$$V_i(A) = \mathbb{E}_\pi\left[\int_0^1 1_{\{Z(u)\in A\}}dY_i(u)\right], \quad i = 1, 2, \tag{5}$$

where $A \subset \mathbb{R}_+^2$ is a Borel set. From (5), one can easily find that V_i defines a finite measure on $\mathbb{R}_+^2$, and has a support on the face $F_i = \{x \in \mathbb{R}_+^2 : x_i = 0\}$. Notice that $V_i(A)$ is the expected fraction of time, during the unit interval, spent in A by the SRBM Z when the "time clock" runs according to the ith reflector Y_i. This is an equivalent quantity to the joint probability vector $\pi_{i,j}$ when the ith component is 0 in the discrete case. Readers may refer to Konstantopoulos, Last and Lin [12] for more interpretations of V_i. Tail probabilities of SRBM models have attracted a lot of interest recently. See, for example, Dupuis and Ramanan [6], Dai and Miyazawa [4] and the references therein. Our focus in this paper is to study the tail behavior of the boundary measures V_i, $i = 1, 2$, in terms of the kernel method. It follows from Dai and Harrison [3], and Harrison and Williams [10] that V_i, $i = 1, 2$, have continuous densities.

There usually exist two types of tail properties, referred to as rough and exact asymptotics. Let $g(x)$ be a positive valued function of $x \in [0, \infty)$. If

$$\alpha = \lim_{x \to \infty} -\frac{1}{x} \log g(x) \tag{6}$$

exists, $g(x)$ is said to have a rough decay rate α. On the other hand, if there exists a function h such that

$$\lim_{x \to \infty} \frac{g(x)}{h(x)} = 1, \tag{7}$$

then $g(x)$ is said to have exact asymptotic $h(x)$. In our case, we are interested in the function defined by the boundary measure V_i.

In order to reach our goal, we use moment generating functions. In the sequel, for $\theta = (\theta_1, \theta_2) \in \mathbb{R}^2$, we define

$$\phi(\theta_1, \theta_2) = \mathbb{E}_\pi e^{\langle \theta, Z \rangle}, \tag{8}$$

$$\phi_1(\theta_2) = \int_{\mathbb{R}^2_+} e^{\theta_2 x_2} V_1(dx) = \mathbb{E}_\pi \int_0^1 e^{\theta_2 Z_2(u)} dY_1(u), \tag{9}$$

and

$$\phi_2(\theta_1) = \int_{\mathbb{R}^2_+} e^{\theta_1 x_1} V_2(dx) = \mathbb{E}_\pi \int_0^1 e^{\theta_1 Z_1(u)} dY_2(u). \tag{10}$$

Functions ϕ and ϕ_i, $i = 1, 2$, are related through the following fundamental form. Let $R = (r_{ij})_{2 \times 2}$ and $\Sigma = (\Sigma_{ij})_{2 \times 2}$. It follows from (2.3) in Dai and Miyazawa [4] that for $\hat{x} = (x, y) \in \mathbb{R}^2$ with $\phi(x, y) < \infty$,

$$\gamma(x, y)\phi(x, y) = \gamma_1(x, y)\phi_1(y) + \gamma_2(x, y)\phi_2(x), \tag{11}$$

where

$$\gamma_1(x, y) = r_{11}x + r_{21}y, \tag{12}$$
$$\gamma_2(x, y) = r_{12}x + r_{22}y, \tag{13}$$

and

$$\gamma(x, y) = -<\hat{x}, \mu> -\frac{1}{2} <\hat{x}, \Sigma\hat{x}>, \tag{14}$$

with $\mu = (\mu_1, \mu_2)$ satisfying (4).

3   Kernel Equation, Branch Points, and Analytic Continuation

In this section, we study the kernel equation:

$$\gamma(x, y) = 0. \tag{15}$$

Specifically, we provide detailed properties on the branch points, and also the function branches defined by the kernel equation. Only elementary mathematics will be involved in obtaining these properties.

We first rewrite the kernel equation in a quadratic form in y with coefficients that are polynomials in x:

$$
\begin{aligned}
\gamma(x, y) &= x\mu_1 + y\mu_2 + \frac{1}{2}\Sigma_{11}x^2 + \Sigma_{12}xy + \frac{1}{2}\Sigma_{22}y^2 \\
&= \frac{1}{2}\Sigma_{22}y^2 + (\mu_2 + \Sigma_{12}x)y + \frac{1}{2}\Sigma_{11}x^2 + x\mu_1 \\
&= ay^2 + b(x)y + c(x) = 0,
\end{aligned} \tag{16}
$$

where

$$a = \frac{1}{2}\Sigma_{22}, \ b(x) = \mu_2 + \Sigma_{12}x \quad \text{and} \quad c(x) = x\mu_1 + \frac{1}{2}\Sigma_{11}x^2.$$

Let

$$D_1(x) = b^2(x) - 4ac(x) \tag{17}$$

be the discriminant of the quadratic form in (16). Therefore, in the complex plane $\mathbb{C}$, for every x, two solutions to (16) are given by

$$Y_{\pm}(x) = \frac{-b(x) \pm \sqrt{b^2(x) - 4ac(x)}}{2a}, \tag{18}$$

unless $D_1(x) = 0$, for which x is called a branch point of Y. We emphasize that in using the kernel method, all functions and variables are usually treated as complex ones.

Symmetrically, when x and y are interchanged, we have

$$\gamma(x, y) = \tilde{a}x^2 + \tilde{b}(y)x + \tilde{c}(y) = 0, \tag{19}$$

where

$$\tilde{a} = \frac{1}{2}\Sigma_{11}, \tilde{b}(y) = \Sigma_{12}y + \mu_1, \quad \text{and} \quad \tilde{c}(y) = \frac{1}{2}\Sigma_{22}y^2 + y\mu_2.$$

Let $D_2(y) = \tilde{b}^2(y) - 4\tilde{a}\tilde{c}(y)$. For each fixed y, two solutions to (19) are given by

$$X_\pm(y) = \frac{-\tilde{b}(y) \pm \sqrt{\tilde{b}^2(y) - 4\tilde{a}\tilde{c}(y)}}{2\tilde{a}}, \tag{20}$$

unless $D_2(y) = 0$, for which y is called a branch point of X.

We have the following properties on the branch points.

Lemma 1. *$D_1(x)$ has two zeros satisfying $x_1 \leq 0 < x_2$ with $x_i, i = 1, 2$ being real numbers. Furthermore, $D_1(x) > 0$ in (x_1, x_2), and $D_1(x) < 0$ in $(-\infty, x_1) \cup (x_2, \infty)$. Similarly, $D_2(y)$ has two zeros satisfying $y_1 \leq 0 < y_2$ with $y_i, i = 1, 2$ being real numbers. Moreover, $D_2(y) > 0$ in (y_1, y_2), and $D_2(y) < \infty$ in $(-\infty, y_1) \cup (y_2, \infty)$.*

Proof. Note that

$$D_1(x) = 4\left[(\Sigma_{12}^2 - \Sigma_{11}\Sigma_{22})x^2 + 2(\Sigma_{12}\mu_2 - \Sigma_{22}\mu_1)x + \mu_2^2\right]. \tag{21}$$

Then it follows from (21) that the discriminant of the quadratic form $D_1(x)$ is given by

$$\Delta = (\mu_2\Sigma_{12} - \Sigma_{22}\mu_1)^2 - (\Sigma_{12}^2 - \Sigma_{11}\Sigma_{22})\mu_2^2. \tag{22}$$

One can verify that $\Delta > 0$. In fact, since $\mu = (\mu_1, \mu_2)$ satisfies conditions (3) and (4), μ_1 or μ_2 is negative. Without loss of generality, we assume that $\mu_2 < 0$. If $\mu_1 > 0$, it follows that since the matrix Σ is positive definite, $\Delta > 0$. If $\mu_1 < 0$, elementary calculations show that $\Delta > 0$. If $\mu_1 = 0$, it is clear that $\Delta > 0$. So, there exist two distinct solutions to $D_1(x) = 0$. We assume $x_1 < x_2$. Now we show that $x_1 \leq 0 < x_2$. If $\mu_1 \neq 0$, then

$$x_1 x_2 = \frac{\mu_2^2}{\Sigma_{12}^2 - \Sigma_{11}\Sigma_{22}} < 0, \tag{23}$$

since $|\Sigma| > 0$, i.e., $\Sigma_{12}^2 - \Sigma_{11}\Sigma_{22} < 0$. Therefore $x_1 < 0 < x_2$. If $\mu_2 = 0$, then we can easily get $x_1 = 0$ and $x_2 > 0$. Since $\Sigma_{12}^2 - \Sigma_{11}\Sigma_{22} < 0$, we get $D_1(x) > 0$ for $x \in (x_1, x_2)$.

Similarly, we can prove the results for $D_2(y)$. $\qquad\square$

It follows from Lemma 1 that $\sqrt{D_1(x)}$ is well defined in $[x_1, x_2]$. Next, we will study the analytic continuation of this function on the cut plane $\mathbb{C} \setminus \{(-\infty, x_1] \cup [x_2, \infty)\}$.

Set $x = u + iv$, where $u, v \in \mathbb{R}$. Then, we can rewrite $D_1(x)$ as:

$$D_1(x) = R(u, v) + I(u, v)i, \tag{24}$$

where

$$R(u, v) = \left(\Sigma_{12}^2 - \Sigma_{11}\Sigma_{22}\right)(u^2 - v^2) + 2\left(\Sigma_{12}\mu_2 - \Sigma_{22}\mu_1\right)u + \mu_2^2,$$

and

$$I(u, v) = 2v\left\{(\Sigma_{12}^2 - \Sigma_{11}\Sigma_{22})u + \Sigma_{12}\mu_2 - \Sigma_{22}\mu_1\right\}.$$

For fixed u and $u \neq \tilde{u} = -\frac{\Sigma_{12}\mu_2 - \Sigma_{22}\mu_1}{\Sigma_{12}^2 - \Sigma_{11}\Sigma_{22}}$, we get from the definition of $I(u, v)$ that

$$I(u, v) = 0 \Longleftrightarrow v = 0. \tag{25}$$

On the other hand,

$$R(u, 0) = D_1(u). \tag{26}$$

It follows from (25), (26) and Lemma 1 that

$$R(u, v) < 0 \Longleftrightarrow u \in (-\infty, x_1) \cup (x_2, \infty). \tag{27}$$

For $u = \tilde{u}$, we have

$$I(u, v) = 0 \Longleftrightarrow v \in \mathbb{R}. \tag{28}$$

Since $x_1 < \tilde{u} < x_2$,

$$D_1(\tilde{u}) = D_1(\tilde{u} + 0i) > 0. \tag{29}$$

Therefore,

$$R(\tilde{u}, 0) = \left(\Sigma_{12}^2 - \Sigma_{11}\Sigma_{22}\right)\tilde{u}^2 + 2\left(\Sigma_{12}\mu_2 - \Sigma_{22}\mu_1\right)\tilde{u} + \mu_2^2 > 0. \tag{30}$$

On the other hand,

$$-\left(\Sigma_{12}^2 - \Sigma_{11}\Sigma_{22}\right)v^2 > 0, \tag{31}$$

since

$$\Sigma_{12}^2 - \Sigma_{11}\Sigma_{12} < 0.$$

It follows from (30) and (31) that

$$R(\tilde{u}, v) = R(\tilde{u}, 0) - \left(\Sigma_{12}^2 - \Sigma_{11}\Sigma_{12}\right)v^2 > 0. \tag{32}$$

Therefore, along the curve

$$C = \{x = u + iv : u = \tilde{u}\},$$

we have

$$R(\tilde{u}, v) = \text{Re}(D_1(x)) > 0.$$

From above arguments, we know that $\sqrt{D_1(x)}$, as the analytic continuation, is analytic in $\mathbb{C} \setminus \{(-\infty, x_1] \cup [x_2, \infty)\}$. For convenience, denote

$$\mathbb{C}_x = \mathbb{C} \setminus \{(-\infty, x_1] \cup [x_2, \infty)\}.$$

$$\mathbb{C}_y = \mathbb{C} \setminus \{(-\infty, y_1] \cup [y_2, \infty)\}.$$

The following lemma is immediate from the above discussion.

Lemma 2. *Both $Y_+(x)$ and $Y_-(x)$ are analytic on $\mathbb{C}_x$. Similarly, both $X_+(y)$ and $X_-(y)$ are analytic on $\mathbb{C}_y$.*

Remark 1. In the SRBM case, the analytic continuation of the lower part of the ellipse $\gamma(x, y) = 0$ coincides with $Y_-(x)$ in the whole cut plane, and the continuation of the upper part with $Y_+(x)$. However, in the discrete case, the analytic continuation, denoted by $Y_0(x)$, of the lower part is not always equal to $Y_-(x)$ or $Y_+(x)$, which is $Y_-(x)$ in some parts of the cut plane and $Y_+(x)$ in other parts. Also, the upper part can only be continued to the whole cut plane meromorphically. One can prove that the meromorphic continuation $Y_1(x)$ has two poles in the cut plane. To be consistent with the discrete case, in the following we use Y_0 and Y_1 instead of Y_- and Y_+. Similarly, we use X_0 and X_1 instead of X_- and X_+.

Based on Lemma 2, we have the analytic continuation of γ_k for $k = 1, 2$.

Lemma 3. *The function $\gamma_2(x, Y_0(x))$ is analytic on $\mathbb{C}_x$. Similarly, the function $\gamma_1(X_0(y), y)$ is analytic on $\mathbb{C}_y$.*

Proof. It follows from the definition of $\gamma_2(\theta)$ that $\gamma_2(x, Y_0(x)) = r_{11}x + r_{21}Y_0(x)$. The analytic continuation is immediate from Lemma 2.   □

Remark 2. Similar results were obtained in [4] based on geometric properties.

4 Interlace Between ϕ_1 and ϕ_2 and Singularity Analysis

For (x, y) satisfying the kernel equation: $\gamma(x, y) = 0$, if $\phi(x, y) < \infty$ then the right-hand side of the fundamental form provides a relationship between the two unknown functions: $\gamma_1(x, y)\phi_1(y) + \gamma_2(x, y)\phi_2(x) = 0$. Through a study of the interlace between the two unknown functions, we will perform a singularity analysis of these functions. For characterizing exact tail asymptotics for the two boundary distributions V_i, $i = 1, 2$, the following are important steps:

(i) analytic continuation of the functions $\phi_1(y)$ and $\phi_2(x)$;
(ii) singularity analysis of the functions $\phi_1(y)$ and $\phi_2(x)$; and
(iii) applications of a Tauberian-like theorem.

The interlace between $\phi_1(y)$ and $\phi_2(x)$ plays a key role in the analysis.

4.1   Analytic Continuation

We first introduce the following notation:

$$\Gamma = \{(x, y) : (x, y) \in \mathbb{R}^2 \text{ such that } \gamma(x, y) < 0\},$$

$$\partial\Gamma = \{(x, y) \in \mathbb{R}^2 : \gamma(x, y) = 0\},$$

$$\Gamma_1 = \{(x, y) : (x, y) \in \mathbb{R}^2, \gamma_1(x, y) \leq 0\},$$

$$\Gamma_2 = \{(x, y) : (x, y) \in \mathbb{R}^2, \gamma_2(x, y) \leq 0\}.$$

We also introduce the following lemma, which is a transformation of Pringsheim's theorem for a generating function (see, for example, Dai and Miyazawa [4] and Markushevich [15]).

Lemma 4. Let $g(\lambda) = \int_0^\infty e^{\lambda x} dF(x)$ be the moment generating function of a probability distribution F on $\mathbb{R}_+$ with real variable λ. Define the convergence parameter of g as

$$C_p(g) = \sup\{\lambda \geq 0 : g(\lambda) < \infty\}. \tag{33}$$

Then, the complex variable function $g(z)$ is analytic on $\{z \in \mathbb{C} : \operatorname{Re}(z) < C_p(g)\}$.

The following lemma is an immediate consequence of the above lemma.

Lemma 5. $\phi_1(z)$ is analytic on $\{z : \operatorname{Re}(z) < \tau_2\}$, and $\phi_2(z)$ is analytic on $\{z : \operatorname{Re}(z) < \tau_1\}$, where $\tau_2 = C_p(\phi_1)$, and $\tau_1 = C_p(\phi_2)$.

The following lemma implies that $\tau_1 > 0$ and $\tau_2 > 0$.

Lemma 6. $\phi_i(z)$, $i = 1, 2$ can be analytically continued up to the region $\{z : \operatorname{Re}(z) < \epsilon\}$ in their respective complex plane, where $\epsilon > 0$.

Proof. In order to simplify the discussion, we let $\mathcal{Q}_i$, $i = 1, 2, 3, 4$ denote the ith quadrant plane. One can easily get that $\gamma(\theta)$ passes through the origin $(0, 0)$. By the proof of Lemma 1, we have that μ_1 or μ_2 is negative. Therefore, without loss of generality, we assume that

$$\Gamma \cap \mathcal{D}_3 \neq \emptyset. \tag{34}$$

Corresponding to (34), without loss of generality, we can further assume that

$$\Gamma \cap \mathcal{D}_2 \neq \emptyset. \tag{35}$$

By (34), for any $(\theta_1, \theta_2) \in \partial\Gamma \cap \mathcal{D}_3$, we have

$$\gamma_1(\theta)\phi_1(\theta_2) + \gamma_2(\theta)\phi_2(\theta_1, \theta_2) = 0. \tag{36}$$

So,

$$\phi_1(\theta_2) = -\frac{\gamma_2(\theta_1, \theta_2)\phi_2(\theta_1)}{\gamma_1(\theta)}. \tag{37}$$

Using $\theta_1 = X_0(\theta_2)$ for $\theta_2 \in [y_2, 0)$ leads to

$$\phi_1(\theta_2) = -\frac{\gamma_2(X_0(\theta_2), \theta_2)\phi_2(X_0(\theta_2))}{\gamma_1(X_0(\theta_2), \theta_2)}. \tag{38}$$

On the other hand, for all $\theta_1 \in [x_1, 0)$ we have

$$\phi_2(\theta_1) < \infty. \tag{39}$$

It follows from (35) that

$$\mathscr{A} = \left\{ \theta_1 : \theta_1 \in (x_1, 0) \text{ such that } Y_0(\theta_1) \leq 0, \text{ and } Y_1(\theta_1) \geq 0 \right\} \neq \emptyset. \tag{40}$$

Let

$$\mathscr{B} = \left\{ \theta_2 : \theta_2 = Y_1(\theta_1) \text{ for any } \theta_1 \in \mathscr{A} \right\}. \tag{41}$$

Then, from (40) and (41) we have that for any $\theta_2 \in \mathscr{B}$,

$$X_0(\theta_2) < 0. \tag{42}$$

By (37), (42) and Lemma 4, we conclude that $\phi_1(\theta_2)$ can be analytically continued to $\{z : \text{Re}(z) < \epsilon\}$ for some $\epsilon > 0$.

Using a similar argument, we can conclude that $\phi_2(\theta_1)$ can be analytically continued to a region $\{z : \text{Re}(z) < \epsilon\}$ with the same $\epsilon > 0$. □

The following property allows us to express $\phi_1(y)$ and $\phi_2(x)$ in terms of each other as a univariate function.

Lemma 7. ϕ_2 *can be analytically continued to the region:* $\{z \in \mathbb{C}_x : \gamma_2(z, Y_0(z)) \neq 0\} \cap \{z \in \mathbb{C}_x : \mathrm{Re}\,(Y_0(z)) < \tau_2\}$, *and*

$$\phi_2(z) = -\frac{\gamma_1(z, Y_0(z))\phi_1(Y_0(z))}{\gamma_2(z, Y_0(z))}. \tag{43}$$

Similarly, ϕ_1 *can be analytically continued to the region:* $\{z \in \mathbb{C}_y : \gamma_1(X_0(z), z) \neq 0\} \cap \{z \in \mathbb{C}_y : \mathrm{Re}\,(X_0(z)) < \tau_1\}$, *and*

$$\phi_1(z) = -\frac{\gamma_2(X_0(z), z)\phi_2(X_0(z))}{\gamma_1(X_0(z), z)}. \tag{44}$$

Proof. Since $\gamma(\theta) = 0$ passes through the origin $(0, 0)$, and $\tau_1 > 0$ and $\tau_2 > 0$, there exists $0 < x_0 < \tau_1$ satisfying the following conditions:

(1) There exists an open neighborhood $U(x_0, \epsilon)$ with $\epsilon > 0$, such that for all $x \in U(x_0, \epsilon)$, $0 < \mathrm{Re}\,(z) < \tau_1$; and
(2) Corresponding to x_0, there exists an open neighborhood $U(Y_0(x_0), \delta)$ such that for all $y \in U(Y_0(x_0), \delta)$, $\mathrm{Re}\,(y) < \tau_2$.

So, we can find a small enough $\delta > \eta > 0$ such that $(x_0, Y_0(x_0) + \eta) \in \Gamma$. On the other hand, $\phi_2(x_0) < \infty$ and $\phi_1(Y_0(x_0) + \eta) < \infty$. So we can get that $\phi(x_0, Y_0(x_0) + \eta) < \infty$. Hence, $\phi(x_0, Y_0(x_0)) < \infty$. By Eq. (11), Eq. (43) holds. Noting that the right-hand side of Eq. (43) is analytic except for the points that $\gamma_2(z, Y_0(z)) = 0$ or $\mathrm{Re}\,(z) \geq \tau_2$, by the uniqueness of analytic continuation, the lemma is now proved. $\square$

Remark 3. Let $\mathbb{D} = \{(x, y) \in \mathbb{R}^2 : \phi(x, y) < \infty\}$. We then have $\phi(z) < \infty$ for any $z \in \mathbb{D} \cap \Gamma$. In fact, Eq. (11) holds as long as $\phi(x, y)$, $\phi_1(y)$ and $\phi_2(x)$ are finite. On the other hand, the function $\gamma(x, y)\phi(x, y)$ is an analytic function of two complex variables x and y for $\mathrm{Re}\,(x) < 0$ and $\mathrm{Re}\,(y) < 0$. This domain can be analytically extended as long as ϕ_1 and ϕ_2 are finite. If both $\phi_1(y)$ and $\phi_2(x)$ are finite, then $\gamma(x, y)\phi(x, y)$ is finite. Moreover, if $\gamma(x, y) \neq 0$, then $\phi(x, y)$ is finite.

4.2 Singularity Analysis

In this subsection, we study properties of singularities of $\phi_2(x)$. We can use the same method to study $\phi_1(y)$, which will not be detailed here. Inspired by Lemmas 4 and 5, in order to determine τ_1, we only need to consider the real number case.

We first introduce a lemma, which will be used later. Let $\tilde{x}$ be the solution of $\tilde{y} = Y_0(x)$ for $\tilde{y} \in (0, y_2]$. Then, we have the following lemma.

Lemma 8. *If* $\tilde{x} = \tau_1$, *then* $\tilde{x} = X_1(\tilde{y})$.

Proof. Since $\tilde{x}$ is a solution $\tilde{y} = Y_0(x)$, $\tilde{x} = \tilde{x}_0 = X_0(\tilde{y})$ or $\tilde{x} = \tilde{x}_1 = X_1(\tilde{y})$. Next, we show that if $\tau_1 = \tilde{x}$, then $\tilde{x} \neq X_0(\tilde{y})$. If $\tilde{x} = X_0(\tilde{y})$, then $\gamma(X_0(\tilde{y}), \tilde{y}) = \gamma_1(X_0(\tilde{y}), \tilde{y}) = 0$.

On the other hand, if $\tilde{x} = \tau_1$, then $\tilde{x} \leq x^*$, which is given in Lemma 11 below. So $\gamma_2(X_0(\tilde{y}), \tilde{y}) \leq 0$. Hence $(X_0(\tilde{y}), \tilde{y}) \in \partial \Gamma_1 \cap \partial \Gamma_2 \cap \Gamma \cap \mathbb{R}^2$. But $\partial \Gamma_1 \cap \partial \Gamma_2 \cap \Gamma \cap \mathbb{R}^2 = \{(0, 0)\}$. It is a contradiction, which proves the lemma.

Lemma 9. *Let* $\tau_1 = \hat{x}$ *be between* 0 *and* x_2. *Then,* $\hat{x}$ *is a zero of* $\gamma_2(x, Y_0(x))$ *or* $Y_0(\hat{x})$ *is a zero of* $\gamma_1(X_0(y), y)$. *Similar results hold for* $\phi_1(y)$.

Proof. If $\tau_1 = \hat{x}$ is not a zero of $\gamma_2(x, Y_0(x))$, then we show that $y^* = Y_0(\hat{x})$ is a zero of $\gamma_1(X_0(y), y)$. It follows from Lemmas 3 and 7 that y^* should be a singular point of $\phi_1(Y_0(x))$.

On the other hand, we have $Y_0(\hat{x}) < y_2$, since $x \in (0, x_2)$. Otherwise, from the definitions of Y_0 and Y_1, we get that $Y_1(\hat{x}) = Y_0(\hat{x})$. Then, $\hat{x} = x_2$, which contradicts the assumption. Similarly, we get that

$$X_0(y^*) < X_1(y^*) < x_2. \tag{45}$$

Since

$$\phi_1(y) = -\frac{\gamma_2(X_0(y), y)\phi_2(X_0(y))}{\gamma_1(X_0(y), y)}, \tag{46}$$

we have

$$\phi_1(Y_0(x)) = -\frac{\gamma_2\big(X_0(Y_0(x)), Y_0(x)\big)\phi_2\big(X_0(Y_0(x))\big)}{\gamma_1\big(X_0(Y_0(x)), Y_0(x)\big)}. \tag{47}$$

By Lemma 8, we have

$$\hat{x} = X_1(y^*). \tag{48}$$

Then, by (45) and (48), we get

$$X_0(y^*) = X_0(Y_0(\hat{x})) < \hat{x}. \tag{49}$$

So, according to (47), (49) and the assumptions, we conclude that y^* is a zero of $\gamma_1(X_0(y), y)$. $\qquad \square$

The following lemma follows directly from Lemmas 7 and 9.

Lemma 10. *The function* $\phi_2(x)$ *is meromorphic on the cut plane* $\mathbb{C}_x$. *Similarly, the function* $\phi_1(y)$ *is meromorphic on the cut plane* $\mathbb{C}_y$.

From Lemma 9, we get that the singular points of $\phi_2(x)$ have a close relationship with the zeros of $\gamma_2(x, Y_0(x))$ and $\gamma_1(X_0(y), y)$. In the sequel, we discuss the zeros of $\gamma_2(x, Y_0(x))$ in detail. Similar results for $\gamma_2(x, Y_0(x))$ can be obtained using a similar argument.

Since $\gamma_2(0,0) = 0$ and $\gamma(0,0) = 0$, $\gamma_2(x, y) = 0$ and $\gamma(x, y) = 0$ must intersect at some point (x_q, y_q) on $\partial\Gamma$ other than $(0,0)$. We claim that $x_q > 0$. In fact, $\gamma_2(x_q, y_q) = 0$ is equivalent to $r_{12}x_q + r_{22}y_q = 0$. This implies that

$$y_q = -\frac{r_{12}}{r_{22}}x_q. \tag{50}$$

On the other hand, since $\gamma(\theta_1, \theta_2) = 0$ for any $\theta = (\theta_1, \theta_2) \in \partial\Gamma$, and Σ is positive definitive, we get that for any $\theta \in \partial\Gamma$ and $\theta \neq 0$,

$$< \theta, \mu > < 0. \tag{51}$$

Combining (50) and (51), we get that $x_q(r_{22}\mu_1 - r_{12}\mu_2) < 0$. It follows from Eq. (4) that $x_q > 0$.

In the next lemma, we will characterize the roots of $\gamma_2(x, Y_0(x)) = 0$.

Lemma 11. x^* is the root of $\gamma_2(z, Y_0(z)) = 0$ in $(0, \quad x_2]$ if and only if $\gamma_2(x_2, Y_0(x_2)) \geq 0$.

Proof. It is obvious that x_2 is a solution of $\gamma_2(z, Y_0(z)) = 0$ if and only if $\gamma_2(x_2, Y_0(x_2)) = 0$.

Next, we assume that $x^* \in (0, \quad x_2)$. We first show that if x^* is a solution of $\gamma_2(x, Y_0(x)) = 0$, then $\gamma_2(x_2, Y_0(x_2)) > 0$. If the statement does not hold, then we have $\gamma_2(x_2, Y_0(x_2)) < 0$. Since $Y_0(x) = Y_1(x)$ at the point x_2, we get

$$\gamma_2(x_2, Y_1(x_2)) < 0. \tag{52}$$

Since, for fixed x, $\gamma_2(x, y)$ is strictly increasing in y, we have

$$\gamma_2(x, Y_0(x)) < \gamma_2(x, Y_1(x)) \tag{53}$$

for $x \in (0, x_2)$.

On the other hand,

$$\gamma_2(x^*, Y_0(x^*)) = 0. \tag{54}$$

By (52), (53) and (54), we know that there exists $\hat{x} \in (x^*, x_2)$ such that

$$\gamma_2(\hat{x}, Y_1(\hat{x})) = 0, \tag{55}$$

which contradicts the fact that the line $\gamma_2(x, y) = 0$ has at least one intersection point with $\gamma(x, y) = 0$ besides the origin $(0, 0)$.

Now we show that if $\gamma_2(x_2, Y_0(x_2)) > 0$, then $\gamma_2(x, Y_0(x)) = 0$ has a root between 0 and x_2. Since $\gamma_2(x, Y_0(x))$ is a continuous function (x_1, x_2), it suffices to show that $\gamma_2(x, Y_0(x)) > 0$ cannot hold for any $x \in (0, \ x_2)$. From the definition of Y_0 and $Y_1(x)$, we get that if $\gamma_2(x, Y_0(x)) > 0$, then $\gamma_2(x, Y_1(x)) > 0$, since $r_{22} > 0$. So, if $\gamma_2(x, Y_0(x)) > 0$ for any $x \in (0, \ x_2)$, then

$$\Sigma_{22}\gamma_2(x, Y_0(x))\gamma_2(x, Y_1(x)) > 0. \tag{56}$$

We can rewrite Eq. (56) as follows:

$$F(x) = x\left((\Sigma_{22}r_{12}^2 - 2r_{11}r_{22}\Sigma_{12} + r_{22}^2\Sigma_{11})x - 2r_{12}r_{22}\mu_2 + 2r_{22}^2\mu_1\right) > 0. \tag{57}$$

On the other hand, it follows from Eq. (4) that

$$2r_{22}^2\mu_1 - 2r_{12}r_{22}\mu_2 < 0.$$

Hence,

$$F'(0) = 2r_{22}^2\mu_1 - 2r_{12}r_{22}\mu_2 < 0. \tag{58}$$

Since $F(0) = 0$, we cannot have $F(x) > 0$ for all $x \in (0, \ x_2)$. From the above arguments, the lemma is proved. $\qquad\square$

Next, we demonstrate how to get the zeros of $\gamma_2(x, Y_0(x))$. For convenience, let $f_0(x) = \gamma_2(x, Y_0(x))$, $f_1(x) = \gamma_2(x, Y_1(x))$ and $f(x) = 2af_0(x)f_1(x)$. Hence, a zero of $f_0(x)$ must be a zero of f. Conversely, any zero of $f(x)$ must be a zero of $f_0(x)$ or $f_1(x)$. From Eq. (16), we have

$$Y_1(x) + Y_0(x) = -2\frac{\mu_2 + \Sigma_{12}x}{\Sigma_{22}}, \tag{59}$$

$$Y_1(x)Y_0(x) = 2\frac{\mu_1 x + \frac{1}{2}x^2\Sigma_{11}}{\Sigma_{22}}. \tag{60}$$

Therefore,

$$\begin{aligned}
f(x) &= 2af_0(x)f_1(x) \\
&= x\left((\Sigma_{22}r_{12}^2 - 2r_{11}r_{22}\Sigma_{12} + r_{22}^2\Sigma_{11})x - 2r_{11}r_{22}\mu_2 + 2r_{22}^2\mu_1\right). \tag{61}
\end{aligned}$$

By Eq. (61), there exist two solutions to $f(x) = 0$, one of which is trivial. We assume that the non-zero solution is x_0, i.e.,

$$x_0 = \frac{2r_{11}r_{22}\mu_2 - 2r_{22}^2\mu_1}{\Sigma_{22}r_{12}^2 - 2r_{11}r_{22}\Sigma_{12} + r_{22}^2\Sigma_{11}}. \tag{62}$$

Remark 4. If x^* is a non-zero solution to $\gamma_2(x, Y_0(x)) = 0$, then $x^* = x_0$.

By the above arguments, we get that the roots of $\gamma_2(x, Y_0(x)) = 0$ are all real. Similarly, $\gamma_1(X_0(y), y) = 0$ has only real roots.

4.3 Asymptotics Behavior of $\phi_2(x)$ and $\phi_1(y)$

In this subsection, we provide asymptotic behavior of the unknown functions $\phi_2(x)$ and $\phi_1(y)$. We only provide details for $\phi_2(x)$, since the behavior for $\phi_1(y)$ can be characterized in the same fashion.

First, we recall the following facts. If $\gamma_2(x, Y_0(x))$ has a zero in $(0, x_2]$, then such a zero is unique, denoted by x^*. If $\gamma_2(x, Y_0(x))$ does not have a zero in $(0, x_2]$, for the convenience of using the minimum function, let $x^* > x_2$ be any number. Similarly, if $\gamma_1(X_0(y), y)$ has a zero in $(0, y_2]$, then such a zero is unique, denoted by y^*. For convenience, if $\gamma_1(X_0(y), y)$ does not have a zero in $(0, y_2]$, we let $y^* > y_2$ be any number. Let $\tilde{x}$ be the solution of $y^* = Y_0(x)$. Then $\tilde{x} = X_0(y^*)$ or $\tilde{x} = X_1(y^*)$. By Lemma 8, for convenience, we let $\tilde{x} > x_2$ be any number, if $\tilde{x} = X_0(y^*)$.

Remark 5. From the above discussion, we know that $\tau_1 \in \{x^*, \tilde{x}, x_2\}$. This same result was obtained by Dai and Miyazawa [4] using a different method by the following four steps: (1) defining $\tau = (\tau_1, \tau_2)$; (2) providing a fixed point equation based on the convergence domain; (3) proving the existence of the solution to the fixed point equation; (4) showing $\tau = (\tau_1, \tau_2)$ is the solution. Our method is in parallel to the kernel method for discrete random walks in Li and Zhao [13, 14].

To state the main theorem, we introduce the following notations for the convenience of expressing the coefficients involved.

(i)

$$
A_1(\tau_1) = \begin{cases}
\dfrac{\Sigma_{22}\gamma_1\big(x^*, Y_0(x^*)\big)\phi_1\big(Y_0(x^*)\big)\gamma_2\big(x^*, Y_1(x^*)\big)}{f^*(x^*)}, & \text{if } \tau_1 = x^* < \min\{\tilde{x}, x_2\}; \\[3ex]
\dfrac{\gamma_1\big(\tilde{x}, Y_0(\tilde{x})\big)\tilde{L}\big(Y_0(\tilde{x})\big)}{\gamma_2\big(\tilde{x}, Y_0(\tilde{x})\big)Y_0'(\tilde{x})}, & \text{if } \tau_1 = \tilde{x} < \min\{x^*, x_2\}; \\[3ex]
\dfrac{\gamma_1\big(x^*, Y_0(x^*)\big)\tilde{L}(Y_0(x^*))\Sigma_{22}}{r_{22}(\Sigma_{11}\Sigma_{22} - \Sigma_{12}^2)(x^* - x_1)}, & \text{if } \tau_1 = x^* = \tilde{x} = x_2,
\end{cases}
$$

where $\tilde{L}(y)$ is given by

$$
\tilde{L}(y) = \frac{\Sigma_{11}\gamma_2(X_0(y), y)\phi_2(X_0(y))\gamma_1(X_1(y), y)}{y\big(r_{11}^2\Sigma_{22} - 2\Sigma_{12}r_{12} + r_{21}^2\big)}, \tag{63}
$$

and $f^*(x)$ is given by

$$f^*(x) = x\left(\Sigma_{22}r_{12}^2 - 2\Sigma_{12}r_{21} + r_{21}^2\right).$$ (64)

(ii)

$$A_2(\tau_1) = \begin{cases} \gamma_1\left(x^*, Y_0(x^*)\right)\dfrac{\Sigma_{22}}{r_{22}\sqrt{\Sigma_{11}\Sigma_{22}-\Sigma_{12}^2(x_2-x_1)}}, & \text{if } \tau_1 = x^* = x_2 < \tilde{x}; \\[4mm] \dfrac{\Sigma_{22}\tilde{L}\left(Y_0(x_2)\right)}{\sqrt{\Sigma_{11}\Sigma_{22}-\Sigma_{12}^2(x_2-x_1)}}\dfrac{\gamma_1\left(x_2, Y_0(x_2)\right)}{\gamma_2\left(x_2, Y_0(x_2)\right)}, & \text{if } \tau_1 = \tilde{x} = x_2 < x^*. \end{cases}$$

(iii)

$$A_3(\tau_1) = \left.\frac{\partial T}{\partial y}\right|_{(x_2,Y_0(x_2))} \tilde{K}(x_2)$$

where $T(x, y)$ is given by

$$T(x, y) = -\frac{\gamma_1(x, y)\phi_1(y)}{\gamma_2(x, y)},$$ (65)

and $\tilde{K}(x_2)$ is given by

$$\tilde{K}(x_2) = \frac{-\sqrt{(\Sigma_{11}\Sigma_{22} - \Sigma_{12}^2)}}{\Sigma_{22}}\sqrt{x_2 - x_1}.$$ (66)

(iv)

$$A_4(\tau_1) = \frac{\gamma_1\left(x^*, Y_0(x^*)\right)\gamma_2\left(X_0(Y_0(x^*)), Y_0(x^*)\right)\phi_2\left(X_0(Y_0(x^*))\right)}{\gamma_2\left(x^*, Y_0(x^*)\right)\gamma_1'\left(X_0(Y_0(x^*)), Y_0(x^*)\right)Y_0'(x^*)}.$$

Theorem 1. *For the function $\phi_2(x)$, a total of four types of asymptotics exist as x approaches to τ_1, based on the detailed properties of τ_1.*

Case 1: If $\tau_1 = x^* < \min\{\tilde{x}, x_2\}$, or $\tau_1 = \tilde{x} < \min\{x^*, x_2\}$, or $\tau_1 = \tilde{x} = x^* = x_2$, then

$$\lim_{x \to \tau_1} (\tau_1 - x)\phi_2(x) = A_1(\tau_1).$$ (67)

Case 2: If $\tau_1 = x^* = x_2 < \tilde{x}$, or $\tau_1 = \tilde{x} = x_2 < x^*$, then

$$\lim_{x \to \tau_1} \sqrt{\tau_1 - x}\phi_2(x) = A_2(\tau_1).$$ (68)

Case 3: If $\tau_1 = x_2 < \min\{\tilde{x}, x^*\}$, *then*

$$\lim_{x \to \tau_1} \sqrt{\tau_1 - x} \phi_2'(x) = A_3(\tau_1).$$ (69)

Case 4: If $\tau_1 = x^* = \tilde{x} < x_2$, *then*

$$\lim_{x \to \tau_1} (\tau_1 - x)^2 \phi_2(x) = A_4(\tau_1).$$ (70)

Proof. We first consider the case that $x^* < \min\{\tilde{x}, x_2\}$. It is obvious that $x^* = x_0$. In such a case, $Y_0(x^*)$ is not a pole of $\phi_1(z)$, and $Y_0(x^*)$ is a simple pole of $\phi_2(z)$. From Eq. (43), we get

$$
\begin{aligned}
\phi_2(x) &= -\frac{\gamma_1(x, Y_0(x))\phi_1(Y_0(x))}{\gamma_2(x, Y_0(x))} \\
&= -\frac{\Sigma_{22}\gamma_1(x, Y_0(x))\phi_1(Y_0(x))\gamma_2(x, Y_1(x))}{2af(x)} \\
&= -\frac{\Sigma_{22}\gamma_1(x, Y_0(x))\phi_1(Y_0(x))\gamma_2(x, Y_1(x))}{(x - x^*)f^*(x)},
\end{aligned}
$$ (71)

where $f^*(x)$ is given by Eq. (64).

Therefore,

$$\lim_{x \to x^*} (x^* - x)\phi_2(x) = \frac{\Sigma_{22}\gamma_1(x^*, Y_0(x^*))\phi_1(Y_0(x^*))\gamma_2(x^*, Y_1(x^*))}{f^*(x^*)}.$$ (72)

Next, we consider the case that $\tilde{x} < \min\{x^*, x_2\}$. In such a case, $\tilde{y} = Y_0(x^*)$ is a zero of $\gamma_1(X_0(y), y)$. Then, using the same argument as the above case, we get that

$$\lim_{y \to \tilde{y}} (\tilde{y} - y)\phi_1(y) = \tilde{L}(\tilde{y}),$$ (73)

where $\tilde{L}(y)$ is given by (63). Hence,

$$
\begin{aligned}
\lim_{x \to \tilde{x}} (\tilde{x} - x)\phi_2(x) &= \lim_{x \to \tilde{x}} (\tilde{x} - x)\frac{\gamma_1(x, Y_0(x))\phi_1(Y_0(x))}{\gamma_2(x, Y_0(x))} \\
&= \lim_{x \to \tilde{x}} (\tilde{x} - x)\frac{\gamma_1(x, Y_0(x))(Y_0(\tilde{x}) - Y_0(x))\phi_1(Y_0(x))}{\gamma_2(x, Y_0(x))(Y_0(\tilde{x}) - Y_0(x))}.
\end{aligned}
$$ (74)

On the other hand, we can rewrite $Y_0(x)$ as follows.

$$
\begin{aligned}
Y_0(x) &= \frac{-(\mu_2 + \Sigma_{12}x) - \sqrt{(\Sigma_{11}\Sigma_{22} - \Sigma_{12}^2)(x - x_1)(x_2 - x)}}{\Sigma_{22}} \\
&= q(x) + p(x),
\end{aligned}
$$ (75)

where

$$p(x) = -\frac{\mu_2 + \Sigma_{12}x}{\Sigma_{22}},$$

and

$$q(x) = -\frac{\sqrt{(\Sigma_{11}\Sigma_{22} - \Sigma_{12}^2)(x - x_1)(x_2 - x)}}{\Sigma_{22}}.$$

So, we easily get that

$$\lim_{x \to \tilde{x}} \frac{p(x) - p(\tilde{x})}{x - \tilde{x}} = -\frac{\Sigma_{12}}{\Sigma_{22}} \neq 0, \qquad (76)$$

and

$$\lim_{x \to \tilde{x}} \frac{q(x) - q(\tilde{x})}{x - \tilde{x}} = q'(x). \qquad (77)$$

By Eqs. (75), (76) and (77),

$$\lim_{x \to \tilde{x}} \frac{Y_0(x) - Y_0(\tilde{x})}{x - \tilde{x}} = q'(\tilde{x}) + p'(\tilde{x}). \qquad (78)$$

Since $\tilde{x} < x_2$, one can get $Y'(\tilde{x}) \neq 0$. Noting that $\gamma_2(\tilde{x}, Y_0(\tilde{x})) \neq 0$, we get from Eqs. (74) and (78) that

$$\lim_{x \to \tilde{x}}(\tilde{x} - x)\phi_2(x) = \frac{\gamma_1(\tilde{x}, Y_0(\tilde{x}))\tilde{L}(Y_0(\tilde{x}))}{\gamma_2(\tilde{x}, Y_0(\tilde{x}))Y_0'(\tilde{x})}. \qquad (79)$$

Now, we consider the case that $x^* = \tilde{x} = x_2$. Since Γ is a convex set, and the curve of γ_1 is above the curve of γ_2, we can easily get that $Y_0(x^*) < y_2$. In such a case, we first have that $\lim_{x \to x^*}(Y_0(x^*) - Y_0(x))\phi_1(Y_0(x)) = \tilde{L}(Y_0(x^*))$. Hence,

$$(x^* - x)\phi_2(x) = -(x^* - x)\frac{\gamma_1(x, Y_0(x))\phi_1(Y_0(x))}{\gamma_2(x, Y_0(x))}$$

$$= \frac{\gamma_1(x, Y_0(x))(Y_0(x^*) - Y_0(x))\phi_1(Y_0(x))}{-\gamma_2(x, Y_0(x)) \setminus \sqrt{x^* - x}} \frac{\sqrt{x^* - x}}{Y_0(x^*) - Y_0(x)}. \qquad (80)$$

By Eq. (75),

$$\lim_{x \to x^*} \frac{Y_0(x^*) - Y_0(x)}{\sqrt{(x^* - x)}} = \frac{\sqrt{(\Sigma_{11}\Sigma_{22} - \Sigma_{22}^2)(x_2 - x_1)}}{\Sigma_{22}}, \qquad (81)$$

and

$$\lim_{x \to x^*} \frac{\gamma_2(x, Y_0(x))}{\sqrt{x^* - x}} = \lim_{x \to x^*} \frac{\gamma_2(x^*, Y_0(x^*)) - \gamma_2(x, Y_0(x))}{\sqrt{x^* - x}}$$

$$= \frac{-r_{22}\sqrt{(\Sigma_{11}\Sigma_{22} - \Sigma_{12}^2)}\sqrt{x_2 - x_1}}{\Sigma_{22}}. \tag{82}$$

So, by Eqs. (80), (81) and (82), we get

$$\lim_{x \to x^*} (x^* - x)\phi_2(x) = \frac{\gamma_1(x^*, Y_0(x^*))\tilde{L}(Y_0(x^*))\Sigma_{22}}{r_{22}(\Sigma_{11}\Sigma_{22} - \Sigma_{12}^2)(x^* - x_1)}. \tag{83}$$

We then consider the case that $x^* = x_2 < \tilde{x}$. In such a case, $\phi_1(Y_0(x))$ is analytic at $y^* = Y_0(x^*)$ and $\gamma_2(x^*, Y_0(x^*)) = 0$. Therefore

$$\lim_{x \to x^*} \sqrt{x^* - x}\phi_2(x) = -\lim_{x \to x^*} \gamma_1(x, Y_0(x))\phi_1(Y_0(x))\frac{\sqrt{x^* - x}}{\gamma_2(x, Y_0(x)) - \gamma_2(x^*, Y_0(x^*))}$$

$$= \gamma_1(x^*, Y_0(x^*))\phi_1(Y_0(x^*))\frac{\Sigma_{22}}{r_{22}\sqrt{\Sigma_{11}\Sigma_{22} - \Sigma_{12}^2}(x_2 - x_1)}. \tag{84}$$

The next case is that $\tilde{x} = x_2 < x^*$. In such a case, $\gamma_2(x^*, Y_0(x^*)) \neq 0$ and $\tilde{y} = Y_0(\tilde{x})$ is a pole of $\phi_1(y)$. Then, we have

$$\lim_{x \to x_2} \sqrt{x_2 - x}\phi_2(x) = \lim_{x \to x_2} \frac{\sqrt{x_2 - x}}{Y_0(x_2) - Y_0(x)}(Y_0(x_2) - Y_0(x))\phi_1(Y_0(x))\frac{\gamma_1(x, Y_0(x))}{\gamma_2(x, Y_0(x))}$$

$$= \frac{\Sigma_{22}\tilde{L}(Y_0(x_2))}{\sqrt{\Sigma_{11}\Sigma_{22} - \Sigma_{12}^2}(x_2 - x_1)}\frac{\gamma_1(x_2, Y_0(x_2))}{\gamma_2(x_2, Y_0(x_2))}. \tag{85}$$

The second last case is $x_2 < \min\{\tilde{x}, x^*\}$. In such a case, we can see that $\tau_2 = y_2$. In fact, if $\tau_2 < y_2$, then, from Remark 5, we get that $\tau_2 = \tilde{y}$ or $\tau_2 = y^*$. If $\tau_2 = \tilde{y}$, then $x_1(\tilde{y})$ is a zero of $\gamma_2(x, Y_0(x))$. But, since $\tau_1 = x_2$, $\tau_1(x, Y_0(x)) \neq 0$ for $x \in (0, x_2)$. So, $\tau_2 \neq \tilde{y}$. If $\tau_2 = y^*$, then $X_1(y^*) = x_2 = \tilde{x}$, which contradicts the assumption of this case. So $\tau_2 = y_2$. Since $x_2 < x^*$, $\phi_1(y)$ is continuous at $y = Y_0(x_2)$. Finally, we have $\gamma_2(x_2, Y_0(x_2)) \neq 0$.

Then, we have

$$\frac{\partial\phi_2(x)}{\partial x} = \frac{\partial T}{\partial x} + \frac{\partial T}{\partial y}\frac{\partial Y_0(x)}{\partial x}, \tag{86}$$

where $T(x, y)$ is given by (65) with

$$\frac{\partial T}{\partial x} = \frac{r_{12}T(x, y) - r_{11}\phi_1(y)}{-\gamma_2(x, y)}, \tag{87}$$

and

$$\frac{\partial T}{\partial y} = \frac{-r_{21}\phi_1(y) - \gamma_1(x, y)\phi_1'(y) + r_{22}T(x, y)}{\gamma_2(x, y)}. \tag{88}$$

From the above argument,

$$\lim_{x \to x_2} \sqrt{x_2 - x}\frac{\partial T}{\partial x} = 0, \tag{89}$$

and

$$\lim_{x \to x_2} \sqrt{x_2 - x}\frac{dY_0(x)}{dx} = \tilde{K}(x_2) \neq 0, \tag{90}$$

where $\tilde{K}(x_2)$ is given by (66).
Combining (89) and (90) leads to

$$\lim_{x \to x_2} (x_2 - x)\phi_2'(x) = \frac{\partial T}{\partial y}\bigg|_{(x_2, Y_0(x_2))} \tilde{K}(x_2). \tag{91}$$

Now, we consider the final case that $x^* = \tilde{x} < x_2$. By (43) and (44),

$$\phi_2(x) = -\frac{\gamma_1(x, Y_0(x))\phi_1(Y_0(x))}{\gamma_2(x, Y_0(x))}$$

$$= -\frac{\gamma_1(x, Y_0(x))\gamma_2(X_0(Y_0(x)), Y_0(x))\phi_2(X_0(Y_0(x)))}{\gamma_2(x, Y_0(x))\gamma_1(X_0(Y_0(x)), Y_0(x))}. \tag{92}$$

On the other hand, from Lemma 8, we have

$$\gamma_1(\tilde{x}, Y_0(\tilde{x}))\gamma_2(X_0(Y_0(x^*)), Y_0(x^*)) \neq 0.$$

We also have

$$\lim_{x \to x^*} \frac{\gamma_1(X_0(Y_0(x)), Y_0(x))}{x^* - x} = \lim_{x \to x^*} \frac{\gamma_1(X_0(Y_0(x)), Y_0(x)) - \gamma_1(X_0(Y_0(x^*)), Y_0(x^*))}{x^* - x}$$

$$= \gamma_1'(X_0(Y_0(x^*)), Y_0(x^*))Y_0'(x^*). \tag{93}$$

Combining Eqs. (76) and (93), we get

$$\lim_{x \to x^*} \frac{(x - x^*)^2}{\gamma_2\big(x, Y_0(x)\big)\gamma_1\Big(X_0\big(Y_0(x)\big), Y_0(x)\Big)}$$

$$= \frac{1}{\gamma_2'\big(x^*, Y_0(x^*)\big)\gamma_1'\Big(X_0\big(Y_0(x^*)\big), Y_0(x^*)\Big)Y_0'(x^*)}. \qquad (94)$$

Finally, it follows from (92) to (94) that

$$\lim_{x \to x^*} (x^* - x)^2 \phi_2(x) = \frac{\gamma_1\big(x^*, Y_0(x^*)\big)\gamma_2\big(X_0(Y_0(x^*)), Y_0(x^*)\big)\phi_2\big(X_0(Y_0(x^*))\big)}{\gamma_2\big(x^*, Y_0(x^*)\big)\gamma_1'\big(X_0(Y_0(x^*)), Y_0(x^*)\big)Y_0'(x^*)}. \qquad (95)$$

It follows from Eq. (78) that $Y_0'(x^*) \neq 0$. By (14) and (13), we have

$$\gamma_2'\big(x^*, Y_0(x^*)\big)\gamma_1'\big(X_0(Y_0(x^*)), Y_0(x^*)\big) \neq 0. \qquad \square$$

5 Tauberian-Like Theorem

Similar to the discrete case, in order to get exact tail asymptotic properties for the boundary measures, we need a technical tool, which is a counterpart to the Tauberian-like theorem used for the discrete reflected random walks. Now, we provide a Tauberian-like theorem for moment generating functions.

We first introduce some notations. Let $g(s)$ be the $\mathbb{L}$-transformation of $f(s)$, i.e.,

$$g(s) = \int_0^\infty e^{st} f(t) dt.$$

Then, $g(s)$ is analytic on the left half-plane. The singularities of $g(s)$ are all in the right half-plane. We now extend the Tauberian-like theorem for a generating function (e.g., Corollary 5.1 in Flajolet and Sedgewick [8]) to that for the continuous case. Denote

$$\Delta(z_0, \epsilon) = \{z \in \mathbb{C} : z \neq z_0, \ |\arg(z - z_0)| > \epsilon\},$$

where $\arg(z) \in (-\pi, \pi]$ is the principal part of the argument of a complex number z.

Theorem 2. *Assume that $g(z)$ satisfies the following conditions:*

(1) The left-most singularity of $g(z)$ is α_0 with $\alpha_0 > 0$. Furthermore, we assume that as $z \to \alpha_0$,

$$g(z) \sim (\alpha_0 - z)^{-\lambda}$$

for some $\lambda \in \mathbb{C} \setminus \mathbb{Z}_{\leq 0}$;

(2) $g(z)$ is analytic on $\Delta(\alpha_0, \epsilon_0)$ for some $\epsilon_0 \in (0, \frac{\pi}{2}]$;

(3) $g(z)$ is bounded on $\Delta(\alpha_0, \epsilon_1)$ for some $\epsilon_1 > 0$.

Then, as $t \to \infty$,

$$f(t) \sim e^{-\alpha_0 t} \frac{t^{\lambda-1}}{\Gamma(\lambda)}, \tag{96}$$

where $\Gamma(\cdot)$ is the Gamma function.

Proof. It follows from the inverse Laplace-transform that

$$f(t) = \frac{1}{2\pi i} \lim_{w \to \infty} \int_{x-iw}^{x+iw} e^{-st} g(s) ds, \tag{97}$$

where x is a constant on the left of α_0. Next, we show that the straight path $[x - iw, \ x + iw]$ of integration can be replaced with the path γ, where γ consists of a circular arc which encircles α_0 on the left, and two beams which are bent with the angle $\pm\theta$ ($\theta > \epsilon_0$) against the positive x-axis. This is valid since we can connect the straight path and the path γ by two large circular arcs γ_1 and γ_2 with radius R above and below, respectively (see Fig. 1 for a picture). All of these paths reside in an area in which, from condition (2), the function $g(z)$ is analytic. Therefore,

$$\int_{x-iw}^{x+iw} e^{-st} g(s) ds + \int_\gamma e^{-st} g(s) ds + \int_{\gamma_1} e^{-st} g(s) ds + \int_{\gamma_2} e^{-st} g(s) ds = 0. \tag{98}$$

Fig. 1 The paths
$[x - iw, \ x + iw]$, γ_1, γ_2 and γ

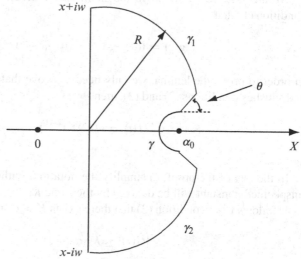

So, we only need to show that for $i = 1, 2$,

$$\lim_{R \to \infty} \int_{\gamma_i} e^{-st} g(s) ds = 0, \tag{99}$$

which can be easily done.

We next assume that $G(z) = (\alpha_0 - z)^{-\lambda}$. We study the asymptotic behavior of

$$\hat{f}(t) = \lim_{R \to \infty} \frac{-1}{2\pi i} \int_{\gamma} e^{-st} G(s) ds. \tag{100}$$

We have

$$\int_{\gamma} e^{-st} G(s) ds = \int_{\gamma} e^{-st} (\alpha_0 - s)^{-\lambda} ds$$

$$= e^{-\alpha_0 t} \int_{\gamma} e^{(\alpha_0 - s)t} (\alpha_0 - s)^{-\lambda} ds$$

$$= e^{-\alpha_0 t} t^{\lambda - 1} \int_{\tilde{\gamma}} e^{-u} (-u)^{-\lambda} du, \tag{101}$$

where $u = (s - \alpha_0)t$ and $\tilde{\gamma}$ is the new curve transformed from γ by μ. It follows from the Hankel's contour integral and (101) that as $R \to \infty$,

$$\hat{f}(t) = \frac{e^{-\alpha_0 t}}{\Gamma(\lambda)} t^{\lambda - 1}. \tag{102}$$

We are now ready to prove the main result of this lemma. It follows from condition (1) that

$$g(z) = (\alpha_0 - z)^{-\lambda} + o\big((\alpha_0 - z)^{-\lambda}\big). \tag{103}$$

In order to prove the lemma, we only need to prove that if $G(z) = o\big((\alpha_0 - z)^{-\lambda}\big)$ and satisfies conditions (2) and (3), then

$$\hat{f}(t) = o\big(e^{-\alpha_0 t} t^{\lambda - 1}\big), \tag{104}$$

as $t \to \infty$.

In the rest of the proof, to simplify the notation, without loss of generality, any unspecified constant will be denoted by the same K.

It follows from condition (3) that there exists $K > 0$ such that

$$|G(z)| \leq K |(\alpha_0 - z)^{-\lambda}| \tag{105}$$

in the whole region Δ. Since $G(z) = o\big((\alpha_0 - z)^{-\lambda}\big)$, there exists $\delta(\epsilon) > 0$ such that for $z \in \Delta$,

$$|\alpha_0 - z| < \delta(\epsilon) \Rightarrow |G(z)| < K\epsilon|(\alpha_0 - z)^{-\lambda}|. \tag{106}$$

In order to prove (104), we need to prove that for some large enough $T(\epsilon) > 0$, we have

$$|\hat{f}(t)| < K\epsilon e^{-\alpha_0 t} t^{\lambda-1}, \quad \text{for } t > T(\epsilon). \tag{107}$$

To prove (107), we choose the contour $\mathscr{D} = \mathscr{D}_1 + \mathscr{D}_2 + \mathscr{D}_3$ as follows:

$$\mathscr{D}_1 = \Big\{z \big| \; |z - \alpha_0| = \frac{1}{t}, \; |\arg(z - \alpha_0)| \geq \epsilon_0 \Big\};$$

$$\mathscr{D}_2 = \Big\{z \big| \frac{1}{t} \leq |z - \alpha_0|, \; |z| \leq \alpha_0 + R, \; |\arg(z - \alpha_0)| = \epsilon_0 \Big\};$$

and

$$\mathscr{D}_3 = \Big\{z \big| \frac{1}{t} \leq |z - \alpha_0|, \; |z| \leq \alpha_0 + R, \; |\arg(z - \alpha_0)| = -\epsilon_0 \Big\}.$$

We proceed to evaluate the contributions to $\hat{f}(t)$ due to each of $\mathscr{D}_i$ separately. For this purpose, we define for $k = 1, 2, 3$,

$$F_k(t) = \frac{1}{2\pi i} \int_{\mathscr{D}_k} e^{-st} G(s) ds.$$

For $k = 1$,

$$
\begin{aligned}
F_1(t) &= \frac{1}{2\pi i} \int_{\mathscr{D}_1} e^{-st} G(s) ds \\
&= e^{-\alpha_0 t} \frac{1}{2\pi i} \int_{\mathscr{D}_1} e^{(\alpha_0 - s)t} G(s) ds \\
&= e^{-\alpha_0 t} \frac{1}{2\pi i} \int_{\tilde{\mathscr{D}}_1} e^{-ut} G(u + \alpha_0) du, \tag{108}
\end{aligned}
$$

where $u = s - \alpha_0$. We can choose $T(\epsilon) > \frac{1}{\delta(\epsilon)}$ such that from (106) we have

$$
\begin{aligned}
|F_1(t)| &= \Big| \frac{1}{2\pi i} e^{-\alpha_0 t} \int_{\tilde{\mathscr{D}}} e^{-ut} G(u + \alpha_0) du \Big| \\
&\leq \frac{1}{2\pi} e^{-\alpha_0 t} t^{\lambda} \epsilon \int_{-\pi}^{\pi} \big| e^{-u(\theta)t} \big| |u'(\theta)| d\theta \\
&\leq K\epsilon e^{-\alpha_0 t} t^{\lambda-1}, \tag{109}
\end{aligned}
$$

where $u(\theta) = \frac{1}{t}(\cos \theta + i \sin \theta)$.

The cases for $k = 2$ and $k = 3$ are similar, so we only provide details for the case of $k = 2$. In this case, let $s = \alpha_0 + \frac{\omega x}{t}$ with $\omega = e^{i\epsilon_0}$. Then, we have

$$\int_{\mathscr{D}_2} e^{-st} G(s) ds = e^{-\alpha_0 t} \int_1^{C(R,t)} e^{\omega s} G(\alpha_0 + \frac{\omega s}{t}) ds, \qquad (110)$$

where $C(R, t)$ is a constant such that

$$\left| \alpha_0 + \frac{\omega C(R, t)}{t} \right| = \alpha_0 + R.$$

Now we decompose the integral (110) as follows:

$$e^{-\alpha_0 t} \int_1^{C(R,t)} e^{\omega s} G(\alpha_0 + \frac{\omega s}{t}) ds = e^{-\alpha_0 t} \int_1^{\log^2 t} e^{\omega s} G(\alpha_0 + \frac{\omega s}{t}) ds$$

$$+ e^{-\alpha_0 t} \int_{\log^2 t}^{C(R,t)} e^{\omega s} G(\alpha_0 + \frac{\omega s}{t}) ds$$

$$= F_{21}(t) + F_{22}(t). \qquad (111)$$

So,

$$|F_2(t)| \leq |F_{21}(t)| + |F_{22}(t)|. \qquad (112)$$

Choose $T_2(\epsilon) > 0$ such that for any $t > T_2(\epsilon)$, we have $\log^2 t/t < \delta(\epsilon)$ and $\log^2 t > 1$.

For F_{21}, by (106), we have

$$\left| e^{-\alpha_0 t} \int_1^{\log^2 t} e^{\omega s} G(\alpha_0 + \frac{\omega s}{t}) ds \right| \leq K \epsilon e^{-\alpha_0 t} t^{\lambda-1} \int_1^{\log^2 t} \left| e^{\omega s} |s^{-\lambda} ds \right|$$

$$= K \epsilon e^{-\alpha_0 t} t^{\lambda-1} \int_1^\infty e^{-\cos \epsilon_0 s} s^{-\lambda} ds. \qquad (113)$$

We also have

$$\int_1^\infty e^{-\cos \epsilon_0 s} s^{-\lambda} ds < \infty \qquad (114)$$

since $\cos \epsilon_0 \geq 0$ for $\epsilon_0 \in (0, \frac{\pi}{2}]$.

For F_{22}, since $|G(s)| \leq K |(\alpha_0 - s)^{-\lambda}|$, we get

$$|F_{22}(t)| \leq \frac{1}{2\pi} e^{-\alpha_0 t} t^{(\lambda-1)} \int_{\log^2 t}^{C(R,t)} |e^{-\omega s}||s^{-\lambda}| ds. \qquad (115)$$

We can also easily get that

$$|e^{-\omega s}| \le e^{-s\cos(\epsilon_0)}. \tag{116}$$

By (116),

$$\int_{\log^2 t}^{C(R,t)} |e^{-\omega s}||s^{-\lambda}|ds \le K\frac{1}{t^2}\int_{\log^2 t}^{C(R,t)} s^{-\lambda}ds \le K\frac{1}{t^2}\int_1^{\infty)} s^{-\lambda}ds. \tag{117}$$

Now, it follows from (111) to (117) that

$$|F_2(t)| < K\epsilon e^{-\alpha_0 t}t^{\lambda-1}, \text{ for some } t > \tilde{T}_2(\epsilon). \tag{118}$$

By (109) and (118), we can easily know that (107) holds. Therefore, the lemma is proved. $\qquad\square$

Remark 6. From the proof of Theorem 2, we can relax conditions (2) and (3) to the following, respectively: for some constant $\beta > \alpha_0$,

(2′) $g(z)$ is analytic on $\Delta(\alpha_0, \epsilon_0)$ for some $\epsilon_0 > 0$ with $\text{Re}(z) < \beta$;
(3′) $g(z)$ is bounded on $\Delta(\alpha_0, \epsilon_1)$ for some $\epsilon_1 > 0$ with $\text{Re}(z) < \beta$.

Remark 7. In [5], Dai and Miyazawa proved another version of the Tauberian-like theorem. It is not difficult to see that the conditions in Theorem 2 are weaker than those assumed in [5].

6 Exact Tail Asymptotics

In this section, we provide the exact tail asympotics for boundary measures. The tail asymptotic property for the boundary measures V_1 and V_2 is a direct consequence of the Tauberian-like theorem and the asymptotic behavior, obtained above, of $\phi_2(x)$ and $\phi_1(y)$.

One may notice that the Tauberian-like theorem is for a density function. However, it is easy to show (e.g., see D.5 of [4]) that the theorem can be applied to the finite measures V_1 and V_2. Specifically, we can show that condition 3 of Theorem 2 is satisfied for ϕ_2 (e.g., see Lemmas 6.6 and 6.7 in Dai and Miyazawa [4]), and therefore by Theorems 1 and 2, we have the following tail asymptotic properties.

Theorem 3. *For the boundary measure* $V_2(x, \infty)$, *we have the following tail asymptotic properties for large x.*

Case 1: If $\tau_1 = x^* < \min\{\tilde{x}, x_2\}$, *or* $\tau_1 = \tilde{x} < \min\{x^*, x_2\}$, *or* $\tau_1 = \tilde{x} = x^* = x_2$, *then*

$$V_2(x, \infty) \sim C_1 e^{-\tau_1 x};$$

Case 2:   *If* $\tau_1 = x^* = x_2 < \tilde{x}$, *or* $\tau_1 = \tilde{x} = x_2 < x^*$, *then*

$$V_2(x, \infty) \sim C_2 e^{-\tau_1 x} x^{-\frac{1}{2}};$$

Case 3:   *If* $\tau_1 = x_2 < \min\{\tilde{x}, x^*\}$, *then*

$$V_2(x, \infty) \sim C_3 e^{-\tau_1 x} x^{-\frac{3}{2}};$$

Case 4:   *If* $\tau_1 = x^* = \tilde{x} < x_2$, *then*

$$V_2(x, \infty) \sim C_4 e^{-\tau_1 x} x;$$

where C_i, $i = 1, 2, 3, 4$ are constants.

Tail asymptotic properties for V_1 can be symmetrically stated.

It is of interest to know why and when one of the above asymptotic types would arise. From an analytical point of view, this solely depends on the type of the dominant singularity of the unknown function for the boundary measure. The above four types of tail asymptotic properties correspond to the following properties of the dominant singularity, respectively:

(i) a simple pole or a double pole and the branch point x_2 simultaneously;
(ii) a simple pole and the branch point x_2 simultaneously;
(iii) the branch point x_2 only; and
(iv) a double pole.

From a practical point of view, it is interesting to know the specific type of the tail asymptotics for a given set (Σ, μ, R) of system parameters (specified numbers). In the following, we provide general steps for the boundary measure V_2 by analyzing the function $\phi_2(x)$.

Step 1.   Based on Eq. (21), evaluate the value of x_2:

$$x_2 = \frac{2(\Sigma_{12}\mu_2 - \Sigma_{22}\mu_1) + \sqrt{\Delta}}{2(\Sigma_{11}\Sigma_{22} - \Sigma_{12}^2)}, \tag{119}$$

where Δ is given by (22).

Step 2.   Recall that

$$Y_0(x) = \frac{-(\mu_2 + \Sigma_{12}x) - \sqrt{D_1(x)}}{\Sigma_{22}} \tag{120}$$

where $D_1(x)$ is given by (21) .

According to Lemma 11, if $\gamma_2\big(x_2, Y_0(x_2)\big) \geq 0$, then, by Remark 4, evaluate the value of x^*:

$$x^* = \frac{2r_{11}r_{22}\mu_2 - 2r_{22}^2\mu_1}{\Sigma_{22}r_{12}^2 - 2r_{11}r_{22}\Sigma_{12} + r_{22}^2\Sigma_{11}}. \tag{121}$$

If $\gamma_2\big(x_2, Y_0(x_2)\big) < 0$, let $x^* > x_2$ be any number.

Step 3. Similar to Steps 1 and 2, evaluate the values of y^* and y_2, respectively.

Step 4. (i) If $y^* > y_2$, then let $\tilde{x} > x_2$ be any number.

(ii) If $y^* \in (0, \, y_2]$, then calculate

$$x^1 = X_1(y^*) = \frac{r_{21}}{r_{11}}y^* - 2\frac{\Sigma_{12}y^* + \mu_1}{\Sigma_{11}}, \tag{122}$$

where

$$y^* = \frac{2r_{11}r_{21}\mu_2 - 2r_{11}^2\mu_2}{r_{21}^2\Sigma_{11} - 2r_{11}r_{21}\Sigma_{12} + \Sigma_{22}r_{11}^2}. \tag{123}$$

Next, verify if

$$Y_0(x^1) = y^*. \tag{124}$$

If (124) is true, then $\tilde{x} = x^1$. Otherwise, let $\tilde{x} > x_2$ be any real number.

Step 5. By the above steps, the values of x^*, x_2 and $\tilde{x}$ are determined. Then, by Theorem 3 the type of tail asymptotic properties of the boundary measure V_2 is determined.

7 A Final Note and Concluding Remarks

The research work in this paper was motivated by Dai and Miyazawa [4], in which the same SRBM was considered, but the exact tail behaviour in boundary probabilities was not reported. Before we completed this paper, Dai and Miyazawa [5] reported the tail behaviour as a continuation of their work in [4]. We therefore present our work as a survey of the kernel method and emphasize the connection of this method to the closely related method in [5]. In [5], the authors used geometric properties of the model to obtain the rough decay and then used an analytic approach for refined tail decay properties. The kernel method is a different (pure analytic) approach. One of the common components in [5] and in this paper is the generalization of the Tauberian-like theorem to the continuous case. The Tauberian-like theorem (Theorem 2) given in this paper has a weaker condition than that given in [5].

The kernel method is a general approach, which could be used for studying tail asymptotics for more general models. For example, for the two-dimensional continuous case, this approach can be a candidate for studying exact tail asymptotic properties of reflected Lévy processes. In fact, this method can be applied to a special Lévy case studied in Miyazawa and Rolski [17]. For high (≥ 3) dimensional models, it could be more challenging since the number of unknown functions in the fundamental form will be increased.

It is noted that the same four types of exact tail asymptotic properties are found for both discrete reflected random walks in the quarter plane and two-dimensional SRBM. This is simply due to the fact that the kernel function in both cases is a quadratic form in two variables. It is interesting to consider whether this is still true for more general two-dimensional continuous models, for example the reflected Lévy process, for which the kernel function in general is not a quadratic form.

Acknowledgements This work was partially supported through NSERC Discovery grants, the National Natural Science Foundation of China (No.11361007), and the Guangxi Natural Science Foundation (No.2012GXNSFBA053010 and 2014GXNSFCA118001). We thank the two reviewers for their comments/suggestions, which improved the quality of the paper.

References

1. Banderier, C., Bousquet-Mélou, M., Denise, A., Flajolet, P., Gardy, D., Gouyou-Beauchamps, D.: Generating functions for generating trees. Discret. Math. **246**, 29–55 (2002)
2. Berman, A., Plemmons, R.J.: Nonnegative Matrices in the Mathematical Sciences. Academic, New York (1979)
3. Dai, J.G., Harrison, J.M.: Reflected Brownian motion in an orthant: numerical methods for steady-state analysis. Ann. Appl. Probab. **2**, 65–86 (1992)
4. Dai, J.G., Miyazawa, M.: Reflecting Brownian motion in two dimensions: exact asymptotics for the stationary distribution. Stoch. Syst. **1**, 146–208 (2011)
5. Dai, J.G., Miyazawa, M.: Stationary distribution of a two-dimensional SRBM: geometric views and boundary measures. Queueng Syst. **74**, 181–217 (2013)
6. Dupuis, P., Ramanan, K.: A time-reversed representation for the tail probabilities of stationary reflected Brownian motion. Stoch. Process. Appl. **98**, 253–287 (2002)
7. Fayolle, G., Iasnogorodski, R., Malyshev, V.: Random Walks in the Quarter-Plane. Springer, New York (1999)
8. Flajolet, P., Sedgewick, R.: Analytic Combinatorics. Cambridge University Press, Cambridge (2009)
9. Harrison, J.M., Hasenbein, J.J.: Reflected Brownian motion in the quadrant: tail behavior of the stationary distribution. Queueing Syst. **61**, 113–138 (2009)
10. Harrison, J.M., Williams, R.J.: Brownian models of open queueing networks with homogeneous customer populations. Stochastic **22**, 77–115 (1987)
11. Knuth, D.E.: The Art of Computer Programming, Fundamental Algorithms, vol. 1, 2nd edn. Addison-Wesley, Massachusetts (1969)
12. Konstantopoulos, T., Last, G., Lin, S.: On a class of Lévy stochastic networks. Queueing Syst. **46**, 409–437 (2004)
13. Li, H., Zhao, Y.Q.: Tail asymptotics for a generalized two-deman queueing model—a kernel method. Queueing Syst. **69**, 77–100 (2011)

14. Li, H., Zhao, Y.Q.: A kernel method for exact tail asymptotics—random walks in the quarter plane. arXiv:1505.04425 [math.PR] (2015)
15. Markushevich, A.I.: Theory of Functions of a Complex Variable, vol. I, II, III, English edn. R.A. Silverman (Trans., ed.). Chelsea Publishing, New York (1977)
16. Miyazawa, M.: Tail decay rates in double QBD processes and related reflected random walks. Math. Oper. Res. **34**, 547–575 (2009)
17. Miyazawa, M., Rolski, T.: Tail asymptotics for a Lévy-driven tandem queue with an intermediate input. Queueng Syst. **63**, 323–353 (2009)
18. Reiman, M.I., Williams, R.J.: A boundary property of semimaritingale reflecting Brownian motions. Probab. Theory Relat. Fields **77**, 87–97 (1988)
19. Taylor, L.M., Williams, R.J.: Existence and uniquess of semimaritingale reflecting Brownian motions in an orthant. Probab. Theory Relat. Fields **96**, 283–317 (1993)
20. Williams, R.J.: Semimartingale reflecting Brownian motions in the orthant. In: Stochastic Networks. IMA Volumes in Mathematics and Its Applications, vol. 71, pp. 125–137. Springer, New York (1995)
21. Williams, R.J.: On the approximation of queueing networks in heavy traffic. In: Kelly, F.P., Zachary, S., Ziedins, I. (eds.) Stochastic Networks: Theory and Applications. Oxford University Press, Oxford (1996)

Central Limit Theorems and Large Deviations for Additive Functionals of Reflecting Diffusion Processes

Peter W. Glynn and Rob J. Wang

1 Introduction

Reflecting diffusion processes arise as approximations to stochastic models associated with a wide variety of different applications domains, including communications networks, manufacturing systems, call centers, finance, and the study of transport phenomena (see, for example, Chen and Whitt [4], Harrison [8], and Costantini [5]). If $X = (X(t) : t \geq 0)$ is the reflecting diffusion, it is often of interest to study the distribution of an additive functional of the form

$$A(t) \overset{\Delta}{=} \int_0^t f(X(s))ds + \Lambda(t),$$

where f is a real-valued function defined on the domain of X, and $\Lambda = (\Lambda(t) : t \geq 0)$ is a process (related to the boundary reflection) that increases only when X is on the boundary of its domain. In many applications settings, the boundary process Λ is a key quantity, as it can correspond to the cumulative number of customers lost in a finite buffer queue, the cumulative amount of cash injected into a firm, and other key performance measures depending on the specific application.

Given such an additive functional $A = (A(t) : t \geq 0)$, a number of limit theorems can be obtained in the setting of a positive recurrent process X.

The Strong Law: Compute the constant α such that

$$\frac{A(t)}{t} \overset{a.s.}{\to} \alpha \tag{1}$$

P.W. Glynn (✉) • R.J. Wang
Department of Management Science and Engineering, Stanford University,
475 Via Ortega, Stanford, CA 94305, USA
e-mail: glynn@stanford.edu; robjwang@stanford.edu

© Springer Science+Business Media New York 2015
D. Dawson et al. (eds.), *Asymptotic Laws and Methods in Stochastics*,
Fields Institute Communications 76, DOI 10.1007/978-1-4939-3076-0_17

as $t \to \infty$. In the presence of (1), we can approximate $A(t)$ via

$$A(t) \overset{\mathscr{D}}{\approx} \alpha t, \tag{2}$$

where $\overset{\mathscr{D}}{\approx}$ means "has approximately the same distribution as" (and no other rigorous meaning, other than that supplied by (1) itself.)

The Central Limit Theorem: Compute the constants α and η such that

$$t^{1/2} \left(\frac{A(t)}{t} - \alpha \right) \Rightarrow \eta N(0, 1) \tag{3}$$

as $t \to \infty$, where $\Rightarrow$ denotes convergence in distribution and $N(0, 1)$ is a normal random variable (rv) with mean 0 and unit variance. When (3) holds, we may improve the approximation (2) to

$$A(t) \overset{\mathscr{D}}{\approx} \alpha t + \eta \sqrt{t} N(0, 1) \tag{4}$$

for large t, thereby providing a description of the distribution of $A(t)$ at scales of order $t^{1/2}$ from αt.

Large Deviations: Compute the rate function $(I(x) : x \in \mathbb{R})$ for which

$$\frac{1}{t} \log P(A(t) \in t\Gamma) \to - \inf_{x \in \Gamma} I(x) \tag{5}$$

as $t \to \infty$, for subsets Γ that are suitably chosen. Given the limit theorem (5), this suggests the (crude) approximation

$$P(A(t) \in \Gamma) \approx \exp\left(-t \inf_{y \in \Gamma} I(y/t) \right) \tag{6}$$

for large t; the approximation (6) is particularly suitable for subsets Γ that are "rare" in the sense that they are more than order $\sqrt{t}$ from αt.

The main contribution of this paper concerns the computation of the quantities α, η, and $I(\cdot)$, when A is an additive functional for a reflecting diffusion that incorporates the boundary contribution Λ. To give a sense of the new issues that arise in this setting, observe that when $\Lambda(t) \equiv 0$ for $t \geq 0$, then α can be easily computed from the stationary distribution π of X via

$$\alpha = \int_S f(x)\pi(dx),$$

where S is the domain of X. However, when Λ is non-zero, this approach to computing α does not easily extend. The key to building a suitable computational theory for reflecting diffusions is to systematically exploit the martingale ideas that (implicitly) underly the corresponding calculations for Markov processes without boundaries; see, for example, Bhattacharyya [2] for a discussion in the central limit setting. In the one-dimensional context, a (more laborious) approach based on the theory of regenerative processes can also be used; see Williams [15] for such a calculation in the setting of Brownian motion. In the course of our development of the appropriate martingale ideas, we will recover the existing theory for non-reflecting diffusions as a special case.

The paper is organized as follows. In Sect. 2, we show how one can apply stochastic calculus and martingale ideas to derive partial differential equations from which the central limit and law of large numbers behavior for additive functionals involving boundary terms can be computed. Section 3 develops the corresponding large deviations theory for such additive functionals. Finally, Sects. 4 and 5 illustrate the ideas in the context of one-dimensional reflecting diffusions.

2 Laws of Large Numbers and Central Limit Theorems

Let S^o be a connected open set in $\mathbb{R}^d$, with S and ∂S denoting its closure and boundary, respectively. We assume that there exists a vector field $\gamma : \partial S \to \mathbb{R}^d$ satisfying

$$\langle \gamma(x), n(x) \rangle > 0$$

for $x \in \partial S$, where $n(x)$ is the unit inward normal to ∂S at x (assumed to exist). Accordingly, $\gamma(x)$ is always "pointing" into the interior of S. Given functions $\mu : S \to \mathbb{R}^d$ and $\sigma : S \to \mathbb{R}^{d \times d}$, we assume the existence, for each $x_0 \in S$, of a pair of continuous processes $X = (X(t) : t \geq 0)$ and $k = (k(t) : t \geq 0)$ (with k of bounded variation) for which

$$X(t) = x_0 + \int_0^t \mu(X(s))ds + \int_0^t \sigma(X(s))dB(s) + k(t), \qquad (7)$$

$$X(t) \in S,$$

$$|k|(t) = \int_0^t I(X(r) \in \partial S)d|k|(r),$$

and

$$k(t) = \int_0^t \gamma(X(s))d|k|(s),$$

where $B = (B(t) : t \geq 0)$ is a standard $\mathbb{R}^d$-valued Brownian motion, and $|k|(t)$ is the (scalar) total variation of k over $[0, t]$; sufficient conditions surrounding existence of such processes can be found in Lions and Snitzman [10]. Note that our formulation permits the direction of reflection to be oblique. Regarding the structure of the boundary process Λ, we assume that it takes the form

$$\Lambda(t) = \int_0^t r(X(s))d|k|(s),$$

for a given function $r : S \to \mathbb{R}$.

We expect laws of large numbers and central limit theorems to hold with the conventional normalizations only when X is a positive recurrent Markov process. In view of this, we assume:

A1: X is a Markov process with a stationary distribution π that is recurrent in the sense of Harris.

Remark. By Harris recurrence, we mean that there exists a non-trivial σ-finite measure ϕ on S for which whenever $\phi(B) > 0$, $\int_0^\infty I(X(s) \in B)ds = \infty$ P_x a.s. for each $x \in S$, where

$$P_x(\cdot) \stackrel{\Delta}{=} P(\cdot \mid X(0) = x).$$

We note that Harris recurrence implies that any stationary distribution must be unique. For a discussion of methods for verification of recurrence in the setting of continuous-time Markov processes, see Meyn and Tweedie [11–13].

The key to developing laws of large numbers and central limit theorems for the additive functional A is to find a function $u : S \to \mathbb{R}$ and a constant α for which

$$M(t) \stackrel{\Delta}{=} u(X(t)) - (A(t) - \alpha t)$$

is a local $\mathscr{F}_t$-martingale, where $\mathscr{F}_t = \sigma(X(s) : 0 \le s \le t)$. In order to explicitly compute u, it is convenient to identify a suitable partial differential equation satisfied by u that can be used to solve for u. Note that if $u \in C^2(S)$, Itô's formula ensures that

$$
\begin{aligned}
dM(t) &= du(X(t)) - (f(X(t)) - \alpha)dt - r(X(t))d|k|(t) \\
&= \nabla u(X(t))dX(t) + \frac{1}{2}\sum_{i,j=1}^d (\sigma\sigma^T)_{ij}(X(t))\frac{\partial^2 u(X(t))}{\partial x_i \partial x_j}dt \\
&\quad - (f(X(t)) - \alpha)dt - r(X(t))d|k|(t) \\
&= \nabla u(X(t))(\mu(X(t))dt + \sigma(X(t))dB(t) \\
&\quad + \gamma(X(t))d|k|(t)) + \frac{1}{2}\sum_{i,j=1}^d (\sigma\sigma^T)_{ij}(X(t))\frac{\partial^2 u(X(t))}{\partial x_i \partial x_j}dt \\
&\quad - (f(X(t)) - \alpha)dt - r(X(t))d|k|(t) \\
&= ((\mathscr{L}u)(X(t)) - (f(X(t)) - \alpha))dt + (\nabla u(X(t))\gamma(X(t)) \\
&\quad - r(X(t)))d|k|(t) + \nabla u(X(t))\sigma(X(t))dB(t),
\end{aligned}
$$

where $\nabla u(x)$ is the gradient of u evaluated at x (encoded as a row vector) and $\mathscr{L}$ is the elliptic differential operator

$$\mathscr{L} = \sum_{i=1}^{d} \mu_i(x)\frac{\partial}{\partial x_i} + \frac{1}{2}\sum_{i,j=1}^{d}(\sigma\sigma^T)_{ij}(x)\frac{\partial^2}{\partial x_i\partial x_j}. \tag{8}$$

The process M can be guaranteed to be a local martingale if we require that u and α satisfy

$$(\mathscr{L}u)(x) = f(x) - \alpha, \ x \in S \tag{9}$$

$$\nabla u(x)\gamma(x) = r(x), \ x \in \partial S,$$

since this choice implies that

$$dM(t) = \nabla u(X(t))\sigma(X(t))dB(t).$$

(We use here the fact that $|k|(t)$ increases only when $X(t) \in \partial S$.) Accordingly, the quadratic variation of M is given by

$$[M, M](t) = \int_0^t \nabla u(X(s))\sigma(X(s))\sigma^T(X(s))\nabla u(X(s))^T ds$$

$$\overset{\Delta}{=} \int_0^t v(X(s))ds.$$

Since v is nonnegative and X is positive Harris recurrent, it follows that

$$\frac{1}{t}\int_0^t v(X(s))ds \to \int_S v(y)\pi(dy) \ \ P_x \ a.s.$$

as $t \to \infty$, for each $x \in S$. Set

$$\eta^2 = \int_S v(y)\pi(dy)$$

$$= \int_S \nabla u(y)\sigma(y)\sigma(y)^T\nabla u(y)\pi(dy),$$

and assume $\eta^2 < \infty$. As a consequence of the path continuity of M, the martingale central limit theorem then implies that for each $x \in S$,

$$t^{-1/2}M(t) \Rightarrow \eta N(0, 1) \tag{10}$$

as $t \to \infty$ under P_x (see, for example, Ethier and Kurtz [7]). In other words,

$$t^{-1/2}(u(X(t)) - (A(t) - \alpha t)) \Rightarrow \eta N(0, 1)$$

as $t \to \infty$ under P_x.

Let $P_\pi(\cdot) = \int_S P_x(\cdot)\pi(dx)$, and observe that X is stationary under P_π. Thus, $u(X(t)) \overset{\mathcal{D}}{=} u(X(0))$ for $t \geq 0$ under P_π (where $\overset{\mathcal{D}}{=}$ denotes equality in distribution), so that

$$t^{-1/2}u(X(t)) \Rightarrow 0 \qquad (11)$$

as $t \to \infty$ under P_π. It follows that

$$\frac{1}{t}A(t) \Rightarrow \alpha$$

as $t \to \infty$ under P_π. Let $E_\pi(\cdot)$ be the expectation operator associated with P_π. If f and r are nonnegative, the Harris recurrence implies that

$$\frac{1}{t}A(t) \to E_\pi A(1) \; P_x \; a.s.$$

as $t \to \infty$, for each $x \in S$. Hence, $E_\pi A(1) = \alpha$, so that

$$\frac{1}{t}A(t) \to \alpha \; P_x \; a.s.$$

as $t \to \infty$, for each $x \in S$. This establishes the desired strong law of large numbers for the additive functional A.

Turning now to the central limit theorem, (10) and (11) together imply that

$$t^{1/2}\left(\frac{A(t)}{t} - \alpha\right) \Rightarrow \eta N(0, 1)$$

as $t \to \infty$ under P_π. Recall that a Harris recurrent Markov process X automatically exhibits one-dependent regenerative structure, in the sense that there exists a non-decreasing sequence $(T_n : n \geq -1)$ of randomized stopping times, with $T_{-1} = 0$, for which the sequence of random elements $(X(T_{n-1} + s) : 0 \leq s < T_n - T_{n-1})$ is identically distributed for $n \geq 1$ and one-dependent for $n \geq 0$; see Sigman [14]. The one-dependence implies that the central limit theorem can be extended from the stationary setting in which $X(0)$ has distribution π to cover arbitrary initial distributions, so that

$$t^{1/2}\left(\frac{A(t)}{t} - \alpha\right) \Rightarrow \eta N(0, 1)$$

as $t \to \infty$ under P_x, for each $x \in S$. We summarize this discussion with the following theorem.

Theorem 1. *Assume A1 and that f and r are nonnegative. If there exists $u \in C^2(S)$ and $\alpha \in \mathbb{R}$ that satisfy*

$$(\mathscr{L}u)(x) = f(x) - \alpha, \ x \in S$$

$$\nabla u(x)\gamma(x) = r(x), \ x \in \partial S,$$

with

$$\eta^2 = \int_S \nabla u(y)\sigma(y)\sigma(y)^T \nabla u(y)\pi(dy) < \infty,$$

then, for each $x \in S$,

$$\frac{1}{t}A(t) \to \alpha \ P_x \ a.s.$$

and

$$t^{1/2}\left(\frac{A(t)}{t} - \alpha\right) \Rightarrow \eta N(0, 1)$$

as $t \to \infty$, under P_x.

The function u satisfying (9) is said to be a solution of the *generalized Poisson equation* corresponding to the pair (f, r).

3 Large Deviations for the Additive Functional A

The key to developing a suitable large deviations theory for A is again based on construction of an appropriate martingale. Here, we propose a one-parameter family of martingales of the form

$$M(\theta, t) = \exp(\theta A(t) - \psi(\theta)t)h_\theta(X(t))$$

for θ lying in some open interval containing the origin, where $\psi(\theta)$ and h_θ are chosen appropriately. As in Sect. 2, we use stochastic calculus to derive a corresponding PDE from which one can potentially compute $\psi(\theta)$ and h_θ analytically. In particular, if $h_\theta \in C^2(S)$, Itô's formula yields

$$dM(\theta, t) \ = \ d(\exp(\theta A(t) - \psi(\theta)t))h_\theta(X(t))$$
$$+ \ \exp(\theta A(t) - \psi(\theta)t)dh_\theta(X(t))$$

$$= \exp(\theta A(t) - \psi(\theta)t)(\theta f(X(t))dt + \theta r(X(t))d|k|(t)$$

$$- \psi(\theta)dt)h_\theta(X(t)) + \exp(\theta A(t) - \psi(\theta)t)\Big[\nabla h_\theta(X(t))\mu(X(t))dt$$

$$+ \nabla h_\theta(X(t))\sigma(X(t))dB(t) + \nabla h_\theta(X(t))\gamma(X(t))d|k|(t)$$

$$+ \frac{1}{2}\sum_{i,j=1}^{d}(\sigma\sigma^T)_{ij}(X(t))\frac{\partial^2 h_\theta(X(t))}{\partial x_i \partial x_j}dt\Big]$$

$$= \exp(\theta A(t) - \psi(\theta)t)\Big[((\mathscr{L}h_\theta)(X(t)) + (\theta f(X(t))$$

$$- \psi(\theta))h_\theta(X(t)))dt + (\nabla h_\theta(X(t))\gamma(X(t))$$

$$+ \theta r(X(t))h_\theta(X(t)))d|k|(t) + \nabla h_\theta(X(t))\sigma(X(t))dB(t)\Big],$$

where $\mathscr{L}$ is the differential operator defined in Sect. 2. If we require that h_θ and $\psi(\theta)$ satisfy

$$(\mathscr{L}h_\theta)(x) + (\theta f(x) - \psi(\theta))h_\theta(x) = 0, \ x \in S \tag{12}$$

$$\nabla h_\theta(x)\gamma(x) + \theta r(x)h_\theta(x) = 0, \ x \in \partial S,$$

then

$$dM(\theta, t) = \nabla h_\theta(X(t))\sigma(X(t))dB(t),$$

and $M(\theta, t) : t \geq 0)$ is consequently a local $\mathscr{F}_t$-martingale. (Again, we use here the fact that $|k|$ increases only when X is on the boundary of S.) Note that (12) takes the form of an eigenvalue problem involving the operator $\mathscr{L} + \theta f \mathscr{I}$, where $\mathscr{I}$ is the identity operator for which $\mathscr{I}u = u$. In this eigenvalue formulation, $\psi(\theta)$ is the eigenvalue and h_θ the corresponding eigenfunction. Since $\mathscr{L} + \theta f \mathscr{I}$ is expected to have multiple eigenvalues, (12) cannot be expected to uniquely determine $\psi(\theta)$ and h_θ. In order to ensure uniqueness, we now add the requirement that h_θ be positive.

Let $(T_n : n \geq 0)$ be the localizing sequence of stopping times associated with the local martingale $(M(\theta, t) : t \geq 0)$, so that

$$E_x \exp(\theta A(t \wedge T_n) - \psi(\theta)(t \wedge T_n))h_\theta(X(t \wedge T_n)) = h_\theta(x) \tag{13}$$

for $x \in S$, where $E_x(\cdot)$ is the expectation operator associated with $P_x(\cdot)$ and $a \wedge b \overset{\Delta}{=} \min(a, b)$ for $a, b \in \mathbb{R}$.

Suppose that S is compact, so that h_θ is then bounded above and below by positive constants (on account of the positivity of h_θ and the fact that $h_\theta \in C^2(S)$). If f and r are nonnegative (as in Sect. 2), it follows that for $\theta \leq 0$,

$$\exp(\theta A(t \wedge T_n) - \psi(\theta)(t \wedge T_n))h_\theta(X(t \wedge T_n))$$

is a bounded sequence of rv's, and thus the Bounded Convergence Theorem implies that

$$E_x \exp(\theta A(t) - \psi(\theta)t)h_\theta(X(t)) = h_\theta(x) \tag{14}$$

for $\theta \leq 0$, and $x \in S$.

On the other hand, if $\theta > 0$, the positivity of h_θ and Fatou's lemma imply that

$$E_x \exp(\theta A(t) - \psi(\theta)t)h_\theta(X(t)) \leq h_\theta(x)$$

for $x \in S$, from which we may obtain the upper bound

$$E_x \exp(\theta A(t)) \leq e^{\psi(\theta)t} \frac{h_\theta(x)}{\inf_{y \in S} h_\theta(y)},$$

and hence $\exp(\theta A(t))$ is P_x-integrable. Since f and r are nonnegative and $\theta > 0$, $\theta A(t \wedge T_n) \leq \theta A(t)$, so

$$\exp(\theta A(t \wedge T_n) - \psi(\theta)(t \wedge T_n))h_\theta(X(t \wedge T_n)) \leq \exp(\theta A(t) + |\psi(\theta)|(t)) \sup_{y \in S} h_\theta(y).$$

The Dominated Convergence Theorem, as applied to (13), then yields the conclusion that

$$E_x \exp(\theta A(t) - \psi(\theta)t)h_\theta(X(t)) = h_\theta(x) \tag{15}$$

for $x \in S$. Since

$$e^{\psi(\theta)t} \frac{h_\theta(x)}{\sup_{y \in S} h_\theta(y)} \leq E_x \exp(\theta A(t)) \leq e^{\psi(\theta)t} \frac{h_\theta(x)}{\inf_{y \in S} h_\theta(y)},$$

it follows that

$$\frac{1}{t} \log E_x \exp(\theta A(t)) \to \psi(\theta)$$

as $t \to \infty$, proving the following theorem.

Theorem 2. *Assume that S is compact and that f and r are nonnegative. If there exists a positive function $h_\theta \in C^2(S)$ and $\psi(\theta) \in \mathbb{R}$ that satisfy*

$$(\mathscr{L}h_\theta)(x) + (\theta f(x) - \psi(\theta))h_\theta(x) = 0, \ x \in S$$

$$\nabla h_\theta(x)\gamma(x) + \theta r(x)h_\theta(x) = 0, \ x \in \partial S,$$

then

$$\frac{1}{t} \log E_x \exp(\theta A(t)) \to \psi(\theta)$$

as $t \to \infty$.

The Gärtner-Ellis Theorem (see, for example, p.45 of Dembo and Zeitouni [6]) then provides technical conditions under which

$$\frac{1}{t} \log P_x(A(t) \in t\Gamma) \to -\inf_{y \in \Gamma} I(y)$$

as $t \to \infty$, where

$$I(y) = \sup_{\theta \in \mathbb{R}}[\theta y - \psi(\theta)].$$

In particular, if $\Gamma = (z, \infty)$, then

$$\frac{1}{t} \log P_x(A(t) \geq tz) \to -(\theta_z z - \psi(\theta_z)),$$

provided that $\psi(\cdot)$ is differentiable and strictly convex in a neighborhood of a point θ_z satisfying $\psi'(\theta_z) = z$. See p.15–16 of Bucklew [3] for a related argument.

4   CLT's for One-dimensional Reflecting Diffusions

We now illustrate these ideas in the setting of one-dimensional diffusions. In this context, we can compute the solution of the generalized Poisson equation corresponding to (f, r) fairly explicitly.

We start with the case where there are two reflecting barriers, at 0 and b, so that $S = [0, b]$. Then, $X = (X(t) : t \geq 0)$ satisfies the stochastic differential equation (SDE)

$$dX(t) = \mu(X(t))dt + \sigma(X(t))dB(t) + dL(t) - dU(t)$$

$$= \mu(X(t))dt + \sigma(X(t))dB(t) + dk(t),$$

with $\gamma(0) = 1$ and $\gamma(b) = -1$; the processes L and U increase only when X visits the lower and upper boundaries at 0 and b, respectively. We consider here the additive functional

$$A(t) = \int_0^t f(X(s))ds + r_0 L(t) + r_b U(t),$$

where $f : [0, b] \to \mathbb{R}$ is assumed to be bounded. In this setting, Theorem 1 leads to consideration of the ordinary differential equation (ODE)

$$\mu(x)u'(x) + \frac{\sigma^2(x)}{2}u''(x) = f(x) - \alpha, \tag{16}$$

$$u'(0) = r_0, \tag{17}$$

$$u'(b) = -r_b. \tag{18}$$

Hence, if $\mu(\cdot)$ and $\sigma^2(\cdot)$ are continuous and $\sigma^2(\cdot)$ positive, (16) can be re-written via the method of integrating factors (see, for example, Karlin and Taylor [9]) as

$$\frac{d}{dx}\left(\exp\left(\int_0^x \frac{2\mu(y)}{\sigma^2(y)}dy\right)u'(x)\right) = \frac{2(f(x) - \alpha)}{\sigma^2(x)}\exp\left(\int_0^x \frac{2\mu(y)}{\sigma^2(y)}dy\right),$$

from which we conclude that

$$u'(x) = \left(u'(0) + \int_0^x \frac{2(f(y) - \alpha)}{\sigma^2(y)}\exp\left(\int_0^y \frac{2\mu(z)}{\sigma^2(z)}dz\right)dy\right) \tag{19}$$

$$\cdot \exp\left(-\int_0^x \frac{2\mu(y)}{\sigma^2(y)}dy\right). \tag{20}$$

But $u'(0) = r_0$ and $u'(b) = -r_b$, and thus

$$-r_b = \left(r_0 + \int_0^b \frac{2(f(y) - \alpha)}{\sigma^2(y)}\exp\left(\int_0^y \frac{2\mu(z)}{\sigma^2(z)}dz\right)dy\right)$$

$$\cdot \exp\left(-\int_0^b \frac{2\mu(y)}{\sigma^2(y)}dy\right).$$

Hence,

$$\alpha = \frac{r_0 + r_b e^{\left(\int_0^b \frac{2\mu(y)}{\sigma^2(y)}dy\right)} + \int_0^b \frac{2f(y)}{\sigma^2(y)}e^{\left(\int_0^y \frac{2\mu(z)}{\sigma^2(z)}dz\right)}dy}{2\int_0^b \frac{1}{\sigma^2(y)}e^{\left(\int_0^y \frac{2\mu(z)}{\sigma^2(z)}dz\right)}dy} \tag{21}$$

By setting $r_0 = r_b = 0$, we conclude that the stationary distribution π of X must satisfy

$$\int_0^b \pi(dx)f(x) = \int_0^b f(x)p(x)dx, \tag{22}$$

where

$$p(x) = \frac{\frac{1}{\sigma^2(x)} \exp\left(\int_0^x \frac{2\mu(z)}{\sigma^2(z)} dz\right)}{\int_0^b \frac{1}{\sigma^2(y)} \exp\left(\int_0^y \frac{2\mu(z)}{\sigma^2(z)} dz\right) dy}.$$

Since (22) holds for all bounded functions f, it follows that $\pi(dx) = p(x)dx$, so that π has now been computed. Furthermore, (19) establishes that

$$u'(x) = \left(r_0 + \int_0^x \frac{2(f(y) - \alpha)}{\sigma^2(y)} \exp\left(\int_0^y \frac{2\mu(z)}{\sigma^2(z)} dz\right) dy\right)$$
$$\cdot \exp\left(-\int_0^x \frac{2\mu(y)}{\sigma^2(y)} dy\right),$$

where α is given by (21). Consequently, we have explicit formulae for both π and u', from which the variance constant

$$\eta^2 = \int_0^b u'(x)^2 \sigma^2(x) p(x) dx$$

of Theorem 1 can now be calculated. We now illustrate these calculations in the context of some special cases, focusing our interest on the boundary processes (by setting $f = 0$).

Example 1 (Two-sided Reflecting Brownian Motion). Here $\mu(x) = \mu$ and $\sigma^2(x) = \sigma^2 > 0$. If $\mu \neq 0$, then, upon setting $\xi = 2\mu/\sigma^2$,

$$\alpha = \frac{\mu(r_0 + r_b e^{\xi b})}{e^{\xi b} - 1}$$

and

$$p(x) = \frac{\xi e^{\xi x}}{e^{\xi b} - 1}.$$

Also,

$$u'(x) = \left(r_0 + \int_0^x -\frac{2\alpha}{\sigma^2} e^{\frac{2\mu}{\sigma^2} y} dy\right) e^{-\frac{2\mu}{\sigma^2} x}$$
$$= \left(\frac{r_0 + r_b}{1 - e^{-\xi b}}\right) e^{-\xi x} - \frac{r_0 e^{-\xi b} + r_b}{1 - e^{-\xi b}}$$

and consequently

$$u'(x)^2 = \left(\frac{r_0 + r_b}{1 - e^{-\xi b}}\right)^2 e^{-2\xi x} - \frac{2(r_0 e^{-\xi b} + r_b)(r_0 + r_b)}{(1 - e^{-\xi b})^2} e^{-\xi x} + \left(\frac{r_0 e^{-\xi b} + r_b}{1 - e^{-\xi b}}\right)^2.$$

Therefore,

$$\eta^2 = \sigma^2 \left[\left(\frac{r_0 + r_b}{1 - e^{-\xi b}} \right)^2 e^{-\xi b} - \frac{(r_0 e^{-\xi b} + r_b)(r_0 + r_b)}{(1 - e^{-\xi b})^2} \frac{2\xi b}{e^{\xi b} - 1} + \left(\frac{r_0 e^{-\xi b} + r_b}{1 - e^{-\xi b}} \right)^2 \right].$$

If $\mu = 0$, then $\alpha = \frac{\sigma^2 (r_0 + r_b)}{2b}$ and $p(x) = \frac{1}{b}$. Also,

$$u'(x) = r_0 - \frac{(r_0 + r_b)}{b} x$$

and therefore

$$\eta^2 = \sigma^2 \int_0^b \frac{\left(\frac{(r_0 + r_b)}{b} x - r_0 \right)^2}{b} dx$$

$$= \frac{\sigma^2 (r_0^3 + r_b^3)}{3(r_0 + r_b)}.$$

Example 2. Two-sided Reflecting Ornstein-Uhlenbeck: For this process, $\mu(x) = -a(x - c)$ and $\sigma^2(x) = \sigma^2 > 0$. We thus have

$$\alpha = \frac{r_0 + r_b e^{-\frac{a(b-c)^2 - ac^2}{\sigma^2}}}{\frac{2}{\sigma^2} \int_0^b e^{-\frac{a(y-c)^2 - ac^2}{\sigma^2}} dy}.$$

Also,

$$u'(x) = \left(r_0 - \frac{2\alpha}{\sigma^2} \int_0^x e^{-\int_0^y \frac{2a(z-c)}{\sigma^2} dz} dy \right) e^{\int_0^x \frac{2a(y-c)}{\sigma^2} dy}$$

$$= r_0 e^{\frac{a(x-c)^2 - ac^2}{\sigma^2}} - \frac{2\alpha}{\sigma^2} \int_0^x e^{-\frac{a(y-c)^2 - a(x-c)^2}{\sigma^2}} dy$$

and

$$p(x) = \frac{e^{-\int_0^x \frac{2a(z-c)}{\sigma^2} dz}}{\int_0^b e^{-\int_0^y \frac{2a(z-c)}{\sigma^2} dz} dy}$$

$$= \sqrt{\frac{2a}{\sigma^2}} \frac{\phi\left((x - c) \sqrt{\frac{2a}{\sigma^2}} \right)}{\Phi\left((b - c) \sqrt{\frac{2a}{\sigma^2}} \right) - \Phi\left((-c) \sqrt{\frac{2a}{\sigma^2}} \right)},$$

where ϕ and Φ are, respectively, the density and cumulative density function (CDF) of a standard normal random variable. From these, one may readily compute

$$\eta^2 = \sigma^2 \int_0^b \left(r_0 e^{\frac{a(x-c)^2 - ac^2}{\sigma^2}} - \frac{2\alpha}{\sigma^2} \int_0^x e^{-\frac{a(y-c)^2 - a(x-c)^2}{\sigma^2}} \, dy \right)^2 p(x) dx$$

numerically when the problem data are explicit.

The diffusions in our examples arise as approximations to queues in heavy traffic, in which $L(t)$ then approximates the cumulative lost service capacity of the server over $[0, t]$, while $U(t)$ describes the cumulative number of customers lost due to blocking (because of arrival to a full buffer); see Zhang and Glynn [16] for details.

Turning now to the setting in which only a single reflecting barrier is present (say, at the origin), S then takes the form $S = [0, \infty)$, and the differential equation for u takes the form

$$\mu(x)u'(x) + \frac{\sigma^2(x)}{2}u''(x) = f(x) - \alpha,$$

$$u'(0) = r_0.$$

Then $u'(x)$ is again given by (19), and

$$\alpha = \frac{r_0 + \int_0^\infty \frac{2f(y)}{\sigma^2(y)} e^{\left(\int_0^y \frac{2\mu(z)}{\sigma^2(z)} dz \right)} dy}{2 \int_0^\infty \frac{1}{\sigma^2(y)} e^{\left(\int_0^y \frac{2\mu(z)}{\sigma^2(z)} dz \right)} dy}, \tag{23}$$

provided that the problem data are such that the integrals in (23) converge and are finite. In particular, X fails to have a stationary distribution if

$$\int_0^\infty \frac{1}{\sigma^2(y)} e^{\left(\int_0^y \frac{2\mu(z)}{\sigma^2(z)} dz \right)} dy = \infty.$$

5   Large Deviations: One-dimensional Reflecting Diffusions

In this setting, we discuss the large deviations theory of Sect. 3, specialized to the setting of one-dimensional diffusions with reflecting barriers at 0 and b. Theorem 2 asserts that the key ODE in this setting requires finding $\psi(\theta) \in \mathbb{R}$ and $h_\theta \in C^2[0, b]$ for which

$$\mu(x)h_\theta'(x) + \frac{\sigma^2(x)}{2}h_\theta''(x) + (\theta f(x) - \psi(\theta))h_\theta(x) = 0, \; 0 \le x \le b \tag{24}$$

$$h_\theta'(0) + \theta r_0 h_\theta(0) = 0,$$

$$-h_\theta'(b) + \theta r_b h_\theta(b) = 0.$$

The above differential equation (24) can be put in the form

$$-\frac{d}{dx}(a(x)h_\theta'(x)) + b(x)h_\theta(x) = \lambda c(x)h_\theta(x) \tag{25}$$

for $0 \leq x \leq b$, where $\lambda = -\psi(\theta)$ and

$$a(x) = \exp\left(\int_0^x \frac{2\mu(y)}{\sigma^2(y)}dy\right),$$

$$b(x) = -\frac{2\theta f(x)}{\sigma^2(x)}\exp\left(\int_0^x \frac{2\mu(y)}{\sigma^2(y)}dy\right),$$

$$c(x) = \frac{2}{\sigma^2(x)}\exp\left(\int_0^x \frac{2\mu(y)}{\sigma^2(y)}dy\right).$$

Suppose that f, μ, and σ^2 are continuous on $[0, b]$, with $\sigma^2(x) > 0$ for $x \in [0, b]$. Because $a(\cdot)$ and $c(\cdot)$ are then positive on $[0, b]$, (25) takes the form of a so-called Sturm-Liouville problem. Consequently, there exist real eigenvalues $\lambda_1 < \lambda_2 < \cdots$ with $\lambda_n \to \infty$ satisfying (25), with corresponding eigenfunction solutions $v_1, v_2, \ldots$. Furthermore, the eigenfunction v_i has the property that it has exactly $i - 1$ roots in $[0, b]$; see, for example, Al-Gwaiz [1] for details on Sturm-Liouville theory. As a consequence, the eigenfunction v_1 is the only eigenfunction that can be taken to be positive over $[0, b]$. Thus, it follows that we should set $\psi(\theta) = -\lambda_1$ and $h_\theta = v_1$.

We now illustrate these ideas in the setting of reflecting Brownian motion in one dimension, again focusing on the boundary process by setting $f = 0$.

Example 3 (Two-sided Reflecting Brownian Motion). Here $\mu(x) = \mu$ and $\sigma^2(x) = \sigma^2 > 0$. The case in which $r_0 = 0$ and $r_b = 1$ was studied in detail in Zhang and Glynn [16]. In particular, consider the parameter spaces given by

$$\mathscr{R}_1 = \{(\theta, \mu, b) : \theta > 0\}$$

$$\mathscr{R}_2 = \{(\theta, \mu, b) : \theta < 0, \mu(\mu + \theta\sigma^2) \leq 0\}$$

$$\mathscr{R}_3 = \{(\theta, \mu, b) : \theta < 0, \mu(\mu + \theta\sigma^2) > 0, b\mu(\mu + \theta\sigma^2) > -\theta\sigma^4\}$$

$$\mathscr{R}_4 = \{(\theta, \mu, b) : \theta < 0, \mu(\mu + \theta\sigma^2) > 0, b\mu(\mu + \theta\sigma^2) < -\theta\sigma^4\}$$

$$\mathscr{B}_1 = \{(\theta, \mu, b) : \theta = 0\}$$

$$\mathscr{B}_2 = \{(\theta, \mu, b) : \theta < 0, \mu(\mu + \theta\sigma^2) > 0, b\mu(\mu + \theta\sigma^2) = -\theta\sigma^4\}.$$

The authors showed that, for $(\theta, \mu, b) \in \mathscr{R}_i$ ($i = 1, 3$), the solutions $\psi = \psi(\theta)$ and $h_\theta(\cdot)$ to

$$(\mathscr{L}h_\theta)(x) = \psi(\theta)h_\theta(x)$$

$$h_\theta'(0) = 0$$

$$h'_\theta(b) = \theta h_\theta(b)$$

$$h_\theta(0) = 1$$

for $0 \le x \le b$ are given by $\psi(\theta) = \frac{\beta(\theta)^2 - \mu^2}{2\sigma^2}$ and

$$h_\theta(x) = \frac{1}{2\beta(\theta)} e^{-\frac{\mu}{\sigma^2}x} \left[(\beta(\theta) - \mu) e^{-\frac{\beta(\theta)}{\sigma^2}x} + (\beta(\theta) + \mu) e^{\frac{\beta(\theta)}{\sigma^2}x} \right],$$

where $\beta(\theta)$ is the unique root in $\mathscr{F}_i$ of the equation

$$\frac{1}{\beta} \log \left(\frac{(\beta - \mu)(\beta + \mu + \theta\sigma^2)}{(\beta + \mu)(\beta - \mu - \theta\sigma^2)} \right) = \frac{2b}{\sigma^2},$$

with $\mathscr{F}_1 = (|\mu| \vee |\mu + \theta\sigma^2|, \infty)$ and $\mathscr{F}_3 = (0, |\mu| \wedge |\mu + \theta\sigma^2|)$. For $(\theta, \mu, b) \in \mathscr{R}_i$ $(i = 2, 4)$, the solutions are given by $\psi(\theta) = -\frac{\xi(\theta)^2 + \mu^2}{2\sigma^2}$ and

$$h_\theta(x) = e^{-\frac{\mu}{\sigma^2}x} \left[\cos\left(\frac{\xi(\theta)x}{\sigma^2} \right) + \frac{\mu}{\xi(\theta)} \sin\left(\frac{\xi(\theta)x}{\sigma^2} \right) \right],$$

where $\xi(\theta)$ is the unique root in $\left(0, \frac{\pi\sigma^2}{b} \right)$ of the equation

$$\frac{b\xi}{\sigma^2} = \arccos \left(\frac{\xi^2 + \mu(\mu + \theta\sigma^2)}{\sqrt{(\xi^2 + \mu(\mu + \theta\sigma^2))^2 + \xi^2\theta^2\sigma^4}} \right).$$

For $(\theta, \mu, b) \in \mathscr{B}_1$, $\psi(\theta) = 0$ and $h_\theta(x) \equiv 1$. Finally, for $(\theta, \mu, b) \in \mathscr{B}_2$, the solutions are given by $\psi(\theta) = -\frac{\mu^2}{2\sigma^2}$ and

$$h_\theta(x) = e^{-\frac{\mu}{\sigma^2}x} \left(\frac{\mu}{\sigma^2}x + 1 \right).$$

The case of arbitrary r_0 and r_b is conceptually similar, but requires even more complicated regions into which to separate the parameter space. For instance, it will be necessary to consider the signs of $\theta(r_0 + r_b)$, $(\mu - \theta r_0\sigma^2)(\mu + \theta r_b\sigma^2)$, and $b(\mu - \theta r_0\sigma^2)(\mu + \theta r_b\sigma^2) + \theta(r_0 + r_b)\sigma^4$, amongst other quantities. It is therefore clear that an explicit description of the solution to (24) will, in general, be very complex.

Acknowledgements The first author gratefully acknowledges the mentorship and friendship of Professor Miklós Csörgő, over the years, and Professor Csörgő's influence on both his research direction and academic career in the years that have passed since his graduation as a student at Carleton University.

References

1. Al-Gwaiz, M.: Sturm-Liouville Theory and Its Applications. Springer, New York (2007)
2. Bhattacharyya, R.N.: On the functional central limit theorem and the law of iterated logarithm for Markov processes. Z. Wahrsch. Verw. Gebiete. **60**, 185–201 (1982)
3. Bucklew, J.: Large Deviation Techniques in Decision, Simulation, and Estimation. Wiley, New York (1990)
4. Chen, H., Whitt, W.: Diffusion approximations for open queueing networks with service interruptions. Queueing Syst.: Theory Appl. **13**(4), 335–359 (1993)
5. Costantini, C.: Diffusion approximation for a class of transport processes with physical reflection boundary conditions. Ann. Probab. **19**(3), 1071–1101 (1991)
6. Dembo, A., Zeitouni, O.: Large Deviation Techniques and Applications, 2nd edn. Springer, New York (1998)
7. Ethier, S.N., Kurtz, T.G.: Markov Processes: Characterization and Convergence, 2nd edn. Wiley, New York (2005)
8. Harrison, J.M.: Brownian Motion and Stochastic Flow Systems. Wiley, New York (1985)
9. Karlin, S., Taylor, H.M.: A Second Course in Stochastic Processes. Academic, New York (1981)
10. Lions, P.L., Snitzman, A.S.: Stochastic differential equations with reflecting boundary conditions. Commun. Pure Appl. **37**, 511–537 (1984)
11. Meyn, S.P., Tweedie, R.L.: Stability of Markovian processes II: continuous-time processes and sampled chains. Adv. Appl. Probab. **25**, 487–517 (1993a)
12. Meyn, S.P., Tweedie, R.L.: Stability of Markovian processes III: Foster-Lyapunov criteria for continuous-time processes. Adv. Appl. Probab. **25**, 518–548 (1993b)
13. Meyn, S.P., Tweedie, R.L.: Exponential and uniform ergodicity of Markov processes. Ann. Probab. **23**(4), 1671–1691 (1995)
14. Sigman, K.: One-dependent regenerative processes and queues in continuous time. Math. Oper. Res. **15**, 175–189 (1990)
15. Williams, R.J.: Asymptotic variance parameters for the boundary local times of reflected Brownian motion on a compact interval. J. Appl. Probab. **29**(4), 996–1002 (1992)
16. Zhang, X., Glynn, P.W.: On the dynamics of a finite buffer queue conditioned on the amount of loss. Queueing Syst.: Theory Appl. **67**(2), 91–110 (2011)

Kellerer's Theorem Revisited

Francis Hirsch, Bernard Roynette, and Marc Yor

1 Introduction

In the following, we shall call a *peacock*, a family $(\mu_t, t \geq 0)$ of probability measures on $\mathbb{R}$ such that:

(i) $\forall t \geq 0, \quad \int |x| \, \mu_t(\mathrm{d}x) < \infty;$

(ii) for every convex function $\psi : \mathbb{R} \longrightarrow \mathbb{R}$, the map:

$$t \geq 0 \longrightarrow \int \psi(x) \, \mu_t(\mathrm{d}x) \in (-\infty, +\infty]$$

is increasing.

An $\mathbb{R}$-valued process $(X_t, t \geq 0)$ will also be called a *peacock*, if the family of its one-dimensional marginal laws is a peacock.

F. Hirsch (✉)
Laboratoire d'Analyse et Probabilités, Université d'Évry-Val d'Essonne,
23 Boulevard de France, F-91037 Evry Cedex, France
e-mail: francis.hirsch@univ-evry.fr

B. Roynette
Institut Elie Cartan, Université Henri Poincaré, B.P. 239, F-54506
Vandœuvre-lès-Nancy Cedex, France

M. Yor
Laboratoire de Probabilités et Modèles Aléatoires, Université Paris VI et VII,
4 Place Jussieu, Case 188, F-75252 Paris Cedex 05, France

Institut Universitaire de France, Paris, France

© Springer Science+Business Media New York 2015
D. Dawson et al. (eds.), *Asymptotic Laws and Methods in Stochastics*,
Fields Institute Communications 76, DOI 10.1007/978-1-4939-3076-0_18

347

The term "peacock" comes from PCOC (pronounced peacock), which is the acronym for the french expression: Processus Croissant pour l'Ordre Convexe (which means: Increasing Process in the Convex Order). We refer to the recent monograph [5] for an introduction to this topic, a description of possible applications, many examples and relevant references.

We say that two $\mathbb{R}$-valued processes are *associated*, if they have the same one-dimensional marginals. A process which is associated with a martingale is called a 1-*martingale*.

Likewise, a family $(\mu_t, t \geq 0)$ of probability measures on $\mathbb{R}$ and an $\mathbb{R}$-valued process $(Y_t, t \geq 0)$ will be said to be *associated* if, for every $t \geq 0$, the law of Y_t is μ_t, i.e. if $(\mu_t, t \geq 0)$ is the family of the one-dimensional marginal laws of $(Y_t, t \geq 0)$.

It is an easy consequence of Jensen's inequality that an $\mathbb{R}$-valued process $(X_t, t \geq 0)$ which is a 1-martingale, is a peacock. A remarkable result due to H. Kellerer [6] states that, conversely, any $\mathbb{R}$-valued process $(X_t, t \geq 0)$ which is a peacock, is a 1-martingale. More precisely, Kellerer's result states that any peacock admits an associated martingale which has the *Markov property*. Note that, in general, it is a difficult challenge to exhibit *explicitly* a martingale associated to a given peacock. The most part of the monograph [5] is devoted to this question.

In the recent paper [4], a new proof of Kellerer's theorem (but without the Markov property) was presented. On the other hand, G. Lowther [8] showed that if $(\mu_t, t \geq 0)$ is a peacock such that the map: $t \longrightarrow \mu_t$ is weakly continuous (i.e. for any $\mathbb{R}$-valued, bounded and continuous function f on $\mathbb{R}$, the map: $t \longrightarrow \int f(x) \ \mu_t(dx)$ is continuous), then $(\mu_t, t \geq 0)$ is associated with a unique *strongly Markov* martingale which moreover is *almost-continuous* (see Sect. 4 for definitions). Actually, this paper [8] partially relies on [7] and [9], but it seems that only [9] was published in a journal.

In this paper, our aim is two-fold:

1. to give a proof of Kellerer's theorem (including the Markov property), following essentially [4] and [9];
2. to present, without proof, results of [7, 8] and [9], which complete and precise Kellerer's theorem on some points.

For the sake of clarity and brevity, we refer here essentially to these papers. Many other references around Kellerer's theorem may be found in [5].

The remainder of this paper is organised as follows:

- Section 2 is devoted to preliminary results about *call functions* (Sect. 2.1), *Lipschitz-Markov property* (Sect. 2.2), *finite-dimensional convergence* (Sect. 2.3) and *Fokker-Planck equation* (Sect. 2.4);
- in Sect. 3, we prove Kellerer's theorem by a two steps approximation. We first consider the *regular case* (Sect. 3.1), then the right-continuous case (Sect. 3.2) and, finally, the general case (Sect. 3.3);
- in Sect. 4, we gather some related results from [7, 8] and [9], concerning notably *strong Markov property*, *almost-continuity* and *uniqueness*.

2 Preliminary Results

In this section, we fix further notation and terminology, and we gather some preliminary results which are essential in the sequel.

2.1 Call Functions

In the following, we denote by $\mathcal{M}$ the set of probability measures on $\mathbb{R}$, equipped with the topology of weak convergence (with respect to the space of $\mathbb{R}$-valued, bounded, continuous functions on $\mathbb{R}$).

We denote by $\mathcal{M}_f$ the subset of $\mathcal{M}$ consisting of measures $\mu \in \mathcal{M}$ such that $\int |x| \, \mu(dx) < \infty$. For $\mu \in \mathcal{M}_f$, we denote by $\mathbb{E}[\mu]$ the expectation of μ, namely:

$$\mathbb{E}[\mu] = \int x \, \mu(dx).$$

We define, for $\mu \in \mathcal{M}_f$, the *call function* C_μ by:

$$\forall x \in \mathbb{R}, \quad C_\mu(x) = \int (y - x)^+ \, \mu(dy).$$

The following easy (and classical) proposition holds (see e.g. [4, Proposition 2.1]).

Proposition 1. *If $\mu \in \mathcal{M}_f$, then C_μ satisfies the following properties:*

(a) *C_μ is a convex, nonnegative function on $\mathbb{R}$.*
(b) *$\lim_{x \to +\infty} C_\mu(x) = 0$.*
(c) *There exists $a \in \mathbb{R}$ such that $\lim_{x \to -\infty}(C_\mu(x) + x) = a$.*

Conversely, if a function C satisfies the above three properties, then there exists a unique $\mu \in \mathcal{M}_f$ such that $C = C_\mu$. This measure μ is the second derivative, in the sense of distributions, of the function C, and $a = \mathbb{E}[\mu]$.

To state the next proposition (which also is classical and whose proof can be found e.g. in [4]), we now recall that a subset $\mathcal{H}$ of $\mathcal{M}$ is said to be *uniformly integrable* if

$$\lim_{c \to +\infty} \sup_{\mu \in \mathcal{H}} \int_{\{|x| \geq c\}} |x| \, \mu(dx) = 0.$$

We remark that, if $\mathcal{H}$ is uniformly integrable, then

$$\mathcal{H} \subset \mathcal{M}_f \quad \text{and} \quad \sup\left\{ \int |x| \, \mu(dx); \ \mu \in \mathcal{H} \right\} < \infty.$$

Proposition 2. *Let I be a set and let $\mathscr{E}$ be a filter on I. Consider a uniformly integrable family $(\mu_i, i \in I)$ in $\mathscr{M}$, and $\mu \in \mathscr{M}$. The following properties are equivalent:*

(1) $\lim_{\mathscr{E}} \mu_i = \mu$ *with respect to the topology on $\mathscr{M}$.*

(2) $\mu \in \mathscr{M}_f$ *and*

$$\forall x \in \mathbb{R}, \quad \lim_{\mathscr{E}} C_{\mu_i}(x) = C_\mu(x).$$

We now fix a family $(\mu_t, t \geq 0)$ in $\mathscr{M}_f$ and we define a function $C(t, x)$ on $\mathbb{R}_+ \times \mathbb{R}$ by:

$$C(t, x) = C_{\mu_t}(x).$$

The following characterization of peacocks is easy to prove and is stated in [5, Exercise 1.7].

Proposition 3. *The family $(\mu_t, t \geq 0)$ is a peacock if and only if:*

1. the expectation $\mathbb{E}[\mu_t]$ does not depend on t,
2. for every $x \in \mathbb{R}$, the function $t \geq 0 \longrightarrow C(t, x)$ is increasing.

We also have (see [5, Exercise 1.1]):

Proposition 4. *Assume that $(\mu_t, t \geq 0)$ is a peacock, and let $T > 0$. Then, the set $\{\mu_t; \ 0 \leq t \leq T\}$ is uniformly integrable.*

2.2 Lipschitz-Markov Property

Following [9] (see also [6, Definition 3]), we now introduce a property, namely the Lipschitz-Markov property, which is stronger than the mere Markov property. (Note that in [9], the Lipschitz-Markov property is simply called *Lipschitz property*.)

If $(X_t, t \geq 0)$ is an $\mathbb{R}$-valued process, we denote by $\mathscr{F}^X$ the filtration generated by X, that is:

$$\forall t \geq 0, \quad \mathscr{F}_t^X = \sigma(X_s; \ s \leq t).$$

On the other hand, for any Lipschitz continuous $f : \mathbb{R} \longrightarrow \mathbb{R}$, we denote by $L(f)$ its Lipschitz constant.

Definition 1 ([9], Definition 4.1). *Let X be an $\mathbb{R}$-valued stochastic process. Then X is said to satisfy the Lipschitz-Markov property if, for all $0 \leq s < t$ and every bounded Lipschitz continuous $g : \mathbb{R} \longrightarrow \mathbb{R}$ with $L(g) \leq 1$, there exists a Lipschitz continuous $f : \mathbb{R} \longrightarrow \mathbb{R}$ with $L(f) \leq 1$ and*

$$f(X_s) = \mathbb{E}[g(X_t)|\mathscr{F}_s^X].$$

The following proposition presents an important example of process satisfying the Lipschitz-Markov property (see also Proposition 9 below, based on [9, Lemma 4.3]). In the sequel, we adopt the following notation: $U = (0, +\infty) \times \mathbb{R}$ and $\overline{U} = \mathbb{R}_+ \times \mathbb{R}$.

Proposition 5. Let $\sigma : (t, x) \in \overline{U} \longrightarrow \sigma(t, x) \in \mathbb{R}$ be a continuous function on $\overline{U}$ such that the derivative σ'_x exists and is continuous on $\overline{U}$. Let X_0 be an integrable random variable and let $(B_t, t \geq 0)$ denote a standard Brownian motion independent of X_0. Then, the stochastic differential equation

$$X_t = X_0 + \int_0^t \sigma(s, X_s) \, dB_s \tag{1}$$

admits a unique strong solution which satisfies the Lipschitz-Markov property.

Proof. It is classical that Eq. (1) admits a unique (non-exploding) strong solution.

Let $s \geq 0$. For any $x \in \mathbb{R}$, we denote by $(X_t^{s,x}, t \geq s)$ the strong solution (for $t \geq s$) of

$$X_t^{s,x} = x + \int_s^t \sigma(u, X_u^{s,x}) \, dB_u.$$

We also denote by $(U_t^{s,x}, t \geq s)$ the process defined by:

$$U_t^{s,x} = \frac{\partial}{\partial x} X_t^{s,x}.$$

We obtain easily:

$$U_t^{s,x} = \exp\left[\int_s^t \sigma'_x(u, X_u^{s,x}) \, dB_u - \frac{1}{2} \int_s^t \sigma'^2_x(u, X_u^{s,x}) \, du\right].$$

In particular, $(U_t^{s,x}, t \geq s)$ is a positive local martingale and hence:

$$U_t^{s,x} \geq 0 \quad \text{and} \quad \mathbb{E}[U_t^{s,x}] \leq 1.$$

Let now g be a bounded Lipschitz continuous function of C^1-class with $|g'| \leq 1$ and $0 \leq s < t$. We define the function $f : \mathbb{R} \longrightarrow \mathbb{R}$ by:

$$f(x) = \mathbb{E}\left[g\left(X_t^{s,x}\right)\right].$$

Then, f is a bounded C^1-function and

$$|f'(x)| = \left|\mathbb{E}\left[g'\left(X_t^{s,x}\right) U_t^{s,x}\right]\right| \leq 1.$$

It is now clear that, if $(X_t, t \geq 0)$ is solution to (1), then

$$\mathbb{E}[g(X_t)|\mathscr{F}_s^X] = \mathbb{E}\left[g\left(X_t^{s,x}\right)\right]\big|_{x=X_s} = f(X_s).$$

The Lipschitz-Markov property follows easily from what precedes. $\qquad\square$

2.3   Finite-dimensional Convergence

Definition 2. Let I be a set and let $\mathscr{E}$ be a filter on I. We consider $\mathbb{R}$-valued stochastic processes $(X^{(i)})_{i \in I}$ and X possibly defined on different probability spaces. Then we shall say that $(X^{(i)})_{i \in I}$ converges (with respect to $\mathscr{E}$) to X in the sense of finite-dimensional distributions if, for any finite subset $\{t_1, t_2, \cdots, t_n\}$ of $\mathbb{R}_+$, the distributions of $(X_{t_1}^{(i)}, X_{t_2}^{(i)}, \cdots, X_{t_n}^{(i)})$ converge weakly to the distribution of $(X_{t_1}, X_{t_2}, \cdots, X_{t_n})$, which means that, for any bounded continuous $f : \mathbb{R}^n \longrightarrow \mathbb{R}$,

$$\lim_{\mathscr{E}} \mathbb{E}[f(X_{t_1}^{(i)}, X_{t_2}^{(i)}, \cdots, X_{t_n}^{(i)})] = \mathbb{E}[f(X_{t_1}, X_{t_2}, \cdots, X_{t_n})].$$

We also shall write:

$$\text{f.d.} \lim_{\mathscr{E}} X^{(i)} = X$$

and we shall say, in short, that $X^{(i)}$ f.d. converges (with respect to $\mathscr{E}$) to X.

The finite-dimensional convergence has important stability properties, in particular with respect to the Lipschitz-Markov property and to the martingale property. This is stated in the next propositions where we consider, like in Definition 2, a set I, a filter $\mathscr{E}$ on I and $\mathbb{R}$-valued stochastic processes $(X^{(i)})_{i \in I}$ and X. The following proposition extends [6, Satz 10].

Proposition 6 ([9], Lemma 4.5). *Suppose that, for every $i \in I$, $X^{(i)}$ satisfies the Lipschitz-Markov property, and that $X^{(i)}$ f.d. converges to X. Then, X satisfies the Lipschitz-Markov property.*

Proof. Let $0 \le s < t$ and a bounded Lipschitz continuous $g : \mathbb{R} \longrightarrow \mathbb{R}$ with $L(g) \le 1$. For any $i \in I$, there exists a Lipschitz continuous $f^{(i)} : \mathbb{R} \longrightarrow \mathbb{R}$ with $L(f^{(i)}) \le 1$ and

$$f^{(i)}(X_s^{(i)}) = \mathbb{E}[g(X_t^{(i)})|\mathscr{F}_s^{X^{(i)}}].$$

Moreover, we may suppose that:

$$\forall i \in I, \quad \sup_{x \in \mathbb{R}} |f^{(i)}(x)| \le \sup_{x \in \mathbb{R}} |g(x)|.$$

Consider an ultrafilter $\mathscr{U}$ on I which refines $\mathscr{E}$. By Ascoli's theorem, there exists a Lipschitz continuous $f : \mathbb{R} \longrightarrow \mathbb{R}$ with $L(f) \le 1$ and

$$\lim_{\mathscr{U}} f^{(i)} = f \quad \text{uniformly on compact sets.}$$

For every $n \in \mathbb{N}$, $0 \le s_1 < s_2 < \cdots < s_n \le s$, for any bounded continuous $h : \mathbb{R}^n \longrightarrow \mathbb{R}$ and for any continuous $\theta : \mathbb{R} \longrightarrow \mathbb{R}$ with compact support, one has, for every $i \in I$,

$$\mathbb{E}[g(X_t^{(i)})\,\theta(X_s^{(i)})\,h(X_{s_1}^{(i)},X_{s_2}^{(i)},\cdots,X_{s_n}^{(i)})]$$
$$= \mathbb{E}[f^{(i)}(X_s^{(i)})\,\theta(X_s^{(i)})\,h(X_{s_1}^{(i)},X_{s_2}^{(i)},\cdots,X_{s_n}^{(i)})].$$

Since

$$\lim_{\mathcal{U}} f^{(i)}\theta = f\theta \quad \text{uniformly}$$

and

$$\text{f.d.}\lim_{\mathcal{U}} X^{(i)} = X,$$

we obtain:

$$\mathbb{E}[g(X_t)\,\theta(X_s)\,h(X_{s_1},X_{s_2},\cdots,X_{s_n})] = \mathbb{E}[f(X_s)\,\theta(X_s)\,h(X_{s_1},X_{s_2},\cdots,X_{s_n})].$$

This holding for any θ with compact support, we also have:

$$\mathbb{E}[g(X_t)\,h(X_{s_1},X_{s_2},\cdots,X_{s_n})] = \mathbb{E}[f(X_s)\,h(X_{s_1},X_{s_2},\cdots,X_{s_n})],$$

which yields the desired result. □

Proposition 7. *Suppose that, for every $i \in I$, $X^{(i)}$ is a martingale, and that $X^{(i)}$ f.d. converges to X. Suppose moreover that, for every $t \geq 0$,*

$$\left\{ X_t^{(i)};\ i \in I \right\} \quad \text{is uniformly integrable.}$$

Then, X is a martingale.

Proof. By Proposition 2, the process X is integrable. We now prove that it is a martingale. We set:

$$\forall p > 0,\ \forall x \in \mathbb{R},\quad \varphi_p(x) = \min[\max(x,-p),p].$$

Then, φ_p is a bounded continuous function, and

$$|\varphi_p(x) - x| \leq |x|\,1_{\{|x|>p\}}.$$

For every $n \in \mathbb{N}$, $0 \leq s_1 < s_2 < \cdots < s_n \leq s \leq t$, for any bounded continuous $h : \mathbb{R}^n \longrightarrow \mathbb{R}$, we have for every $i \in I$,

$$\mathbb{E}[h(X_{s_1}^{(i)},\cdots,X_{s_n}^{(i)})\,X_t^{(i)}] = \mathbb{E}[h(X_{s_1}^{(i)},\cdots,X_{s_n}^{(i)})\,X_s^{(i)}].$$

We set: $\| h \|_\infty = \sup\{|h(x)|;\ x \in \mathbb{R}^n\}$. Then,

$$\left| \mathbb{E}[h(X_{s_1}, \cdots, X_{s_n}) \, \varphi_p(X_t)] - \mathbb{E}[h(X_{s_1}, \cdots, X_{s_n}) \, X_t] \right|$$

$$\leq \| h \|_\infty \, \mathbb{E}\left[|X_t| \, 1_{\{|X_t| > p\}} \right], \quad \text{for every } p > 0,$$

$$\left| \mathbb{E}[h(X_{s_1}^{(i)}, \cdots, X_{s_n}^{(i)}) \, \varphi_p(X_t^{(i)})] - \mathbb{E}[h(X_{s_1}^{(i)}, \cdots, X_{s_n}^{(i)}) \, X_t^{(i)}] \right|$$

$$\leq \| h \|_\infty \, \mathbb{E}\left[|X_t^{(i)}| \, 1_{\{|X_t^{(i)}| > p\}} \right], \quad \text{for every } i \in I \text{ and every } p > 0,$$

and likewise, replacing t by s. Moreover,

$$\lim_{\mathscr{E}} \mathbb{E}[h(X_{s_1}^{(i)}, \cdots, X_{s_n}^{(i)}) \, \varphi_p(X_t^{(i)})] = \mathbb{E}[h(X_{s_1}, \cdots, X_{s_n}) \, \varphi_p(X_t)],$$

and likewise, replacing t by s. Finally, we obtain, for $p > 0$,

$$\left| \mathbb{E}[h(X_{s_1}, \cdots, X_{s_n}) \, X_t] - \mathbb{E}[h(X_{s_1}, \cdots, X_{s_n}) \, X_s] \right|$$

$$\leq 2 \, \| h \|_\infty \, \left(\sup_{i \in I} \mathbb{E}\left[|X_t^{(i)}| \, 1_{\{|X_t^{(i)}| > p\}} \right] + \sup_{i \in I} \mathbb{E}\left[|X_s^{(i)}| \, 1_{\{|X_s^{(i)}| > p\}} \right] \right),$$

and the desired result follows, letting p go to ∞. $\square$

2.4 Fokker-Planck Equation

We now state M. Pierre's uniqueness theorem for a Fokker-Planck equation, which plays an important role in our proof of Kellerer's theorem in the regular case. This theorem is stated and proved in Subsection 6.1 of [5].

Theorem 1 ([5], Theorem 6.1). *Let $a : (t, x) \in \overline{U} \longrightarrow a(t, x) \in \mathbb{R}_+$ be a continuous function such that $a(t, x) > 0$ for $(t, x) \in U$, and let $\mu \in \mathscr{M}_f$. Then there exists at most one family of probability measures $(p(t, \mathrm{d}x), t \geq 0)$ such that*

(FP1) $t \geq 0 \longrightarrow p(t, \mathrm{d}x)$ *is weakly continuous,*
(FP2) $p(0, \mathrm{d}x) = \mu(\mathrm{d}x)$ *and*

$$\frac{\partial p}{\partial t} - \frac{\partial^2}{\partial x^2}(a p) = 0 \ \ in \ \mathscr{D}'(U)$$

   (i.e. in the sense of Schwartz distributions in the open set U).

3 Kellerer's Theorem

In this section, we shall give a proof of Kellerer's theorem. Following [4], we shall proceed by a two steps approximation, starting with the regular case.

3.1 The Regular Case

Theorem 2. *Let $(X_t, t \geq 0)$ be an $\mathbb{R}$-valued integrable process such that $\mathbb{E}[X_t]$ is independent of t, and let $C : \overline{U} \longrightarrow \mathbb{R}_+$ the corresponding call function (see Sect. 2.1):*

$$C(t, x) = \mathbb{E}[(X_t - x)^+].$$

We assume:

(i) C is a C^∞-function on $\overline{U}$.
   We set:

$$\forall (t, x) \in \overline{U}, \quad p(t, x) = \frac{\partial^2 C}{\partial x^2}(t, x).$$

   Thus, the law of X_t is $p(t, x)\mathrm{d}x$.

(ii) $p > 0$ on $\overline{U}$ and $\dfrac{\partial C}{\partial t} > 0$ on U.

We set:

$$\forall (t, x) \in \overline{U}, \quad \sigma(t, x) = \left(2 \frac{\frac{\partial C}{\partial t}(t, x)}{p(t, x)} \right)^{1/2}.$$

Then, the stochastic differential equation

$$Y_t = Y_0 + \int_0^t \sigma(s, Y_s) \, \mathrm{d}B_s$$

(where Y_0 is a random variable with law $p(0, x)\mathrm{d}x$, independent of the Brownian motion $(B_s, s \geq 0)$) admits a unique strong solution, which is a martingale associated to X and satisfying the Lipschitz-Markov property.

Proof. (1) We first prove that Y is associated to X. Set:

$$a = \frac{1}{2} \sigma^2 = \frac{\frac{\partial C}{\partial t}}{p}.$$

We have;

$$\frac{\partial^2}{\partial x^2}(a p) = \frac{\partial^2}{\partial x^2} \frac{\partial}{\partial t} C = \frac{\partial}{\partial t} p$$

on $\overline{U}$. In particular, the family $(p(t, x)\mathrm{d}x, t \geq 0)$ satisfies (FP1) and (FP2) in Theorem 1. On the other hand, for any C^2-function $\varphi : \mathbb{R} \longrightarrow \mathbb{R}$ with compact support, we have by Itô's formula:

$$\mathbb{E}[\varphi(Y_t)] = \mathbb{E}[\varphi(Y_0)] + \int_0^t \mathbb{E}[\varphi''(Y_s) \, a(s, Y_s)] \, \mathrm{d}s$$

and therefore, denoting, for every $t \geq 0$, by $q(t, dx)$ the law of Y_t,

$$\frac{d}{dt} \int \varphi(x) \, q(t, dx) = \int \varphi''(x) \, a(t, x) \, q(t, dx).$$

Thus, the family $(q(t, dx), t \geq 0)$ also satisfies properties (FP1) and (FP2). Hence, by Theorem 1,

$$\forall t \geq 0, \quad q(t, dx) = p(t, x) dx$$

and Y is associated to X.

(2) Obviously, Y is a local martingale. We now prove that it is a true martingale. Let ϕ be a C^2-function on $\mathbb{R}$ such that $\phi(x) = 1$ for $|x| \leq 1$, $\phi(x) = 0$ for $|x| \geq 2$, and $0 \leq \phi(x) \leq 1$ for all $x \in \mathbb{R}$. We set, for $k > 0$, $\phi_k(x) = x \phi(k^{-1} x)$. Fix now $0 \leq s_1 \leq \cdots \leq s_n \leq s \leq t$ and a bounded continuous $h : \mathbb{R}^n \longrightarrow \mathbb{R}$. We set:

$$\theta_k = \mathbb{E}[h(Y_{s_1}, Y_{s_2}, \cdots, Y_{s_n}) \, \phi_k(Y_t)] - \mathbb{E}[h(Y_{s_1}, Y_{s_2}, \cdots, Y_{s_n}) \, \phi_k(Y_s)]$$

and $m = \sup_{x \in \mathbb{R}^n} |h(x)|$. By dominated convergence,

$$\lim_{k \to \infty} \theta_k = \mathbb{E}[h(Y_{s_1}, Y_{s_2}, \cdots, Y_{s_n}) \, Y_t] - \mathbb{E}[h(Y_{s_1}, Y_{s_2}, \cdots, Y_{s_n}) \, Y_s].$$

On the other hand, since Y is associated to X, Itô's formula yields:

$$|\theta_k| \leq \frac{m}{2} \int_s^t \mathbb{E}\left[|\phi_k''(Y_u)| \, \sigma^2(u, Y_u) \right] \, du = m \int \int_s^t |\phi_k''(x)| \frac{\partial C}{\partial u}(u, x) \, du \, dx.$$

Besides,

$$\int |\phi_k''(x)| \, dx = \int |x \phi''(x) + 2\phi'(x)| \, dx$$

and $\phi_k''(x) = 0$ for $|x| \notin [k, 2k]$. Therefore, there exists a constant $\tilde{m}$ such that:

$$|\theta_k| \leq \tilde{m} \, \sup\{C(t, y) - C(s, y); \; k \leq |y| \leq 2k\}.$$

Thus, since by hypothesis $\mathbb{E}[X_s] = \mathbb{E}[X_t]$, Proposition 1 entails: $\lim_{k \to \infty} \theta_k = 0$, which yields the desired result.

(3) Finally, by Proposition 5, Y satisfies the Lipschitz-Markov property.

$\square$

Remark 1. By now, the formula giving σ in terms of the derivatives of C, in the statement of Theorem 2, is common in Mathematical Finance, where it is referred to Dupire [3] or Derman-Kani [2].

3.2 The Right-continuous Case

We now present our proof of Kellerer's theorem for right-continuous peacocks.

Theorem 3. *Let $(\mu_t, t \geq 0)$ be a peacock such that the map:*

$$t \geq 0 \longrightarrow \mu_t \in \mathcal{M}$$

is right-continuous. Then there exists a càdlàg martingale associated to $(\mu_t, t \geq 0)$ and satisfying the Lipschitz-Markov property.

Proof. We set, as in Sect. 2.1, $C(t, x) = C_{\mu_t}(x)$. We shall regularize, in space and time, $p(t, dx) := \mu_t(dx)$ considered as a distribution on U. Thus, let α be a density of probability on $\mathbb{R}$, of C^∞-class, with compact support contained in $[0, 1]$. We set, for $\varepsilon \in (0, 1)$ and $(t, x) \in \mathbb{R}_+ \times \mathbb{R}$,

$$p_\varepsilon(t, x) = \frac{1 - \varepsilon}{\varepsilon} \int \alpha(u) \left[\int \alpha\left(\frac{y - x}{\varepsilon}\right) \mu_{t + \varepsilon u}(dy) \right] du + \varepsilon \, g(t, x)$$

with

$$g(t, x) = \frac{1}{\sqrt{2\pi(1 + t)}} \exp\left(-\frac{x^2}{2(1 + t)}\right).$$

Lemma 1. *The function p_ε is of C^∞-class on $\mathbb{R}_+ \times \mathbb{R}$ and $p_\varepsilon(t, x) > 0$ for any (t, x). Moreover,*

$$\int p_\varepsilon(t, x) \, dx = 1 \quad and \quad \int |x| \, p_\varepsilon(t, x) \, dx < \infty.$$

The proof is straightforward.

We now set:

$$\mu_t^\varepsilon(dx) = p_\varepsilon(t, x) \, dx.$$

By Lemma 1, $\mu_t^\varepsilon \in \mathcal{M}_f$ and we set:

$$C_\varepsilon(t, x) = C_{\mu_t^\varepsilon}(x).$$

Lemma 2. *For any $t \geq 0$, the set $\{\mu_t^\varepsilon; 0 < \varepsilon < 1\}$ is uniformly integrable.*

Proof. Let $a = \int y \alpha(y) \, dy$. A simple computation yields:

$$\int_{\{|x| \geq c\}} |x| \, \mu_t^\varepsilon(dx) \leq \int \alpha(u) \left[\int_{\{|y| \geq c - 1\}} (|y| + a) \, \mu_{t + \varepsilon u}(dy) \right] du$$

$$+ \int_{\{|x| \geq c\}} |x| \, g(t, x) \, dx$$

and the result follows from the uniform integrability of $\{\mu_v; 0 \leq v \leq t + 1\}$ (Proposition 4). $\qquad \square$

Lemma 3. *One has:*

$$C_\varepsilon(t, x) = (1 - \varepsilon) \int \int \alpha(u)\, \alpha(y)\, C(t + \varepsilon u, x + \varepsilon y)\; \mathrm{d}y\, \mathrm{d}u$$

$$+ \varepsilon \int_x^{+\infty} (y - x)\, g(t, y)\; \mathrm{d}y.$$

The function C_ε is of C^∞-class on $\mathbb{R}_+ \times \mathbb{R}$. Moreover, for any $(t, x) \in \mathbb{R}_+ \times \mathbb{R}$,

$$\frac{\partial C_\varepsilon}{\partial t}(t, x) > 0 \quad and \quad \frac{\partial^2 C_\varepsilon}{\partial x^2}(t, x) = p_\varepsilon(t, x).$$

Proof. The above expression of C_ε follows directly from the definitions. We deduce therefrom that C_ε is of C^∞-class on $\mathbb{R}_+ \times \mathbb{R}$. Now, by property 2 in Proposition 3,

$$\frac{\partial C_\varepsilon}{\partial t}(t, x) \geq \varepsilon \frac{\partial}{\partial t} \left[\int_x^{+\infty} (y - x)\, g(t, y)\; \mathrm{d}y \right] = \frac{\varepsilon}{2}\, g(t, x) > 0.$$

Finally, the equality:

$$\frac{\partial^2 C_\varepsilon}{\partial x^2}(t, x) = p_\varepsilon(t, x)$$

holds, since, by Proposition 1, it holds in the sense of distributions, and both sides are continuous. $\square$

The following lemma is an easy consequence of the right-continuity of $(\mu_t, t \geq 0)$.

Lemma 4. *For $t \geq 0$,*

$$\lim_{\varepsilon \to 0} \mu_t^\varepsilon = \mu_t \quad in\ \mathcal{M}.$$

By Theorem 2, there exists a martingale $(M_t^\varepsilon, t \geq 0)$ satisfying the Lipschitz-Markov property, which is associated to $(\mu_t^\varepsilon, t \geq 0)$. For every $n \in \mathbb{N}$ and $\tau_n = (t_1, \cdots, t_n) \in \mathbb{R}_+^n$, we denote by $\mu_{\tau_n}^{(\varepsilon, n)}$ the law of $(M_{t_1}^\varepsilon, \cdots, M_{t_n}^\varepsilon)$, a probability on $\mathbb{R}^n$.

Lemma 5. *For every $n \in \mathbb{N}$ and $\tau_n \in \mathbb{R}_+^n$, the set of probability measures: $\{\mu_{\tau_n}^{(\varepsilon, n)};\ 0 < \varepsilon < 1\}$, is tight.*

Proof. Let $n \in \mathbb{N}$ and $\tau_n = (t_1, \cdots, t_n) \in \mathbb{R}_+^n$. For $x = (x_1, \cdots, x_n) \in \mathbb{R}^n$, we set $|x| := \sup_{1 \leq j \leq n} |x_j|$. Then, for $c > 0$,

$$\mu_{\tau_n}^{(\varepsilon, n)}(|x| \geq c) = \mathbb{P}\left(\sup_{1 \leq j \leq n} |M_{t_j}^\varepsilon| \geq c \right) \leq \frac{1}{c} \mathbb{E}\left[\sup_{1 \leq j \leq n} |M_{t_j}^\varepsilon| \right]$$

$$\leq \frac{1}{c} \sum_{j=1}^n \mathbb{E}\left[|M_{t_j}^\varepsilon| \right] = \frac{1}{c} \sum_{j=1}^n \int |x|\, \mu_{t_j}^\varepsilon(\mathrm{d}x).$$

Now, by Lemma 2, for $1 \le j \le n$,

$$\sup_{0 < \varepsilon < 1} \int |x| \, \mu_{t_j}^{\varepsilon}(dx) < \infty.$$

Thus,

$$\lim_{c \to +\infty} \sup_{0 < \varepsilon < 1} \mu_{\tau_n}^{(\varepsilon, n)}(|x| \ge c) = 0,$$

which yields the tightness of $\{\mu_{\tau_n}^{(\varepsilon, n)}; \ 0 < \varepsilon < 1\}$. $\qquad\qquad\square$

Let $\mathcal{U}$ be an ultrafilter on $(0, 1)$ such that $\lim_{\mathcal{U}} \varepsilon = 0$. As a consequence of the previous lemma, the family of probabilities on $\mathbb{R}^n$: $(\mu_{\tau_n}^{(\varepsilon, n)}, \varepsilon > 0)$, weakly converges (with respect to $\mathcal{U}$) to a probability which we denote by $\mu_{\tau_n}^{(n)}$. We remark that, by Lemma 4, for any $t \ge 0$, $\mu_{(t)}^{(1)} = \mu_t$. By Kolmogorov's extension theorem, there exists a process $(M_t, t \ge 0)$ such that, for every $n \in \mathbb{N}$ and every $\tau_n = (t_1, \cdots, t_n) \in \mathbb{R}_+^n$, the law of $(M_{t_1}, \cdots, M_{t_n})$ is $\mu_{\tau_n}^{(n)}$. Since for any $t \ge 0$, $\mu_{(t)}^{(1)} = \mu_t$, the process $M = (M_t, t \ge 0)$ is associated to $(\mu_t, t \ge 0)$. Moreover, by Proposition 6, M satisfies the Lipschitz-Markov property, and by Lemma 2 and Proposition 7, M is a martingale. By the classical theory of martingales (see, for example, [1]), almost surely, for every $t \ge 0$,

$$\tilde{M}_t = \lim_{s \to t, s \in \mathbb{Q}, s > t} M_s$$

is well defined, and $(\tilde{M}_t, t \ge 0)$ is a right-continuous martingale which, by the right-continuity of $(\mu_t, t \ge 0)$, is associated to $(\mu_t, t \ge 0)$. Besides, it is easy to see that $\tilde{M}$ still satisfies the Lipschitz-Markov property. Modifying $\tilde{M}$ on a negligible set, we may assume that $\tilde{M}$ is a càdlàg process and, moreover,

$$\forall t \in \mathbb{Q}_+, \quad \tilde{M}_t = M_t \ \text{a.s.}$$

$\qquad\qquad\square$

3.3 The General Case

We now obtain, by approximation, a proof of Kellerer's theorem in the general case.

Theorem 4. *Let $(\mu_t, t \ge 0)$ be a peacock. Then there exists a martingale associated to $(\mu_t, t \ge 0)$ and satisfying the Lipschitz-Markov property.*

Proof. We consider a peacock $(\mu_t, t \ge 0)$ and we set $C(t, x) = C_{\mu_t}(x)$.

Lemma 6. *There exists a countable set $D \subset \mathbb{R}_+$ such that the map:*

$$t \longrightarrow \mu_t \in \mathcal{M}$$

is continuous at any $s \notin D$.

Proof. By property 2 in Proposition 3, there exists a countable set $D \subset \mathbb{R}_+$ such that, for every $x \in \mathbb{Q}$, the map:

$$t \longrightarrow C(t, x)$$

is continuous at any $s \notin D$. Since

$$\forall x, y, t, \quad |C(t, y) - C(t, x)| \leq |y - x|,$$

this continuity property holds for every $x \in \mathbb{R}$. It suffices then to apply Proposition 2, taking into account Proposition 4. □

We may write $D = \{d_n; \ n \in \mathbb{N}\}$. For $p \in \mathbb{N}$, we denote by $(k_n^{(p)}, n \geq 0)$ the increasing rearrangement of the set:

$$\{k\, 2^{-p}; \ k \in \mathbb{N}\} \cup \{d_j; \ 0 \leq j \leq p\}.$$

We define $(\mu_t^{(p)}, t \geq 0)$ by:

$$\mu_t^{(p)} = \frac{k_{n+1}^{(p)} - t}{k_{n+1}^{(p)} - k_n^{(p)}} \, \mu_{k_n^{(p)}} + \frac{t - k_n^{(p)}}{k_{n+1}^{(p)} - k_n^{(p)}} \, \mu_{k_{n+1}^{(p)}} \qquad \text{if } t \in [k_n^{(p)}, k_{n+1}^{(p)}].$$

Lemma 7. *The following properties hold:*

(i) *$(\mu_t^{(p)}, t \geq 0)$ is a peacock and the map: $t \longrightarrow \mu_t^{(p)} \in \mathcal{M}$ is continuous.*

(ii) *For any $t \geq 0$, the set $\{\mu_t^{(p)}; \ p \in \mathbb{N}\}$ is uniformly integrable.*

(iii) *For $t \geq 0$, $\lim_{p \to \infty} \mu_t^{(p)} = \mu_t$ in $\mathcal{M}$.*

Proof. Properties (i) and (iii) are clear by construction. Property (ii) follows directly from Proposition 4. □

By Theorem 3, there exists, for each p, a martingale $(M_t^{(p)}, t \geq 0)$ which is associated to $(\mu_t^{(p)}, t \geq 0)$ and satisfies the Lipschitz-Markov property. For any $n \in \mathbb{N}$ and $\tau_n = (t_1, \cdots, t_n) \in \mathbb{R}_+^n$, we denote by $\mu_{\tau_n}^{(p,n)}$ the law of $(M_{t_1}^{(p)}, \cdots, M_{t_n}^{(p)})$, a probability measure on $\mathbb{R}^n$. The proof of the following lemma is quite similar to that of Lemma 5, hence we omit this proof.

Lemma 8. *For every $n \in \mathbb{N}$ and $\tau_n \in \mathbb{R}_+^n$, the set of probability measures $\{\mu_{\tau_n}^{(p,n)}; \ p \geq 0\}$, is tight.*

Let now $\mathscr{U}$ be an ultrafilter on $\mathbb{N}$, which refines Fréchet's filter.[1] As a consequence of the previous lemma, for every $n \in \mathbb{N}$ and every $\tau_n \in \mathbb{R}_+^n$, $\lim_{\mathscr{U}} \mu_{\tau_n}^{(p,n)}$ exists in $\mathscr{M}$ and we denote this limit by $\mu_{\tau_n}^{(\infty,n)}$. By property (iii) in Lemma 7, $\mu_{(t)}^{(\infty,1)} = \mu_t$. By Kolmogorov's extension theorem, there exists a process $M = (M_t, t \geq 0)$ such that, for every $n \in \mathbb{N}$ and every $\tau_n = (t_1, \cdots, t_n) \in \mathbb{R}_+^n$, the law of $(M_{t_1}, \cdots, M_{t_n})$ is $\mu_{\tau_n}^{(\infty,n)}$. In particular, this process $(M_t, t \geq 0)$ is associated to $(\mu_t, t \geq 0)$. Moreover, by Proposition 6, M satisfies the Lipschitz-Markov property, and by property (ii) in Lemma 7 and Proposition 7, M is a martingale.   □

4 Related Results

The following definition was first introduced by Lowther in [7], and also plays a central role in [8, 9].

Definition 3. Let $X = (X_t, t \geq 0)$ be an $\mathbb{R}$-valued stochastic process. Then:

1. X is *strong Markov* if for every bounded, measurable $g : \mathbb{R} \longrightarrow \mathbb{R}$ and every $t \geq 0$, there exists a measurable $f : \mathbb{R}_+ \times \mathbb{R} \longrightarrow \mathbb{R}$ such that

$$f(\tau, X_\tau) = \mathbb{E}[g(X_{\tau+t}) | \mathscr{F}_\tau]$$

   for every stopping time τ.

2. X is *almost-continuous* if it is càdlàg, continuous in probability, and given any two independent, identically distributed càdlàg processes Y, Z with the same distribution as X and for every $0 \leq s < t$ we have

$$\mathbb{P}(Y_s < Z_s ,\ Y_t > Z_t \text{ and } Y_u \neq Z_u \text{ for every } u \in (s,t)) = 0.$$

3. X is an *almost-continuous diffusion* (abbreviated to ACD) if it is strong Markov and almost-continuous.

The main result of [8] is the following theorem:

Theorem 5 ([8], Theorem 1.3). *Let $(\mu_t, t \geq 0)$ be a peacock such that the map:*

$$t \geq 0 \longrightarrow \mu_t \in \mathscr{M}$$

is continuous. Then there exists an ACD martingale which is associated to $(\mu_t, t \geq 0)$, and it is unique in law.

On the other hand, the Lipschitz-Markov property entails the strong Markov property:

---

[1] At this place, it seems that the use of filters rather than sequences is necessary, since we have to consider a convergence in a uncountable product of spaces of probabilities, namely: $\prod_{n \in \mathbb{N}} (\mathscr{M}(\mathbb{R}^n))^{\mathbb{R}_+^n}$.

Proposition 8 ([9], Lemma 4.2). *Let X be a càdlàg, $\mathbb{R}$-valued process which satisfies the Lipschitz-Markov property. Then it is strong Markov.*

As a consequence, the martingale appearing in the statement of Theorem 3 is strong Markov. Note that a kind of converse of Proposition 8 holds (see also [7, Theorem 1.5]):

Proposition 9 ([9], Lemma 4.3). *If X is an ACD local martingale, then X satisfies the Lipschitz-Markov property.*

The following result concerns the stability with respect to the f.d. convergence.

Theorem 6 ([9], Theorem 1.2). *If $(M^{(i)})_{i \in I}$ is a family of ACD martingales which f.d. converges on a dense subset of $\mathbb{R}_+$ to a process X which is càdlàg and continuous in probability, then X is an ACD.*

It follows from what precedes that if $(\mu_t, t \geq 0)$ is a peacock such that the map:

$$t \geq 0 \longrightarrow \mu_t \in \mathcal{M}$$

is continuous, the martingale $\tilde{M}$ "constructed" in the proof of Theorem 3 is the *unique* ACD martingale associated to $(\mu_t, t \geq 0)$ (and, in particular, it is independent of the ultrafilter $\mathcal{U}$ and of the regularization process). About the continuity of $\tilde{M}$, we may use the following remarkable result, the proof of which relies on [7].

Theorem 7 ([9], Lemma 1.4). *Let X be an almost-continuous process. If the support of the law of X_t is connected for every t in $\mathbb{R}_+$ outside of a countable set, then X is continuous.*

Finally, we obtain from what precedes:

Theorem 8. *Suppose that $(\mu_t, t \geq 0)$ is a peacock such that the map:*

$$t \geq 0 \longrightarrow \mu_t \in \mathcal{M}$$

is continuous, and such that the support of μ_t is connected for every t in $\mathbb{R}_+$ outside of a countable set. Then there exists one and only one continuous, strongly Markov martingale which is associated to $(\mu_t, t \geq 0)$.

References

1. Dellacherie, C., Meyer, P.-A.: Probabilités et potentiel, Chapitres V à VIII, Théorie des martingales, Hermann (1980)
2. Derman, E., Kani, I.: Riding on a smile. Risk **7**, 32–39 (1994)
3. Dupire, B.: Pricing with a smile. Risk Mag. **7** , 18–20 (1994)
4. Hirsch, F., Roynette, B.: A new proof of Kellerer's theorem. ESAIM: PS **16**, 48–60 (2012)
5. Hirsch, F., Profeta, C., Roynette, B., Yor, M.: Peacocks and Associated Martingales, with Explicit Constructions. Bocconi & Springer Series, vol. 3. Springer, Milan (2011)

6. Kellerer, H.G.: Markov-Komposition und eine Anwendung auf Martingale. Math. Ann. **198**, 99–122 (1972)
7. Lowther, G.: Properties of expectations of functions of martingale diffusions. http://arxiv.org/abs/0801.0330v1 (2008)
8. Lowther, G.: Fitting martingales to given marginals. http://arxiv.org/abs/0808.2319v1 (2008)
9. Lowther, G.: Limits of one-dimensional diffusions. Ann. Proba. **37**(1), 78–106 (2009)

Part VII
Statistics: Theory and Methods

Empirical Likelihood and Ranking Methods

Mayer Alvo

1 Introduction

In the parametric setting, when the joint distribution of the observations is known up to one or more parameters, likelihood methods in statistics have been effectively used to provide tests of hypotheses and confidence intervals. On the other hand, estimating equation methods have been used in estimation problems when the complete probability distribution is not specified. Boos [4] discusses the use of score tests in this more general context. Aitchison and Silvey [1] considered maximum likelihood estimation and hypothesis testing subject to constraints on the parameters. In cases when the likelihood function is misspecified, serious errors in inference can result. Empirical likelihood is a nonparametric technique that is entirely data driven. Owen [11] provides a thorough and excellent treatment of the subject and discusses its relation to the bootstrap. Qin and Lawless [12] have extended empirical likelihood methods to deal with constraints on parameters. DiCiccio, Hall and Romano [5] have shown that unlike the bootstrap, empirical likelihood is Bartlett-correctable, thus yielding second order approximations.

In a series of articles, Alvo and Cabilio generalized ranking methods to deal with tests of trend and the analysis of two-way layouts. The test statistics developed were motivated by notions of distance between permutations. Liu et al. [9] have described a rank-based empirical likelihood approach for inference on population medians. The goal of the present article is to apply empirical likelihood methods to various problems in two-way layouts involving the use of ranks and to compare with the previous results of Alvo and Cabilio. In Sect. 2 we introduce the usual problem of testing for concordance and place it in the context of empirical likelihood. In Sect. 3, we discuss generalizations and applications to the multi-sample situation.

M. Alvo (✉)
Department of Mathematics and Statistics, University of Ottawa, Ottawa, ON K1N 6N5, Canada
e-mail: malvo@uottawa.ca

© Springer Science+Business Media New York 2015   367
D. Dawson et al. (eds.), *Asymptotic Laws and Methods in Stochastics*,
Fields Institute Communications 76, DOI 10.1007/978-1-4939-3076-0_19

2 Empirical Likelihood Methods

Suppose that Y_{ij} for $0 \leq j \leq r, 1 \leq i \leq n$, represent the jth response of $k = r + 1$ treatments in the ith replication. Let Y_{ij} have a continuous cumulative distribution function F_{ij}. We would like to test the hypothesis

$$H_0 : F_{i0} = \ldots = F_{ir}, i = 1, \ldots, n \tag{1}$$

against the alternative

$$H_1 : F_{ij}(x) = F_i(x - \theta_j), 0 \leq j \leq r, i = 1, \ldots, n. \tag{2}$$

Let R_{ij} denote the rank of Y_{ij} among the k responses $\{Y_{i0}, Y_{i1}, \ldots, Y_{ir}\}$. Since

$$R_{ij} = 1 + \sum_{t=0}^{r} I_{[Y_{ij} > Y_{it}]}$$

it follows that

$$ER_{ij} = 1 + \sum_{t=0}^{r} P(Y_{ij} > Y_{it})$$

which clearly depends on $\boldsymbol{\theta} = (\theta_1, \ldots, \theta_k)'$. We shall further assume that the k–dimensional distributions $\{F_i(\cdot, \ldots, \cdot), i = 1, \ldots, n\}$ are independent and identical to some cumulative distribution function $F(\cdot, \ldots, \cdot)$. As well, we suppose that $F(\cdot, \ldots, \cdot)$ is differentiable and symmetric in its arguments. In that case, it is possible to drop the first subscript and consider the random vector of ranks $\mathbf{R} = \{R_0, R_1, \ldots, R_r\}$. Under the null hypothesis (1), all rankings are equally likely in the space $\mathscr{P} = \{\varpi_i\}$ of all possible $k!$ permutations, where ϖ is a column vector permutation of the integers $(1, 2, \ldots, k)$. It follows that for any components R, R'

$$P(R = i) = \frac{1}{k}, P(R = i, R' = j) = \frac{1}{k(k-1)}, i \neq j$$

and consequently

$$ER = \frac{(k+1)}{2}, VarR = \frac{k^2 - 1}{12}, Cov(R, R') = -\frac{k+1}{12}.$$

The covariance matrix

$$Cov(\mathbf{R}) = \begin{cases} \frac{k^2-1}{12} & \text{on the diagonal} \\[2mm] -\frac{k+1}{12} & \text{off the diagonal} \end{cases}$$

is a $k \times k$ singular matrix of rank $(k-1)$. Let $\mathbf{1}_k$ be a k-dimensional vector of ones and let T be the $k \times k!$ matrix given by

$$T = \left(\varpi_1 - \frac{k+1}{2}\mathbf{1}_k, \ldots, \varpi_{k!} - \frac{k+1}{2}\mathbf{1}_k \right).$$

Noting that

$$\sum_{i=1}^{k!} \varpi_i = (k-1)! \left[\frac{k(k+1)}{2} \right] \mathbf{1}_k$$

and

$$\mathbf{R} = \sum_{i=1}^{k!} \varpi_i I_{[\mathbf{R}=\varpi_i]}$$

it follows that

$$ER = \sum_{i=1}^{k!} \varpi_i P\left(\mathbf{R} = \varpi_i\right)$$

$$= \sum_{i=1}^{k!} \varpi_i \pi_i$$

$$= T\pi + \frac{k+1}{2}\mathbf{1}_k \tag{3}$$

where $\pi = (\pi_1, \ldots, \pi_{k!})'$.

In view of the fact that the covariance matrix of $\mathbf{R}$ is singular, we shall consider instead the reduced ranking $\mathbf{S} = \{R_1, \ldots, R_r\}$. The covariance matrix of $\mathbf{S}$ is nonsingular of full rank r with

$$Cov\,(\mathbf{S}) = \begin{cases} \frac{k^2-1}{12} & \text{on the diagonal} \\[2mm] -\frac{k+1}{12} & \text{off the diagonal} \end{cases}.$$

The inverse takes the form of

$$(Cov\,(\mathbf{S}))^{-1} = \frac{12}{k(k+1)} \begin{cases} 2 & \text{on the diagonal} \\[2mm] 1 & \text{off the diagonal} \end{cases}$$

$$= \frac{12}{k(k+1)} (J_r + I_r)$$

where J_r is an $r \times r$ matrix of ones and I_r is the identity matrix of order r.

Set $\mu = ES$. Let us consider an empirical likelihood approach for testing hypotheses about the mean μ [11]. Suppose that we observe a random sample of n rankings $S_1, \ldots, S_n$ and suppose that $\Psi(Y_{ij}, 0 \leq j \leq r, \mu)$ is a vector valued function. Let μ_0 be a fixed point of μ for which the variance-covariance matrix of $\Psi(Y_{ij}, 0 \leq j \leq r, \mu_0)$ is finite and has rank $q > 0$. Let w_i be the probability mass placed at $\{Y_{ij}, 0 \leq j \leq r\}$. If μ_0 satisfies $E\left(\Psi(Y_{ij}, 0 \leq j \leq r, \mu_0)\right) = 0$, then

$$l_E(\mu_0) = -2 \log \mathscr{R}(\mu_0) \rightarrow \chi_q^2 \tag{4}$$

where the profile empirical likelihood ratio function is given by

$$\mathscr{R}(\mu) = \max\left\{ \Pi_{i=1}^n nw_i \mid \sum_{i=1}^n w_i \Psi(Y_{ij}, 0 \leq j \leq r, \mu) = 0, w_i \geq 0, \sum_{i=1}^n w_i = 1 \right\}. \tag{5}$$

We now choose

$$\Psi(Y_{ij}, 0 \leq j \leq r, \mu) = (S_i - \mu), q = r, \mu_0 = \frac{k+1}{2} 1_r.$$

The results of the maximization adapted to this choice show that an empirical log likelihood-ratio statistic $l_E(\mu)$ for the mean can be obtained with

$$\hat{w}_i = \frac{1}{n} \frac{1}{\{1 + t'(S_i - \mu_0)\}}, i = 1, \ldots, n \tag{6}$$

and t is a $(r-1) \times 1$ vector of Lagrange multipliers satisfying

$$\sum_{i=1}^n \frac{(S_i - \mu_0)}{1 + t'(S_i - \mu_0)} = 0.$$

Owen (Theorem 3.2 p.219) has shown that $\|t\| = O_p\left(n^{-\frac{1}{2}}\right)$, and

$$l_E(\mu_0) = n(\bar{S} - \mu_0)'(Cov(S))^{-1}(\bar{S} - \mu_0) + o_p(1) \tag{7}$$

where $\bar{S} = n^{-1}(S_1 + \ldots + S_n)$ and that hence $l_E(\mu_0) \rightarrow_d \chi_{k-1}^2$. The empirical likelihood ratio test would then reject the null hypothesis whenever

$$l_E(\mu_0) = -2 \sum \log(n\hat{w}_i) > \chi_{k-1}^2(\alpha)$$

where $\chi_{k-1}^2(\alpha)$ is the upper $100(1 - \alpha)\%$ point of a chi-square distribution with $(k-1)$ degrees of freedom and the $\{\hat{w}_i\}$ are given in (6).

As in Owen, a $100(1 - \alpha)\%$ confidence region may be written as

$$C_{r,n} = \left\{ \sum_{i=1}^{n} \hat{w}_i S_i \,|\, \Pi_{i=1}^{n} n\hat{w}_i \geq r \right\}$$

with $r = \exp\left(-\frac{\chi^2_{k-1}(\alpha)}{2}\right)$.

We may now relate (7) to the vector of relative frequencies $\hat{\pi}$. Setting $\bar{R}_0 = n^{-1} \sum_i R_{0i}$, we have

$$(\bar{S} - \mu_0)' (Cov\,(S))^{-1} (\bar{S} - \mu_0)$$

$$= \frac{12}{k(k+1)} (\bar{S} - \mu_0)' (J_r + I_r) (\bar{S} - \mu_0)$$

$$= \frac{12}{k(k+1)} (\bar{S} - \mu_0)' (1_r 1_r' + I_r) (\bar{S} - \mu_0)$$

$$= \frac{12}{k(k+1)} \left\{ (\bar{S} - \mu_0)' (1_r 1_r') (\bar{S} - \mu_0) + (\bar{S} - \mu_0)' (\bar{S} - \mu_0) \right\}$$

$$= \frac{12}{k(k+1)} \left\{ \left(\bar{R}_0 - \frac{k+1}{2} \right)^2 + (\bar{S} - \mu_0)' (\bar{S} - \mu_0) \right\}$$

$$= \frac{12}{k(k+1)} \left\{ (\bar{R} - \mu_0)' (\bar{R} - \mu_0) \right\}$$

$$= \frac{12}{k(k+1)} (T\hat{\pi})' (T\hat{\pi}) \tag{8}$$

since $T\mu_0 = \frac{k+1}{2} T1_r = 0$. We recognize that (8) is the usual Friedman statistic (see Alvo, Cabilio and Feigin [3]) and that in view of (3)

$$(T\hat{\pi})' (T\hat{\pi}) = \sum_{i=0}^{k} \left(\bar{R}_i - \frac{(t+1)}{2} \right)^2$$

where $\bar{R}_i$ is the average of the ranks assigned to object i. Hence, the empirical log likelihood-ratio statistic is asymptotically equivalent in distribution to the Friedman statistic. As such, the Friedman statistic shares, at least asymptotically, many of the properties of the empirical likelihood ratio statistic. Specifically, we can obtain narrower confidence intervals for the mean [11].

The expression for the power function may be derived from Sen [13]. Specifically, it was shown that the power function for the Friedman statistic is given by the non central chi square with noncentrality parameter

$$\lambda = \frac{nk}{k+1} \left[\int_{-\infty}^{\infty} f(x,x)\, dx \right]^2 \sum_{j=0}^{r} \theta_j^2$$

where $f(\cdot,\cdot)$ is the joint marginal density function derived from $F(\cdot,\cdot,\ldots,\cdot)$ of any pair of elements.

3 General Block Designs

The methods of empirical likelihood may be generalized to deal with general block designs. Suppose that t objects are ranked k_h at a time $2 \le k_h \le t$ by b judges (blocks) independently, $h = 1,\ldots,b$, in such a way that each object is presented to r_i judges and each pair of objects (i,j) is presented together to λ_{ij} of these judges, $i,j = 1,\ldots,t$. For a balanced incomplete block design (BIBD), $k_h = k, r_i = r$, $\lambda_{ij} = \lambda$ and we must have that

$$bk = rt$$

$$\lambda(t-1) = r(k-1) .$$

In the complete ranking situation $k_h = t, r = b = \lambda$. We would like to test the hypothesis of no treatment effect; that is each judge selects the ranking at random from the space of $k_h!$ permutations of the integers $(1,\ldots,k_h)$. In order to consider the asymptotics in this situation we shall allow n replications of such basic designs. Alvo and Cabilio [2] introduced the notion of compatibility in order to deal with precisely such a situation. We recall

Definition 1. The complete ranking μ of t objects is said to be compatible with an incomplete ranking μ^* of a subset of k of these objects, $2 \le k \le t$, if the relative ranking of every pair of objects ranked in μ^* coincides with their relative ranking in μ.

As an example, the incomplete ranking $\mu^* = (2,-,1)$ is compatible with each of the rankings in the class

$$C(\mu^*) = \{(2,3,1),(3,2,1),(3,1,2)\} .$$

We may arrange the complete rankings in a fixed but arbitrary order and then specify by means of a matrix all the compatible classes corresponding to the missing pattern. Let $\mu_1 = (1,2,3), \mu_2 = (1,3,2), \mu_3 = (2,1,3), \mu_4 = (2,3,1), \mu_5 = (3,2,1), \mu_6 = (3,2,1)$. Then the compatibility matrix corresponding to rankings whereby only objects one and three are ranked

$$C = \begin{array}{c} \mu_1 \\ \mu_2 \\ \mu_3 \\ \mu_4 \\ \mu_5 \\ \mu_6 \end{array} \begin{pmatrix} 1 & 0 \\ 1 & 0 \\ 1 & 0 \\ 0 & 1 \\ 0 & 1 \\ 0 & 1 \end{pmatrix}.$$

Here the columns of the matrix correspond to the incomplete rankings $(1, -, 2)$, $(2, -, 1)$ respectively. Alvo and Cabilio [2] have shown that for a specific pattern of missing observations for each of the b blocks, the matrix of scores is given by

$$T^* = (T_1^* | T_2^* | \ldots | T_b^*)$$

$$= T \left(\frac{k_1!}{t!} C_1 \Big| \frac{k_2!}{t!} C_2 \Big| \ldots \Big| \frac{k_b!}{t!} C_b \right)$$

where C_i is the compatibility matrix for block i. This decomposition suggests that we may define an empirical likelihood test for each block separately. In view of the independence of the observations in each block, the profile empirical likelihood ratio function for the general block design will be the sum of the individual profile empirical likelihood ratio function for each block. Consequently,

$$l_E(\mu_0) = n \sum_{h=1}^{b} \left(\bar{\mathbf{S}}_h^* - \mu_{h0} \right)' \left(Cov \left(\mathbf{S}_h^* \right) \right)^{-1} \left(\bar{\mathbf{S}}_h^* - \mu_{h0} \right) + o_p(1)$$

$$= n \left(T^* \hat{\pi} \right)' \Gamma^- \left(T^* \hat{\pi} \right) + o_p(1) \rightarrow \chi^2_{rank(\Gamma)}$$

where $\bar{\mathbf{S}}_h^*$ is the vector whose components are the averages of the ranks in block h,

$$\hat{\pi} = (\hat{\pi}_1 | \hat{\pi}_2 | \ldots | \hat{\pi}_b), \mu_0 = (\mu_{10} | \mu_{20} | \ldots | \mu_{b0})$$

and

$$\Gamma = \sum_{h=1}^{b} \frac{1}{k_h!} \left(\frac{k_h!}{t!} C_h \right) \left(\frac{k_h!}{t!} C_h \right)'.$$

In the special case of the BIBD, it can be shown that

$$l_E(\mu_0) \rightarrow \chi^2_{t-1}.$$

Specific results for general designs may be obtained from Alvo and Cabilio [2] who obtained the eigenvalues of Γ for general cyclic designs and group divisible designs.

4 Two Sample Problems

Consider now the two sample problem. Suppose that Y_{lij} for $0 \leq j \leq r, 1 \leq i \leq n_l, l = 1, 2$ represent the jth response of $k = r + 1$ treatments in the ith replication, lth population. Let Y_{lij} have a continuous cumulative distribution function F_{lij}. We assume that $F_{lij}(x) = F_{lj}(x - \theta_l)$ for all $1 \leq i \leq n_l$. We would like to test the null hypothesis

$$H_0 : F_{lij}(x) = F_j(x)$$

against the alternative

$$H_1 : F_{lij}(x) = F_j(x - \theta_l) .$$

Let R_{lij} denote the rank of Y_{lij} among the k responses $\{Y_{li0}, \ldots, Y_{lik}\}$ and let $\mathbf{S}_{li} = (R_{li1}, \ldots, R_{lik})$ be the ith vector of rankings in the lth population. We shall suppose that we observe a random sample of n_l rankings $\{\mathbf{S}_{li}\}, l = 1, 2$. Set $\boldsymbol{\gamma}_l = E\mathbf{S}_{li}, \boldsymbol{\gamma} = \boldsymbol{\gamma}_2 - \boldsymbol{\gamma}_1$ and using the Neyman-Scott [10] parametrization, set

$$\boldsymbol{\gamma}_1 = \boldsymbol{\mu} - \frac{n_2}{n_1}\boldsymbol{\gamma}, \boldsymbol{\gamma}_2 = \boldsymbol{\mu} - \frac{n_1}{n_2}\boldsymbol{\gamma}.$$

Under the null hypothesis, $\boldsymbol{\gamma} = \mathbf{0}$. Suppose that $\Psi_l \left(Y_{lij}, 0 \leq j \leq r, 1 \leq i \leq n_l, \boldsymbol{\mu}, \boldsymbol{\gamma}\right)$ is a vector valued function. Let $\boldsymbol{\gamma}_0$ be a fixed point of $\boldsymbol{\gamma}$ for which the variance-covariance matrix of $\Psi_l \left(Y_{lij}, 0 \leq j \leq r, 1 \leq i \leq n_l, \boldsymbol{\mu}, \boldsymbol{\gamma}\right)$ is finite and has rank $q > 0$. Let w_{li} be the probability mass placed at

$$\Psi_l \left(Y_{lij}, 0 \leq j \leq r, 1 \leq i \leq n_l, \boldsymbol{\mu}, \boldsymbol{\gamma}\right) .$$

If $\boldsymbol{\gamma}_0$ satisfies

$$E \left(\Psi_l \left(Y_{lij}, 0 \leq j \leq r, 1 \leq i \leq n_l, \boldsymbol{\mu}, \boldsymbol{\gamma}_0\right)\right) = 0,$$

then we shall show that

$$l_E \left(\boldsymbol{\mu}, \boldsymbol{\gamma}_0\right) = -2 \log \mathscr{R} \left(\boldsymbol{\mu}, \boldsymbol{\gamma}_0\right) \to \chi_q^2$$

where the profile empirical likelihood ratio function is given by

$$\mathscr{R}(\boldsymbol{\mu}, \boldsymbol{\gamma})$$

$$= \max \left\{ \Pi_{l=1}^2 \Pi_{i=1}^{n_l} n_l w_{li} \Big| \sum_{l=1}^2 \sum_{i=1}^{n_l} w_{li} \Psi_l \left(Y_{lij}, 0 \leq j \leq r, 1 \leq i \leq n_l, \boldsymbol{\mu}, \boldsymbol{\gamma}\right) = \mathbf{0}, \right.$$

$$\left. w_{ij} \geq 0, \sum_{i=1}^{n_l} w_{ij} = 1 \right\} .$$

The maximization is solved by using Lagrange multipliers. The Lagrangian is

$$L = \sum_l \sum_i \log w_{li} - \sum_{l=1}^{2} v_l \left(\sum_{i=1}^{n_l} w_{li} - 1 \right)$$
$$- \sum_l n_l t_l \sum_i w_{li} \Psi_l \left(Y_{lij}, 0 \leq j \leq r, 1 \leq i \leq n_l, \boldsymbol{\mu}, \boldsymbol{\gamma} \right)$$

where $\{v_l, t_l\}$ are the Lagrange multipliers. Following Liu et al. [9], and setting

$$\Psi_l \left(Y_{lij}, 0 \leq j \leq r, 1 \leq i \leq n_l, \boldsymbol{\mu}, \boldsymbol{\gamma} \right) = (\mathbf{S}_{li} - \boldsymbol{\gamma}_l), q = r,$$

we see that

$$\hat{w}_{li} = \frac{1}{n_l} \frac{1}{\{1 + \mathbf{t}_l' (\mathbf{S}_{li} - \boldsymbol{\gamma}_l)\}}, i = 1, \ldots, n_l, l = 1, 2$$

and $\mathbf{t}_l$ is a $(r-1) \times 1$ vector of Lagrange multipliers satisfying

$$\sum_{i=1}^{n_l} \frac{(\mathbf{S}_{li} - \boldsymbol{\gamma}_l)}{1 + \mathbf{t}_l' (\mathbf{S}_{li} - \boldsymbol{\gamma}_l)} = \mathbf{0}, l = 1, 2.$$

Owen (Theorem 3.2 p.219) has shown that $\|\mathbf{t}_l\| = O_p\left(n_l^{-\frac{1}{2}}\right)$, for each $l = 1, 2$. Moreover, it can be seen that under the null hypothesis

$$l_E(\boldsymbol{\mu}) = \sum_{l=1}^{2} n_l (\bar{\mathbf{S}}_l - \boldsymbol{\mu})' (Cov(\mathbf{S}))^{-1} (\bar{\mathbf{S}}_l - \boldsymbol{\mu}) + o_p(1)$$

where $\bar{\mathbf{S}}_l = n_l^{-1} \sum_{i=1}^{n_l} \mathbf{S}_{li}$. Estimating $\boldsymbol{\mu}$ by

$$\hat{\boldsymbol{\mu}} = \frac{n_1 \bar{\mathbf{S}}_1 + n_2 \bar{\mathbf{S}}_2}{n_1 + n_2}$$

we have that

$$l_E(\hat{\boldsymbol{\mu}}) = \frac{n_1 n_2}{n_1 + n_2} (\bar{\mathbf{S}}_1 - \bar{\mathbf{S}}_2)' (Cov(\mathbf{S}))^{-1} (\bar{\mathbf{S}}_1 - \bar{\mathbf{S}}_2) + o_p(1)$$

from which it follows that $l_E(\hat{\boldsymbol{\mu}}) \to_d \chi_{k-1}^2$ as $n \to \infty$, with $n_l/n \to \lambda_l, l = 1, 2$.

It can be seen that the empirical likelihood ratio test is asymptotically equivalent to the Feigin and Alvo [6] two-sample test. The $Cov(\mathbf{S})$ can be estimated as described in Feigin and Alvo [6].

5 Other Test Statistics

Test statistics other than the mean ranking can be considered using the empirical likelihood approach. Specifically we may consider the Kendall score function $sgn\left(Y_{iq_2} - Y_{iq_1}\right)$ which represents the level of agreement between the ranking of judge i and the complete natural ordering $\{1, 2, \ldots, t\}$ with respect to the paired comparison of the objects q_1, q_2. The parameter of interest in that case is

$$\tau = \int \int sgn\left(y_{11} - y_{21}\right) sgn\left(y_{12} - y_{22}\right) dF\left(\mathbf{y}_1\right) dF\left(\mathbf{y}_2\right)$$

where $\mathbf{y}_1 = \left(y_{11}, y_{12}\right), \mathbf{y}_2 = \left(y_{21}, y_{22}\right)$ have common distribution F. For independent variables $\mathbf{Y}_1, \mathbf{Y}_2$, the parameter τ represents the covariance of the sign of the difference between the first coordinates and the sign of the difference between the second coordinates. This shows that the empirical likelihood approach is much more generally applicable to new situations.

6 Simulations

It is well known that the chi square distribution is not a good approximation to the null distribution of the Friedman statistic for small and even moderate sample sizes [7]. Unfortunately, a Bartlett correction for this test statistic is not possible since the condition

$$\lim_{\|t\| \to \infty} \sup \left\| Ee^{itX} \right\| < \infty$$

is not satisfied for lattice random variables. Jensen [8] obtained an $O(\frac{1}{n^{\frac{1}{2}}})$ approximation. In view of the asymptotic equivalence of the Friedman statistic to the empirical likelihood ratio test, we may consider a calibration mentioned by Owen (p.33–35) which involves using the bootstrap. Specifically, for $b = 1, \ldots, B$ and $i = 1, \ldots, n$ let $\mathbf{S}_i^{*b}$ be independent random vectors sampled from among the rankings $\mathbf{S}_1, \ldots, \mathbf{S}_n$. This resampling can be implemented by drawing nB random integers $J(i, b)$ independently from the uniform distribution on $(1, \ldots, n)$ and setting $\mathbf{S}_i^{*b} = \mathbf{S}_{J(i,b)}$. Now let

$$C^{*b} = -2 \log \mathscr{R}^{*b}\left(\bar{\mathbf{S}}\right)$$

where

$$\mathscr{R}^{*b}\left(\bar{\mathbf{S}}\right) = \max \left\{ \Pi_{i=1}^n n w_i \mid \sum_{i=1}^n w_i \left(\mathbf{S}_i^{*b} - \bar{\mathbf{S}}\right) = \mathbf{0}, w_i \geq 0, \sum_{i=1}^n w_i = 1 \right\}.$$

Define the order statistics $C^{(1)} \leq C^{(2)} \leq \ldots \leq C^{(B)}$. We may now compare the 95 % critical value $C^{(0.95B)}$ with that of the appropriate Chi square. We do not pursue this further in this paper.

7 Conclusion

In this paper, we applied the methods of empirical likelihood to various non-parametric problems involving ranking data. Specifically, it was shown that the Friedman statistic has an empirical likelihood interpretation. This should enable us to construct narrower confidence intervals for the treatment means. As well, it was shown that empirical likelihood methods can be applied to the two sample problem as well as to various block design situations.

Acknowledgements Work supported by Natural Sciences and Engineering Council of Canada Grant OGP0009068.

References

1. Aitchison, J., Silvey, S.D.: Maximum-likelihood estimation of parameters subject to restraints. Ann. Math. Stat. **29**, 813–828 (1958)
2. Alvo, M., Cabilio, P.: A general rank based approach to the analysis of block data. Commun. Stat. Theory Methods **28**, 197–215 (1999)
3. Alvo, M., Cabilio, P., Feigin, P.D.: Asymptotic theory for measures of concordance with special reference to average Kendall tau. Ann. Stat. **10**, 1269–1276 (1982)
4. Boos, D.D.: On generalized score tests. Am. Stat. **46**, 327–333 (1992)
5. DiCiccio, T.J., Hall, P., Romano, J.: Empirical likelihood is Bartlett-correctable. Ann. Stat. **19**, 1053–1061 (1991)
6. Feigin, P.D., Alvo, M.: Intergroup diversity and concordance for ranking data: an approach via metrics for permutations. Ann. Stat. **14**, 691–707 (1986)
7. Iman, R.L., Davenport, J.M.: Approximations of the critical region of the Friedman statistic. Commun. Stat. Theory Methods **A9**(6), 571–595 (1980)
8. Jensen, D.R.: On approximating the distribution of Friedman's χ^2. Metrika **24**, 75–85 (1977)
9. Liu, T., Yuan, X., Lin, N., Zhang, B.: Rank-based empirical likelihood inference on medians of k populations. J. Stat. Plan. Inference **142**, 1009–1026 (2012)
10. Neyman, J., Scott, E.: Note on techniques of evaluation of single rain stimulations experiments. In: Proceedings of the Fifth Berkeley Symposium vol. 5, pp. 327–350. University of California Press, Berkeley (1967)
11. Owen, A.B.: Empirical Likelihood. Chapman & Hall/CRC, Boca Raton (2001)
12. Qin, J., Lawless, J.F.: Estimating equations, empirical likelihood and constraints on parameters. Can. J. Stat. **23**, 145–159 (1995)
13. Sen, P.K.: Asymptotically efficient tests by the method of n rankings. J. R. Stat. Soc. B **30**, 312–317 (1968)

Asymptotic and Finite-Sample Properties in Statistical Estimation

Jana Jurečková

1 Introduction

Consider first the problem of estimating the shift parameter θ based on observations $X_1, \ldots, X_n$, distributed according to distribution function $F(x - \theta)$. Parallel problem consists of estimating the regression parameter in model $Y_i = \mathbf{x}_i^\top \boldsymbol{\beta} + e_i, i = 1, \ldots, n$. Many estimators of θ are asymptotically normally distributed, which is proven with the aid of the central limit theorem. The word "central" is suitable, because it approximates well the central part, but less accurately the tails of the true distribution of the estimator. The leading idea of robust estimators was their assumed resistance to heavy-tailed distributions and to the gross errors. However, while they are often asymptotically normal, we can show that they themselves can be heavy-tailed for any finite n.

Another interesting fact is that though many estimators are asymptotically admissible with respect to quadratic or generally to convex risk functions, some of them are not finite-sample admissible for any distribution at all, and cannot be even Bayesian. This is true mainly for trimmed estimators, as the median, trimmed mean or the trimmed least squares estimator. Generally this is true for many estimators with bounded influence functions; cf. [6, 7].

If we do not know F exactly, we usually take recourse to robust estimators, less sensitive to the outlying observations and to the gross errors. Well-known are the classes of M-, L- and R-estimators, each of which containing elements, asymptotically normal and efficient for specific distributions. In the family of symmetric contaminated distributions, $\mathscr{F} = (1 - \varepsilon)F + \varepsilon H$, $H \in \mathscr{H}$ with unimodal central distribution F, any of these classes contains an element with the mini-maximally optimal asymptotic variance over $\mathscr{F}$. Under a fixed F, we can

J. Jurečková (✉)
Faculty of Mathematics and Physics, Charles University in Prague, Prague, Czech Republic
e-mail: jurecko@karlin.mff.cuni.cz

© Springer Science+Business Media New York 2015

379

D. Dawson et al. (eds.), *Asymptotic Laws and Methods in Stochastics*,
Fields Institute Communications 76, DOI 10.1007/978-1-4939-3076-0_20

obtain the M-, L- and R-estimators with identical influence functions by a suitable transformation (dependent on F) of the respective score (weight) function. However, the influence function characterizes the statistical functional rather than its finite-sample estimator, and the M-, L- and R-estimators can behave differently for finite n.

The asymptotic approach often stretches the truth; when the number of observations is finite, the distribution of a robust estimator is far from normal, and it inherits the tails from the parent distribution F. From this point of view, the estimator is non-robust. Our purpose in the present paper is to illustrate some distinctive differences between the asymptotic and finite-sample properties of robust estimators. We shall devote attention to the tail-behavior of M-estimators and of their one-step versions, and generally to the tail-behavior of equivariant estimators. Concerning the one-step version $T_n^{(1)}$ of estimator T_n, starting with an initial estimator $T_n^{(0)}$, it is interesting though not well known that while asymptotic properties of $T_n^{(1)}$ depend on those of non-iterated T_n, its finite-sample properties rather depend on the initial $T_n^{(0)}$. The finite-sample properties of an estimator depend on its finite sample distribution; we shall illustrate the exact finite-sample densities of some equivariant estimators. However, to calculate the density numerically requires a multiple numerical integration, for which a very good approximation is needed. We recommend the saddle-point approximation, which is very precise even for a very small n.

2   Tail-Behavior of Equivariant Estimators

2.1   Estimation of Shift Parameter, i.i.d. Observations

Let $X_1, \ldots, X_n$ be a random sample from an unknown distribution function $F(x - \theta)$, where F is absolutely continuous with positive density f. For the sake of identifiability of θ, assume that f is symmetric around 0, or another condition guaranteeing the identifiability. Suppose that F is heavy-tailed in the sense that

$$\lim_{x \to \infty} \frac{-\ln(1 - F(x))}{m \ln x} = 1, \quad \text{for some } m > 0. \tag{1}$$

Then, for $x > 0$,

$$1 - F(x) = x^{-m} L(x) \tag{2}$$

where $L(x)$ is slowly varying at infinity, i.e. $\lim_{x \to \infty} \frac{L(ax)}{L(x)} = 1 \ \forall a > 0$.

For that, we should verify that $L_m(x) = x^m(1 - F(x))$ is slowly varying at infinity. Indeed, for $x > 0$ and any $a > 0$ fixed, under (1)

$$\ln\left(\frac{L_m(ax)}{L_m(x)}\right) = m \ln a + \ln(1 - F(ax)) - \ln(1 - F(x))$$

$$= m \ln a + \left(\frac{\ln(1 - F(ax))}{m \ln(ax)}\right) \cdot m \ln(ax) - \left(\frac{\ln(1 - F(x))}{m \ln x}\right) \cdot m \ln x \to 0$$

as $x \to \infty$, and it confirms (2). In that case F belongs to domain of attraction of the Fréchet distribution. Conversely, (2) implies (1).

Let $T_n = T_n(X_1, \ldots, X_n)$ be a translation equivariant estimator of θ, further satisfying the following natural condition:

$$\min_{1 \le i \le n} X_i > 0 \Rightarrow T_n(\mathbf{X}) > 0, \quad \max_{1 \le i \le n} X_i < 0 \Rightarrow T_n(\mathbf{X}) < 0. \tag{3}$$

Tail-behavior of T_n can be characterized by means of a measure proposed in [3]:

$$B(a, T_n) = \frac{-\ln P_\theta(|T_n - \theta| \ge a)}{-\ln(1 - F(a))} = \frac{-\ln P_0(|T_n| \ge a)}{-\ln(1 - F(a))} \tag{4}$$

and its values for $a \gg 0$. If T_n satisfies (3), then under any fixed n

$$1 \le \liminf_{a \to \infty} B(a, T_n) \le \limsup_{a \to \infty} B(a, T_n) \le n$$

(see [3] for the proof). Particularly, if $\lim_{a \to \infty} B(a, T_n) = \lambda_n > 0$ and F is heavy-tailed with tail index m, then

$$P_0(T_n \ge a) = a^{-m\lambda_n} L_1(a), \quad L_1 \text{ slowly varying at infinity,}$$

hence T_n is also heavy-tailed. Specifically, it applies also to median $\tilde{X}_n$ and to the M-estimator M_n with bounded ψ-function, where $\lambda_n = \frac{n}{2}$. It means that $\tilde{X}_n$ and M_n are heavy-tailed with the tail index $\frac{mn}{2}$. It is finite for every n, though increasing with n, which classifies the distribution of these estimates as heavy-tailed for any finite n. The distribution of estimates is light-tailed (normally, exponentially tailed) only under $n = \infty$. The sample mean $\bar{X}_n$ has $\lambda_n \equiv 1$; thus $\bar{X}_n$ is heavy-tailed with the tail index m for any $n < \infty$.

2.2 Estimation of Shift Parameter, Non-identically Distributed Observations

Let us now consider the case where the X_i, $i = 1, \ldots, n$ are independent, but non-identically distributed, X_i having continuous distribution function $F_i(x - \theta)$, symmetric around θ, and heavy-tailed in the sense that

$$1 - F_i(x) = x^{-m_i} L_i(x), \quad 0 < m_i < \infty, \quad L_i \text{ slowly varying at infinity, } i = 1, \ldots, n.$$

Denote

$$m_* = \min\{m_i, \ 1 \le i \le n\} \quad m^* = \max\{m_i, \ 1 \le i \le n\}.$$

If we are not aware of the difference between $F_1, \ldots, F_n$, we automatically use an equivariant estimate T_n satisfying (3) as before. Then even its tail behavior cannot be exponentially-tailed. In fact, as proven in [8],

$$a^{-m^*} L(a) \le P_\theta(T_n - \theta > a) \le a^{-m_*} L(a) \quad \text{for} \ a > a_0,$$

where $L(\cdot)$ is slowly varying at infinity. Particularly, if $X_1, \ldots, X_n$ are heteroscedastic in the sense that $F_i(x) = F(x/\sigma_i)$, $i = 1, \ldots, n$, then $m_1, \ldots, m_n$ coincide. Hence, the heteroscedasticity does not affect the tail index of T_n, which is always equal to m.

2.3   Estimation of Regression Parameter

Consider the linear model $\mathbf{Y}_n = \mathbf{X}_n \boldsymbol{\beta} + \mathbf{e}_n$ with a fixed (nonrandom) design matrix $\mathbf{X}_n$ of order $n \times p$ and of rank p, with the rows $\mathbf{x}_i^\top$, $i = 1, \ldots, n$. The vector of errors $\mathbf{e}_n$ consists of n independent components, identically distributed with a symmetric distribution function F such that $0 < F(z) < 1$, $z \in \mathbb{R}^1$. Let $\mathbf{T}_n$ be an estimator of $\boldsymbol{\beta}$, regression equivariant in the sense

$$\mathbf{T}_n(\mathbf{Y} + \mathbf{Xb}) = \mathbf{T}_n(\mathbf{Y}) + \mathbf{b}, \quad \forall \mathbf{b} \in \mathbb{R}^p.$$

He et al. [2] extended the tail measure (4) to $\mathbf{T}_n$ in the linear model in the following way:

$$B(a, \mathbf{T}_n) = \frac{-\ln P\left(\max_i |\mathbf{x}_i^\top (\mathbf{T}_n - \boldsymbol{\beta})| > a\right)}{-\ln(1 - F(a))}, \ a \gg 0. \tag{5}$$

The same authors showed that if there exists at least one non-positive and one non-negative residual $r_i = Y_i - \mathbf{x}_i^\top \mathbf{T}_n$, then $\limsup_{a \to \infty} B(a, \mathbf{T}_n) \le n$. The properties of this measure were further studied by Mizera and Müller [12] and Portnoy and Jurečková [13], and this measure was extended to multivariate models by Zuo ([15, 16] and [17]). Jurečková, Koenker and Portnoy [11] studied the tail behavior of the least-squares estimator with random (possibly heavy-tailed) matrix $\mathbf{X}$.

It is traditionally claimed that robust estimators are insensitive to outliers in $\mathbf{Y}$ and to heavy-tailed distributions of model errors. However, we can show that an equivariant estimator $\mathbf{T}_n$ in the linear model is still heavy-tailed for any finite n provided the distribution function F is heavy-tailed, even if $\mathbf{X}$ is non-random. More precisely, if $\mathbf{T}_n$ is a regression equivariant estimator of $\boldsymbol{\beta}$ such that there exists at least one non-negative and one non-positive residual $r_i = Y_i - \mathbf{x}_i^\top \mathbf{T}_n$, $i = 1, \ldots, n$, then

$$P_\beta \left(\|\mathbf{T}_n - \boldsymbol{\beta}\| > a \right) \geq a^{-m(n+1)} L(a)$$

where $L(\cdot)$ is slowly varying at infinity. Hence, the distribution of $\|\mathbf{T}_n - \boldsymbol{\beta}\|$ is heavy-tailed under every finite n (see [8] for the proof).

2.4 Tail-Behavior of M-Estimator of Regression Parameter

The class of M-estimators defined as

$$\mathbf{T}_n = \arg \min_{\mathbf{b} \in \mathbb{R}^p} \left\{ \sum_{i=1}^n \rho(Y_i - \mathbf{x}_i^\top \mathbf{b}) \right\}$$

covers the Huber estimator and some redescending M-estimators. Assume that F is symmetric with nondegenerate tails (heavy or light) and such that

$$\lim_{a \to \infty} \frac{-\ln(1 - F(a + c))}{-\ln(1 - F(a))} = 1 \quad \text{for } \forall c > 0.$$

Following [12], we suppose that ρ satisfies the conditions (discussed in [12] in detail):

(i) ρ is absolutely continuous, nondecreasing on $[0, \infty)$, $\rho(z) \geq 0$, $\rho(z) = \rho(-z)$, $z \in \mathbb{R}^1$.
(ii) $\rho(z)$ is unbounded and its derivative $\psi(z)$ is bounded for $z \in \mathbb{R}^1$.
(iii) ρ is subadditive in the sense that there exists $L > 0$ such that $\rho(z_1 + z_2) \leq \rho(z_1) + \rho(z_2) + L$ for $z_1, z_2 \geq 0$.

Define

$$m_* = m_*(n, \mathbf{X}, \rho)$$

$$= \min \left\{ \text{card } \mathcal{M} : \sum_{i \in \mathcal{M}} \rho(\mathbf{x}_i^\top \mathbf{b}) \geq \sum_{i \notin \mathcal{M}} \rho(\mathbf{x}_i^\top \mathbf{b}) \text{ for some } \mathbf{b} \neq \mathbf{0} \right\}$$

where $\mathcal{M}$ runs over subsets of $\mathcal{N} = \{1, 2, \ldots, n\}$. Then it is proven in [5] that

$$\liminf_{a \to \infty} B(a, \mathbf{T}_n) \geq m_*.$$

It means that m_* is the lower bound for the tail behavior of M-estimator generated by ρ and it coincides with the lower bound derived in [12] for the finite-sample breakdown point of the M-estimator $\mathbf{T}_n$.

3   One-Step Version of an Estimator, Its Tail-Behavior and Breakdown Point

A broad class of estimators $\mathbf{T}_n$ of $\boldsymbol{\beta}$ admit a representation

$$\mathbf{T}_n(\mathbf{Y}) = \boldsymbol{\beta} + \frac{1}{\gamma}(\mathbf{X}_n^\top \mathbf{X}_n)^{-1} \sum_{i=1}^{n} \mathbf{x}_i \psi(Y_i - \mathbf{x}_i^\top \boldsymbol{\beta}) + \mathbf{R}_n,$$

$$\|\mathbf{R}_n\| = o_p(\|\mathbf{X}_n^\top \mathbf{X}_n\|^{-1/2}) \tag{6}$$

with a suitable function ψ and a functional $\gamma = \gamma(\psi, F)$.

The one-step version of $\mathbf{T}_n$ is defined as the one-step Newton-Raphson iteration of the system of equations $\sum_{i=1}^{n} \mathbf{x}_i \psi(Y_i - \mathbf{x}_i^\top \mathbf{b}) = \mathbf{0}$, even when the estimator is not a root of this system (as in the case of L_1-estimator or of other M-estimators with discontinuous ψ).

Let us start with a consistent initial estimator $\mathbf{T}_n^{(0)}$ of $\boldsymbol{\beta}$, satisfying $n^{1/2}(\mathbf{T}_n^{(0)} - \boldsymbol{\beta}) = O_p(1)$. The one-step version of $\mathbf{T}_n$ is defined as

$$\mathbf{T}_n^{(1)} = \begin{cases} \mathbf{T}_n^{(0)} + \frac{1}{n\hat{\gamma}_n}(\mathbf{Q}_n^*)^{-1} \sum_{i=1}^{n} \mathbf{x}_i \psi(Y_i - \mathbf{x}_i^\top \mathbf{T}_n^{(0)}) \ \ldots \ \text{if } \hat{\gamma}_n \neq 0 \\ \mathbf{T}_n^{(0)} \qquad\qquad\qquad\qquad\qquad\qquad\qquad \ldots \ \text{otherwise} \end{cases}$$

where $\mathbf{Q}_n^* = n^{-1}\mathbf{X}_n^\top \mathbf{X}_n$. The two-step or the k-step versions of $\mathbf{T}_n$ are defined analogously for $k = 2, 3, \ldots$. Here we assume that $\gamma \neq 0$ and that $\hat{\gamma}_n$ is a consistent estimator of γ such that $1 - (\gamma/\hat{\gamma}_n) = O_p(n^{-1/2})$. For possible regression invariant estimates of γ we refer the reader to [9].

While the asymptotic properties of $\mathbf{T}_n^{(1)}$ depend on those of the non-iterated estimator $\mathbf{T}_n$, its finite-sample breakdown point depends on that of initial $\mathbf{T}_n^{(0)}$ (see [13]). There is a conjecture that even more finite sample properties of $\mathbf{T}_n^{(1)}$ depend solely on the initial estimator. We shall illustrate this phenomenon at least in the special case of location model:

3.1   One-Step Version in the Location Model

Let T_n be an equivariant estimator of a location parameter and $T_n^{(0)}$ be an equivariant initial estimator. Consider a modified one-step version of T_n :

$$T_n^{(1)} = \begin{cases} T_n^{(0)} + \hat{\gamma}_n^{-1} W_n \ \ldots \ \text{if } |\hat{\gamma}_n^{-1} W_n| \leq c, \ 0 < c < \infty \\ T_n^{(0)} \qquad\qquad \ldots \ \text{otherwise} \end{cases}$$

where $W_n = n^{-1} \sum_{i=1}^{n} \psi(Y_i - T_n^{(0)}) = O_p(n^{-1/2})$. Then $T_n^{(1)} - T_n = o_p(n^{-1/2})$ and $T_n^{(1)}$ is also equivariant. Surprisingly, the tail behavior of $T_n^{(1)}$ and of $T_n^{(k)}$ depends more on that of $T_n^{(0)}$ than on the tail-behavior of non-iterative T_n. The following theorem is proven in [5]:

Theorem 1. *Let* $Y_1, \ldots, Y_n$ *be a sample from a population with distribution function* $F(y - \theta)$, F *symmetric and increasing on the set* $\{x : 0 < F(x) < 1\}$. *Let* T_n *be an equivariant estimator of* θ *admitting the representation*

$$T_n(\mathbf{Y}) = \theta + \frac{1}{n\gamma} \sum_{i=1}^{n} \psi(Y_i - \theta) + R_n, \quad R_n = o_p(n^{-1/2})$$

with a bounded skew-symmetric non-decreasing ψ. *Then, for* $k = 1, 2, \ldots$

$$\liminf_{a \to \infty} B(T_n^{(0)}, a) \leq \liminf_{a \to \infty} B(T_n^{(k)}, a)$$

$$\leq \limsup_{a \to \infty} B(T_n^{(k)}, a) \leq \limsup_{a \to \infty} B(T_n^{(0)}, a).$$

Example 1. (i) Let $T_n^{(0)} = \tilde{X}_n$ be the sample median, n odd. Let T_n be an equivariant estimator and $T_n^{(k)}$ its k-step version starting with $\tilde{X}_n$. Then, under the conditions of Theorem 1,

$$\lim_{a \to \infty} B(T_n^{(k)}, a) = \frac{n+1}{2} \quad \text{for } k = 1, 2, \ldots.$$

(ii) Let $T_n^{(0)} = \bar{X}_n$ be the sample mean. Let T_n be an equivariant estimator and $T_n^{(k)}$ its k-step version starting with $\bar{X}_n$. Then, under the conditions of Theorem 1,

$$\lim_{a \to \infty} B(T_n^{(k)}, a) = \begin{cases} n & \text{if } F \text{ is of type I (exponentially tailed)} \\ 1 & \text{if } F \text{ is of type II (heavy tailed)} \end{cases}$$

for $k = 1, 2, \ldots$, where the types I or II of F mean that its tails satisfy

$$\lim_{a \to \infty} \frac{-\ln(1 - F(a))}{ba^r} = 1, \quad b > 0, \quad r \geq 1$$

$$\lim_{a \to \infty} \frac{-\ln(1 - F(a))}{m \ln a} = 1, \quad m > 0,$$

respectively (see [3] for more details).

4 Finite-Sample Density of Equivariant Estimators

The finite-sample properties of estimator T_n, including the moments, depend on its entire scope, not only on its central part. The finite sample density can be sometimes derived, though it does not have a simple form. For instance, let $X_1, \ldots, X_n$ be a sample from the distribution with distribution function $F(x - \theta)$ where F has a continuously differentiable density f and finite Fisher information. Denote by $g_\theta(t)$ the density of a translation equivariant estimator T_n of θ. Then (see [10])

$$g_\theta(t) = \int_{T(x_1,\ldots,x_n) \leq t} \cdots \int \sum_{i=1}^{n} \frac{f'(x_i - \theta)}{f(x_i - \theta)} \prod_{k=1}^{n} f(x_k - \theta) dx_1 \ldots dx_n$$

$$= \mathrm{E}_0 \left\{ \sum_{i=1}^{n} \frac{f'(X_i)}{f(X_i)} I\Big[T(X_1, \ldots, X_n) \leq t - \theta \Big] \right\}.$$

If T_n is a solution of the equation $\sum_{i=1}^{n} \psi(X_i - t) = 0$ with monotone ψ, then $g_\theta(t)$ can be rewritten as

$$g_\theta(t) = \mathrm{E}_0 \left\{ \sum_{i=1}^{n} \frac{f'(X_i)}{f(X_i)} I\Big[\sum_{j=1}^{n} \psi(X_j - (t - \theta)) \leq 0 \Big] \right\}.$$

To calculate it numerically means an n-fold integration, and we recommend to use a saddle point approximation as it is more precise.

This density is numerically compared in [10] with its saddle-point approximation, developed in [1], for the Huber and maximum likelihood estimators, and for various parent distributions, including the Cauchy. The numerical comparisons demonstrate that the saddle-point approximations are very precise even for small sample sizes, and thus can be recommended in applications. A similar approach applies to the density of a regression quantile, derived in [4], and its saddle-point approximation, computed in [14].

Acknowledgements Research was supported by the Grant GAČR 15-00243S. The author would like to thank the Editor for organizing the volume, and the Referee for his/her valuable comments.

References

1. Field, C.A., Ronchetti, E.: Small Sample Asymptotics. Lecture Notes–Monograph Series. Institute of Mathematical Statistics, Hayward (1990)
2. He, X., Jurečková, J., Koenker, R., Portnoy, S.: Tail behavior of regression estimators and their breakdown points. Econometrica **58**, 1195–1214 (1990)
3. Jurečková, J.: Tail behavior of location estimators. Ann. Stat. **9**, 578–585 (1981)

4. Jurečková, J.: Finite sample distribution of regression quantiles. Stat. Probab. Lett. **80**, 1940–1946 (2010)
5. Jurečková, J.: Tail-behavior of estimators and of their one-step versions. Journal de la Société Francaise de Statistique **153/1**, 44–51 (2012)
6. Jurečková, J., Klebanov, L.B.: Inadmissibility of robust estimators with respect to L1 norm. In: Dodge, Y. (ed.) L_1-Statistical Procedures and Related Topics. Lecture Notes–Monographs Series, vol. 31, pp. 71–78. Institute of Mathematical Statistics, Hayward (1997)
7. Jurečková, J., Klebanov, L.B.: Trimmed, Bayesian and admissible estimators. Stat. Probab. Lett. **42**, 47–51 (1998)
8. Jurečková, J., Picek, J.: Finite-sample behavior of robust estimators. In: Chen, S., Mastorakis, N., Rivas-Echeverria, F., Mladenov, V. (eds.) Recent Researches in Instrumentation, Measurement, Circuits and Systems pp. 15–20. ISBN: 978-960-474-282-0. ISSN: 1792-8575
9. Jurečková, J., Portnoy, S.: Asymptotics for one-step M-estimators in regression with application to combining efficiency and high breakdown point. Commun. Stat. A **16**, 2187–2199 (1987)
10. Jurečková, J., Sabolová, R.: Finite-sample density and its small sample asymptotic approximation. Stat. Probab. Lett. **81**, 1311–1318 (2011)
11. Jurečková, J., Koenker, R., Portnoy, S.: Tail behavior of the least-squares estimator. Stat. Probab. Lett. **55**, 377–384 (2001)
12. Mizera, I., Mueller, C.H.: Breakdown points and variation exponents of robust M-estimators in linear models. Ann. Stat. **27**, 1164–1177 (1999)
13. Portnoy, S., Jurečková, J.: On extreme regression quantiles. Extremes **2**(3), 227–243 (1999)
14. Sabolová, R.: Small sample inference for regression quantiles. KPMS Preprint 70, Charles University in Prague (2012)
15. Zuo, Y.: Finite sample tail behavior of the multivariate trimmed mean based on Tukey-Donoho halfspace depth. Metrika **52**, 69–75 (2000)
16. Zuo, Y.: Finite sample tail behavior of Hodges-Lehmann type estimators. Statistics **35**, 557–568 (2001)
17. Zuo, Y.: Finite sample tail behavior of multivariate location estimators. J. Multivar. Anal. **85**, 91–105 (2003)

Publications of Miklós Csörgő

Research Books, Monographs

[A1] M. Csörgő and P. Révész, *Strong Approximations in Probability and Statistics*. Akadémiai Kiadó, Budapest 1981 – Academic Press, New York 1981 (284 pages).

[A2] M. Csörgő, *Quantile Processes with Statistical Applications*. CBMS-NSF Regional Conference Series in Applied Mathematics **42**, SIAM Philadelphia 1983 (156 pages).

[A3] M. Csörgő, S. Csörgő and L. Horváth, *An Asymptotic Theory for Empirical Reliability and Concentration Processes*. Lecture Notes in Statistics, **33**, Springer-Verlag, Berlin, Heidelberg 1986 (171 pages).

[A4] M. Csörgő and L. Horváth, *Weighted Approximations in Probability and Statistics*. Wiley, Chichester 1993 (436 pages).

[A5] M. Csörgő and L. Horváth, *Limit Theorems in Change-Point Analysis*. Wiley, Chichester 1997 (414 pages).

Books Edited

[B1] *Statistics and Related Topics* (M. Csörgő, D.A. Dawson, J.N.K. Rao, A.K.M. E. Saleh, Eds.), North-Holland, Amsterdam 1981.

[B2] *Proceedings of the Theme Term Changepoint Analysis–Empirical Reliability*, Technical Report Series of the Laboratory for Research in Statistics and Probability, No.224–June 1993, Carleton University–University of Ottawa.

[B3] *Limit Theorems in Probability and Statistics, Volumes I.,II.* (I. Berkes, E. Csáki, M. Csörgő, Eds.), János Bolyai Mathematical Society, Budapest 2002.

© Springer Science+Business Media New York 2015
D. Dawson et al. (eds.), *Asymptotic Laws and Methods in Stochastics*,
Fields Institute Communications 76, DOI 10.1007/978-1-4939-3076-0

Volumes in Honour of

[V1] *Asymptotic Methods in Probability and Statistics: A Volume in Honour of Miklós Csörgő* (B. Szyszkowicz, Ed.). Elsevier Science B.V., Amsterdam 1998.

[V2] *Asymptotic Methods in Stochastics: Festschrift for Miklós Csörgő* (L. Horváth, B. Szyszko-wicz, Eds.). Fields Institute Communications, Volume 44, AMS 2004.

– Fields Institute - Supplements:
  http://www.fields.utoronto.ca/publications/supplements
– *Path Properties of Forty Years of Research in Probability and Statistics: In Conversation with Miklós Csörgő.*
– *Miklós Csörgő List of Publications.*

[V3] *Asymptotic Laws and Methods in Stochastics: A Volume in Honour of Miklós Csörgő* (Donald Dawson, Rafal Kulik, Mohamedou Ould Haye, Barbara Szyszkowicz, Yiqiang Zhao, Eds.), Fields Institute Communications, Volume 76, Springer, New York.

Papers

1962
[1] On the empty cell test. *Technometrics* **4** (1962), 235–247 (with I. Guttman).

1964
[2] On the consistency of the two sample empty cell test. *Canadian Math. Bull.* **7** (1964), 57–63 (with I. Guttman).

1965
[3] Exact and limiting probability distributions of some Smirnov type statistics, *Canadian Math. Bulletin* **8** (1965), 93–103.

[4] Some Rényi type limit theorems for empirical distribution functions. *Annals Math. Statist.* **36** (1965), 322–326.

[5] Some Smirnov type theorems of probability theory. *Annals. Math. Statist.* **36** (1965), 1113–1119.

[6] *K*-sample analogues of Rényi's Kolmogorov-Smirnov type theorems. *Bull. Amer. Math. Soc.* **71** (1965), 616–618.

[7] Exact probability distribution functions of some Rényi type statistics. *Proc. Amer. Math. Soc.* **16** (1965), 1158–1167.

1966
[8] Some *K*-sample Kolmogorov-Smirnov-Rényi type theorems for empirical distribution functions. *Acta Math. Acad. Sci. Hungarica* **17** (1966), 325–334.

1967

[9] A new proof of some results of Rényi and the asymptotic distribution of the range of his Kolmogorov-Smirnov type random variables. *Canad. J. Math.* **19** (1967), 550–668.

[10] On some limit theorems involving the empirical distribution function. *Canad. Math. Bull.* **10** (1967), 739–741.

1968

[11] On the strong law of large numbers and the central limit theorem for martingales. *Trans. Amer. Math. Soc.* **131** (1968), 259–275.

1969

[12] Tests for the exponential distribution using Kolmogorov type statistics, *J. Roy. Statist. Soc.* **31** (1969), 499–509 (with V. Seshadri, M.A. Stephens).

1970

[13] Departure from independence: the strong law, standard and random-sum central limit theorems. *Acta Math. Acad. Sci. Hung.* **21** (1970), 105–114 (with R. Fischler).

[14] An invariance principle for the empirical process with random sample size. *Bull. Amer. Math. Soc.* **76** (1970), 704–710 (with S. Csörgő).

[15] Distribution results and power functions for Kac statistics. *Annals Inst. Statist. Math.* **22** (1970), 257–259 (with M. Alvo).

[16] On the problem of replacing composite hypotheses by equivalent simple ones. *Review Int. Statist. Inst.* **38** (1970), 351–368 (with V. Seshadri).

[17] On a law of iterated logarithm for strongly multiplicative systems. *Acta Math. Acad. Sci. Hung.* **21** (1970), 315–321.

1971

[18] Characterizing the Gaussian and Exponential laws via mappings onto the unit interval. *Z. Wahrscheinlichkeitstheorie und verw. Gebiete* **18** (1971), 333–339 (with V. Seshadri).

[19] Characterizations of the Brehrens-Fisher and related problems (a goodness of fit point of view). *Teoriya Veroiystnostei i ee Primeneniya* **XVI** (1971), 20–33 (with V. Seshadri).

[20] On mixing and the random-sum central limit theorem. *Tohoku Math. J.* **23** (1971), 139–145 (with R. Fischler).

1972

[21] Distribution results for distance functions based on the modified empirical distribution functions of M. Kac. *Annals Inst. Statist. Math.* **24** (1972), 101–110.

[22] On the problem of replacing composite hypotheses by equivalent simple ones (a characterization approach to goodness-of-fit). *Colloquia Mathematica Societatis János Bolyai* **9**, European Meeting of Statisticians, Budapest (Hungary), 1972, 159–180.

1973

[23] Some examples and results in the theory of mixing and random-sum central limit theorems. A. Rényi memorial volume of *Periodica Mathematics Hungarica* **3** (1973) 41–57 (with R. Fischler).

[24] On weak convergence of randomly selected partial sums. *Acta Sci. Math. Szegediensis* **34** (1973) 53–60 (with S. Csörgő).

[25] Some exact tests for normality in the presence of unknown parameters. *J. Roy. Statist. Soc. Ser. B.* **35** (1973), 507–522 (with V. Seshadri, M. Yalowsky).

[26] Glivenko-Cantelli type theorems for distance functions based on the modified empirical distribution function of M. Kac and for the empirical process with random sample size in general. *Lecture Notes in Mathematics* **296**, Probability and Information Theory II, Springer-Verlag 1973, 149–164.

1975

[27] A new method to prove Strassen type laws of invariance principle, I. *Z. Wahrscheinlichkeitstheorie und verw. Gebiete* **31** (1975), 255–259 (with P. Révész).

[28] A new method to prove Strassen type laws of invariance principle, II. *Z. Wahrscheinlichkeitstheorie und verw. Gebiete* **31** (1975), 261–269 (with P. Révész).

[29] Some notes on the empirical distribution function and the quantile process. *Colloquia Mathematica Societatis János Bolyai* **11**, Limit Theorems of Probability Theory (Keszthely, Hungary 1974), North Holland 1975, 59–71 (with P. Révész).

[30] Applications of characterizations in the area of goodness of fit. *Statistical distributions in scientific work* **2**: Model Building and Model Selection, 1975, 79–90 (with V. Seshadri, M. Yalovsky).

[31] Random-indexed limit theorems with the help of strong invariance theorems (in Hungarian). *Matematikai Lapok* **26** (1975), 39–66 (with S. Csörgő, R. Fischler, P. Révész).

[32] A strong approximation of multivariate empirical process. *Studia Sci. Math. Hungaricae* **10** (1975), 427–434 (with P. Révész).

1976

[33] Weak approximations of the empirical process when parameters are estimated. *Lecture Notes in Mathematics*, Empirical Distributions and Processes (Oberwolfach 1976) **566** (1976), 1–16, (with M.D. Burke).

[34] On the Erdős-Rényi increments and the P. Lévy modulus of continuity of a Kiefer process. *Lecture Notes in Mathematics* **566** (1976), 17–33 (with A.H.C. Chan).

1977

[35] On conditional medians and a law of iterated logarithm for strongly multiplicative systems. *Acta Math. Acad. Sci. Hung.* **29** (1977), 309–311, (with D.L. McLeish).

[36] On the empirical process when parameters are estimated. *Transactions of the Seventh Prague conference on Probability and Statistics*, 1974. Academia, Prague 1977, 87–97 (with J. Komlós, P. Major, P. Révész, G. Tusnády).

[37] Recent developments in the theory of strong invariance approximations. *La Gazette des Sciences Mathématiques du Québec* **1** (1977), 3–20 and 41–45 (Solicited).

1978

[38] How big are the increments of a multi-parameter Wiener process? *Z. Wahrscheinlichkeitstheorie verw. Gebiete* **42** (1978), 1–12 (with P. Révész).

[39] Strassen type limit points for moving averages of a Wiener process. *The Canadian Journal of Statistics* **6** (1978), 57–75 (with A.H.C. Chan and P. Révész).

[40] Strong approximations of the quantile process. *The Annals of Statistics* **6** (1978), 882–894 (with P. Révész).

1979

[41] On the standardized quantile process. *Optimizing Methods in Statistics* (Ed. J.S. Rustagi), 125–140. Academic Press 1979 (with P. Révész).

[42] Strong approximations of the Hoeffding, Blum, Keifer, Rosenblatt multivariate empirical process. *Journal of Multivariate Analysis* **9** (1979), 84–100.

[43] How big are the increments of a Wiener process? *The Annals of Probability* **7** (1979), 731–737 (with P. Révész).

[44] How small are the increments of a Wiener process? *Stochastic Processes and their Applications* **8** (1979), 119–129 (with P. Révész).

[45] Approximations of the empirical process when parameters are estimated. *The Annals of Probability*, **7** (1979), 790–810 (with M.D. Burke, S. Csörgő, P. Révész).

[46] Brownian motion–Wiener process. *Canadian Math. Bull.* **22** (1979) 257–280.

1980

[47] Asymptotics for the multisample, multivariate Cramér-von Mises statistic with some possible applications. *Proceedings of the 1st International Conference on Statistical Climatology* (Inter-University Seminar House, Tokyo, Japan, 1979), 67–83, Elsevier, New York, 1980 (with D.S. Cotterill).

[48] Weak convergence of sequences of random elements with random indices. *Math. Proc. Cambridge Philos. Soc.* **88** (1980) 171–174 (with Z. Rychlik).

1981

[49] On a test for goodness-of-fit based on the empirical probability measure of Foutz and testing for exponentiality. Analytical methods in Probability Theory (Proceedings of the Conference held at Oberwolfach, 1980), 25–34. *Lecture Notes in Math.* **861**, Springer, Berlin 1981.

[50] On the nondifferentiability of the Wiener sheet. *Contributions to Probability* (A Collection of Papers Dedicated to Eugene Lukács), 143–150, Academic Press, New York, 1981 (with P. Révész).

[51] Gaussian processes, strong approximations: An Interplay. *Colloques Internationaux du Centre National de la Recherche Scientifique*, No. **307**, Aspects Statistiques et Aspects Physiques des Processus Gaussiens (Saint-Flour 22–29 juin 1980), Editions du CNRS, Paris 1981, 131–229.

[52] Improved Erdős-Rényi and strong approximation laws for increments of partial sums. *The Annals of Probability* **9** (1981), 988–996 (with J. Steinebach).

[53] Asymptotic properties of randomly indexed sequences of random variables. *The Canadian Journal of Statistics* **9** (1981), 101–107 (with Z. Rychlik).

[54] Quantile processes and sums of weighted spacings for composite goodness-of-fit. *Statistics and Related Topics* (Proceedings of the International Symposium on Statistics and Related Topics, Ottawa, Canada, 1980), 69–87, North–Holland, Amsterdam, 1981 (with P. Révész).

[55] On the asymptotic distribution of the multivariate Cramér-Von Mises and Hoeffding-Blum-Kiefer-Rosenblatt independence criteria. *Statistical Distributions in Scientific Work* **5** Inference Problems and Properties, 141–156, Reidel–Dordrecht, 1981.

1982

[56] On the limiting distribution of and critical values for the multivariate Cramér-Von Mises statistic. *The Annals of Statistics* **10** (1982), 233–244 (with D.S. Cotterill).

[57] An invariance principle for *NN* empirical density functions. In *Colloquia Mathematica Societatis János Bolyai* - **32**. Nonparametric Statistical Inference (B.V. Gnedenko, M.L. Puri and I. Vincze eds.), 151–170, North–Holland, Amsterdam, 1982 (with P. Révész).

1983

[58] How big are the increments of the local time of a Wiener process? *The Annals of Probability* **11** (1983), 593–608 (with E. Csáki, A. Földes, P. Révész).

[59] An *NN*-estimator for the score function. In *6th International Summer School on Problems of Model Choice and Parameter Estimation in Regression Analysis* (Sellin, 1983), 62–82, *Seminarberichte* **51**, Humboldt Univ., Berlin, 1983 (with P. Révész).

1984

[60] Three strong approximations of the local time of a Wiener process and their applications to invariance. In *Colloquia Mathematica Societatis János Bolyai* **36**. *Limit Theorems in Probability and Statistics* (P. Révész, ed.), 223–254, North–Holland, Amsterdam, 1984 (with P. Révész).

[61] Quantile processes for composite goodness-of-fit. In *Colloquia Mathematica Societatis János Bolyai* **36**. *Limit Theorems in Probability and Statistics* (P. Révész, ed.), 255–304, North–Holland, Amsterdam, 1984 (with P. Révész).

[62] Two approaches to constructing simultaneous confidence bounds for quantiles. *Probability and Mathematical Statistics* **4** (1984), 221–236 (with P. Révész).

[63] Invariance Principles for Empirical Processes. In *Handbook of Statistics* **4**, Elsevier Science Publishers (1984), 431–462.

[64] On weak and strong approximations of the quantile process. In *Proc. of the 7th Conference on Probability Theory*, Brasov (Aug. 29 – Sept. 4, 1982) (M. Josifescu, ed.), 81–95, Editura Academici Republicii Socialiste România, Bucuresti, 1984 (with S. Csörgő, L. Horváth, P. Révész).

[65] On the parameters estimated quantile process. In *Proc. of the First Saudi Symposium on Statistics and its Applications*, Riyadh (May 2–5, 1983), Developments in Statistics and its Applications (Aboummoh et al. eds.), 421–444, King Saud University Press, Riyadh, Saudi Arabia, 1984 (with E.-E.A.A. Aly).

1985

[66] On the limiting distribution of and critical values for the Hoeffding-Blum-Kiefer-Rosenblatt independence criterion. *Statistics & Decisions* **3** (1985), 1–48 (with D.S. Cotterill).

[67] Strong approximations of the quantile process of the product-limit estimator. *Journal of Multivariate Analysis* **16** (1985), 185–210 (with E.E.A.A. Aly, L. Horváth).

[68] Multivariate Cramér-Von Mises Statistics. In *Encyclopedia of Statistical Sciences* **6** (S. Kotz, N.L. Johnson and C.B. Read, eds.), 35–39, Wiley, New York 1985.

[69] On the stability of the local time of a symmetric random walk. *Acta Scientiarum Mathematicarum Szegedienses* **48** (1985), 85–96 (with P. Révész).

[70] Quadratic nuisance-parameter-free goodness-of-fit tests in the presence of location and scale parameters. *The Canadian Journal of Statistics* **13** (1985), 53–70 (with E.E.A.A. Aly).

[71] On strong invariance for local time of partial sums. *Stochastic Processes and their Applications* **20** (1985), 59–84 (with P. Révész).

[72] On the asymptotic distribution of weighted empirical and quantile processes in the middle and on the tails. *Stochastic Processes and their Applications* **21** (1985), 119–132 (with D.M. Mason).

1986

[73] Quantile Processes. In *Encyclopedia of Statistical Sciences* **7** (S. Kotz, N.L. Johnson and C.B. Read, eds.), 412–424, Wiley, New York 1986.

[74] Weighted empirical and quantile processes. *The Annals of Probability* **14** (1986), 31–85 (with S. Csörgő, L. Horváth, D.M. Mason).

[75] Normal and stable convergence of integral functions of the empirical distribution function. *The Annals of Probability* **14** (1986), 86–118 (with S. Csörgő, L. Horváth, D.M. Mason).

[76] A nearest neighbour estimator for the score function. *Probability Theory and Related Fields* **71** (1986), 293–305 (with P. Révész).

[77] Mesure du voisinage and occupation density. *Probability Theory and Related Fields* **73** (1986), 211–226 (with P. Révész).

[78] How large must be the difference between local time and mesure du voisinage of Brownian motion. *Statistics & Probability Letters* **4** (1986), 161–166 (with L. Horváth, P. Révész).

[79] Weighted empirical spacings processes. *Canadian Journal of Statistics* **14** (1986), 221–232 (with L. Horváth).

[80] Approximations of weighted empirical and quantile processes. *Statistics & Probability Letters* **4** (1986), 275–280 (with L. Horváth).

[81] Strong approximations for renewal processes. *C.R. Mathematical Reports of the Academy of Science, Canada* **8** (1986), 151–154 (with L. Horváth, J. Steinebach).

[82] Bootstrapped confidence bands for percentile lifetime. *Annals of the Institute of Statistical Mathematics* **38** (1986), Part A, 429–438 (with B. Barabás, L. Horváth, B.S. Yandell).

[83] Sup-norm convergence of the empirical process indexed by functions and applications. *Probability and Mathematical Statistics* **7** (1986), 13–26 (with S. Csörgő, L. Horváth, D.M. Mason).

1987

[84] Approximation of intermediate quantile processes. *Journal of Multivariate Analysis* **21** (1987), 250–262 (with L. Horváth).

[85] Asymptotic distributions of pontograms. *Math. Proc. Camb. Phil. Soc.* **101** (1987), 131–139 (with L. Horváth).

[86] P–P plots, rank processes, and Chernoff-Savage theorems. In *New Perspectives in Theoretical and Applied Statistics* (M. Puri, J.P. Vilaplana and W. Wertz, eds.), Part **3**, Nonparametric Theory, 135–156, Wiley, New York, 1987 (with E.-E.A.A. Aly, L. Horváth).

[87] Estimation of total time on test transforms and Lorenz curves under random censorship. *Statistics* **18** (1987), 77–97 (with S. Csörgő, L. Horváth).

[88] On the distribution of the supremum of weighted empirical processes. In *Contributions to Stochastics in Honour of the 75th Birthday of Walther Eberl, Sr.* (W. Sendler, ed.) 1–18, Physica-Verlag Heidelberg, 1987 (with L. Horváth, J. Steinebach).

[89] On the optimality of estimating the tail index and a naive estimator. *The Austrian Journal of Statistics* **29** (1987), 166–178 (with L. Horváth, P. Révész).

[90] An approximation of stopped sums with applications in queuing theory. *Advances in Applied Probability.* **19** (1987), 674–690 (with P. Deheuvels, L. Horváth).

[91] Estimation of percentile residual life. *Operations Research.* **35** (1987), 598–606 (with S. Csörgő).

[92] Invariance principles for renewal processes. *The Annals of Probability* **15** (1987), 1441–1460 (with L. Horváth, J. Steinebach).

[93] Nonparametric tests for the change-point problem. *Journal of Statistical Planning and Inference* **17** (1987), 1–9 (with L. Horváth).

[94] Detecting change in a random sequence. *Journal of Multivariate Analysis* **23** (1987), 119–130 (with L. Horváth).

[95] Stability and instability of local time of random walk in random environment. *Stochastic Processes and their Applications* **25** (1987), 185–202 (with L. Horváth, P. Révész).

[96] Rates of convergence for random walk summation. *Bulletin of the London Mathematical Society* **19** (1987), 531–536 (with L. Horváth).

1988

[97] Asymptotic representations of self-normalized sums. *Probability and Mathematical Statistics* **9** (1988), 15–24 (with L. Horváth).

[98] Asymptotics for L_p-norms of kernel estimators of densities. *Computational Statistics and Data Analysis* **6**, (1988), 241–250 (with L. Horváth).

[99] Convergence of the empirical and quantile distributions to Poisson measures. *Statistics and Decisions* **6** (1988), 129–136 (with L. Horváth).

[100] On the distributions of L_p norms of weighted uniform empirical and quantile processes. *The Annals of Probability* **16** (1988), 142–161 (with L. Horváth).

[101] A note on strong approximations of multivariate empirical processes. *Stochastic Processes and their Applications* **27** (1988), 101–109 (with L. Horváth).

[102] A law of the iterated logarithm for infinite dimensional Ornstein–Uhlenbeck processes. *C.R. Mathematical Reports of the Academy of Science, Canada* **10** (1988), 113–118 (with Z.Y. Lin).

[103] On moduli of continuity for Gaussian and χ^2 processes generated by Ornstein–Uhlenbeck processes. *C.R. Mathematical Reports of the Academy of Science, Canada* **10** (1988), 203–207 (with Z.Y. Lin).

[104] Rate of convergence of transport processes with an application to stochastic differential equations. *Probability Theory and Related Fields* **78** (1988), 379–387 (with L. Horváth).

[105] Nonparametric methods for changepoint problems. In *Handbook of Statistics* **7**, Quality Control and Reliability (P.R. Krishnaiah, P.K. Sen, eds.) Ch. 20, 403–425, North–Holland, Amsterdam 1988 (with L. Horváth).

[106] Invariance principles for changepoint problems. *Journal of Multivariate Analysis* **27** (1988), Krishnaiah Memorial Volume, 151–168 (with L. Horváth).

[107] Central limit theorems for L_p-norms of density estimators. *Probability Theory and Related Fields* **80** (1988), 269–291 (with L. Horváth).

1989

[108] Brownian local time approximated by a Wiener sheet. *The Annals of Probability* **17** (1989), 516–537 (with E. Csáki, A. Földes, P. Révész).

[109] On best possible approximations of local time. *Statistics & Probability Letters* **8** (1989), 301–306 (with L. Horváth).

[110] Invariance principles for changepoint problems. In *Multivariate Statistics and Probability, Essays in Memory of P.R. Krishnaiah* (C.R. Rao, M.M. Rao, eds.), 151–168 (with L. Horváth), Academic Press, San Diego, 1989.

[111] Comments on "Asymptotics via empirical processes" by D. Pollard. *Statistical Science* **4** (1989), 360–365 (with L. Horváth).

[112] On confidence bands for the quantile function. In *Statistical Applications in the Earth Sciences* (eds. F.P. Agterberg and G.F. Bonham-Carter); Geological Survey of Canada. Paper **89–9**, 221–231, 1989 (with L. Horváth).

[113] On confidence bands for the quantile function of a continuous distribution function. In *Colloquia Mathematica Societatis János Bolyai* **57**. *Limit Theorems in Probability and Statistics* (P. Révész, ed.) Pécs (Hungary), 1989, 96–106, North–Holland, Amsterdam (with L. Horváth).

[114] Path properties of infinite dimensional Ornstein–Uhlenbeck processes. In *Colloquia Mathematica Societatis János Bolyai* **57**. *Limit Theorems in Probability and Statistics* (P. Révész, ed.) Pécs (Hungary), 1989, 107–135, North–Holland, Amsterdam (with Z.Y. Lin).

1990

[115] On moduli of continuity for Gaussian and l^2–norm squared processes generated by Ornstein–Uhlenbeck processes. *Canadian Journal of Mathematics* **42** (1990), 141–158 (with Z.Y. Lin).

[116] Confidence bands for quantile function under random censorship. *Annals of the Institute of Statistical Mathematics* **42** (1990), 21–36 (with C. -J.F. Chung, L. Horváth).

[117] On the distributions of L_p norms of weighted quantile processes. *Annales de l'Institut Henri Poincaré* – Probabilités et Statistiques **26** (1990), 65–90 (with L. Horváth).

[118] Asymptotic tail behavior of uniform multivariate empirical processes. *The Annals of Probability* **18** (1990), 1723–1738 (with L. Horváth).

[119] Fernique type inequalities for not necessarily Gaussian processes. *C.R. Mathematical Reports of the Academy of Science, Canada* **12** (1990), 149–154 (with E. Csáki).

[120] On the distribution of the supremum of weighted quantile processes. *Studia Scientiarum Mathematicarum Hungarica* **25** (1990), 353–375 (with L. Horváth).

1991

[121] On the estimation of the adjustment coefficient in risk theory via intermediate order statistics. *Insurance: Mathematics and Economics* **10** (1991), 37–50 (with J. Steinebach).

[122] Criteria for limit inferior of small increments of Banach space valued stochastic processes. *C.R. Mathematical Reports of the Academy of Science, Canada* **13** (1991), 173–178 (with Q-M. Shao).

[123] Estimating the quantile–density function. In *Nonparametric Functional Estimation and Related Topics* (G. Roussas, Ed.), Proceedings of the NATO Advanced Study Institute on Nonparametric Functional Estimation and Related Topics, Spetses, Greece, 1990, 213–223, Kluwer Academic Publishers, Netherlands 1991 (with P. Deheuvels, L. Horváth).

[124] Path properties of kernel generated two-time parameter Gaussian processes. *Probability Theory and Related Fields* **89** (1991), 423–445 (with Z.Y. Lin).

[125] On infinite series of independent Ornstein -Uhlenbeck processes. *Stochastic Processes and their Applications* **39** (1991), 25–44 (with E. Csáki, Z.Y. Lin, P. Révész).

[126] Central limit theorems for L_p distances of kernel estimators of densities under random censorship. *The Annals of Statistics* **19** (1991), 1813–1831 (with E. Gombay, L. Horváth).

[127] A note on local and global functions of a Wiener process and some Rényi-type statistics. *Studia Scientiarum Mathematicarum Hungarica* **26** (1991), 239–259 (with Q.-M. Shao, B. Szyszkowicz).

1992

[128] Rényi-type empirical processes. *Journal of Multivariate Analysis* **41** (1992), 338–358 (with L. Horváth).

[129] Strong approximation of a multi -time parameter Poisson process. In *Nonparametric Statistics and Related Topics*. (A.K.Md.E. Saleh, ed.) 365–370, North-Holland, Amsterdam 1992 (with L. Horváth).

[130] Inequalities for increments of stochastic processes and moduli of continuity. *The Annals of Probability* **20** (1992), 1031–1052 (with E. Csáki).

[131] Long random walk excursions and local time. *Stochastic Processes and their Applications* **41** (1992), 181–190 (with P. Révész).

[132] Invariance principles for logarithmic averages. *Mathematical Proceedings of Cambridge Philosophical Society.* **112** (1992), 195–205 (with L. Horváth); Corrigendum **ibid.** (1994).

[133] Fernique type inequalities and moduli of continuity for l^2-valued Ornstein-Uhlenbeck Processes. *Ann. Inst. Henri Poincaré, Probabilités et Statistiques* **28** (1992), 479–517 (with E. Csáki, Q.-M. Shao).

[134] Strong approximation of additive functionals. *Journal of Theoretical Probability* **5** (1992), 679–706 (with E. Csáki, A. Földes, P. Révész).

1993

[135] Convergence of integrals of uniform empirical and quantile processes. *Stochastic Processes and their Applications* **45** (1993), 283–294 (with L. Horváth, Q.-M. Shao).

[136] Randomization moduli of continuity for l^2-norm squared Ornstein-Uhlenbeck processes. *Canadian Journal of Mathematics* **45** (1993), 269–283 (with Z.-Y. Lin, Q.-M. Shao).

[137] On the weighted asymptotics of partial sums and empirical processes of independent random variables. In *Contemporary Mathematics* **149** (1993), 139–147, *Doeblin and Modern Probability*, Harry Cohn, Editor, American Mathematical Society 1993 (with L. Horváth, Q.-M. Shao, B. Szyszkowicz).

[138] Strong limit theorems for large and small increments of ℓ^p-valued Gaussian processes. *The Annals of Probability* **21** (1993), 1958–1990 (with Q. - M. Shao).

1994

[139] Kernel generated two-time parameter Gaussian processes and some of their path properties. *Canadian Journal of Mathematics* **46** (1994), 81–119 (with Z.-Y. Lin, Q.-M. Shao).

[140] A new proof on the distribution of the local time of a Wiener process. *Statistics & Probability Letters* **19** (1994), 285–290 (with Q.-M. Shao).

[141] Studentized increments of partial sums. *Science in China* (Series A) **37** (1994), 365–376 (with Z.-Y. Lin, Q.-M. Shao).

[142] Bahadur-Kiefer representations on the tails. In *Recent Advances in Statistics and Probability*, 255–261, J. Pérez Vilaplana and M.L. Puri (Eds.), VSP 1994 (with L. Horváth).

[143] A self-normalized Erdős-Rényi type strong law of large numbers, *Stochastic Processes and their Applications* **50** (1994), 187–196 (with Q.-M. Shao).

[144] On almost sure limit inferior for B-valued stochastic processes and applications, *Probability Theory and Related Fields* **99** (1994), 29–54 (with Q.-M. Shao).

[145] Path properties for l^∞-valued Gaussian processes, *Proc. Amer. Math. Soc.* **121** (1994), 225–236 (with Z.-Y. Lin, Q.-M. Shao).

[146] Empirical and partial sum processes with sample paths in Banach function spaces, In *Probability Theory and Mathematical Statistics*, 143–158, B. Grigelionis et al. (Eds), 1994 VSP/TEV (with R. Norvaiša).

[147] Applications of multi-time parameter processes to change-point analysis, In *Probability Theory and Mathematical Statistics*, 159–222, B. Grigelionis et al. (Eds), 1994 VSP/TEV (with B. Szyszkowicz).

[148] Weighted multivariate empirical processes and contiguous change-point analysis, In *IMS Lecture Notes-Monograph Series* **23** (1994), 93–98 (with B. Szyszkowicz).

1995

[149] On the distance between smoothed empirical and quantile processes, *The Annals of Statistics* **23** (1995), 113–131 (with L. Horváth).

[150] Moduli of continuity for ℓ^p-valued Gaussian processes, *Acta Sci. Math. (Szeged)* **60** (1995), 149–175 (with E. Csáki, Q.-M. Shao).

[151] On moduli of continuity for local times of Gaussian processes, *Stochastic Processes and their Applications* **58** (1995), 1–21 (Z.-Y. Lin, Q.-M. Shao).

[152] On additive functionals of Markov chains, *Journal of Theoretical Probability* **8** (1995), 905–919 (with E. Csáki).

[153] Global Strassen-type theorems for iterated Brownian motions, *Stochastic Processes and their Applications* **59** (1995), 321–341 (with E. Csáki, A. Földes, P. Révész).

[154] Invariance principles in Banach function spaces, *Studia Scientiarum Mathematicarum Hungarica* **30** (1995), 63–92 (with R. Norvaiša).

1996

[155] A note on the change-point problem for angular data, *Statistics & Probability Letters* **27** (1996), 61–65 (with L. Horváth).

[156] The local time of iterated Brownian Motion, *Journal of Theoretical Probability* **9** (1996), 717–743 (with E. Csáki, A. Földes and P. Révész).

[157] Strassen's LIL for the Lorenz curve, *Journal of Multivariate Analysis* **59** (1996), 1–12 (with R. Zitikis).

[158] Mean residual life processes, *The Annals of Statistics* **24** (1996), 1717–1739 (with R. Zitikis).

[159] Weak approximations for quantile processes of stationary sequences, *The Canadian Journal of Statistics* **24** (1996), 403–430 (with Hao Yu).

1997

[160] On confidence bands for the Lorenz and Goldie curves, In *Advances in the Theory and Practice of Statistics: A Volume in Honor of Samuel Kotz*, 261–281, Edited by Norman L. Johnson and N. Balakrishnan, Wiley, New York 1997 (with R. Zitikis).

[161] Almost sure summability of partial sums, *Studia Scientiarum Mathematicarum Hungarica* **33** (1997), 43–74 (with L. Horváth, Q.-M. Shao).

[162] On the rate of strong consistency of Lorenz curves, *Statistics & Probability Letters* **34** (1997), 113–121 (with R. Zitikis).

[163] Estimation of total time on test transforms for stationary observations, *Stochastic Processes and their Applications* **68** (1997), 229–253 (with Hao Yu).

[164] On the occupation time of an iterated process having no local time, *Stochastic Processes and their Applications* **70** (1997), 199–217 (with E. Csáki, A. Földes, P. Révész).

[165] Integral tests for suprema of Kiefer processes with application, *Statistics & Decisions* **15** (1997), 365–377 (with L. Horváth, B. Szyszkowicz).

1998

[166] Standardized sequential empirical processes, *Studia Scientiarum Mathematicarum Hungarica* **34** (1998), 51–69 (with R. Norvaiša).

[167] Sequential Quantile and Bahadur–Kiefer Processes, In *Handbook of Statistics* **16**, *Order Statistics: Theory and Methods* (N. Balakrishnan and C.R. Rao, eds.) Ch.21, 631–688, 1998 Elsevier Science B.V. (with B. Szyszkowicz).

[168] On the rate of strong consistency of the total time on test statistics, *Journal of Multivariate Analysis* **66** (1998), 99–117 (with R. Zitikis).

[169] Asymptotic confidence bands for the Lorenz and Bonferroni curves based on the empirical Lorenz curve, *Journal of Statistical Planning and Inference* **74** (1998), 65–91 (with J.L. Gastwirth, R. Zitikis).

1999

[170] Some asymptotic properties of the local time of the uniform empirical process, *Bernoulli* **5** (1999), 1035–1058 (with Z. Shi, M. Yor).

[171] Convergence of weighted partial sums when the limiting distribution is not necessarily Radon, *Stochastic Processes and their Applications* **81** (1999), 81–101 (with R. Norvaiša, B. Szyszkowicz).

[172] On the Vervaat and Vervaat–error processes, *Acta Applicandae Mathematicae* **58** (1999), 91–105 (with R. Zitikis).

[173] Random walking around financial mathematics, In *Random Walks, Bolyai Society Mathematical Studies*, **9** (Pál Révész, Bálint Tóth, eds.), Budapest 1999, 59–111.

[174] Weak approximations for empirical Lorenz curves and their Goldie inverses of stationary observations, *Advances in Applied Probability* **31** (1999), 698–719 (with Hao Yu).

2000

[175] Weighted quantile processes and their applications to change-point analysis, In *Stochastic Models* (Luis G. Gorostiza, B. Gail Ivanoff, eds.), *Canadian Mathematical Society Conference Proceedings* **26**, 2000, 67–84 (with B. Szyszkowicz).

[176] Asymptotic properties of integral functions of geometric stochastic processes, *Journal of Applied Probability* **37** (2000), 480–493 (with E. Csáki, A. Földes, P. Révész).

[177] Approximation for bootstrapped empirical processes, *Proceedings of the American Mathematical Society* **128** (2000), 2457–2464 (with L. Horváth, P. Kokoszka).

[178] Increment sizes of the principal value of Brownian local time, *Probability Theory and Related Fields* **117** (2000), 515–531 (with E. Csáki, A. Földes, Z. Shi).

[179] Be n fixed, or go to infinity: difficulties are welcome and almost surely overcome. A tribute to Endre Csáki, *Periodica Mathematica Hungarica* **41/1–2** (2000), 1–25.

[180] Pre-super Brownian motion, *Periodica Mathematica Hungarica* **41/1–2** (2000), 71–102 (with P. Révész).

2001

[181] A functional modulus of continuity for a Wiener process, *Statistics & Probability Letters* **51** (2001), 215–223 (with B. Chen).

[182] Path properties of Cauchy's principal values related to local time, *Studia Scientarium Mathematicarum Hungarica* **38** (2001), 149–169 (with E. Csáki, A. Földes, Z. Shi).

[183] An L^p–view of a general version of the Bahadur-Kiefer process, *Journal of Mathematical Sciences* **105** (2001), 2534–2540 (with Zhan Shi).

[184] The Vervaat process in L_p spaces, *Journal of Multivariate Analysis* **78** (2001), 103–138 (with R. Zitikis).

2002

[185] Pointwise and uniform asymptotics of the Vervaat error process, *Journal of Theoretical Probability* **15** (2002), 845–875 (with E. Csáki, A. Földes, Z. Shi, R. Zitikis).

[186] Life is a random walk with a local time: a tribute to Pál Révész, In *Limit Theorems in Probability and Statistics I.* (I. Berkes, E. Csáki, M. Csörgő, eds.), János Bolyai Mathematical Society, Budapest 2002, 21–99.

[187] On a class of additive functionals of two-dimensional Brownian motion and random walk, In *Limit Theorems in Probability and Statistics I*. (I. Berkes, E. Csáki, M. Csörgő, eds.), János Bolyai Mathematical Society, Budapest 2002, 321–345 (with E. Csáki, A. Földes, Z. Shi).

[188] On the general Bahadur-Kiefer, quantile, and Vervaat processes: old and new. In *Limit Theorems in Probability and Statistics I*. (I. Berkes, E. Csáki, M. Csörgő, eds.), János Bolyai Mathematical Society, Budapest 2002, 389–426 (with R. Zitikis).

[189] A glimpse of the impact of Pál Erdős on probability and statistics, *The Canadian Journal of Statistics* **30** (2002), 493–556.

2003

[190] Darling–Erdős theorem for self-normalized sums, *The Annals of Probability* **31** (2003), 676–692 (with B. Szyszkowicz, Q. Wang).

[191] Donsker's theorem for self-normalized partial sums processes, *The Annals of Probability* **31** (2003), 1228–1240 (with B. Szyszkowicz, Q. Wang).

2004

[192] On weighted approximations and strong limit theorems for self-normalized partial sums processes. In *Asymptotic Methods in Stochastics: Festschrift for Miklós Csörgő* (L. Horváth, B. Szyszkowicz, eds.), Fields Institute Communications, **44**, AMS 2004, 489–521 (with B. Szyszkowicz, Q. Wang).

[193] Weighted invariance principle for Banach space valued random variables, *Lietuvos Matematikos Rinkinis*, **44** (2004), 139–175 (with R. Norvaiša).

2005

[194] Joint path properties of two of my most favourite friends in mathematics: a tribute to Endre Csáki and Pál Révész, *Periodica Mathematics Hungarica*, **50** (2005), 1–27.

[195] An L^p-view of the Bahadur-Kiefer theorem, *Periodica Mathematica Hungarica*, **50** (2005),79–98 (with Z. Shi).

[196] Limit theorems for nearest-neighbour density estimation under long-range dependence, *Journal of Statistical Research*, **39/1** (2005), 121–138 (with B. Szyszkowicz, L. Wang).

2006

[197] Strong invariance principles for sequential Bahadur–Kiefer and Vervaat error processes of long-range dependent sequences, *The Annals of Statistics*, **34** (2006), 1013–1044 (with B. Szyszkowicz, L. Wang).

[198] Change in the mean in the domain of attraction of the normal law, *Austrian Journal of Statistics*, **35** (2006), 93–103 (with B. Szyszkowicz, L. Wang).

2007

[199] Large deviations for two-parameter Gaussian processes related to change-point analysis. In **ALM2** *Advanced Lectures in Mathematics, Asymptotic Theory in Probability and Statistics with Applications* (Tze Leung Lai, Lianfen Qian, QiMan Shao, eds.), Higher Education Press, Beijing and International Press, Cambridge MA 2007, 177–197 (with Bin Chen).

[200] Correction Note: Strong invariance principles for sequential Bahadur-Kiefer and Vervaat error processes for long-range dependent sequences, *The Annals of Statistics*, **35** (2007), 2815–1817 (with B. Szyszkowicz, L. Wang).

[201] A glimpse of the KMT (1975) approximation of empirical processes by Brownian bridges via quantiles, *Acta Scientiarum Mathematicarum (Szeged)*, **73** (2007), 349–366.

[202] Path properties of ℓ^p-valued Gaussian random fields, *Science in China Series A: Mathematics*, **50** (2007), 1501–1520 (with Y.-K. Choi).

[203] On Vervaat and Vervaat-error type processes for partial sums and renewals, *Journal of Statistical Planning and Inference*, **137** (2007), 953–966 (with E. Csáki, Z. Rychlik, J. Steinebach).

2008

[204] Asymptotics of Studentized U-type processes for changepoint problems, *Acta Mathematica Hungarica*, **121** (2008), 333–357 (with B. Szyszkowicz, Q. Wang).

[205] On weighted approximations in D[0,1] with applications to self-normalized partial sum processes, *Acta Mathematica Hungarica*, **121** (2008), 307–332 (with B. Szyszkowicz, Q. Wang).

[206] Reduction principles for quantile and Bahadur-Kiefer processes of long-range dependent linear sequences, *Probability Theory and Related Fields*, **142** (2008), 339–366 (with R. Kulik).

[207] Weak convergence of Vervaat and Vervaat error processes of long-range dependent sequences, *Journal of Theoretical Probability*, **21** (2008), 672–686 (with R. Kulik).

[208] Limsup results and LIL for partial sum processes of a Gaussian random field, *Acta Math. Sin. (Engl. Ser.)*, **24** (2008), 1497–1506 (with Y.K. Choi).

[209] Asymptotic properties of increments of L^∞-valued Gaussian random fields, *Canadian Journal of Mathematics*, **60** (2008), 313–333 (with Y.K. Choi).

2009

[210] Strong limit theorems for a simple random walk on the 2-dimensional comb, *Electronic Journal of Probability*, **14** (2009), 2371–2390 (with E. Csáki, A. Földes, P. Révész).

[211] Random walk local time approximated by a Brownian sheet combined with an independent Brownian motion, *Annales de l'Institut Henri Poincaré-Probabilités et Statistiques*, **45** (2009), 515–544 (with E. Csáki, A. Földes, P. Révész).

2010

[212] Random walk and Brownian local times in Wiener sheets: a tribute to my almost surely most visited 75 years young best friends, Endre Csáki and Pál Révész, *Periodica Mathematica Hungarica*, **61** (2010), 1–21.

[213] On the supremum of iterated local time, *Publicationes Mathematicae Debrecen*, **76** (2010), 255–270 (with E. Csáki, A. Földes, P. Révész).

[214] On Vervaat processes for sums and renewals in weakly dependent cases, *Dependence in probability, analysis and number theory*, 145–156, Kendrick Press, Heber City, UT 2010, (with E. Csáki, R. Kulik).

2011

[215] On the local time of a random walk on the 2-dimensional comb, *Stochastic Processes and their Applications*, **121**, (2011),1290–1314. Available online: 1 February 2011 (with E. Csáki, A. Földes, P. Révész).

[216] Functional central limit theorems for self-normalized least squares processes in regression with possibly infinite variance data, *Stochastic Processes and their Applications*, **121** (2011), 2925–2953. Available online: 16 August 2011 (with Yu. V. Martsynyuk).

2012

[217] Strassen-type law of the iterated logarithm for self-normalized increments of sums, *Journal of Mathematical Analysis and Applications*, **393/1** (2012), 45–55 (with Zh. Hu, H. Mei).

[218] Random walk on half-plane half-comb structure, *Annales Mathematicae et Informaticae*, **39** (2012), 29–44 (with E. Csáki, A. Földes, P. Révész).

2013

[219] Strong limit theorems for anisotropic random walks on $\mathbf{Z}^2$, *Periodica Mathematica Hungarica*, **67**, (2013), 71–94 (with E. Csáki, A. Földes, P. Révész).

[220] A strong approximation of self-normalized sums, *Science China Mathematics*, **56/1** (2013), 149–160 (with Zh. Hu).

[221] Strassen-type law of the iterated logarithm for self-normalized sums, *Journal of Theoretical Probability*, **26** (2013), 311–328 (with Zh. Hu, H. Mei).

[222] Asymptotics of randomly weighted u- and v-statistics: application to bootstrap, *Journal of Multivariate Analysis*, **121** (2013), 176–192 (with M.M. Nasari).

2014

[223] Change in the mean in the domain of attraction of the normal law via Darling-Erdős Theorems, *Brazilian Journal of Probability and Statistics*, **28** (2014), 538–560 (with Zh. Hu).

[224] Another look at bootstrapping the Student t-statistic. *Mathematical Methods of Statistics*, **23** (2014), No. 4, 256–278 (with Yu. V. Martsynyuk and M. M. Nasari).

2015

[225] Inference from small and big data sets with error rates. *Electronic Journal of Statistics*, **9** (2015), 535–566 (with M. M. Nasari).

A Substantial Length Review

[226] Review of *Empirical Processes with Applications to Statistics*, by Galen R. Shorack and Jon A. Wellner, Wiley, New York, 1986, in *Bulletin AMS* **17** (1987), 189–200, by Miklós Csörgő.

Papers to Appear

[227] Weak convergence of self-normalized partial sums processes. In [V3], this volume (with Zh. Hu).

[228] On Bahadur-Kiefer type proesses for sums and renewals in dependent cases. In *Mathematical Statistics and Limit Theorems: Conference in Honor of Paul Deheuvels* (Hallin, M., Mason, D., Pfeifer, D., Steinebach, J.G., Eds.), Springer, New York, (with E. Csáki).

Papers Under Review for Publication

[229] Strong approximations for long memory sequences based partial sums, counting and their Vervaat processes. [arXiv:1302.3740v] [math. PR] (with E. Csáki and R. Kulik).

[230] Randomized pivots for means of short and long memory linear processes. http://arxiv-web3.library.cornell.edu/pdf/1309.4158.v2.pdf. (with M. M. Nasari and M. Ould Haye).

[231] Some limit theorems for heights of random walks on a spider. http://arxiv.org/abs/1501.00466 (with E. Csáki, A. Földes, P. Révész).

Printed in the United States
By Bookmasters